Electromagnetic Transients of Power Electronics Systems

Zhengming Zhao · Liqiang Yuan
Hua Bai · Ting Lu

Electromagnetic Transients of Power Electronics Systems

清华大学出版社
TSINGHUA UNIVERSITY PRESS

Springer

Zhengming Zhao
Department of Electrical Engineering
Tsinghua University
Beijing, China

Liqiang Yuan
Department of Electrical Engineering
Tsinghua University
Beijing, China

Hua Bai
University of Tennessee
Knoxville, TN, USA

Ting Lu
Department of Electrical Engineering
Tsinghua University
Beijing, China

ISBN 978-981-10-8811-7   ISBN 978-981-10-8812-4   (eBook)
https://doi.org/10.1007/978-981-10-8812-4

Jointly published with Tsinghua University Press, Beijing, China

The print edition is not for sale in China Mainland. Customers from China Mainland please order the print book from: Tsinghua University Press.
ISBN of the China Mainland edition: 978-7-302-46634-5

Library of Congress Control Number: 2018957247

This Springer imprint is published by the registered company Springer Nature Singapore Pte Ltd.
The registered company address is: 152 Beach Road, #21-01/04 Gateway East, Singapore 189721, Singapore

Preface

The transient analysis of power electronics systems, successor of the motor dynamic analysis, and power system transient analysis, is shaping the perspective, forming the new mindset and booming new ideas, methodology, and techniques of the electrical engineering, which could ultimately influence the dynamic analysis of modern electrical engineering.

Different from the machinery and power system, power electronics was defined as an interdisciplinary engineering branch since it was born, involving power semiconductor, power conversion circuit, control of the switch and circuit, electromagnetic field, thermal analysis, mechanical assembly, etc. Especially with the introduction of fully controlled semiconductor switches and the pulse width modulation (PWM), the continuous electromagnetic energy is translated into quasi-discrete and controllable energy pulse sequences, obstructing the conventional large-timescale electromagnetic transient analysis in machinery and power system from being applied to power electronics. Along with puzzles of pulsed electromagnetic transients came the evolutionary cognition of the electromagnetic energy. From this perspective, the development of power electronics deepens and widens the content of the whole electrical engineering.

A power electronics system consists of the semiconductor switch, electronics circuit, control, etc. Although all elements in the circuit display different characteristics, at the full-system level their harmonious interaction yields the power electronics system as an organic combination. Generally, the power electronics system is a combination of the software and hardware, an interaction of the energy and information, an effective transformation between the linearity and nonlinearity, a mixture of the discrete and continuity, and a coordination of multiple timescales. All such special and comprehensive characteristics exhibit the property of power electronics. The related power electronics technique is in charge of the controllable energy transformation under such characteristics, balances the electromagnetic energy in transients, and deals with the relationship between the device and equipment, between the control and the main circuit, and between stray parameters and lumped parameters.

Presently, the power electronics theory and techniques are still staying at the application level, synthesizing power semiconductors, electronic circuits, and control algorithms. Its theoretical framework is still progressing. The perspectives of system integration, energy conversion, and electromagnetic transients provide a promising direction for the next-step theoretical study of power electronics. In the last two decades, our research team carried out in-depth research and engineering applications in domains of motor drive, grid-tied photovoltaic system, high-power multilevel converter, electric router for the energy internet and wireless charger. In this process, the electromagnetic transient analysis is one of the core topics. Based on such cognition combined with our research exploration and applications in the past multiple years, we composed this book as an early attempt at exploring rules and analysis methodology of electromagnetic transients. Hope this book can be inspiring for whole power electronics community.

Overall ten chapters are included in this book. Chapter 1 is the introduction, beginning with the analysis and synthesis of high-power converters to summarize and recognize the structure and property of power electronics systems. This chapter, centered with the core of power electronics systems, i.e., energy balancing in the transient commutation process, elaborates characteristics of the power electronics conversion, e.g., multi-timescale, quasi-discrete, and strong nonlinearity. After illustrating characteristics of energy transients in various high-power converters, this chapter listed typical confusions and problems in the power electronics researches and applications. Chapter 2 is the foundation of the argument, depicting various transients in different commutation loops and timescales of power electronics systems. It provides mathematical models and modeling methodology for these transient processes, details the difference and impacts of various timescale transients, and analyzes the mathematical expression and differences of the electromagnetic pulses and pulse sequences. Chapter 3, based on the internal mechanism and external impact factors of the power semiconductor, analyzes its transient behaviors when equipped in actual power electronics systems. Chapter 4 is focused on the transient commutation process and related stray parameters, including the impact, extraction, and reduction of stray parameters in various converters. Based on the switch characteristics and interactions among components inside the power electronics converter, Chap. 5 defines the safe operation area. The conflict between the system potential and reliability is discussed based on the system-level energy conversion and full-time electromagnetic transients. Chapter 6 emphasizes the structure, content, and function of the sampling system. It describes the difference between energy and information signals, analyzes the influence of the sampling delay and error, and optimizes the sampling system. Chapter 7 is about the difference and interaction of the information pulse, gate-drive pulse, and power pulse. Based on the comparison of those pulses the interaction of the information and energy in the power electronics system is discussed. Chapter 8 from the control aspect analyzes the high-performance closed-loop control and related constraints. The structure of the closed-loop control, characteristics of the conventional control with limitations, particularly the cause and impact of invalid and abnormal pulses are the focus. Based on the transient energy flow and distribution, Chap. 9 sets forth the concept and fundamental of the transient energy balancing control, demonstrates the transient energy balance based control algorithm, and analyzes the control stability and robustness. Chapter 10 aims at the transient analysis in the dynamic balancing process of the IGBT-series-connection based converter and the SiC-based high-switching-frequency power amplifier.

This book summarizes the exploration and application of our research group in the power electronics domain over the past 10 years. Many scholars and students who worked in our group made significant contributions to this book. They are Haitao Zhang, Rong Yi, Yingchao Zhang, Yongchang Zhang, Yulin Zhong, Sideng Hu, Lu Yin, Gaoyu Zou, Fanbo He, Kainan Chen, Shiqi Ji, Junjie Ge, et al. We would show our great gratitude toward their effort. Meanwhile, during the composition of this book, other faculty members and students in our research group made a great effort to assist and edit, for example, Kai Li, Xiaoying Sun, Xing Weng, Ye Jiang, Yatao Ling, Sizhao Lu, et al. They all deserve our thanks. In addition, tremendous amount of articles were referred and cited during the book composition. We would like to thank the authors of these references.

Part of this book was under the sponsorship of the Major Project of National Science Foundation of China (NSFC), multiple-timescale dynamic behaviors and operational mechanism of high-power power electronics mixture systems (51490680, 51490683), which emphasizes the comprehensive system analysis of power electronics, modeling of electromagnetic transients, comparison of the information and power pulses, comparison and analysis based on multi-timescales, and the control algorithm based on the energy balancing control. We are very grateful for the great NSFC support.

This book provides the reference for power electronics engineers especially those in the high-power converter research, development, and application. It can also benefit the faculty members and graduate students in the same domain.

Due to the limitation of our knowledge and insights, and the continual progress of the research of power electronics transients, our work is still quite preliminary. Some flaws and even mistakes might exist in the book. Comments and criticisms are more than welcome.

Your Sincerely

Beijing, China   Zhengming Zhao
April 2018

Contents

Chapter 1
Introduction

The major difference between power electronics and microelectronics is the power rating. All the power electronics systems in this book are rated beyond tens of kilowatts up to multiple megawatts, with the voltage level above hundreds of volts and current over tens of amperes and even thousands of amperes, facing various applications.

Power electronics with related applications has a three-level industrial chain, i.e., devices (bottom level), power electronic converters (middle level) and associated applications (top level). Such industry covers nearly all the key technologies related to the economic development and military defense, e.g., materials and manufacturing, information and communication, aviation and transportation, energy and environments, etc.

Power electronics is an interdisciplinary engineering branch, covering semiconductor devices, power conversion circuits and associated control strategies. Specifically, it uses weak signals to control strong power, adapts software to control hardware, embeds device into system, and bridges information with energy. Back to 1973, Dr. William E. Newell from Westinghouse Electric pointed out that power electronics is an interdisciplinary domain containing microelectronics, electricity and control. Thus, the essence of power electronics is the interaction of devices, circuits and control thereby realizing the effective electromagnetic power conversion.

To realize the expected power conversion, power electronics systems usually adopt the methods such as pulse-width modulation, which outputs the power in forms of energy pulses and their pulse sequences. Such pulses are the basic behaviors of power electronics systems, which involve the critical electromagnetic dynamic processes. Time constants of these electromagnetic transients are usually between nano- and micro-seconds, and these short timescale transient processes determine the system reliability. On one hand, such processes are the fundamental of power conversion. On the other hand, without effective control the damages and failures of devices will occur. Therefore, short timescale electromagnetic transients are the cores of power electronics systems.

© Tsinghua University Press and Springer Nature Singapore Pte Ltd. 2019
Z. Zhao et al., *Electromagnetic Transients of Power Electronics Systems*,
https://doi.org/10.1007/978-981-10-8812-4_1

1.1   Decomposition of Power Electronics Systems

Multiple factors together form the power electronics systems. To detail the electro-magnetic transients, an analysis of those building blocks (power devices, circuitry and control, etc.) is a must, such as their structure, characteristics, functions and limitations.

1.1.1   Power Semiconductor Devices

Semiconductor devices are the fundamental of power electronics systems. Since 1950s when the first thyristor (also known as silicon controlled rectifier, SCR) came to the world, power electronics witnessed the evolution of devices from uncontrolled to controlled, current controlled to voltage controlled, hybrid integrated modules to wide-bandgap (WBG) devices, e.g., metal oxide semiconductor field effect transistors (MOSFETs) in 1970s, insulated gate bipolar transistors (IGBTs) in 1980s, integrated gate commutation thyristors (IGCTs) in 1990s, and WBG devices such as silicon carbide (SiC) device in 2000s. In different development periods, high power, high voltage and high switching frequency are always the pursuits of power semi-conductor devices. As the successors of SCRs and gate turn-off thyristors (GTOs), IGBTs become the mainstream of power electronics devices. Tradeoffs thereby opti-mizations of the on-state voltage drop and the switching power losses are the ultimate goals of IGBTs in high-frequency and high-power applications. Meanwhile, multi-series and multi-parallel dies within one package is the must. Nowadays IGBTs have become the main game player in high-voltage and high-power applications. Further improvement for higher frequency, higher power and easier drive is anticipated.

WBG semiconductor devices have become the hot spots of power electronics. The surge of WBG materials such as SiC and its related devices, e.g., SiC schottky diodes, SiC JFETs and SiC MOSFETs is witnessed in recent years. Compared to the con-ventional Si-based devices, WBG devices exhibit superior thermal performance and higher dielectrics, which provide enormous potentials to improve the performance and enhance the capability of power electronics systems.

Any single semiconductor device in power electronic converters acts as a power switch, which is the source of power pulses. However, its on-state and off-state char-acteristics are much different from conductors and insulators, which results from its PN junctions inside. PN junctions are the base of semiconductor devices. Compre-hending the characteristics of PN junctions is very important for the development and applications of power electronics. In the following sections review on the char-acteristics of PN junctions will be extended.

1. **On-state and off-state of single PN junction**

It is well known that free electrons can be activated inside the intrinsic semiconductors under specific temperatures. Same amount of holes emerge to fill the vacancy of such

Fig. 1.1 The PN junction structure

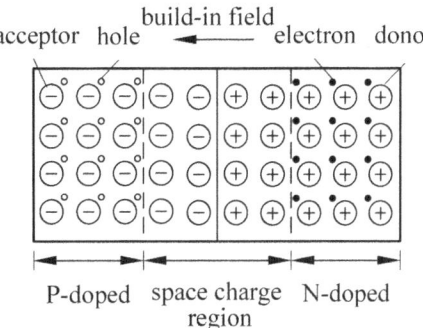

free electrons, which is the major difference between semiconductors and conductors solely using electrons as the current carriers. Intrinsic semiconductors contain holes and electrons as the current carriers, whose concentration varies significantly with the temperature. It directly results in that its conductance changes with the operational temperature, a main characteristic of semiconductor switches.

On the other hand, using appropriate techniques to dope impurity inside the semiconductor, e.g., alloying, diffusing, extending, ion implanting, the conductance of semiconductors varies drastically. Based upon the valence of impurity, two types of semiconductor appear, i.e., N-type using electrons and P-type with holes. The difference of the carrier concentration exists around the contact surface of N-type and P-type semiconductors, yielding a concentration gradient, which in turn results in the directional diffusion of current carriers from high-concentration areas to low-concentration domains, instead of random movements. Specifically, electrons will diffuse from N-type to P-type and holes will flow in the opposite direction, forming the diffusion current. It is obvious that such diffusion current does not comply with the Ohm's law. The diffusion generates immobile positive and negative space charges around the P-N junction boundary, forming an electric field pointing from N towards P, i.e., built-in field shown as Fig. 1.1. Such field obstructs the further diffusion. Meanwhile some electrons and holes drift back to the N-type and P-type semiconductor, respectively. Hence the drifting direction is opposite to the diffusing. Such drifting current meets the Ohm's law given that drifting carriers are moved by the electric field.

In summary, the diffusing and drifting current together are major internal behaviors of semiconductors. Such two currents are independent and meanwhile interacting with each other. A dynamic balance will be ultimately realized when a zone full of stabilized space charges is formed. We name such area as the space-charge region, i.e., depletion region, inside which positive charge is equal to negative charge thereby maintaining neutral. This region is called PN junction.

The major behavior of the PN junction is its unidirectional current flow, which will only be exhibited when the external voltage is imposed. A forward voltage U_D, with its positive terminal connected to P and its negative terminal connected to N, leads to an electric field opposite to the built-in field, which is named as the forward bias

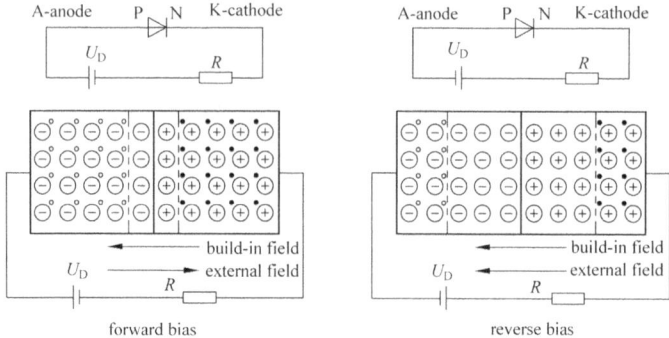

Fig. 1.2 Forward bias and reverse bias of a PN junction

of the PN junction. When a negative voltage is imposed, i.e., the positive terminal of U_D is connected to N while the negative is connected to P, an electric field aligned with the built-in field is created, i.e., the reverse biased, as shown in Fig. 1.2.

When forward biased, the diffusing current is dominant over the drifting current under the external field, swiping out space charges and narrowing the depletion region. This is also called minority-carrier injection. Current appears in the external circuit flowing towards the P region, i.e., forward current I. Increasing U_D further weakens the built-in field thereby enhancing the forward current. A low voltage drop appears across the PN junction when the current flows through, named as forward voltage.

When reverse biased, the external field halts the diffusion, while the drifting effect is enhanced for minor carriers. More current carriers leave the PN junction, widening the depletion region. Such phenomenon is known as minority-carrier withdrawal. The whole PN junction withstands the external voltage with barely any current flowing through. This is called reverse bias, determining the voltage withstanding capability of power switches.

2. **Interactions of multi-PN junctions**

Majority of power semiconductor devices contain more than one PN junction, for instance, a bipolar junction transistor (BJT) contains two PN junctions, while a GTO has three. Interactions among PN junctions, and those among PN junctions and peripheral circuits, determine the characteristics of power switches. Take the IGBT as an example, a typical hybrid switch made of MOSFET and BJT. Such device is a fully controlled switch, consisting of a voltage-controlled PNP part and two PN junctions, as shown in Fig. 1.3.

Such a three-layer two-junction structure is not formed through circuitry interconnection, but through a complex doping technology. The bottom P layer of the device is the emitter of the BJT part, injecting current carriers into the base. The middle base is made of two N-type layers, one of which is lightly doped with the other heavily doped. Such base is to transport and control carriers. The collector is one P-type layer

Fig. 1.3 The structure of PN-junctions in IGBT

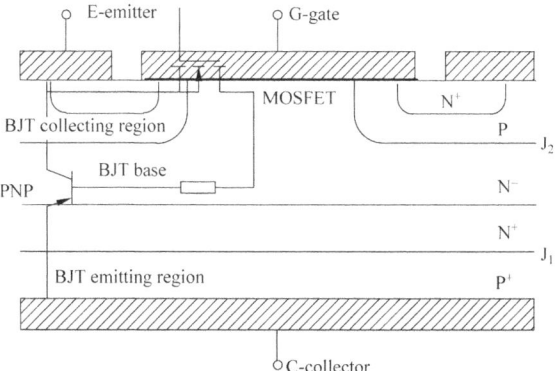

collecting current carriers. PN junctions are formed at the boundaries of different layers. Here we name the PN junction between the base and emitter as J_1, and that between the collector and base as J_2.

A large enough forward voltage imposed between the gate and emitter of the IGBT, when collector-emitter is forward biased, will turn on the IGBT. At this state, J_1 is forward biased while J_2 is reverse biased. Controlled by the gate, an N-channel appears between gate and the emitter, allowing electrons to flow from N + zone underneath the emitter towards the base. Due to the reverse biased J_2, part of electrons in base diffuse into the collector. On the other hand, because of the existence of electrons flowing from the emitter to the base, some holes flowing from the collector will combine with electrons in the base, while left holes drift to the emitter due to the built-in field of J_1, forming the current from the collector to the emitter. Such holes through J_1 shrink the space charges, yielding the IGBT turn-on state. When V_{GE} is lower than the threshold voltage, the N-channel is not sustainable, yielding no electrons supplied from the emitter to the base. Therefore the interaction between the emitter and collector is significantly weakened. Given the reverse biased PN junction has no extra current carriers flowing through, IGBT remains off.

In summary, a single PN junction based device, such as diode, has its on and off states totally determined by the external circuit, while a multi-PN-junction device, such as IGBT, relies on the interactions among PN junctions to turn on or off, which are ultimately determined by the state of the control terminal. Understanding that a multi-junction device is a controllable switch by manipulating the control terminal thereby building interactions among PN junctions is the fundamental of power electronics. Given that turn-on/off processes are all related to variations of space-charge zones of the PN junction, the controllable switches have turn-on, turn-off, conducting and blocking characteristics, the former two of which are transient while the latter two are static. An ideal switch will ignore transient processes of turn-on/off, the forward voltage drop in the conducting state and the leakage current in the blocking state. To emphasize the switching function of power semiconductor devices, we will name it as power switches in following sections.

1.1.2 Power Conversion Circuit

Interconnections between power switches and other related components thereby realizing the effective power conversion, such as voltage, current, frequency and waveform transformation, forms the power conversion circuits, namely topology, the premise of power electronics systems. Topology in conventional power electronics study is focused on the interaction between ideal switches and passive components represented by lumped parameters, targeting the macroscopic electrical transformation, e.g., input/output waveform and values. The rectifier was proposed in 1920s, DC-DC converters emerged in 1960s, and DC-AC inverters were created in 1970s. Nowadays various hybrid topologies become the focus. All such effort is to enhance the capability of power conversion, e.g., high voltage, high current, low harmonics and high efficiency. Typical high-voltage and high-power electronics topologies include: neutral-point-clamped (NPC) structure, flying capacitor multi-level topology, cascaded H-bridge circuits, and modular multi-level converters, as shown in Fig. 1.4.

All such topologies share some common points, i.e., (1) each power switch only undertakes part of the DC-bus voltage during the off state, allowing low-voltage switches to be combined together for high-voltage and high-power applications without using any dynamic voltage balancing circuit; (2) more voltage levels improve the output voltage waveform; (3) a lower switching frequency can be adopted to realize the same output as conventional two-level converters, resulting in lower switching loss thereby higher efficiency; (4) more voltage levels reduce du/dt under the same DC-bus voltage, preventing the breakdown of associated devices and improving the electromagnetic capability (EMC).

It is obvious that the description above is based on ideal switches and lumped parameters, without considering the transient process of non-ideal switches and stray parameters along the circuitry interconnections. In real practice, such ideal topology based power electronics design, analysis and control tend to discount the power converting capability and reliability, especially in high-power converters.

1.1.3 Pulse Control

The pulse control is critical for power electronics controllability, which essentially is the control of power switches. To effectively control the output voltage, current, frequency and waveform, modulation techniques such as pulse-width or pulse-amplitude modulation are adopted, i.e. transforming the continuous input into the discrete pulse sequences through switching on or off semiconductor switches. The reference voltage, current or frequency can be approached then through varying the pulse width, amplitude or period.

Pulse control is one of main characteristics of power electronics, including two aspects. One is the hardware, mainly the signal processing unit. From the analogue

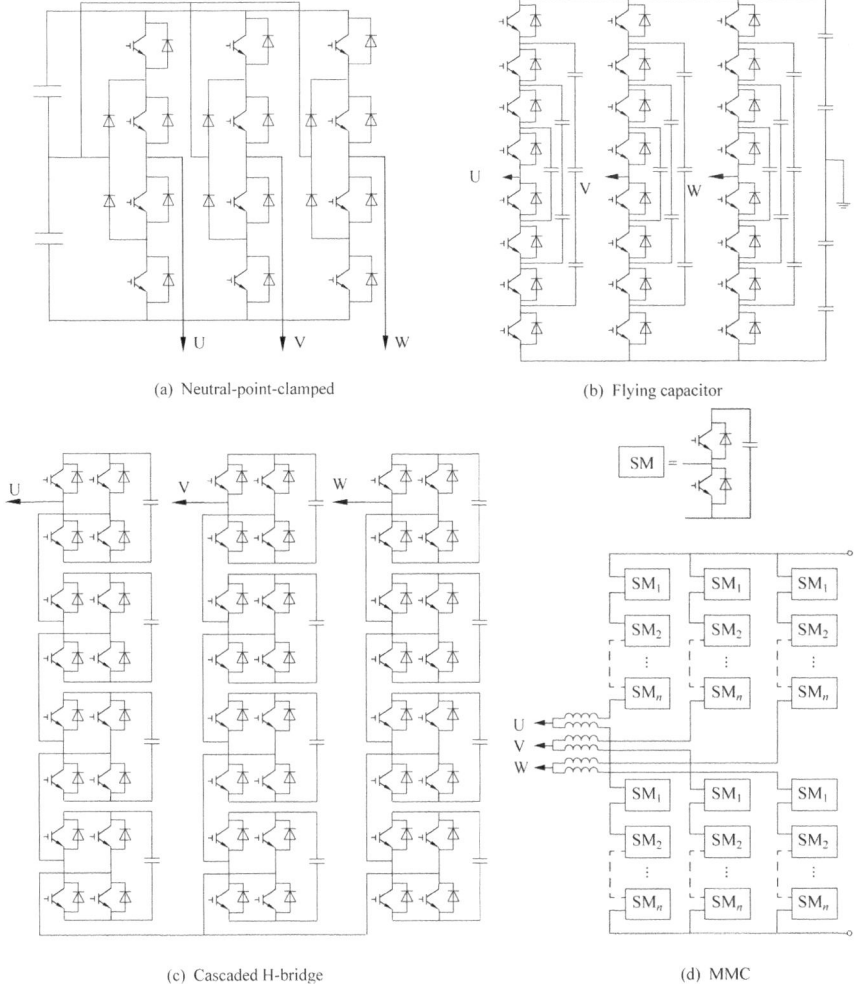

(a) Neutral-point-clamped

(b) Flying capacitor

(c) Cascaded H-bridge

(d) MMC

Fig. 1.4 Typical topologies used in high-voltage and high-power power electronics systems

circuits in 1950s, single board computer in 1970s, single-chip microcontroller in 1980s, digital signal processors (DSPs) in 1990s, to multi-core DSPs in this century, the computation speed, accuracy and storage have been improved drastically. The other is the software, mainly pulse control techniques, including pulse modulations and system control strategies.

Multiple pulse modulation techniques have been proposed, e.g., signal comparators, hysteresis control, space-vector control, selected harmonics elimination and one-cycle control. For high-voltage and high-power systems, main modulation schemes include multi-level sinusoidal pulse-width modulation (MSPWM), space vector pulse width modulation (SVPWM) and its related variations. Shown in Fig. 1.5 are the typical five-level SPWM and three-level SVPWM. Employing such PWM

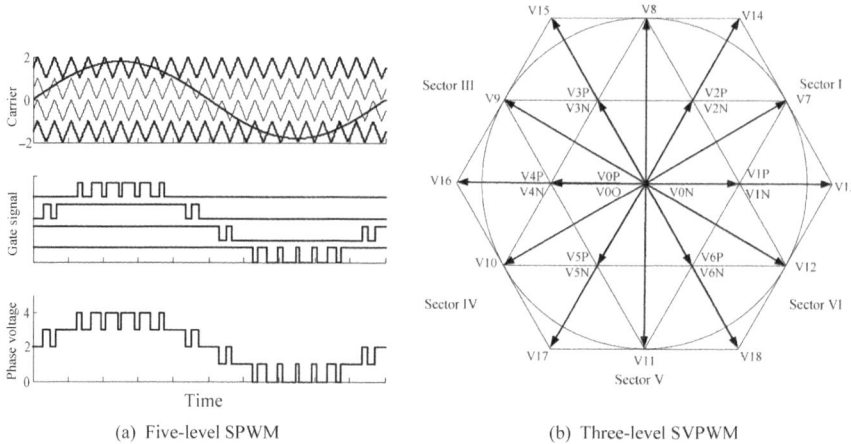

Fig. 1.5 Schemes of five-level SPWM and three-level SVPWM

schemes with the closed-loop control generates various closed-loop PWM control methods, e.g., vector control (VC) and direct torque control (DTC) in motor drive systems.

Implementing bottom-level pulse modulation techniques in the control hardware and software expedites the development of power electronics systems. Particularly for high-voltage and high-power applications, future trends include (1) combination of parameter control and reference control (inner loop with outer loop), (2) the multi-core chip with a field-programmable gate array (FPGA), and (3) signal modulated combined with direct power control. However, it is worthwhile pointing out that previous pulse-width modulations are based upon ideal switches and linear circuits. The problem caused by the non-ideal switch and non-linear circuits is the signal transportation delay and waveform distortion. Shown in Fig. 1.6 are the signal generated from CPU (top plot), the gate signal generated by the gate-drive circuit (middle plot) and the voltage/current waveform of the power switch (power pulse shown in the bottom plot), respectively. Time delay and distortion exist in the real practice, especially due to the existence of the "blind zone" in switching on/off processes, i.e., non-controllable switching on/off intervals, which obstruct the control strategy, generate abnormal pulses, distort the waveforms and even damage power switches and systems.

Fig. 1.6 Experimental
curves among control, driver
and power pulses

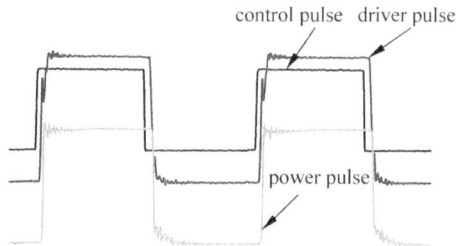

1.2 Synthesis of Power Electronics Systems

Different functional factors co-exist in power electronics systems, which from the system aspect are integrated systematically thereby exhibiting a high synthesis. This section is to help understand interactions among factors thereby systematically comprehending the electromagnetic transients of power conversions. In summary, power electronics embraces the integration of the software and hardware, interaction between the information and energy, transfer between the linearity and nonlinearity, mixture of continuity and discreteness, and co-ordinations of multi-timescale subsystems.

1.2.1 Integration of Software and Hardware

An exemplary power electronics system is shown as Fig. 1.7, the hardware of which includes: (1) a rectifier made of power diodes (or fully controllable switches), converting AC into DC; (2) the DC bus made of capacitors and interconnections, filtering the voltage ripples thereby generating a high-quality DC source; (3) an inverting stage made of fully controlled switches, inverting the DC component into needed AC waveforms; (4) a control system comprised of sensors, control chips, communications and gate-drive circuits, acting as the carrier and channels of information signals.

Therefore the hardware system mainly consists of electric components (various power switches, passive components, signal processing units, etc.) and their interconnections. An ideal hardware system treats power switches as ideal, i.e., no time interval is taken for switching on/off actions. Thus, there is no need to consider switching transients but only the on-state current conducting capability and off-state voltage blocking capability. Passive components mainly are inductors and capacitors, treated as ideal current sources and voltage sources, respectively, i.e., no need to consider the component loss or stray parameters but only lumped parameters. For signal processing units, only the signal processing capability is focused on, instead of the time delay or signal distortion. Interconnections among components incudes the connector and cable materials, connecting methodology (topology). Ideally no stray parameters of the connectors and cables but only the topology are considered. All the above bottleneck the hardware system of a typical power converter.

Software systems include the signal sampling, I/O communication, signal processing, system control, protection, system management, etc. Such functions are realized by the software code inside micro-controllers or analogue circuits. Ideally neither delay nor distortion of the software system is considered.

Both software and hardware systems are necessities of power electronics systems, i.e., the hardware system embodies its controllability through the software, while software displays its function through the hardware. Assuming both systems are ideal, we could completely utilize the computer to emulate the whole power conversion

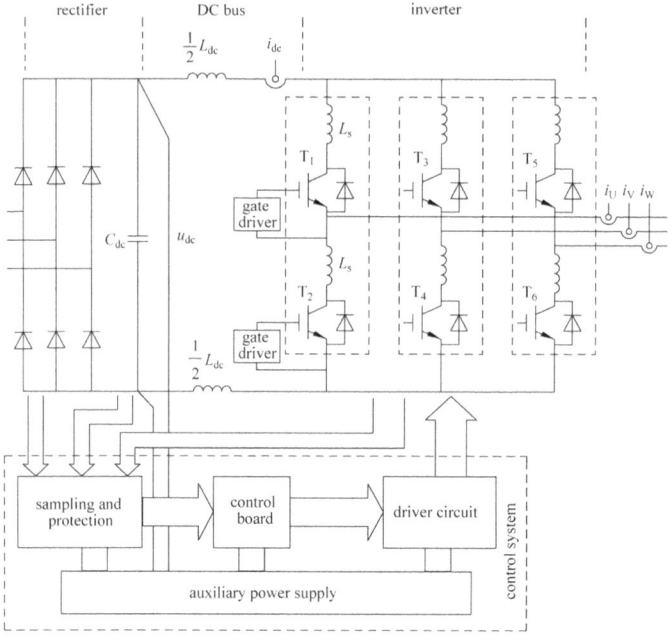

Fig. 1.7 Exemplary power electronics system

system without considering characteristics of actual software and hardware, i.e., ideal operation or computer aided simulation. However, neither software nor hardware is ideal. For instance, both switching-on and switching-off processes of power switches take time and generate the loss, stray parameters exist along the connectors, and the signal processing creates the time delay and signal distortion. All these non-ideal factors emerge in the actual operation among each firmware component and software behavior, e.g., difference exists between IGBTs and IGCTs even though they are both power electronics devices employing the same connecting method. Power converters using these two switches behave differently as well. Hence, an effective power conversion needs address such non-ideal factors and match parameters of actual software and hardware.

1.2.2 Interaction Between Information and Energy

The hardware and software carry the energy and information inside the power conversion system, respectively. Here the energy is electromagnetic type, provided by external sources, transformed and transferred by the power converter and delivered to the load. The information is carried by sensor signals, control commands and communication, resulting from sensors, external control circuit and logic operations

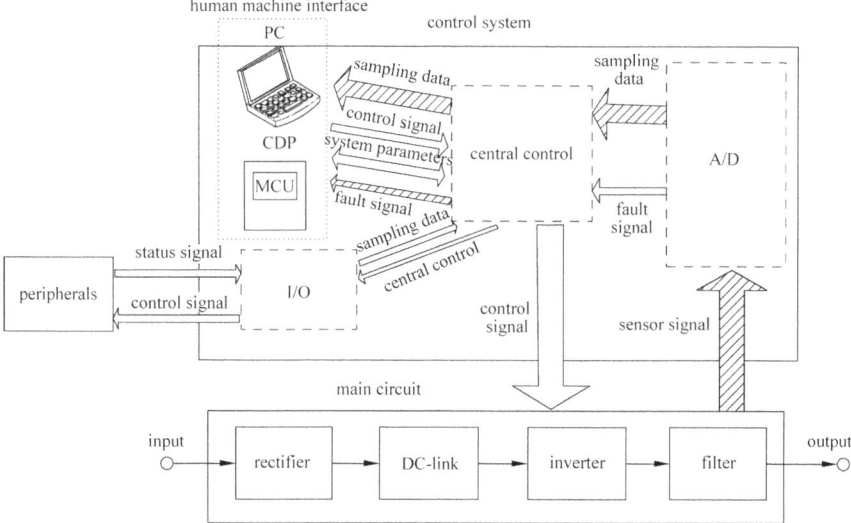

Fig. 1.8 Scheme of energy flow and information flow

inside the signal processing unit, respectively. Both energy and the information signals have electromagnetic energy, however, are totally different in terms of the order of magnitude and functions. Energy usually has the high order of magnitude to do the work for the load. Information signals usually have small magnitudes, carrying the information for control and communication during the system operation. The interaction between energy and information inside one power electronics system is illustrated as Fig. 1.8.

It can be seen from Fig. 1.8 that energy passes through the main circuit, information exists in the control system, peripherals and main circuits, and the effective transformation and transfer of the energy flow depends on the information flow.

Magnitude, distribution and direction are the three characteristics of the energy flow. It is the function of both one-dimension time and three-dimension space. It can be not only bidirectional but also multidirectional. It is not only conducted along main components and connection cables, but also radiated into the space. It is worthwhile pointing out that energy flow must comply with energy conservation without sudden changes. Therefore ideal switching characteristics of power switches never exist. Electromagnetic energy change in the switching process is always continuous, exhibited as one short-timescale transient. This is one of the major characteristics of power electronics transients.

Information flow is, on the other hand, multivariate, e.g., sensed information of the main-circuit status, external command signals, analysis/calculation carried inside the microcontroller. It can be the single-amplitude pulse or continuous analogue signal. It can be real time or past data. The most important is the switching control information of power semiconductors, which meanwhile is assisted and determined by other surrounding information signals.

Interactions between energy and information determine the effectiveness of the power conversion, as shown in Fig. 1.8. In this figure, energy flow is determined by the switching control information for fully controlled power switches, which ultimately is generated by the external command, sensing information and logic information inside the micro-controller altogether. In one word, the switching control information determines the energy transformation and transfer characteristics, while other information determines the switching control information. All information signals coexist and function together, forming an inseparable unity.

1.2.3 Transfer Between Linearity and Non-linearity

Non-linearity characteristics of power electronics components result in various blind zones and uncontrollability during the interaction of information and energy. Here linearity or non-linearity is defined as the relationship between input and output during the power conversion and information control. When the input and output of the system meets the homogeneity and superposition, linearity becomes true, otherwise non-linearity. Table 1.1 illustrates the non-linearity inside power electronics systems.

It can be seen that the non-linearity roots from hardware materials, component operations and connecting ways. For a power electronics system, nonlinearity is a norm while linearity is a special case. Mathematical description of those non-linearities is the pre-condition to comprehend the operation of power electronics, which facilitates the understanding of interactions between energy and information through involving the hardware material, interconnections and working modes. How-

Table 1.1 Non-linearity in a power electronics system

Element	Feature	Illustration	Formula	Effect
Power semi-conductor	Switching characteristics		$u_{out} = \left\{ 1, 0 \right\}$	Switching mode
Inductor	V-A characteristics		$i = f(u)$	Magnetic saturation
Capacitor	Q-V characteristics		$q = f(u)$	Variable dielectric coefficient
Connector	Parasitic parameter		$u = i \times \sum_{k=1}^{n} z_k$	High frequency oscillation
Circuit	Multi state variable loops		$i^{(n)} = f(i^{(n-1)}, \ldots, i', i)$	High order differential equation

ever, given the actual nonlinearities are extremely complicated, we usually simplify a non-linear system into a linear one thereby minimizing the complexity of energy-information interactions. The advantage of this operation is to simplify the system design, analysis and control, however it meanwhile results in the inaccurate description of electromagnetic processes, mismatches among parameters and imprecision of the control implementation, especially for fault diagnosis and system protections.

More importantly, some abnormal nonlinear behaviors emerge in power electronics systems, for example, abnormal pulses during switching actions, spurs in the gate-drive output, electromagnetic oscillation on the interconnection bus bars. In general, such abnormal behaviors of power electronics systems can be traced back to inner and outer factors. Optimization of those factors could help eliminate such abnormal behaviors, however, such parameter variation usually is caused by the component aging and change of the operational environment. Therefore, control of such abnormal transients needs adapt the parameter variation, model the system with varied parameters, set appropriate boundary conditions and optimize the controller, without altering the original topology.

1.2.4 Mixture of Continuity and Discreteness

The conventional mixture system contains both continuous and discrete signals, where two types of signals and events mix and interact with each other thereby forming one dynamic unity. Compared to a single time continuous or discrete system, the mixture system has following specialties: two different types of signals coexist in one system, one is the time continuous while the other is the discrete type; the system behavior is determined by both continuous and discrete signals; when the continuous signal exceed the threshold, discrete events are triggered.

In power electronics systems, digital control is adopted (ADC sampling, DSP operations, etc.) along with the pulse modulation for power switches, making both information and energy the mixture of discrete and continuous signals. Its discrete characteristics are revealed through the information pulse and energy pulse, i.e., discrete information pulse comes from the micro-controller, shown as the standard pulse sequence; the discrete energy pulse comes through power switches during switching on/off actions. However, energy cannot change abruptly, which makes the energy pulse not a real discrete signal, but a continuous signal fast changing within a short time interval. In this book, *discrete* and *continuous* correspond to information pulses and energy pulses, respectively. Shown in Fig. 1.9 is the relationship between discreteness and continuity in power electronics systems, i.e., the discrete determines the continuous, and information signals determine energy pulses.

Information pulse sequences are usually close to ideal in power electronics systems, i.e., standard square waveforms. However, in the transportation process, signals might be delayed or distorted due to stray parameters in the loop, for instance, fiber signals and gate signals all exhibit the delay and distortion. Therefore digitizing the information pulse is not accurate either. For the energy pulse, its transition is

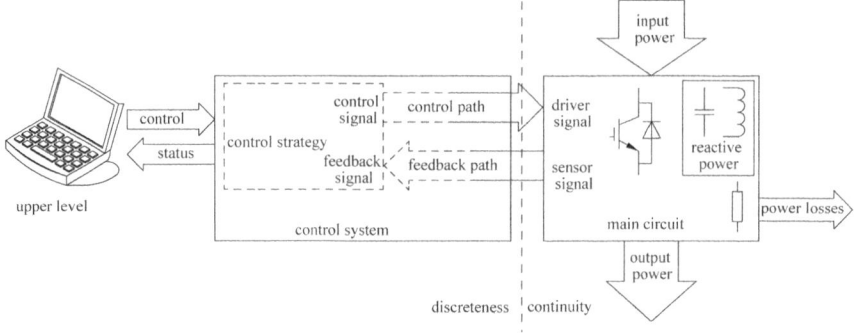

Fig. 1.9 Relationship between discreteness and continuity in power electronics systems

ultra-fast, e.g., at the nano-second scale. Such high energy transformation within ultra-short timescale is one of major characteristics of electromagnetic transients in power electronics systems. Therefore the fundamental of power electronics lies in that information controls energy and the discrete controls the continuous. The characteristics of discrete signals are expressed by the starting moment, lasting interval, and ending moment, while the continuous characteristics are described by amplitude, changing trend and lasting interval. Quantifying the relationships among those variables is the key to realize the optimal control of the continuous-discrete mixture system.

1.2.5 Coordination of Multi-timescale Subsystems

Various electromagnetic loops exist in power electronics systems with complex nonlinear dynamics. Mixed with discrete and continuous signals, the overall system exhibits a comprehensive electromagnetic dynamic process, with each subsystem having its own transients and timescales. The basic structure of a power electronics circuit is shown as Fig. 1.10.

Due to the different inductance, capacitance and resistance in different loops, their time constants are also different. Therefore electromagnetic transients of various timescales coexist in a power electronics system. The transition speed of energy pulses is not a constant. Electromagnetic energy flow, holes flow and electron flow are all travelling at different speeds. In such a system, each subsystem has its own time constant for energy conversion. For example, main passive components and their circuits have time constants of milliseconds, power switches and related control circuits have time constants of microseconds, and some high-switching-frequency soft-switching circuits and the digital control system have nanosecond-level time constants. When driving the motor, the power electronics converter has time constant of seconds for its mechatronics conversion. Subsystems with different time constants

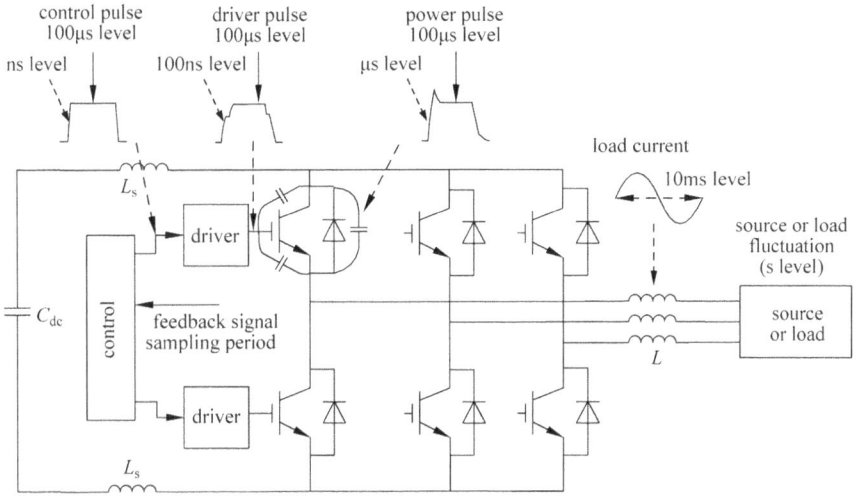

Fig. 1.10 The basic structure of a power electronics circuit

form the overall power conversion system, where the key is to reach the dynamic balance of the electromagnetic energy during the transformation, transfer and storage.

Real applications indicated that most failures of components and systems occur during the electromagnetic transients, i.e., energy transition from one steady state to another. During such transients, energy distribution tends to be imbalanced especially when multiple subsystems co-function together with different time constants. Destructive local energy converge might appear. Previous research on topology and circuits is mostly focused on the second-level or millisecond level, e.g., investigation of the voltage/current total harmonics distortion (THD) and electromagnetic interference (EMI). The component failure and EMI caused by short-timescale transients behave differently at different power ratings, which become more severe in high-power power electronics systems due to higher energy pulses and faster energy transitions. Various parts inside one power electronics system have different electromagnetic parameters, resulting in different time constants in the dynamic processes thereby yielding a mixture of multi-timescale transient processes. Therefore coordination among such transients is highly demanded through matching and optimizing each subsystem, which is a key characteristic of power electronics systems.

It is worthwhile pointing out that for different timescale transients, the modeling methodology, analyzing method, parameter matching and control strategies need be differentiated. If the whole-time-domain modelling and analysis of electromagnetic transients is required, we need bridge all different timescale processes together. To obtain the consistency of the theoretical modeling, mathematical analysis and control precision, we need classify all these transients based upon their own timescales.

As shown above, the transformation from the information signal flow to the energy flow in power electronics is hard to be explained by the conventional electromagnetics. It is also difficult to apply the conventional topology to the switching process of power semiconductors. We need stand on top of the energy pulse sequences, analyze based upon the microscopic characteristics, such as non-ideal switching processes and energy pulse transients, and measure with short timescales (ns–μs).

Power electronics systems are used to control the energy conversion and transfer the energy in the defined direction. In this process, regardless of the shape and timescales of the energy pulse or pulse sequence, energy conservation and no sudden energy change must be strictly complied with, which is the theoretical precondition and fundamental of power electronics analysis.

1.3   Applications of Power Electronics Systems

With the development of the device, topology and control, a surge of power electronics equipment and systems has been witnessed with a fast-growing demand, especially those high-voltage and high-power power electronics systems, where the IGCT, IGBT and IEGT (injection enhanced gate transistor) are main switches, cascaded H-bridge, NPC multi-level and modular multi-level are the main topologies and multi-level SPWM and SVPWM are the main modulation schemes. Majority of high-voltage and high-power systems are focused on flexible alternative current transmission systems (FACTS) or voltage-source converter based high-voltage direct current transmission systems (VSC-HVDC), high-power traction systems, variable frequency motor drive systems, wind energy generation, grid-tied solar systems, all-electric ships, etc. Different switches, topologies and control strategies are adopted in different applications, leading to different electromagnetic transients, which are detailed below through several typical high-voltage and high-power power electronics systems.

1.3.1   Flexible AC or DC Current Transmission

1. FACTS

Power electronics is the core of FACTS, to increase the power capability and enhance the system controllability for AC transmissions. Power electronics technology plays a critical role in FACTS, such as the unified power flow controller (UPFC), the interline power flow controller (IPFC), static synchronous compensator (STATCOM), static synchronous series compensator (SSSC), static var compensator (SVC), static var generator (SVC), thyristor switched series capacitor (TSSC) and convertible static compensator (CSC). All of the above employ high-power semiconductors as main switches, combine all switches effectively, and form high-current and high-voltage electromagnetic energy conversion systems.

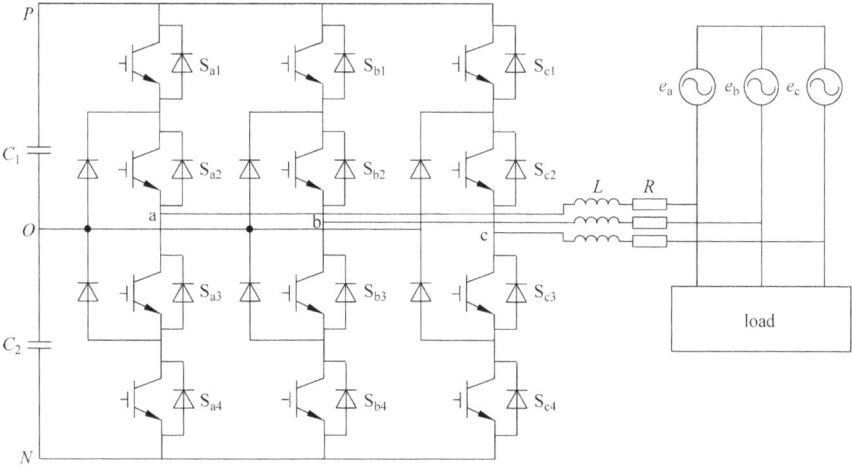

Fig. 1.11 The circuit schematic of STATCOM

Take the STATCOM as an example. Its basic operation principle is to form bridge circuits via high-power switchable devices such as IGBTs and GTOs, connect the power inductor to the grid and adjust the amplitude and phase of the bridge-circuit output voltage or directly control the output current, which alters the reactive current generated to or from the grid thereby ultimately compensating the reactive power. The circuit schematic is shown as Fig. 1.11.

The control objective of the STATCOM is the amplitude and phase of a constant-frequency AC system. The goal is to improve the power system performance, including the dynamic voltage control during the electricity transmission and delivery, damp the oscillation in the transmission system, enhance the system stability and mitigate the voltage fluctuation and flicker, all of which are high-power and large-timescale electromagnetic transients with characteristics of high voltage, high voltage variation and fast dynamic response. Therefore the focus of the STATCOM includes power-stage topology, modulation schemes of multi-level inverter, reactive power compensation and energy storage, and control strategies, which emphasize the transients of sub-milliseconds to millisecond timescales.

2. VSC-HVDC

Such systems are also named as HVDC Light, using power electronics voltage source converters (VSCs) to transmit the DC power. It has smaller impedance thereby exhibiting faster dynamic response compared to AC systems. According to the bridge-circuit difference, HVDC can be classified into three types, i.e., controllable switching topology, controllable power-source topology and controllable switching power-source topology.

(1) Controllable switching topology, which is the two-level or three-level converter. Such topology has relatively simple structure and control, a high switching fre-

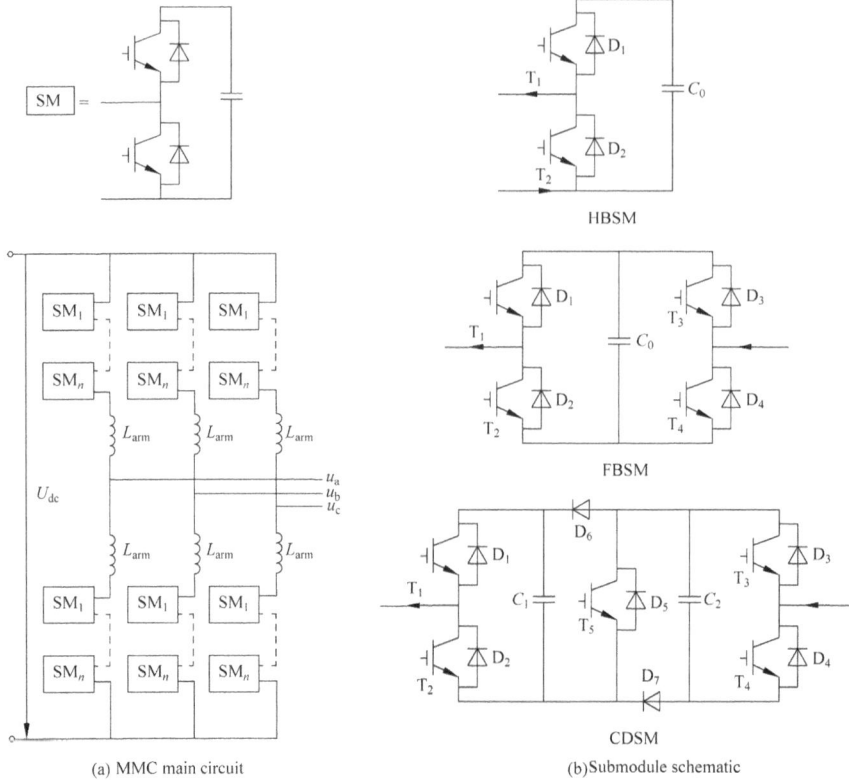

Fig. 1.12 MMC topology using various sub-modules

quency thereby a high power loss. In addition, series connected switches are demanded for high-voltage applications, which however faces the challenge of dynamic voltage balancing, a typical short-timescale transient control. Therefore such topology is rarely seen in high-voltage applications.

(2) Controllable power-source topology, where MMC is an exemplary candidate. Such topology cascades sub modules thereby avoiding the switch series connection. No dynamic voltage balancing or synchronous gate triggering is required. Therefore such topology is particularly suitable for the high-voltage DC transmission. Based upon the structure of sub-modules, the MMC could be classified into three basics, i.e., half-bridge sub-module (HBSM) MMC (H-MMC), full-bridge sub-module (FBSM) MMC (F-MMC), and clamped double sub-module (CDSM) MMC (C-MMC), as shown in Fig. 1.12b. Compared to other two topologies, H-MMC cannot self-heal DC-side short-circuit faults, however, it is more widely used due to its simplest structure, least switch amount and least complex electromagnetic transients.

(3) Controllable switching power-source topology includes hybrid cascaded multi-level converters (HCMCs) and alternate-arm multi-level converters (AAMCs), as shown in Fig. 1.13.

The cascaded H-bridge sub-modules in HCMC is aimed to shape the two level output voltage to form a high-quality sinusoidal voltage waveform at the AC side. Since the required number of sub-modules is about half of that of MMC, and the sub-module capacitor is the key factor to decide the converter size, this structure is beneficial to build a more compact converter station.

The conducting switch in AAMC bears a part of the DC voltage, which can reduce the number of sub-modules. Under the extreme condition, the conducting switch can withstand half of the DC voltage, thus reducing the number of submodules to half of the F-MMC. At the same time, if the number of sub-modules is large enough, there is no need to use the conducting switch to withstand the DC voltage. At this point, AAMC is equivalent to F-MMC.

From the topology and operation function above, mixture of device combination and modular structure is the main solution of power electronic converters in the flexible DC transmission systems, and the electromagnetic transient process is focused on the microsecond to sub-millisecond timescales.

1.3.2 Power Electronic Systems in Grid-Tied Renewable Energy Generation

Renewable energy generation such as wind and photovoltaic is one of the most important application areas of power electronics. Power electronic converters are the key for those renewable energy resources to be converted to electricity.

1. Photovoltaic generation system

Photovoltaic inverters are categorized as single-stage type without the DC/DC conversion, as shown in Fig. 1.14a, and two-stage type with the DC/DC conversion, as shown in Fig. 1.14b. The single-stage type only contains one DC/AC inverter, which is responsible for the AC current and DC voltage control simultaneously meanwhile realizing the maximum power point tracking (MPPT). The two-stage type consists of one DC/DC converter taking care of the MPPT, and one DC/AC inverter controlling the DC and AC power.

When constructing the photovoltaic systems, multiple combinations are adopted, i.e., centralized, string-type, multi-string type and decentralized.

The centralized type, as shown in Fig. 1.15a, combines all solar panels together, connects overall terminals to one single solar inverter, and sends all the power to the AC grid.

The string type, as shown in Fig. 1.15b, series connects multiple solar panels into a string and connects each such string to a DC/AC inverter to supply energy to the grid.

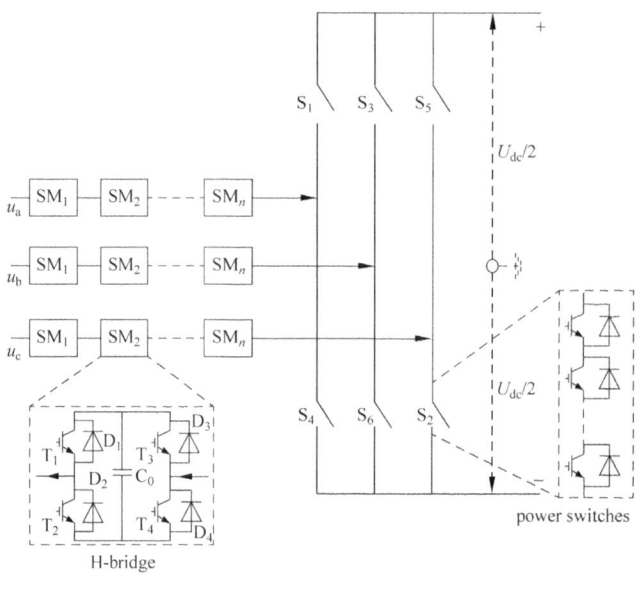

(a) HCMC

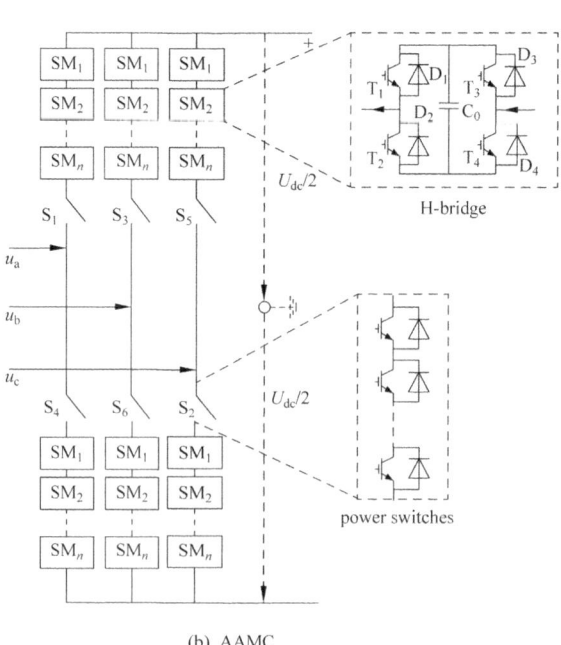

(b) AAMC

Fig. 1.13 Topologies of controllable switching power-source converters

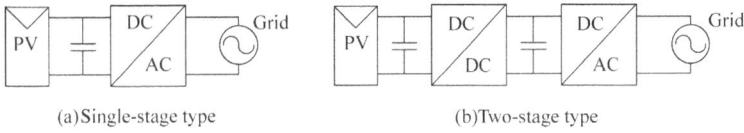

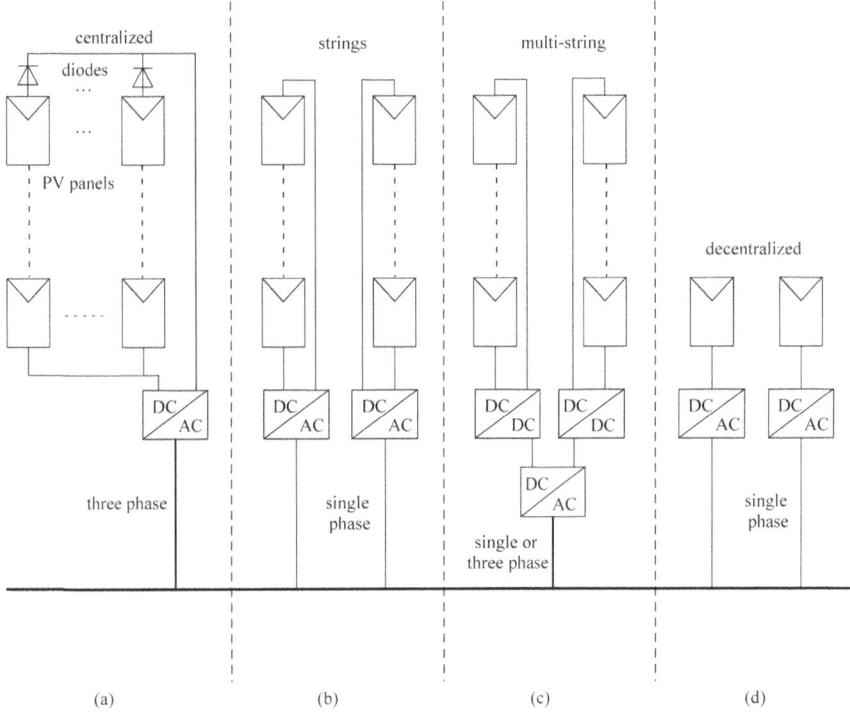

Fig. 1.14 Photovoltaic inverters

Fig. 1.15 Inverter combinations of the PV generation system

The multi-string type, as shown in Fig. 1.15c, series connects multiple solar panels into a string, which then is connected to a DC/DC converter. Multiple such DC/DC converters are paralleled together to connect with one single DC/AC inverter, forwarding the power to the grid.

The decentralized type, as shown in Fig. 1.15d, connects each solar panel with one DC/AC inverter, exhibiting a high flexibility.

Another grid-tied solar system is shown in Fig. 1.16, which covers various combinations thereby named as multi-combination type.

Regardless of the solar inverter type, the common point is to smooth the electromagnetic energy conversion and suppress the transient energy variation. Especially with the variation of the solar intensity, the randomness of the source energy becomes

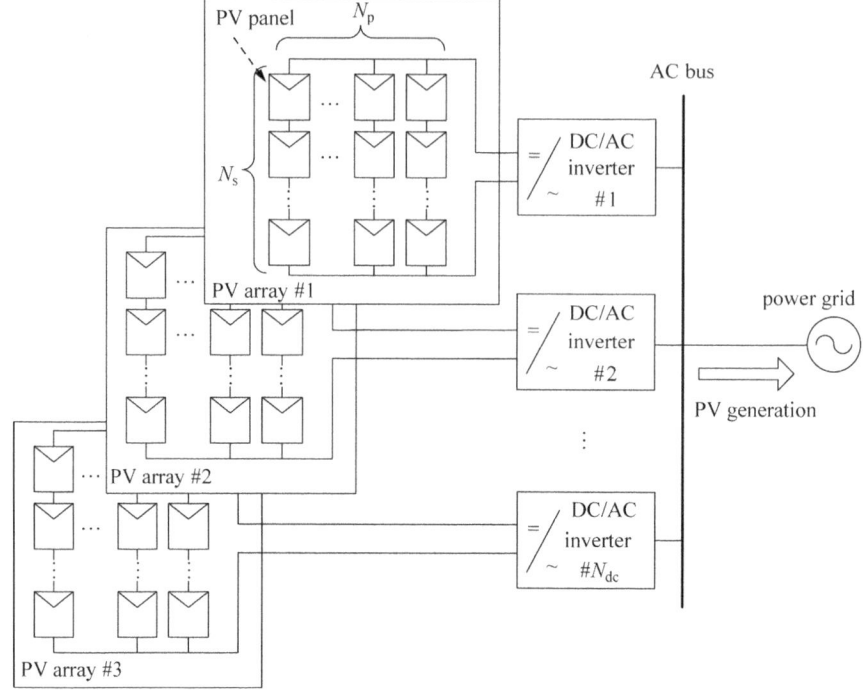

Fig. 1.16 General combination of the PV generation system

obvious, which highly impacts electromagnetic transients inside the photovoltaic inverter. This is one important feature of the solar energy conversion.

2. **Wind power system**

Different from grid-tied solar systems, the power rating of wind mills is much higher with stricter constraints of the input and output. A typical direct-drive wind-power power converter is shown in Fig. 1.17, which is made of multiple AC/DC and DC/AC converters.

The specialties of the wind-power converter include:

(1) The direct-drive type is the mainstream. The power and voltage rating increment per windmill is witnessed;
(2) High power density, general-purpose modular design, with the power density still gradually increasing;
(3) Various and flexible topology selections, e.g., IGCT based four-quadrant NPC three-level converter, IGBT based fly-capacitor four-level converter, and IGCT based fly-capacitor five-level converter;
(4) The higher the power rating, the more severe impact of its low-voltage ride-through capability on the grid, which can only be alleviated by power electronics systems.

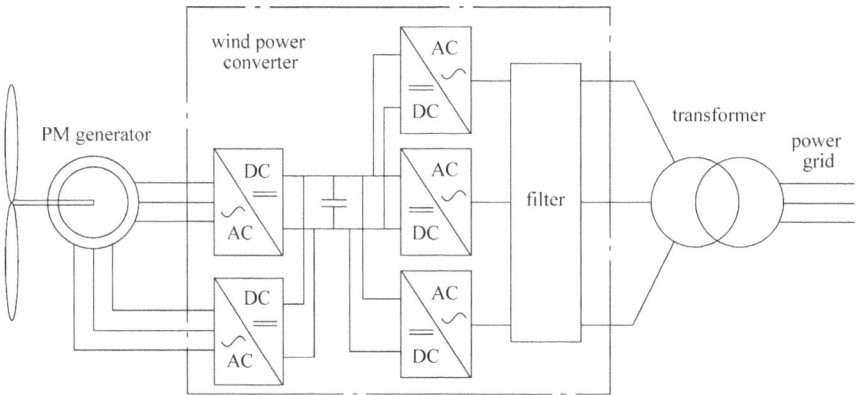

Fig. 1.17 Structure of the direct drive wind power converter

In addition to inherited electromagnetic transients inside the power converter, solar and wind power systems share one common point, i.e., high randomness and high instability of primary energy sources, which worsens electromagnetic transients thereby complicating the control of power electronics. An extra energy storage system is usually required to smooth such transients and maintain the stability of the output, which in return requires the bidirectional power flow control of the storage system.

1.3.3 Traction System

The traction systems include high-speed trains, metro and light rail transits, electric vehicle drive, et al., all of which adopt the power electronics technology to spin the motor under the speed and torque control. The load of power electronics systems is the motor, which faces the complicated operation modes such as driving, braking and varying the speed. High power density and high control accuracy are demanded, which all require the accurate modulation of electromagnetic transients.

Take the motor drive system in the high-speed train as an example. Such system contains two parts, one is a bidirectional rectifier, and the other is a motor-drive inverter, forming a back-to-back topology. A typical three-level NPC back-to-back topology is shown as Fig. 1.18. Four-quadrant operations are guaranteed with the enhancement of the power quality at both sides. The problem is the bulky DC-link capacitor and the coupling of electromagnetic characteristics at both sides.

To regulate electromagnetic transients accurately, weakening the electromagnetic coupling between both sides is required, which further led to a solid-state transformer. The heavy DC-link capacitor is replaced by a DC-DC power electronic transformer, as shown in Fig. 1.19. In addition to a three-level topology, a conventional two-level topology also applies to the rectifier and inverter, where a solid-state transformer transforms the primary DC-voltage into a high-frequency AC, reflects

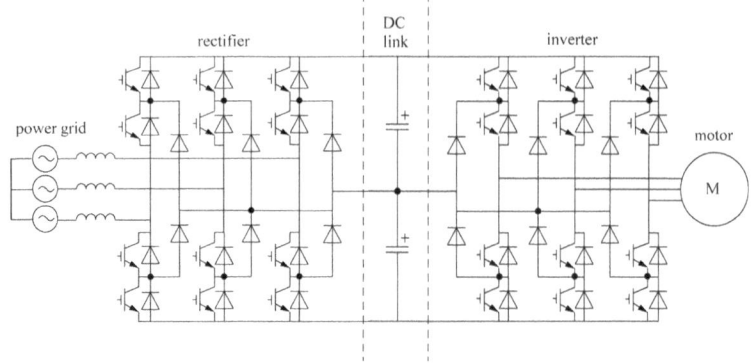

Fig. 1.18 A typical three-level back-to-back NPC topology

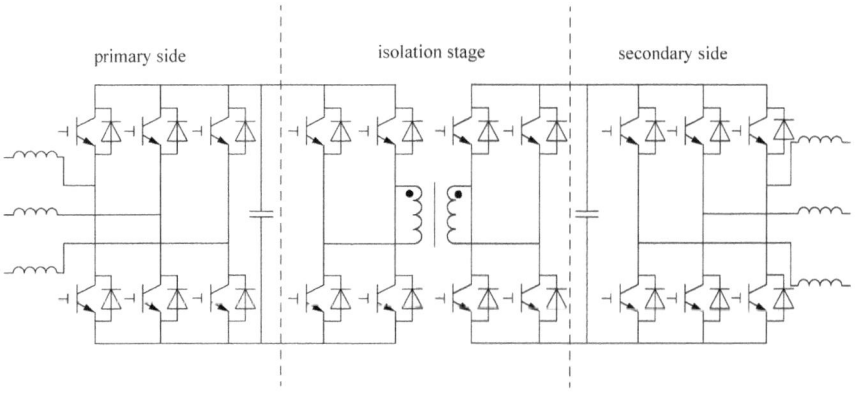

Fig. 1.19 Scheme of the power electronic solid-state transformer based traction system

the corresponding AC to the secondary side, and rectifies it into a DC or AC. A high-frequency transformer replaces the original line-frequency transformer, shrinking the size and weight of the overall system. On the other hand, due to the specialty of the traction system, the drive inverter could be directly connected to the output of the solid-state transformer, which simplifies the system structure, reduces the weight and further improves the transformer efficiency. From the transient aspect, introducing the high-frequency transformer decouples the electromagnetic characteristics at both sides thereby enhancing the control accuracy.

Despite various advantages of the solid-state transformer, multiple problems exist in its transient processes, i.e., (1) adding a high-frequency DC-DC converter expands the power loop, expedites transient processes, and makes the energy balance more difficult to reach; (2) compared to the conventional design, a solid-state transformer based topology employs more switches, adds the system complexity and deteriorates the system reliability.

1.4 Existing Challenges in Power Electronics Systems

Even though power electronics technologies have been widely used, obstacles to its future development still exist, especially for high-power applications. Three major challenges are detailed as below:

(1) To increase the power capability. Due to the limited current-conducting and voltage-blocking capability of a single switch, the conventional two-level converter using single-switch power modules does not meet the high-voltage and high-current requirement. Even in the future one single switch reaches enough voltage and current capability, due to its high switching loss and high inrush voltage and current, using single switch is not an appropriate option. In general, multiple switches or switching units need be in series/parallel connection to form a hybrid converter, which however has the dynamic current/voltage balancing as the potential bottleneck.

(2) To optimize the system design. Since the power electronic converter is always working at the PWM mode, various nonlinearities and uncertainties exhibit, which are hard to describe through mathematical modelling and even measure. Therefore the quantification of the design is difficult, no mention the optimization. Usually the design with sufficient allowance and overkills is recommended, like a big horse pulling a small carriage, or totally based on the empirical parameters using experience plus statistics, which yields the low device utilization, low economic performance, mismatched parameters and poor adaption.

(3) To enhance the system reliability, which is the focus and also one of the biggest bottlenecks of power electronics. For example, the annual faulty rate of HVDC systems is much higher than that of regular transmission systems, most of which are caused by power electronic systems. Hence enhancing the reliability of power electronic equipment is the must to secure the reliability of the overall system. As shown before, failure and damage of devices and equipment usually occurs in electromagnetic transients, when the energy transformed exceeds limits. Therefore, comprehending the principles of transient processes inside power electronic systems, refining mathematical models, and furthermore enhancing the control accuracy are essential to improve the overall reliability.

Going deep into converters, in terms of the electromagnetic transients, we could categorize all existing problems as below.

1.4.1 Misunderstanding the Short-Timescale Switching Process of Power Switches

Power switches are the base of power electronic converters. As shown above, majority of power electronics especially the high-power applications utilize SCRs, GTOs, IGBTs, IGCTs and IEGTs. Differences among those switches result in various com-

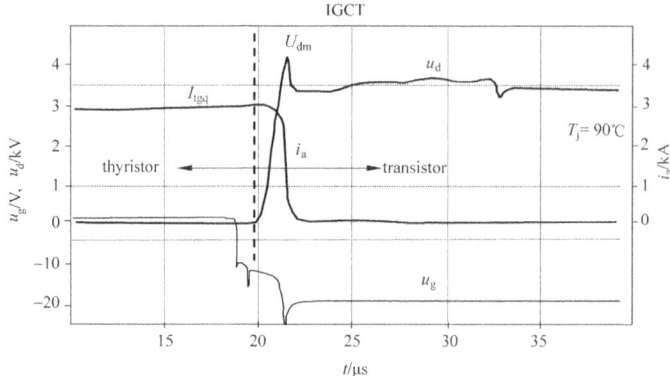

Fig. 1.20 Measured voltage and current during IGCT turning-off

plex transients thereby obstructing the establishment of models for such transient processes, which further leads to not insured system reliability.

The root is the fast movement of current carriers inside semiconductors, yielding that transients complete within ns–μs time intervals. Therefore to understand its moving principles is difficult. Shown in Fig. 1.20 is the waveform of one IGCT switching-off voltage and current, which reveals that the transition from the thyristor to the transistor only takes 1 μs. Analytical methods are hard to apply to such complex transients. Moreover, the diversity of device parameters complicates the characteristics when combining components altogether. A tiny diversity might lead to a significant impact.

Therefore, accurately modelling the short-timescale electromagnetic transient and its interaction with other system components is rewarding. On one hand, simulation models of semiconductor switches are pursued based on either the lumped charge or numerical solutions. To obtain related parameters is difficult and its description of the transient processes is inaccurate, making it not suitable for the analysis of high-voltage power electronic systems yet. Other endeavors include building the functional model of power switches, which trades off between the simulation speed and accuracy and is found useful in real practice. On the other hand, measurements of switch characteristics are demanded, e.g., special equipment for the IGCT evaluation. The problem is, however, not being able to bridge switches with other peripheral components in the system.

1.4.2   Idealization of Power-Conversion Topology for Transient Study

Power electronics topology is the fundamental to realize the energy conversion. To obtain the excellent output performance with limited switch options, variation of

Fig. 1.21 A buck converter including stray parameters

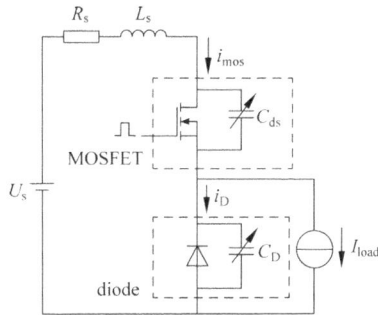

Fig. 1.22 Measured turn-on voltage and current of MOSFET in the buck converter

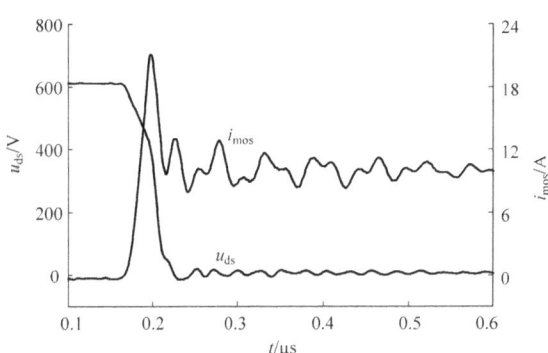

the circuitry structure emerges as the applicable power electronics topologies, for instance, the NPC topology, cascaded H-bridge and conventional two-level using series connected devices. Uniformed topology is always the pursuit of power electronics researchers, during the search of which various topology appears. Graph theory has also been adopted to analyze the topology and trace the current commutation. Such attempts are beneficial for low-power systems, but face challenges when applied to the high-power systems, i.e., no consideration of the electromagnetic transients while only covering on-state and off-state of the circuit, no consideration of stray parameters of commutation loops and connectors while only emphasizing lumped-parameter components.

The target output waveform and amplitude are approached by fast switching power semiconductors, exhibiting the high du/dt and di/dt externally. Such fast switching behaviors yield fast transitions of electromagnetic energy, which in return creates surge voltage and current at ns~µs level through interactions with stray parameters. Different time constants are displayed in different loops with different stray parameters, which obstructs the energy balancing. Shown in Fig. 1.21 is a buck converter considering stray parameters with the turn-on voltage and current shown in Fig. 1.22. It can be seen that a current pulse is overlapped on the MOSFET turn-on current with oscillations.

Therefore the transient commutation topology embracing stray parameters of the main-power loop, which allows to further modelling the dynamic energy imbalance in different time-constant subsystems, is a must for the research of power electronics topology. The focus then is shifted to the extraction of stray parameters and their impact evaluations.

In fact, due to fast electromagnetic transients in the commutation process, the system is highly influenced by the transient behavior of the power switch and stray parameters. Take IGCT as an example. A voltage spike across the switch will appear due to the forward voltage drop of the diode and stray inductance, which causes EMI, waveform distortion, increment of the electric stress and even the switch failure. Obviously stray parameters cannot be ignored. However, the commutation topology will be drastically changed when considering such parameters. The idealization of power electronics topology has become a major defect of conventional topology study.

1.4.3 Unrecognizing the Difference Between Information Pulses and Energy Pulses

Majority of power electronic systems adopt modulation strategies such as PWM to approach the target output waveforms. The essence of conventional PWM strategies is to control the power conversion macroscopically using ideal switches without considering switching processes. However, as shown above, all switching processes take time especially in high-power converters, where the energy transition is more drastic. Ultimately, the conventional PWM control has the discounted performance, with the introduction of the digital error, control delay and even the switch failure.

In the real practice, the non-linear impact factors of the switches and power loops yield differences between the information pulse and energy pulse, e.g., distortion and delay. Therefore theoretical control strategies are difficult to implement, with many abnormal and even destructive pulses emerging. As shown in Fig. 1.23, various irregular pulses appear on the line-line output voltage of a three-level NPC inverter.

Revealing the generation and transfer of energy pulses, and furthermore actively controlling such energy pulses and pulse sequences are the necessity for the future power electronics development. The key is to comprehend abnormal pulses and difference between information pulses and energy pulses. In high-power applications, energy pulses are prominent and inevitable, and the cognition of principles of such short-timescale transients still faces challenges.

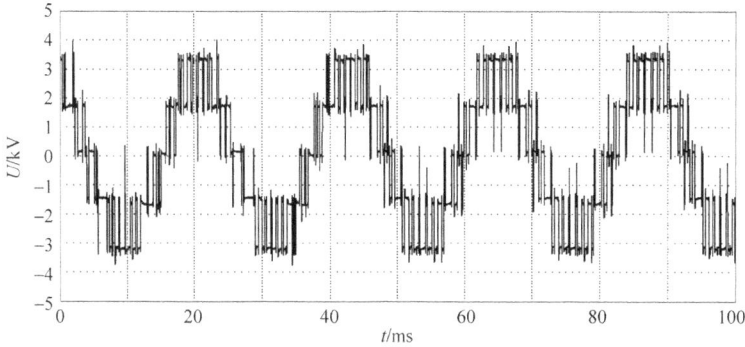

Fig. 1.23 Measured output line-line voltage of a three-level NPC inverter

1.4.4 Misidentifying Electromagnetic Transients

Conventional circuit theory is employed in previous power electronics analysis and system design, treating the components and connections all ideal, which encounters the barrier in the actual system design.

To integrate all parts of the system for electromagnetic transient analysis is one of the core challenges of power electronics. Given a power electronic converter is a unity of energy transformation, switch modulation and energy storage, such a system design should not only focus on the circuit design but also be treated as a comprehensive design emphasizing the multi-field coupling and time-space variation. A more optimal design, more accurate analysis and more precise control could be anticipated once the energy conversion and flow are investigated from the electromagnetic-energy-flow point of view. In particular, it is worthwhile pointing out the energy interface, e.g., some connectors and terminals. In during the transient process the voltage and current at the two sides of the interface might behave differently.

A power electronic converter is a mixture of multi-timescale transient processes. Such transients especially those ultra-short-timescale ones, e.g., at ns level, is critical for the energy conversion. Nonlinear component models and stray parameters need be taken into account to establish multi-dynamic equation sets thereby analyzing the transients more accurately.

As a summary, electromagnetic energy pulses are essential for power electronics transient processes. When such transients are close to the ns levels, the circuit theory and methodology used in conventional power electronics study are revoked. A new perspective and change of the power electronics theory and techniques becomes a rewarding research.

Based on the analysis above, it can be concluded that short-timescale electromagnetic transient behaviors are cores of high power electronic systems, which are key factors to improve the overall performance.

For high power electronic converters, the controllable electromagnetic energy conversion must be dealt with properly, concerning electromagnetic transient processes

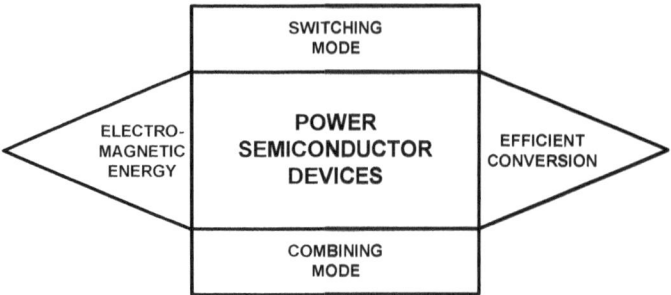

Fig. 1.24 Reconsideration frame on power electronics

and energy balance. Also, relations of devices and equipment, of control circuits and main circuits, and of distributed parameters and lumped parameters must be dealt with properly. The key to satisfy all these requirements is the switching characteristics. Such switching characteristics, like high power rating, low loss, high frequency and fast response, have always been pursued in power electronic converters. Developments of devices, circuits and control strategies are all focused on improving the switching characteristics.

Based on the understanding above, a reconsideration of power electronics can be concluded that power electronics is associated with the efficient conversion of electromagnetic energy based on the switching and combining mode of power semiconductors. Four key words are involved, (1) power semiconductor is the carrier and foundation of electromagnetic energy conversion; (2) switching and combining modes are the basic means of conversion; (3) electromagnetic energy is the target object; and (4) efficient conversion is the goal of power electronic systems. Such reconsideration aims to expand the content of power electronics, from electronics on electron movement to efficient conversion of electromagnetic energy; from the simple combination of microelectronics, electricity and control to controllable behavior patterns of electromagnetic energy in switching and combining modes.

The reconsideration can be summarized as one center and two basic points which have two meanings, (1) to consider power semiconductor as the center, and electromagnetic energy and efficient conversion as two basic points; (2) to consider power semiconductor as the center, and switching and combining mode as two basic points. The relationship of the two meanings of the reconsideration is graphically illustrated in Fig. 1.24. From left to right, it shows the electromagnetic energy can be converted efficiently by the controllable power semiconductor, which is a description from the perspective of technology and application with the object and goal. From up to down, it shows two function characteristics of the power semiconductor, the switching and combining characteristics. Switching characteristic mainly reflects the characteristic of a single power semiconductor device, while the combining characteristic mainly focuses on the whole circuit composed of multiple power semiconductor devices. It is a description from the perspective of the inherent features of power electronics. The two meanings are both independent from and interdependent on each other, forming an organic whole of power electronics.

Although the characteristics of the elements in a power electronics system vary from each other, from a holistic perspective, the organic integration of the independent elements creates a unified system which shows its scientific nature. In summary, the content of power electronic systems should include the unique characteristics as mentioned in Sect. 1.2: the integration of hardware and software, the interaction of power and information, the transfer of linearity and nonlinearity, the mixture of continuity and discreteness, and the coordination of multi-timescales.

Chapter 2
Electromagnetic Transients and Modelling

Transients are the norms and the musts of the energy conversion in power electronics systems. Previous research is focused on the large-timescale electromagnetic transient processes, while neglecting short-timescale switching processes and stray parameters. When failure mechanisms of components and the enhancement of the power-conversion capability are to be emphasized, such short-timescale transients must be taken into account. To fully understand electromagnetic transients, establishing the mathematical model of full-timescale transients is the fundamental.

2.1 Electromagnetic Transients of Power Electronics Systems

Transients in power electronic systems are complex. As described before, a power electronic system is made of power switches, connecting cables, control units and other auxiliaries, as shown in Fig. 2.1.

According to the power rating, such system could be divided into the main-power circuit, gate-drive circuit and the control circuit. The control circuit manages the main-power loop through the gate-drive circuit, which realizes the effective energy conversion and the accurate control of the energy flow via the information flow. As introduced in Chap. 1, a power electronic system embraces the unity of hardware and software, interaction between the energy and information, transfer between linear and nonlinear, mixture of the continuity and discreteness, and coordination of multi-timescales. Subsystems exhibit different electromagnetic parameters, with different characteristics during electromagnetic transients.

© Tsinghua University Press and Springer Nature Singapore Pte Ltd. 2019   33
Z. Zhao et al., *Electromagnetic Transients of Power Electronics Systems*,
https://doi.org/10.1007/978-981-10-8812-4_2

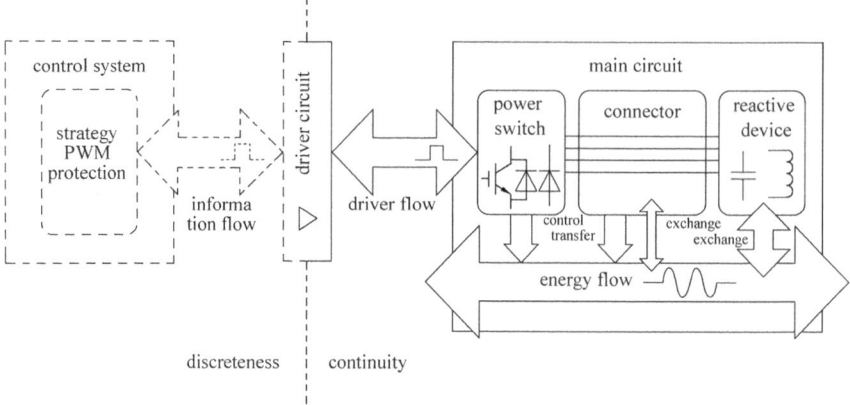

Fig. 2.1 Scheme of a typical power electronic system

2.1.1 Electromagnetic Transients in the Main-Power Loop

Power electronics main-power loop is the path for the power transformation and transfer. It realizes the power transformation and carries the energy. Two major building
blocks construct the main-power loop, i.e., lumped-parameter components, such as
inductors, capacitors and switches, and connections, such as bus bars, connectors
and cables. Shown in Fig. 2.2a is one single-phase main-power loop of a three-level
NPC inverter. Figure 2.2b illustrates the physical circuit of three-phase main power
loop.

Comparison of Fig. 2.2a and b reveals a visible difference between the equivalent
and actual circuits. In the equivalent circuit, lumped-parameter components and
connections are main players. Components are related to the energy transformation
and storage, e.g., resistors for the energy consumption, inductors for the magnetic
energy storage, capacitors for the electric energy storage, and switches for the energy
transformation from the continuous to the discontinuous. Connections in the circuit
represent the linkage of various components, which in reality act as the energy
conducting paths. Therefore the actual main-power circuit is treated as the mixture of
lumped-parameter components and connections, where components act as the bridge
of the energy flow while connections have their own "components" as well, e.g.,
stray inductance, capacitance and resistance. The existence of such stray parameters
complicates electromagnetic transients in the main-power loop.

1. Electromagnetic transients inside the power switches

Power switches are critical components of the energy conversion. With the switching
on and off, drastic changes of voltage and current across the switch occur, yielding
the pulse-like energy variation in the main-power loop, defined as electromagnetic
pulse transients.

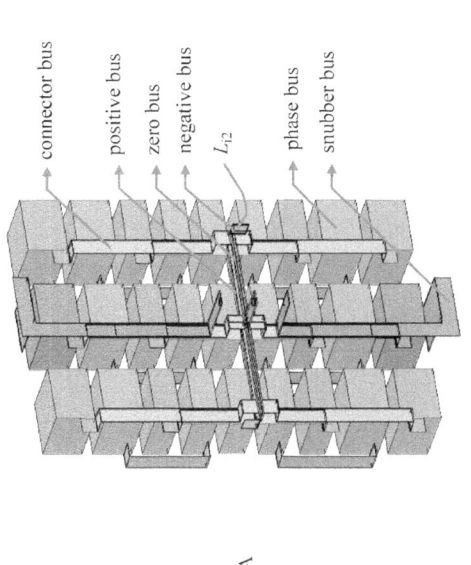

(b) Physical structure of three-level NPC inverter

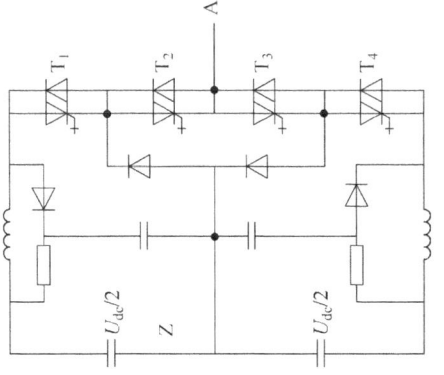

(a) Circuit of three-level NPC bridge

Fig. 2.2 Main-power-loop structure of one power electronic system

Such pulse transients impose high electric stress, such as di/dt and du/dt, which could be coupled with stray inductance and capacitance, respectively thereby generating high voltage spike and high inrush current in the switching process. The switching-off voltage waveform of one IGCT is shown in Fig. 2.3. It can be seen that the transition from the on-state to the off-state takes several μs. Such a voltage spike overlapped with the original DC-bus voltage has the potential to breakdown the switch and even damage the system.

In the transient process, over voltage and current will be induced by the stray inductance and capacitance, especially those introduced by leads of the switches and connectors, which might be more critical for transients than others. Figure 2.4 shows the turn-off voltage across one IGCT measured at different timescales. It is concluded that different waveform is shown at different measure timescales. At the μs scale, more voltage pulses with different amplitudes and widths are observed. Transients in this process are extremely complex with atrocious amount of coupling among stray parameters. In general, the voltage spike amplitude needs be paid special attention on during the power electronic system design, analysis and control.

2. **Electromagnetic transients of power connections**

The transmission line of the main-power loop is usually made of good conductors, forming various cables or bus bars. Though their location, shape and even materials are different, the rules of the electromagnetic energy flow are the same, complying with the same diffusion rules. The differential equations for the voltage and current of the bus-bar conductors are

$$k_u \frac{\partial^2 u}{\partial x^2} = \tau_u \frac{\partial^2 u}{\partial t^2} + \frac{\partial u}{\partial t} \text{ and } k_i \frac{\partial^2 i}{\partial x^2} = \tau_i \frac{\partial^2 i}{\partial t^2} + \frac{\partial i}{\partial t} \tag{2.1}$$

where u and i are the conductor voltage and current, k_u and k_i represent the diffusion coefficients, τ_u and τ_i are corresponding time constants. All these parameters are related to the bus-bar material, structure, shape and even operational modes. Take the voltage u as an example. Analytic solution of u in Eq. (2.1) is shown as below

Fig. 2.3 Measured turn-off voltage of one IGCT

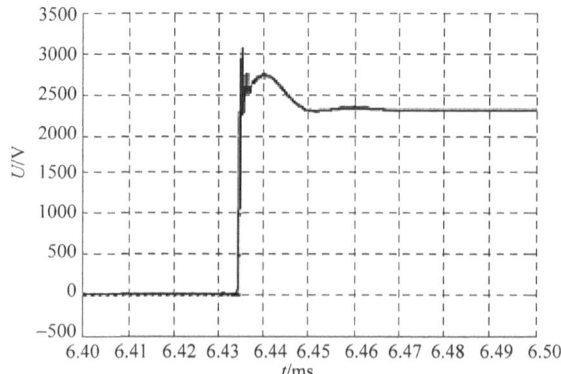

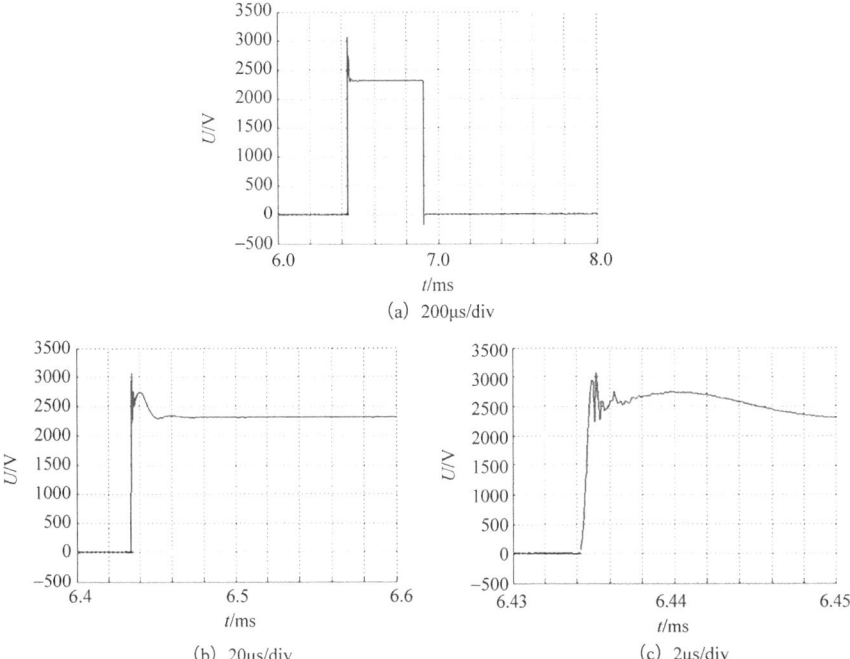

Fig. 2.4 Turn-off voltage across one IGCT measured at different timescales

$$u(x,t) = \begin{cases} 0, & t < x/c \\ u(0, t - x/c)e^{-\frac{x/c}{2\tau_u}} + \Delta U(x,t), & t > x/c \end{cases} \tag{2.2}$$

Here $c = \sqrt{\frac{\tau_u}{k_u}}$, $\Delta U(x,t) = \frac{x}{2\sqrt{k_u t_u}} \int_{x/c}^{t} \frac{I_1(\frac{1}{2\tau_u}\sqrt{\sigma^2 - \frac{x^2}{c^2}})}{\sqrt{\sigma^2 - \frac{x^2}{c^2}}} e^{-\frac{\sigma}{2\tau_u}} u(0, t - \sigma)d\sigma$. I_1 is the first-order Bessel function of imaginary argument. $i(x, t)$ has a similar form as Eq. (2.2), which indicates that voltage and current transients are both time-space variables.

In actual bus bars, other various stray inductances and capacitances exist, which induce other shorter-timescale transients than the original ones. Shown in Fig. 2.5 is the terminal voltage and current of the IGBT, which contains ripples in addition to the voltage and current spikes. The root is traced back to stray parameters along the transmission line.

Differences of current, voltage, magnetic strength and thermal distribution at different points of the current-commutation loop are instructive to the placement of key components, which requires the accurate extraction of stray parameters of the power loop. Conventional power electronics analysis using lumped linear parameters and ideal switches will lose the simulation fidelity for such short-timescale transients.

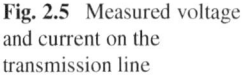

Fig. 2.5 Measured voltage and current on the transmission line

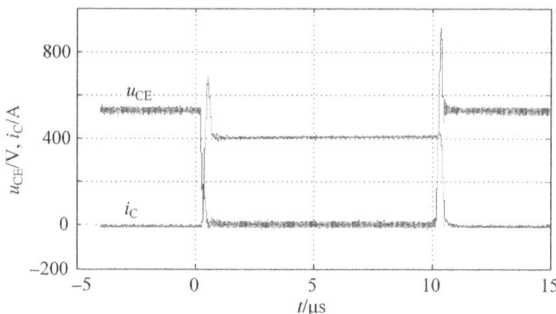

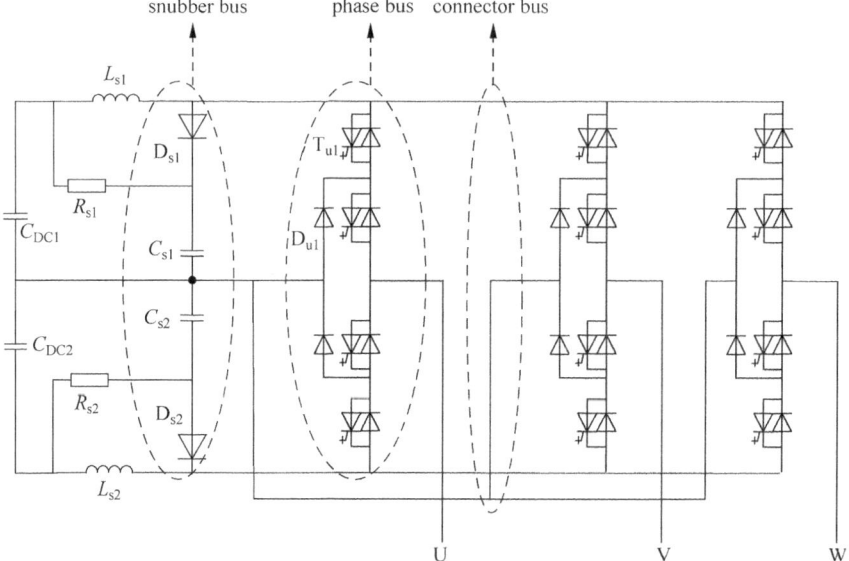

Fig. 2.6 Categories of buses in a three-level inverter

3. **Similarities and differences of the main-power loop and the lumped-parameter topology**

As shown above, the actual main-power loop involves characteristics of the material, energy storage and stray parameters. Such characteristics are not considered in the lumped-parameter based circuit, which only covers the linkage among ideal lumped-parameter components.

Consider an IGCT based three-level NPC inverter, as shown in Fig. 2.5. We could categorize all bus bars as snubber-circuit bus bars, connecting bus bars and phase bus bars based on their locations. Here the snubber-circuit bus bar is to connect the snubber capacitor (C_s) and the snubber diode (D_s). The connecting bus bar is used to bridge the snubber circuit with three-phase legs. The phase bus bar is employed to connect IGCTs with clamping diodes and heatsinks in each phase.

Fig. 2.7 3-D structure of buses

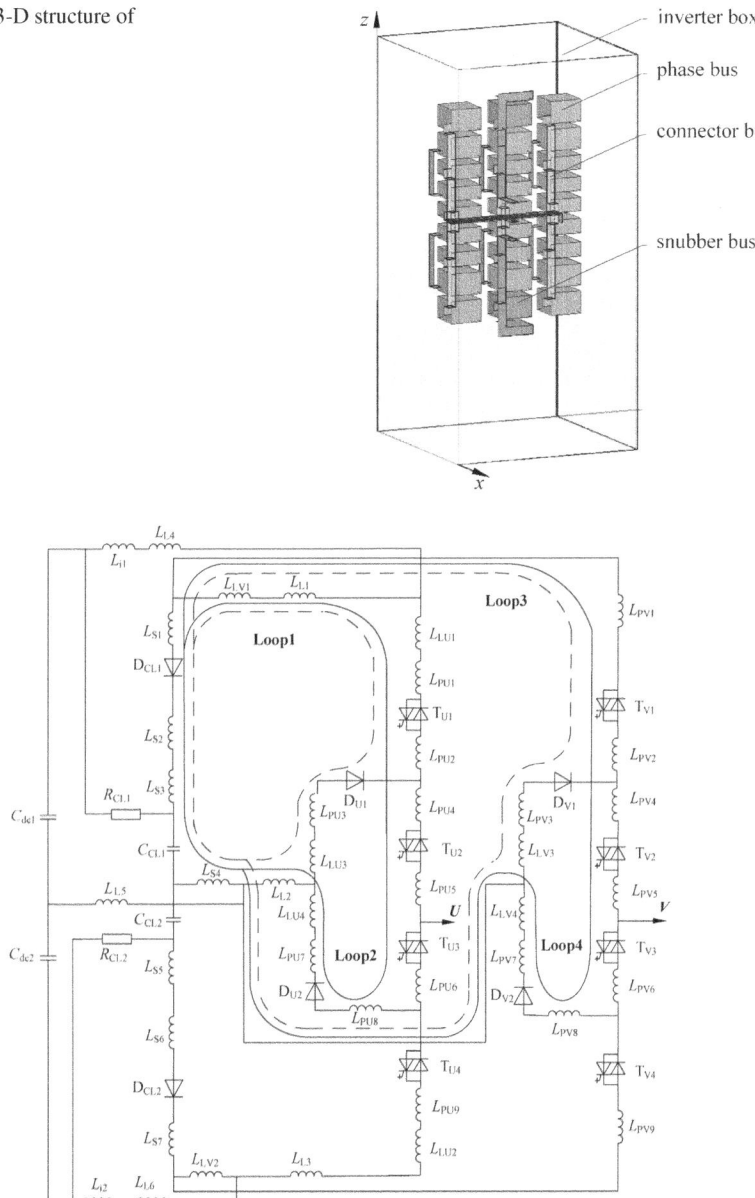

Fig. 2.8 The equivalent circuit of phase U and V

Physical bus bars are largely dimensioned, complicatedly connected, irregularly shaped and highly diversified. Influenced by the skin effect and the proximity effect, stray parameters vary with the switching frequency, i.e., nonlinear characteristics.

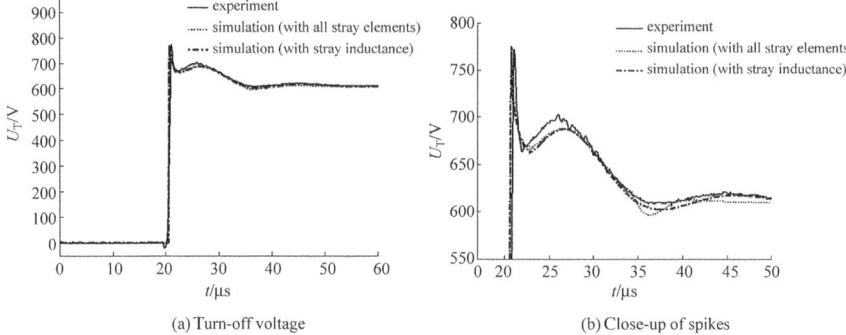

(a) Turn-off voltage   (b) Close-up of spikes

Fig. 2.9 Comparison of the turn-off voltage by experiment and simulation

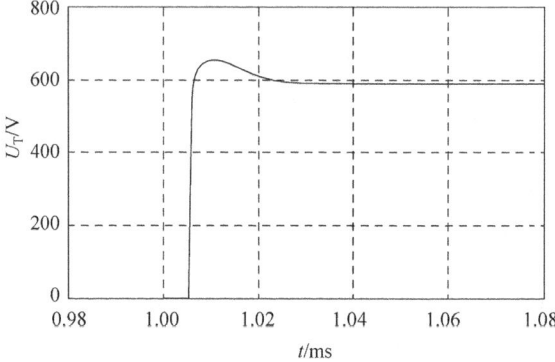

Fig. 2.10 Turn-off voltage across the power switch simulated with only lumped parameters

To fully unveil transients in the main-power loop, stray parameters of bus bars must be taken into account. Bus bars used in Fig. 2.6 have a 3D physical model shown as Fig. 2.7, which contain tens of units to form connecting bus bars, phase bus bars and snubber-circuit bus bars. The inverter cabinet is made of steel sheets coated with Aluminum-Zinc alloy, which is connected to the ground as the EMI shielding unit.

The equivalent circuit including all stray parameters is shown in Fig. 2.8. Due to the symmetry of the topology and mechanical structure, only the U-phase and V-phase equivalent circuits are given. To maintain the circuit simplicity, only self-inductance of each bus bar unit is considered. Here L_{L1}-L_{L6} are the connecting-bus-bar stray inductance; L_{S1}-L_{S7} are the inductance of the snubber-circuit bus bars; L_{PU1}-L_{PU9} correspond to the U-phase inductance; L_{PV1}-L_{PV9} correspond to the V-phase inductance; Loop1 ~ Loop4 represent transient commutation loops of switch T_{U1}, T_{U3}, T_{V1}, T_{V3}, respectively. An obvious difference is revealed between transient commutation loops and original ideal loops.

Equivalent circuits of bus bars are complicated due to numerous stray parameters. To generalize such equivalent circuits, a reasonable simplification needs be carried out when building a circuit-emulation oriented transient model. For instance, we could combine bus bar units thereby reducing the number of overall units, which will be detailed in Sect. 4.4.2.

We name such main-power-loop topology as the transient commutation topology, which targets at electromagnetic transients of the main-power loop. As shown in Fig. 2.9, using such topology yields nearly the same waveforms as the experimental ones. With only lumped parameters, however, no first voltage spike is observed, as shown in Fig. 2.10. Therefore the transient commutation topology is a must for the short-timescale transient analysis.

2.1.2 Electromagnetic Transients in the Gate-Drive Loop

Transients in the gate-drive loop of power electronics systems represent the control of electric signals over the electric power, which is also the power amplification. However, such amplification is different from that of transistors in analog electronics, which work in the active region to amplify the signal. In power electronics, switches are working at the switching mode to amplify the signal. Here gate-drive loops merge the information flow into the power flow and control the electromagnetic energy through triggering switches.

Compared to main-power loops, gate-drive loops are part of information signal loops. Such loops, however, are coupled with main-power loops, resulting in transients as well. In general, the voltage-controlled switch requires a lower gate-drive power and gets triggered faster, yielding a higher switching frequency. The current controlled device utilizes the current signal to vary the switch states, which requires a higher drive power with a slower response, resulting in a lower switching frequency.

The basic gate-drive circuit of IGBTs is shown as Fig. 2.11. The gate resistance R_G directly determines transients in gate-drive loop. This circuit has a typical flyback power supply. A microcontroller is employed to set the reference of output voltage through adjusting the duty cycle of the active switch.

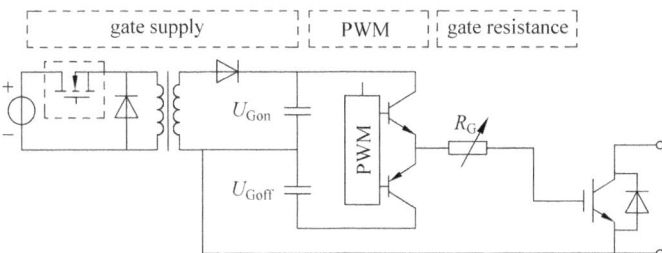

Fig. 2.11 The basic gate-drive circuit of IGBTs

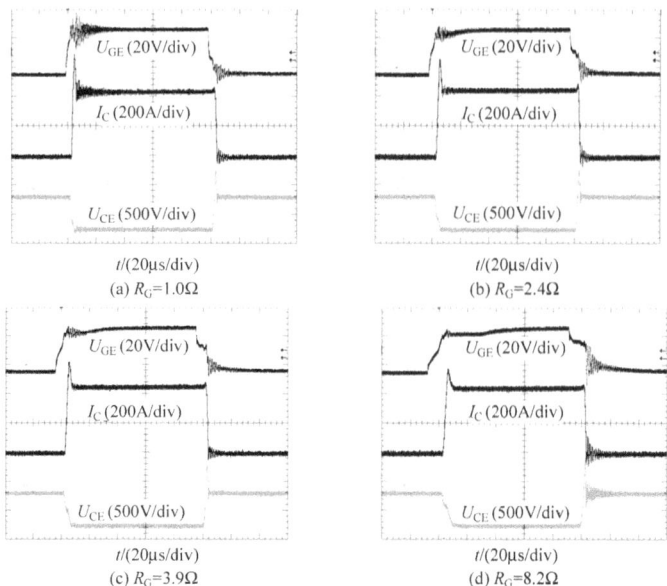

Fig. 2.12 Gate-resistance impact on the switching performance

The gate resistance has little influence on the IGBT forward saturation voltage drop. However, it highly affects the dynamic performance of the IGBT, e.g., turn-on delay, turn-off delay, current rise time and fall time.

At the ambient temperature of 30 °C, DC-bus voltage of 600 V and the current of 450 A, the IGBT switching process with different gate resistances is shown as Fig. 2.12.

It indicates that a too small value of the gate resistance induces the turn-on current oscillation while a too large value of the gate resistance increases the switching time and loss. An optimized gate-resistance value secures the switching dynamic characteristics.

A frequently encountered problem of the gate-drive circuit is the triggering failure. As shown in Fig. 2.13, narrow pulses are detected across the collector and emitter of an IGBT, all of which are caused by some weak and under-threshold gate signals. Essentially, any gate-drive pulses narrower than the minimum pulse width, which is determined by the switching delay and switching duration time, will cause such abnormal transients.

Over-voltage and over-current are two most common faults when switching IGBTs. Protections against them are required, e.g., removing the gate signal, turning off the main power input and discharging the DC-bus capacitor. Particularly, we need differentiate the random over-current case from the destructive surging current, mainly due to the different lasting time. The surging current could be suppressed by lowering the gate-drive voltage, given the IGBT on-state resistance has the negative correlation with the gate-drive voltage. For example, when the over-current occurs,

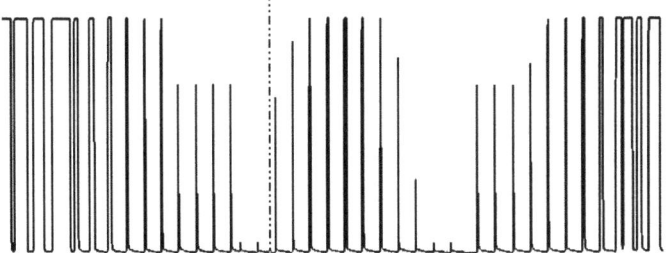

Fig. 2.13 Measured voltage pulses across the IGBT in one actual converter

halving the gate-drive voltage will significantly increase the IGBT resistance thereby reducing the current.

Therefore parameters and electromagnetic transients of the gate-drive loop highly impact characteristics of the main-power loop.

2.1.3 Electromagnetic Transients in the Control Loop

All output energy pulses of power electronics systems, i.e., voltage and current pulses in the main-power loop, are triggered by the control circuit, with the typical structure shown as Fig. 2.14. Take the voltage pulse in a VSC as an example. The ideal information pulses are generated based upon the control algorithm, turned into control pulses via the PWM control unit, further shaped by logic units (e.g., interlocking, dead-band setting and minimum pulse width limitation, etc., to be detailed in future chapters) and imposed on the actual gate of the power switch through the gate-drive circuit. Energy pulses are then generated, combined and superposed to form the ultimate output energy pulse waveforms.

Figure 2.14 reveals processing characteristics of the control loop, i.e., the effect of each sector on pulses is gradually accumulated, which creates the difference between

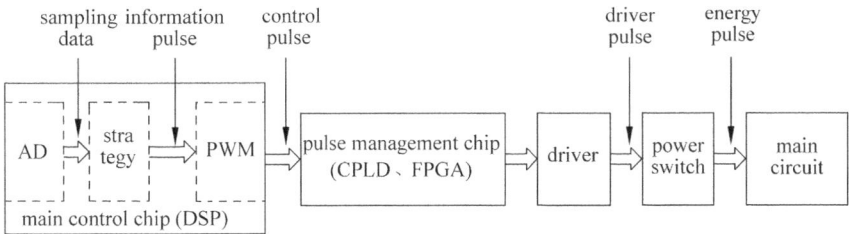

Fig. 2.14 Typical structure of the control circuit in power electronic converters

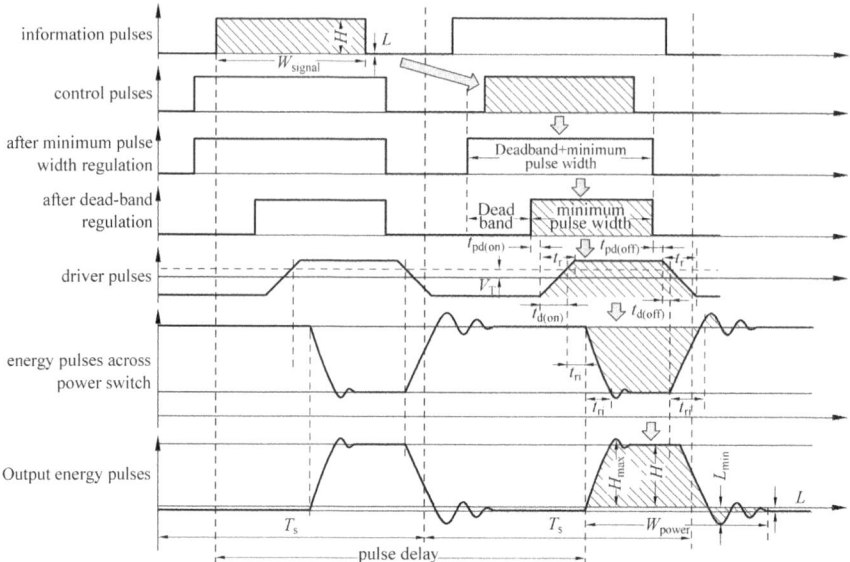

Fig. 2.15 Transitions from information pulses to energy pulses

the information pulses of the control strategy and energy pulses generated by the power electronic converter.

In the transition from information to energy pulses, in addition to the variation of energy features, such as high/low levels of the voltage and current, other distortions of the pulse rising edge, falling edge and even overall shape also happen. A qualitative analysis is carried out in Fig. 2.15.

In the process of generating voltage pulses, from the top to bottom of Fig. 2.15, it can be seen that original information pulses are stored inside the memory as internal variables, such as the duty cycle or comparator registers. At this stage, pulses are ideal. Due to the reloading mechanism of the digital signal processor (DSP), its output pulses usually have a delay of half or one control period, compared to ideal pulses.

Setting the dead band and minimum pulse width is processed inside the control chip or pulse management unit, after which pulse rising and falling edges are varied. Note that control pulses are the weak-electricity signals in digital circuits. Input/output (IO) characteristics of the digital chips determine that control pulses are close to the ideal. An assumption could be made that edges of control pulses are ideal without transient processes, even though the occurrence moments might be varied.

Control pulses, with constraints of the dead-band and minimum pulse width, pass through the gate-drive circuit and are finally transformed into gate signals of the power switch. Under the impact of the gate-drive loop and the switch gate terminal, energy features of pulses show differences. Non-ideal edges appear on the gate, even though no overshoot or oscillation happens.

When gate-drive pulses are poised to switch power semiconductors, energy pulses emerge in the main-power loop. This process is affected by switch characteristics and main-power-loop parameters. In addition to the essential change from weak-electricity signals to strong-electricity signals and from the information level to the power level, pulse shapes vary significantly. Non-ideal edges emerge across power switches, with potential overshoot and oscillations in transients.

Multiple energy pulses in power electronics systems are restrained by main-power loop structures and parameters, forming final output pulses through combinations and superpositions. The system output pulses have similar shapes of the switch output, however, are determined by multiple switch output pulses together in terms of the time sequence and shape features.

As shown above, the original information pulse generated by the control unit, after passing through all sections of the control loop, becomes significantly different from ideal pulses in terms of the energy feature, e.g., pulse edges and pulse shapes. Previous literatures on power electronics idealized energy pulses and ignored variations of pulse edges and shapes between information and energy pulses. Majority of non-ideal characteristics of energy pulses cannot be discovered by using the ideal concept. Therefore the ideal pulse based analysis fails to unveil electromagnetic transients thereby becoming the bottleneck of further improving the performance of power electronic systems.

To quantify the impact of the control-loop non-ideal characteristics on energy pulse sequences, shapes and even the performance of power electronic systems, a mathematical model of such pulses is in need. In order to quantify actual energy pulses and compensate the control error between ideal information and non-ideal energy pulses, such a mathematical model should include information pulses, gate-drive pulses, energy pulses and pulse sequences, based upon characteristic parameters of actual pulses.

The control part is an information-level system, with transients of ns ~μs. The control units in modern power electronics are mainly DSP or multi-core CPU centered digital systems, with the code implementation time of ns ~μs. Its peripheral gate-drive circuit has the signal distortion and delay caused by parasitics with the similar timescale. Such transients highly impact the gate-drive and non-ideal switching performance. Nearly all abnormal waveforms existing in the output voltage and current of power converters are caused by such transients. Shown in Fig. 2.16 are those abnormal pulses measured in a 1.25 MW/6 kV power inverter, i.e., within one dead-band time when current crosses zero (Fig. 2.16a) or when switching actions between two phases are too close (Fig. 2.16b). Waveform fidelity will be sacrificed and even the failure of the current commutation occurs.

Besides, due to limitations of the switching characteristics, such as the dead-band and minimum pulse width, some macroscopic control strategies will fail to be accurately implemented, especially when the output voltage is low. For example, at the low-speed operation of a variable speed motor drive system, the duty cycle of switches is small. However, the minimum pulse width further limits such duty cycle, yielding the output distortion and even failure. Voltage vector distribution under different minimum-pulse-width settings are observed in Fig. 2.17, when the motor

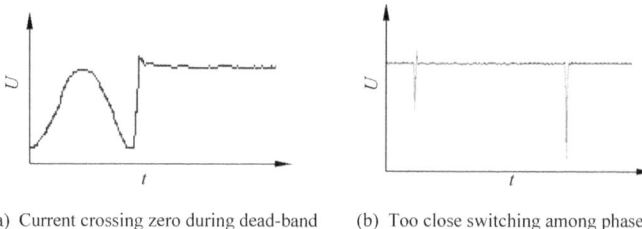

(a) Current crossing zero during dead-band   (b) Too close switching among phases

Fig. 2.16 Abnormal pulses measured in an actual high power inverter

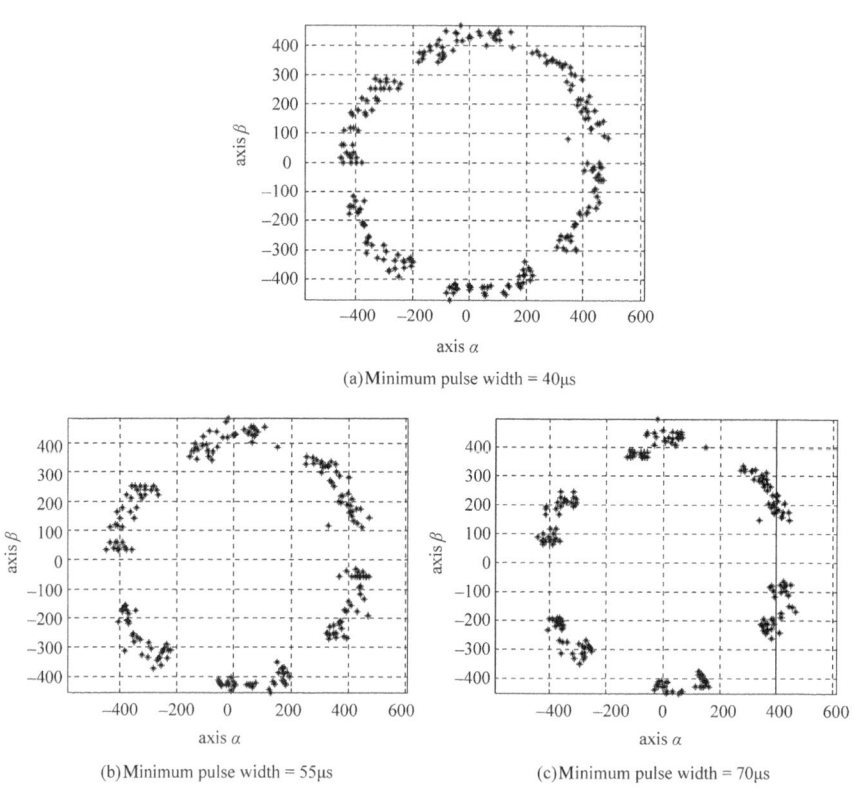

Fig. 2.17 Voltage vector distribution under different minimum-pulse-width settings

drive system runs at the low speed. Note the ideal distribution should be a perfect circle. The larger the minimum pulse width, the more distortion of space vectors.

As a summary, pulses and pulse sequences in the control loop determine related transient processes. Most of control strategies are mainly based upon the ideal assumption, which encounters difficulties with the non-ideal and non-linear factors in real practice.

2.2 Mathematical Models of Electromagnetic Transients

Electromagnetic transients in power electronic systems occur in the main-power loop, switches, energy storage components, control circuits and load. The most frequently used mathematical model to describe such transients are discrete iteration equations based on the state-space variables, piecewise smooth differential equations based upon switching modes, and the time-continuous average models based on the piecewise smooth state-space equations. Transient processes vary with circuit parameters. Details of such modelling methods and circuit transients are shown below.

2.2.1 Modelling Electromagnetic Transients

Such transients of power electronics combine the discreteness with the continuity, integrate the linearity with the non-linearity, and contain both information and energy. While using one single mathematical modelling method is not practical, following non-linear dynamic modeling methods are recommended.

1. **Discrete iteration equations based upon the system state-space variables**

Discrete modelling is also known as discrete time mapping, i.e., mapping the transient process in the present sampling period to the next sampling period in a discrete way. Based upon the system structure, parameters and actual working conditions, such discrete mapping could be applied to one-dimension, two-dimension or multi-dimension systems. The basic is to formulate piecewise linear differential equations, solve the state-space transfer matrix, and obtain the final solution.

Take a buck converter shown in Fig. 2.18 as an example, with following assumptions made:

(1) All components are ideal not considering switching transitions and inductor/capacitor loss;
(2) The cutoff frequency of the RLC low pass filter is far lower than the switching frequency, and the voltage U_0 across the output capacitor C is a constant;
(3) The current hysteresis control is applied.

In this way the circuit is turned into a first-order differential system. The inductor current i_L linearly increases when the switch turns on and linearly drops when the

Fig. 2.18 The circuit schematics of a buck converter

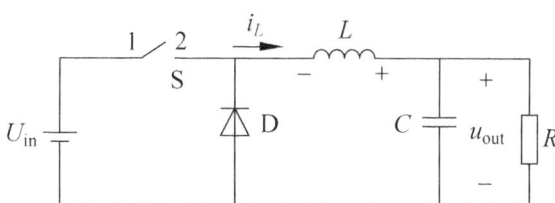

switch is off. Assume $i_n = i_L(nt)$, $i_{n+1}=i_L((n + 1)t)$. Here t is the sampling period and n is the nth sampling action. The one-dimension discrete time mapping model is shown below.

$$i_{n+1} = \begin{cases} i_n + m_1 t & i_n < I_{b1} \\ (1 + \frac{m_2}{m_1})I_{ref} - m_2 t - \frac{m_2}{m_1} i_n, & I_{b1} < i_n < I_{b2} \\ 0 & i_n > I_{b2} \end{cases} \qquad (2.3)$$

Here $m_1 = (U_{in}-U_{out})/L$, the inductor current slope when the switch is on. $m_2 = U_{out}/L$, the inductor current sliding slope when the switch is off. I_{b1} and I_{b2} are upper and lower limits of the current hysteresis, respectively. When $i_n < I_{b1}$, the switch remains on during the next sampling period. When $i_n > I_{b2}$, the inductor current in the next period drops to zero, making the circuit work in the discontinuous mode.

As shown above, such discrete iteration method is very easy to model all transient processes comprehensively. Its demerit, however, is to simplify and approximate the system and differential equations in order to derive the non-linear characteristics of discrete iterations, which creates a major difference between modelled and actual transients.

2. Piecewise smooth differential equations

This is also named as state-space variable analytical method, which directly applies Kirchhoff Current and Voltage Laws to the circuit thereby deriving the state-space differential equations.

Take a voltage-closed-loop-control buck converter as an example, as shown in Fig. 2.19.

When the switch is on, based on KCL, KVL and Ohm's Law with the ideal amplifier characteristics, we have

Fig. 2.19 A voltage-closed-loop-control buck converter

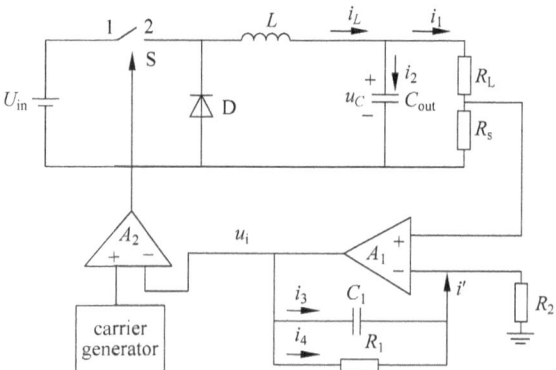

$$U_{\text{in}} = L\frac{di_L}{dt} + (R_{\text{L}} + R_{\text{s}})i_1 \tag{2.4}$$

$$i_1 + i_2 = i_L \tag{2.5}$$

$$i_1 = \frac{u_C}{R_{\text{L}} + R_{\text{s}}} \tag{2.6}$$

In addition, $i_2 = C_{\text{out}}\frac{du_C}{dt}$, $i' = i_3 + i_4 = \frac{u_{R_2}}{R_2}$, $u_{R_2} = i_1 R_{\text{s}}$, $i_4 = \frac{u_i - u_{R_2}}{R_1}$, $i_3 = C_1\frac{d(u_i - u_{R_2})}{dt}$.

Based on the above, we have

$$\begin{cases} \dfrac{di_L}{dt} = -\dfrac{u_C}{L} + \dfrac{U_{\text{in}}}{L} \\[2mm] \dfrac{du_C}{dt} = \dfrac{1}{C_{\text{out}}}i_L - \dfrac{u_C}{C_{\text{out}}(R_{\text{L}} + R_{\text{s}})} \\[2mm] \dfrac{du_i}{dt} = \dfrac{R_{\text{s}}}{C_{\text{out}}(R_{\text{L}} + R_{\text{s}})}i_L + \left[\dfrac{R_{\text{s}}(R_1 + R_2)}{C_1 R_1 R_2 (R_{\text{L}} + R_{\text{s}})} - \dfrac{R_{\text{s}}}{C_{\text{out}}(R_{\text{L}} + R_{\text{s}})^2}\right]u_C - \dfrac{1}{C_1 R_1}u_i \end{cases} \tag{2.7}$$

Similarly, when the switch is off, state-space equations are turned into

$$\begin{cases} \dfrac{di_L}{dt} = -\dfrac{u_C}{L} \\[2mm] \dfrac{du_C}{dt} = \dfrac{1}{C_{\text{out}}}i_L - \dfrac{u_C}{C_{\text{out}}(R_{\text{L}} + R_{\text{s}})} \\[2mm] \dfrac{du_i}{dt} = \dfrac{R_{\text{s}}}{C_{\text{out}}(R_{\text{L}} + R_{\text{s}})}i_L + \left[\dfrac{R_{\text{s}}(R_1 + R_2)}{C_1 R_1 R_2 (R_{\text{L}} + R_{\text{s}})} - \dfrac{R_{\text{s}}}{C_{\text{out}}(R_{\text{L}} + R_{\text{s}})^2}\right]u_C - \dfrac{1}{C_1 R_1}u_i \end{cases} \tag{2.8}$$

Together with (2.7) and (2.8), we have

$$\begin{cases} \dfrac{di_L}{dt} = -\dfrac{u_C}{L} + \dfrac{U_{\text{m}}}{L}s \\[2mm] \dfrac{du_C}{dt} = \dfrac{1}{C_{\text{out}}}i_L - \dfrac{u_C}{C_{\text{out}}(R_{\text{L}} + R_{\text{s}})} \\[2mm] \dfrac{du_i}{dt} = \dfrac{R_{\text{s}}}{C_{\text{out}}(R_{\text{L}} + R_{\text{s}})}i_L + \left[\dfrac{R_{\text{s}}(R_1 + R_2)}{C_1 R_1 R_2 (R_{\text{L}} + R_{\text{s}})} - \dfrac{R_{\text{s}}}{C_{\text{out}}(R_{\text{L}} + R_{\text{s}})^2}\right]u_C - \dfrac{1}{C_1 R_1}u_i \end{cases} \tag{2.9}$$

where s represents the switch state $(1, 0)$, yielding i_L, u_C and u_i all nonlinear functions. Hence the analytical solution is difficult to get.

As shown above, no simplification or approximation is needed for such method, all equations precisely describe the right behavior of the circuit, and the solution reveals the actual physical characteristics. The problem is that the numerical instead of analytical solution of the transients is generated, without considering switching transients as well.

3. State-space average modelling

This is a widely used modelling method of power electronics circuits. Such method is simple and does not contain time variants thereby facilitating the analysis and design. It will average system variables within one switching period, i.e., obtain the average value based on the switch duty cycle, derive state-space average equations and solve transients of state-space variables (Fig. 2.20).

Assume the switching period is T_s with the on time as $T_{on} = dT_s$. Here d is the duty cycle. Its off time is $T_{off} = (1-d)T_s$. For the Cuk converter, the energy storage and transfer are processed in these two time intervals (T_{on} and T_{off}) and two power loops in parallel, as shown in Fig. 2.21.

After multiple switching periods, the circuit reaches the steady state. During T_{on}, as shown in Fig. 2.21a, the switch is on, closing both the input and output circuits thereby leading to the diode reverse biased. During this period, the inductor current i_{L1} stores the energy in L_1. i_{L2} discharges C_1, stores the energy in L_2 and provides the energy to the load. The switch current $i = i_{L1} + i_{L2}$. During T_{off}, the circuit is shown as Fig. 2.21b, when the switch S is off while diode D is forward biased, closing the input and output loops. Here the input current and inductor discharging current, i_{L1}, charges C_1. The discharging current of L_2, i_{L2}, supplies the load power consumption. The diode current is the sum of the input and output current. State-space equations of the Cuk converter are

$$\begin{cases} \dfrac{di_{L_1}}{dt} = -\dfrac{(1-s)u_{C_1}}{L} + \dfrac{U_{in}}{L} \\[2mm] \dfrac{di_{L_2}}{dt} = \dfrac{su_{C_1}}{L} - \dfrac{u_{C_2}}{L} \\[2mm] \dfrac{du_{C_2}}{dt} = \dfrac{i_{L_2}}{C} - \dfrac{u_{C_2}}{CR} \\[2mm] \dfrac{du_{C_1}}{dt} = \dfrac{(1-s)i_{L_1}}{C} - \dfrac{si_{L_2}}{C} \end{cases} \qquad (2.10)$$

Here $L_1 = L_2 = L$, $C_1 = C_2 = C$, s represents the switch status, which is 1 when the switch is on, otherwise 0. Due to the voltage feedback,

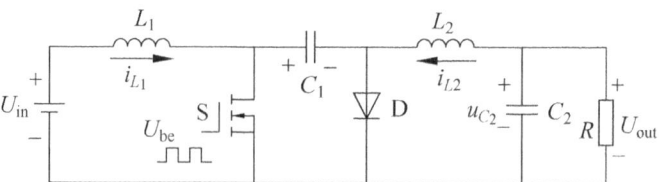

Fig. 2.20 Circuit schematics of a Cuk converter

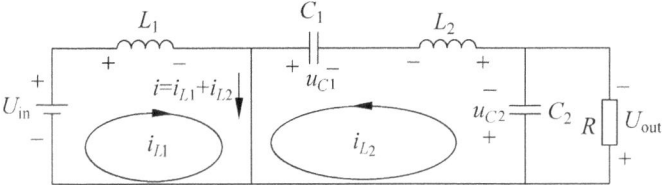

(a) When the switch is on

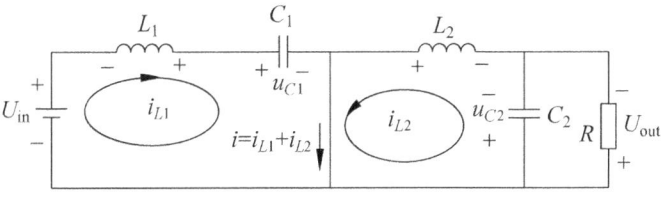

(b) When the switch is off

Fig. 2.21 Current and voltage distribution of a Cuk converter

$$i_{L_1} + i_{L_2} = g(u_{C_2}) \tag{2.11}$$

Here i_{L1} and i_{L2} are the current of inductor L_1 and L_2, respectively, u_{C2} is the output voltage, g(.) is the transfer function. For the purpose of simplicity, a proportional function is used, i.e.,

$$\Delta(i_{L_1} + i_{L_2}) = -k\Delta u_{C_2} \tag{2.12}$$

where k is the gain. Equivalently, Eq. (2.12) can be converted as

$$i_{L_1} + i_{L_2} = k_0 - k_1 u_{C_2} \tag{2.13}$$

Here k_0 and k_1 represent the control parameters. Replacing s in Eq. (2.10) with the duty cycle d, we can get the system average model. Since i_{L1} and i_{L2} are functions of u_{C2}, the system order could be reduced by one after considering Eq. (2.13), i.e.,

$$\begin{cases} \dfrac{di_{L_2}}{dt} = \dfrac{du_{C_1}}{L} - \dfrac{u_{C_2}}{L} \\[2mm] \dfrac{du_{C_2}}{dt} = \dfrac{i_{L_2}}{C} - \dfrac{u_{C_2}}{CR} \\[2mm] \dfrac{du_{C_1}}{dt} = \dfrac{(1-d)(k_0 - k_1 u_{C_2})}{C} - \dfrac{i_{L_2}}{C} \end{cases} \tag{2.14}$$

At the same time, derived from Eq. (2.13) is

$$\frac{d(i_{L_1} + i_{L_2})}{dt} = -k_1 \frac{du_{C_2}}{dt} \tag{2.15}$$

Substituting Eq. (2.10) and (2.14)–(2.15) yields to

$$d = -\frac{1}{2} \frac{\frac{k_1 L}{C} i_{L_2} - (1 + \frac{k_1 L}{CR}) u_{C_2} + U_{in}}{2 u_{C_1}} \tag{2.16}$$

Substituting Eqs. (2.16)–(2.14) generates the state-space average model of the Cuk converter

$$\begin{cases} \dfrac{di_{L_2}}{dt} = -\dfrac{k_1 i_{L_2}}{2C} - (1 - \dfrac{k_1 L}{CR}) \dfrac{u_{C_2}}{2L} + \dfrac{u_{C_1}}{2L} - \dfrac{U_{in}}{2L} \\[2mm] \dfrac{du_{C_2}}{dt} = \dfrac{i_{L_2}}{C} - \dfrac{u_{C_2}}{CR} \\[2mm] \dfrac{du_{C_1}}{dt} = -\dfrac{i_{L_2}}{C} + (\dfrac{k_0 - k_1 u_{C_2}}{2C})\{1 + \dfrac{1}{u_{C_1}} [\dfrac{k_1 L}{C} i_{L_2} - (1 + \dfrac{k_1 L}{CR}) u_{C_2} + U_{in}]\} \end{cases} \tag{2.17}$$

It can be seen that the average model is an approximation which is only true when the output fundamental frequency is much lower than the switching frequency. Only low-frequency characteristics of the circuit will be maintained while all high-frequency features are lost. Besides, due to the existence of the large disturbance, the converter is usually working at the large-signal mode. Using the prediction based on the small-signal transient model will create large errors.

Therefore, all three models have their own merits and demerits. Their application scopes are different, and they all treat switches as ideal without considering stray parameters of components and connections, which limits their applications especially for the complex converter circuit.

In the real practice, nonlinear factors inside power electronic converters are complicated, which will be detailed as follows.

2.2.2 Transient Model of the Main-Power Loop

Take a typical PWM rectifier as an example, as shown in Fig. 2.22. Such system is made of the AC power source, rectifying circuit and DC bus bars. Here e_a, e_b and e_c are the three-phase grid voltage, L_r is the input inductance, R_r is the equivalent line resistance, i_a, i_b, i_c are the three-phase input currents, C_{out} is the output capacitance, u_C is the capacitor and output voltage and i_L is the output current.

Assume all power switches are ideal with only lumped parameters being considered. With the symmetric three-phase grid, d-axis and q-axis components of the grid voltage and current could be derived with the d-q transformation, i.e., e_d and e_q, i_d and i_q, respectively. The mathematical model of such PWM rectifier under the d-q coordinates is

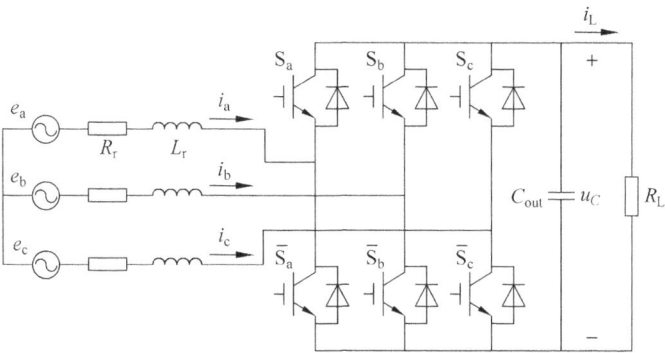

Fig. 2.22 The main-circuit structure of a typical PWM rectifier

$$\begin{cases} L_\mathrm{r}\dfrac{\mathrm{d}i_d}{\mathrm{d}t} = \omega L_\mathbf{r} i_q + e_d - i_d R_\mathrm{r} - u_d \\[2mm] L_\mathrm{r}\dfrac{\mathrm{d}i_q}{\mathrm{d}t} = -\omega L_\mathbf{r} i_d + e_q - i_q R_\mathrm{r} - u_q \\[2mm] C_\mathrm{out}\dfrac{\mathrm{d}u_C}{\mathrm{d}t} = \dfrac{3(u_d i_d + u_q i_q)}{2u_C} - i_\mathrm{L} \end{cases} \qquad (2.18)$$

Here ω is the rotatory angular velocity. u_d and u_q are the inverter d-axis and q-axis voltage with the SVPWM control, respectively, both of which are average values. Therefore the transient model of such PWM rectifier is essentially an average model.

2.2.3 Transient Models of Electric Components

All models above do not take transient processes of electric components into account. They treat either the power switches or passive components as ideal. In fact, due to the existence of stray parameters, all electric parts have their transients, even though the timescale is short. When considering the high-frequency performance, all such transients cannot be neglected.

Electric components in power electronic converters are mainly power switches (IGBT, etc.) and passive parts (inductors, capacitors, etc.). In following sections, we will analyze their electromagnetic transients based upon their structure and operational mechanisms.

1. **Power switches.**

Take IGBTs, the fully controlled power switches using MOSFETs to control BJTs, as an example. Compared to MOSFETs, IGBTs have one extra P + layer, which forms the base current of the BJT controlled by the MOSFET gate voltage. The goal is to inject current carriers from the collector and utilize the PN-junction conductance modulation to lower the on-state resistance and the forward voltage. By imposing

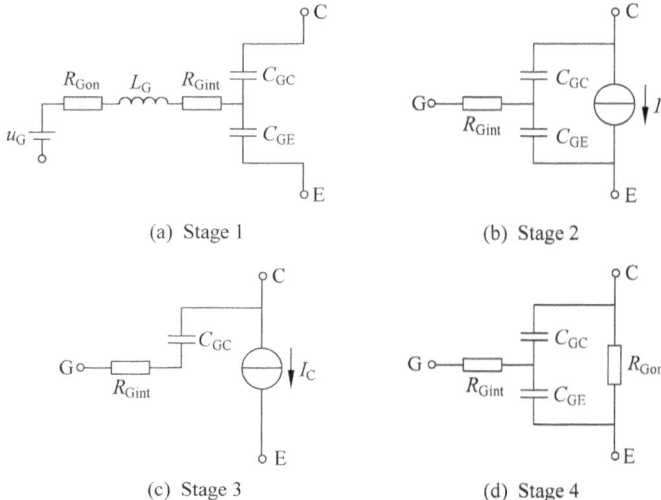

(a) Stage 1   (b) Stage 2

(c) Stage 3   (d) Stage 4

Fig. 2.23 The turn-on circuit model of the IGBT

a positive voltage on the gate, a channel is formed to provide the base current of the BJT thereby turning on the IGBT. Otherwise, a negative voltage on the gate eliminates the channel thereby turning off the IGBT. Different from the MOSFET, the existence of the PNP BJT results in a time delay to combine carriers even after the channel diminishes, creating the tailing current.

The parasitic junction capacitance of the IGBT greatly impacts the switching process. As shown in Fig. 2.23, there are four stages of the IGBT turn-on process. Here R_{Gint} is the gate internal resistance. C_{GC} and C_{GE} are both junction parasitic capacitances. Its detailed turn-on process is shown as below.

Stage 1: imposed with the gate-drive voltage u_{G}, IGBT remains off until the gate-emitter voltage u_{GE} rises to the threshold $u_{\text{GE(th)}}$. u_{CE} and I_{C} remain unchanged.

Stage 2: the gate continues being charged. The collector current I_{C} rises with u_{GE}. The collector-emitter voltage u_{CE} drops quickly close to u_{GE};

Stage 3: u_{CE} further drops close to the on-state saturation voltage. Due to the miller effect, its input capacitance C_{in}, made of C_{GC} and C_{GE}, becomes very large. The miller plateau appears with nearly all the gate current going through C_{GC}.

Stage 4: u_{CE} drops to the on-state saturation voltage and remains constant. The miller effect disappears when the IGBT stays on. The gate-drive voltage u_{G} keeps charging the gate.

With the gate voltage u_{GE} as the state variable, all four stages above could be treated as the gate-drive voltage u_{G} charges C_{in} through the gate-drive resistance R_{G} and gate-drive-loop inductance L_{G}. Therefore, the transient model of the IGBT gate-drive loop is

$$L_{\text{G}} C_{\text{in}} \frac{\text{d}^2 u_{\text{GE}}}{\text{d}t^2} + R_{\text{G}} C_{\text{in}} \frac{\text{d}u_{\text{GE}}}{\text{d}t} + u_{\text{GE}} = u_{\text{Gh}} \qquad (2.19)$$

Here u_{Gh} is the high-level of u_G. R_G is the gate resistance where $R_G = R_{Gon} + R_{Gint}$. Note that R_{Gon} is the gate external resistance and R_{Gint} is the gate internal resistance. L_G is the gate-loop inductance. The equivalent input capacitance C_{in} is

$$C_{in} = \frac{dQ_G}{du_{GE}} = \frac{C_{GE}du_{GE} + C_{GC}(du_{GE} - du_{CE})}{du_{GE}} = C_{GE} + (1 - \frac{du_{CE}}{du_{GE}})C_{GC} \quad (2.20)$$

In the turn-on process, all parasitic capacitance will vary greatly, leading to a significant change of C_{in}. However, back to each single stage shown in Fig. 2.23, C_{in} does not change much, which allows us to linearize C_{in} locally for each stage.

The transient process of u_{GE} for each stage could be derived through (2.19).

(1) When $R_G^2 C_{in} > 4L_G$

$$u_{GE} = \begin{cases} u_{Gh} + (u_{G0} - u_G)e^{-(t-t_0)/(R_G C_{in})}, & L_G = 0 \\ k_1 e^{\lambda_1(t-t_0)} + k_2 e^{\lambda_2(t-t_0)} + u_{Gh}, & L_G \neq 0 \end{cases} \quad (2.21)$$

Here u_{G0} is the gate initial voltage and t_0 is the initial moment. Meanwhile

$$\begin{cases} \lambda_1 = -\frac{R_G}{2L_G} + \frac{\sqrt{(R_G C_{in})^2 - 4L_G C_{in}}}{2L_G C_{in}} \\ \lambda_2 = -\frac{R_G}{2L_G} - \frac{\sqrt{(R_G C_{in})^2 - 4L_G C_{in}}}{2L_G C_{in}} \\ k_1 = (u_{G0} - u_E)\lambda_2/(\lambda_2 - \lambda_1) \\ k_2 = (u_{G0} - u_G)\lambda_1/(\lambda_1 - \lambda_2) \end{cases} \quad (2.22)$$

(2) When $R_G^2 C_{in} = 4L_G$

$$u_{GE} = m_1 e^{\lambda(t-t_0)} + m_2(t - t_0)e^{\lambda(t-t_0)} + u_{Gh} \quad (2.23)$$

Here

$$\begin{cases} \lambda = -R_G/(2L_G) \\ m_1 = u_{G0} - u_G \\ m_2 = -m_1\lambda \end{cases} \quad (2.24)$$

(3) When $R_G^2 C_{in} < 4L_G$, we have

$$u_{GE} = n_1 e^{\omega_1(t-t_0)} \cos \omega_2(t - t_0) + n_2 e^{\omega_1(t-t_0)} \sin \omega_2(t - t_0) + u_{Gh} \quad (2.25)$$

Here

$$\begin{cases} n_1 = u_{G0} - u_G \\ n_2 = -\omega_1(u_{G0} - u_G)/\omega_2 \\ \omega_1 = -R_G/(2L_G) \\ \omega_2 = \sqrt{4L_G C_{in} - (R_G C_{in})^2}/(2L_G C_{in}) \end{cases} \tag{2.26}$$

From Eq. (2.25), a too small R_G ($R_G^2 C_{in} < 4L_G$) results in a potential gate-voltage oscillation, leading to the system instability. To secure the system stability, an appropriate R_G is required.

At the stage 2 and 3, i_C and u_{GE} comply with the device transfer characteristics, which could be obtained through the datasheet or experiments and approximated by a quadratic function, i.e.,

$$i_C = a(u_{GE} - u_{GE\_th})^2 + b(u_{GE} - u_{GE\_th}) \tag{2.27}$$

Here a and b are coefficients obtained through the least-square method.

The IGBT turn-off process is opposite to the turn-on process. It also has four stages, as shown in Fig. 2.23. In this process, the gate voltage u_{GE} complies with

$$L_G C_{in}\frac{d^2 u_{GE}}{dt^2} + R_G C_{in}\frac{du_{GE}}{dt} + u_{GE} = u_{GI} \tag{2.28}$$

Here u_{GI} is the low-level gate-drive voltage. Difference from the turn-on process is the stage 4 of the turn-off process, i.e., stage 1 in Fig. 2.23. When u_{GE} drops below u_{GE_th}, I_C does not disappear right away. Instead, it appears as the tailing current to drop to zero gradually. The tailing current is determined by the IGBT manufacturing technologies and operational conditions, which in theory could be approached by an exponential function

$$i_C = \alpha i_{C\_on}e^{-(t-t_0)/\tau} \tag{2.29}$$

Here α is the amplification coefficient of the BJT, i_{C_on} is the IGBT on-state current, t_0 is the initial moment of the tailing current, and τ is the time constant of the tailing current.

From the above, we can see that electromagnetic transients exist in both turn-on and turn-off processes of the IGBT, determined by equivalent parameters of the device. Small parameter values lead to short time constants, much smaller than those of the main-power loop transients. In Sect. 2.3, differences and influences of these time constants will be discussed.

2. Capacitor transient models

In power electronic circuits, capacitor functions are mainly filtering, storing energy, decoupling and DC blocking. Different types of capacitors have vastly different capacitance values, frequency and loss characteristics, thereby used for different purposes. The commonly used capacitors in power electronic systems are:

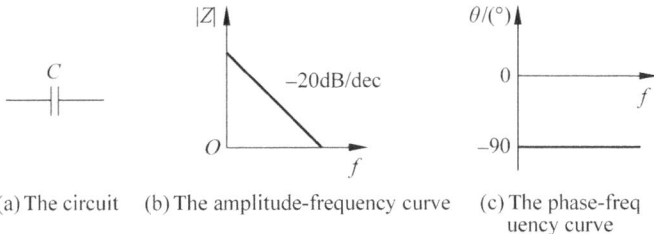

(a) The circuit   (b) The amplitude-frequency curve   (c) The phase-freq
                                                          uency curve

Fig. 2.24 The ideal impedance characteristics of the capacitor

(1) Electrolytic capacitors, which have a high capacitance-size ratio, however, large internal parasitic inductance and resistance. Electrolytic capacitors have two types, aluminum type and tantalum type. The latter one has better capacitance performance, which however is only true within the low-frequency range and has a higher cost. Due to such limitations, such capacitors are mainly used for filtering in power electronic converters, seldom used for noise reduction;

(2) Paper capacitors, which have a wide range of the capacitance and voltage ratings. Its equivalent series resistance (ESR) is much smaller than that of the electrolytic capacitors, however still has quite large equivalent series inductance (ESL). Various paper capacitors along with metallized paper capacitors are widely used in power electronic systems. Due to its better performance and high reliability, it is also used for the grid filtering;

(3) Ceramic capacitors. Different dielectrics have different energy density and the thermal coefficients. The characteristics of ceramic capacitors vary with the time, temperature and voltage. The voltage transients are easy to destroy such capacitors. However, such capacitors have small dimensions, excellent high-frequency performance and low ESR, making them widely applied in the printed circuit boards (PCBs);

(4) Film capacitors, which are categorized as polyester capacitors, polypropylene capacitors, polystyrene capacitors and polycarbonate capacitors, etc. It has a low ESR, which is subject to its coiling method. It exhibits a very low energy density as well. Among all capacitors, polycarbonate capacitors have very low electrolytic loss and very stable capacitance-frequency characteristics.

The ideal impedance of the capacitor is shown in Fig. 2.24.

The actual capacitor will show inductance at the high-frequency operation. The actual impedance of a film capacitor is shown as Fig. 2.25a.

As shown in Fig. 2.25, this capacitor shows capacitance at the low frequency and acts as the inductance at the high frequency. Its equivalent circuit is shown as Fig. 2.25b, which could be modeled as

$$u_{AB} = R_l + L_l \frac{di}{dt} + u_C \tag{2.30}$$

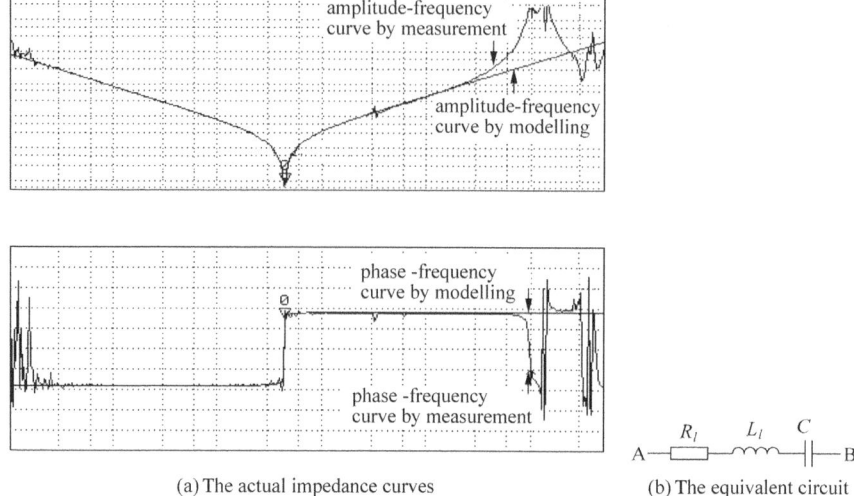

(a) The actual impedance curves   (b) The equivalent circuit

Fig. 2.25 Capacitor measurement and modelling

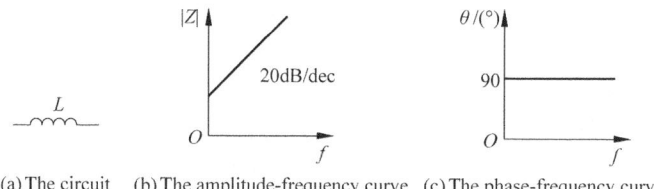

(a) The circuit   (b) The amplitude-frequency curve (c) The phase-frequency curve

Fig. 2.26 Impedance characteristics of an ideal inductor

3. Inductor transient models

The impedance characteristics of an ideal inductor are shown in Fig. 2.26.

Inductors are usually wound as coils. Based upon the magnetic core utilized, an inductor could be classified as air-core type and magnetic-core type. Power loss exists in both the winding and the core. Stray capacitance exists between turns, between the insulating layer and bobbins, and between layers if a multi-layer winding is adopted. The measured impedance of an air-core inductor is shown in Fig. 2.27a with the equivalent circuit shown as Fig. 2.27b.

The transient model of an inductor is shown as below

$$u_{C_l} = R_l + L\frac{\mathrm{d}i}{\mathrm{d}t} \tag{2.31}$$

The combination of Eqs. (2.19), (2.28), (2.30), (2.31) and (2.18) generates the overall differential equation set of the main-power loop, the solution of which describes main-power-loop electromagnetic transients.

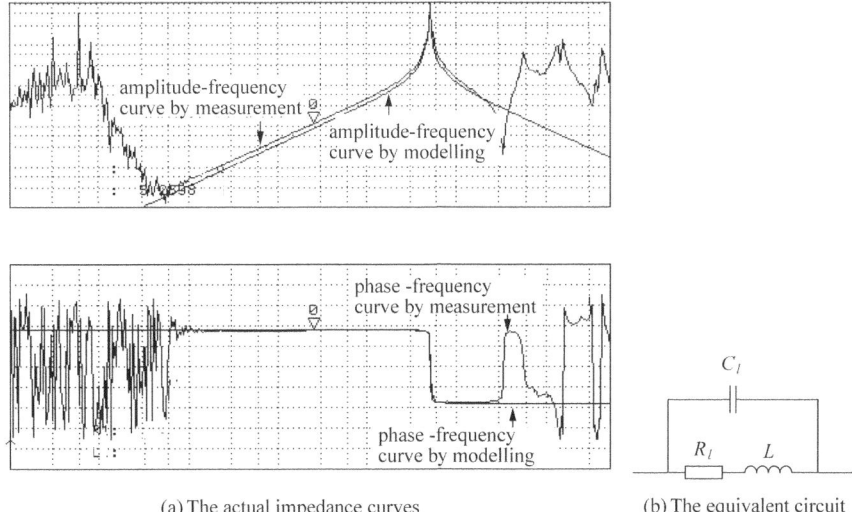

(a) The actual impedance curves (b) The equivalent circuit

Fig. 2.27 Inductor measurement and modelling

2.2.4 Transients Model of Gate-Drive and Control Circuits

1. Gate-drive circuit

Different switches have different gate-drive circuits. Take the IGBT as an example. Its gate-drive circuit is shown in Fig. 2.28, where R_G highly impacts the switching performance. A too large R_G slows down the switching speed, increases the switching loss and even fails triggering the switch. A too small R_G increases du_{CE}/dt, induces the EMI and potentially creates the mis-trigger. The reason for the mis-triggering is that at the switching-off moment, u_{CE} rises to the DC-bus voltage from the saturation forward voltage. A large du_{CE}/dt induces a high reverse current flowing through R_G thereby mis-switching on the device. A negative turn-off voltage could alleviate the impact of the reverse current thereby preventing mis-triggering.

Fig. 2.28 Typical gate-drive circuit for IGBT

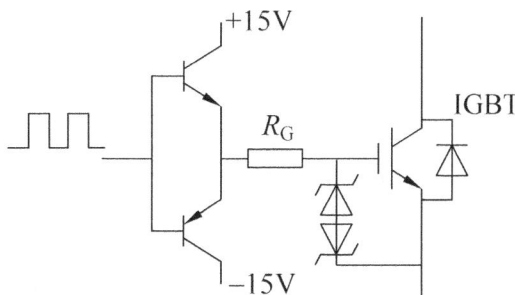

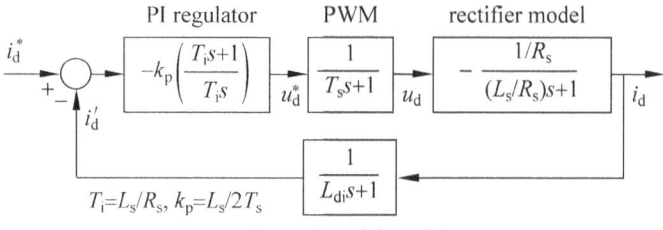

Fig. 2.29 Linearized transient model of field oriented control considering the sampling delay

The model of transient processes in the IGBT gate-drive loop is shown as Eqs. (2.19) and (2.28). As a major parameter of the gate-drive-loop transient processes, R_G varies the loop time constant.

2. **Control circuit**

The control circuit includes the sensing circuit, signal conditioning circuit, AD conversion and modulation circuit, communication circuit and control chip. Even though the control circuit has voltage and current as small signals, transient processes still exist. For instance, in the PWM rectifier, an exemplary linearized transient model of field-oriented control considering the sampling delay is shown in Fig. 2.29. Here R_s, L_s, T_i, T_{di} and T_s are corresponding parameters of the control loop.

Its open-loop transfer function is

$$G_0(s) = \frac{1}{2T_s^2 T_{di} \cdot s(s + \frac{1}{T_s})(s + \frac{1}{T_{di}})} \tag{2.32}$$

Its closed-loop transfer function is

$$G(s) = \frac{T_{di}s + 1}{2T_s^2 T_{di}s^3 + 2T_s(T_s + T_{di})s^2 + 2T_s s + 1} \tag{2.33}$$

Such equations are found useful when transients in the loop need be considered.

As a summary, when considering transient processes in the components, gate-drive loop and the control loop, their state-space equations need be incorporated with the whole system, which drastically increases the system orders. Furthermore, time constants of transients in each subsystem are so different that the coefficients of equations form a stiffness matrix, impossible to solve. To build a full-frequency-domain mathematical model of those transients in power electronic systems, link different-time-constant subsystems, address the relationship between the discrete and continuous variables and apply such models to the real design, analysis and control, remains as a problem to be solved for the transient modelling and applications.

2.3 Timescale Difference and Impact

Different parts of power electronic systems have different time constants thereby different timescale electromagnetic transients, e.g., ms for passive components in the main-power loop, μs for power switches and part of the control loop with stray parameters considered, and tens of ns for the high-frequency soft-switching and digital control system. If a mechanical load is connected, such as the motor drive system, its timescale will be multiple seconds. Various subsystems with different time constants form overall transients in the power electronic systems, during which energy balancing in the transformation, transfer and storage is critical. Majority of electrical components and equipment fail during transients. When subsystems with different time constants work together, imbalance of the energy transformation, distribution and transmission might occur, leading to the local energy concentration thereby destroying devices.

2.3.1 Comparison of Different Time-Scale Transients

A conventional three-phase two-level converter, as shown in Fig. 2.30 and also pictured as Fig. 1.10, from the left to right includes the control circuit, gate-drive circuit, three-phase two-level inverter (or rectifier) and power supply or load. The timescales are from nano-second level to second level. A quantitative comparison of these subsystems is detailed below.

The definition of the time constant could be traced back to the circuit theories, i.e., a constant representing the reaction speed of electromagnetic processes. For a value fading in the exponential function, the time taken to drop from 100% to 1/e (e = 2.71828) is defined as the time constant. The meaning of the time constant varies with

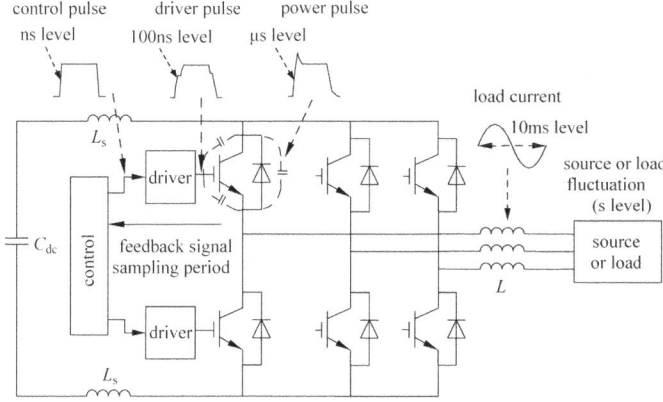

Fig. 2.30 The basic structure of a three-phase two-level inverter

applications. For an *RC* circuit, the time constant is the multiplication of the resistance and capacitance. With the unit of the capacitance as μF and that of the resistance as MΩ, the unit of time constant τ is s. Assume a constant current *I* is flowing through. The time period for the capacitor voltage to reach 1-1/e of *IR*, its maximum value, is the time constant τ. When open circuit, the time constant is the period for the capacitor voltage to reach 1/e, i.e., 0.37 of its peak value. In an *RC* circuit, when its capacitor voltage u_C always monotonically decreases from the initial value $u_C(0)$ to 0 in an exponential way, its time constant is *RC*. All other circuit components except the capacitor can be treated as a two-terminal active electric networks. In an *RL* circuit, its inductor current i_L monotonically decreases from the initial value $i_L(0)$ to 0 following an exponential function, with its time constant as *L/R*.

For a complex circuit containing multiple inductors and capacitors, multiple differential equations could be set for the final solution of multiple time constants. In the real practice, the time constant could be obtained through measuring experimental waveforms in transient processes. Sections below will discuss time constants in different loops.

1. Switching loop.

Take the IGBT as an example. As shown previously, its equivalent circuit contains multiple inductors and capacitors. Experimental methods are preferred to extract its time constant. We could use a device tester to locate its turn-on and turn-off commutating loops, as illustrated in Fig. 2.31a. The DC power supply is connected to an IGBT bridge, where the top device is a diode and the load is an inductor. During the test, the duty-cycle-controllable pulses are employed to switch the IGBT, with which the IGBT switching-on/off voltage and current waveforms could be probed. The time constant can be further derived.

Shown in Fig. 2.31b is the large-signal model of the switching loop with the small-signal model shown as Fig. 2.31c, which indicates that the switching-loop impedance is made of IGBT output capacitance C_{oss}, diode resistance R_D and loop stray inductance L_s. Here L_s includes the switch internal and external inductance.

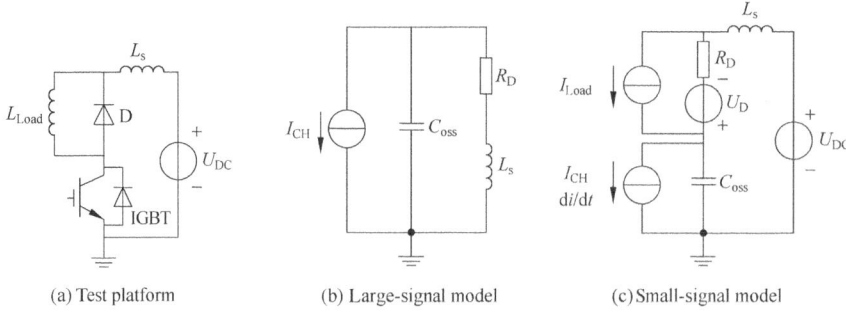

(a) Test platform   (b) Large-signal model   (c) Small-signal model

Fig. 2.31 Switch test platform, loop large-signal and small-signals models

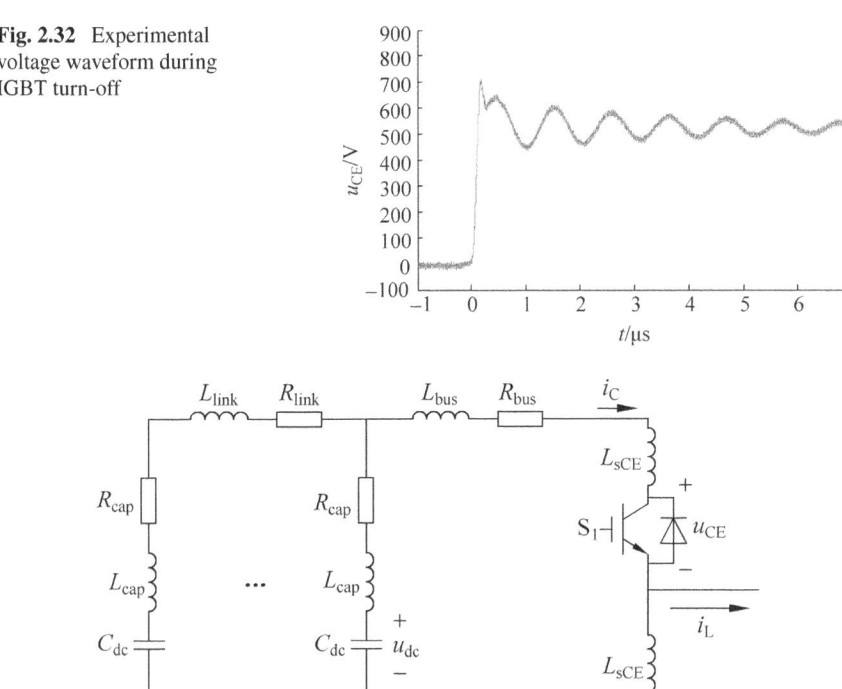

Fig. 2.32 Experimental voltage waveform during IGBT turn-off

Fig. 2.33 Phase equivalent circuit without snubbers

Experimental voltage waveform of the IGBT turn-off shown in Fig. 2.32 indicates that the time constant is ~100 ns.

2. Switching loop with bus-bar parasitics

The tester above is to directly connect the DC power supply to the switch leg. Once connected with bus bars, the DC-bus capacitance (including the capacitor ESR and ESL), stray inductance, capacitance and the resistance of bus bars need be taken into account, as shown in Fig. 2.33.

Based upon the equivalent circuit of Fig. 2.33, in the switching-off process, the voltage peak value of the IGBT and the DC-bus voltage comply with

$$u_{\text{CE - peak}} = u_{\text{dc}} + \left(L_{\text{cap}} + L_{\text{bus}} + L_{\text{sCE}}\right)\frac{\text{d}i_C}{\text{d}t} \tag{2.34}$$

The objective IGBT (FF30R17ME4) has the turn-off voltage and current shown as Fig. 2.34. During $t_2 \sim t_3$, the IGBT current drops from i_L to the tailing current i_{tail}, using the time interval of t_{fi}. Such time interval remains nearly constant at different turn-off currents and junction temperature, with the time constant of ~200 ns.

3. **Switching loop with the snubber circuit**

To limit the voltage spike during the IGBT turn-off, snubber circuits are usually employed, which typically have the charging-discharging type and non-discharging type. The former one include *RC* snubber, *RCD* snubber and *C* snubber, as shown in Fig. 2.35. The latter one includes two types of *RCD* circuits, as shown in Fig. 2.36. In this section, the *RC* circuit is used to calculate the time constant of the snubber.

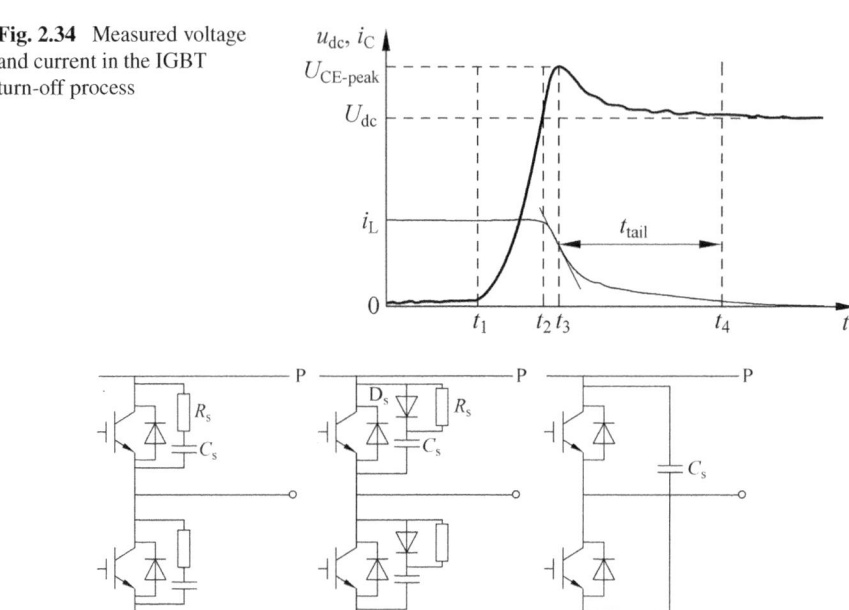

Fig. 2.34 Measured voltage and current in the IGBT turn-off process

Fig. 2.35 Charging-discharging snubber

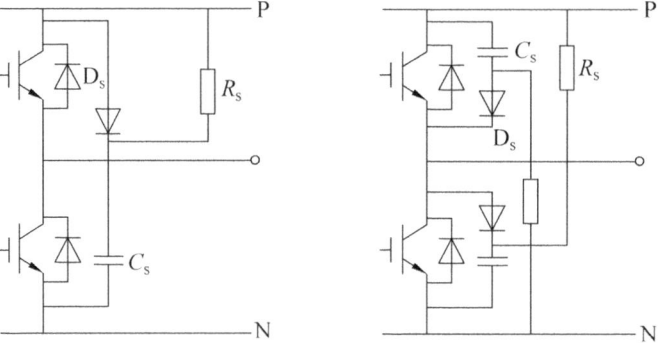

Fig. 2.36 Blocking-discharging snubber

For different topology-oriented snubber circuits, the maximum time to discharge C_s is different. A time constant of $1/6 \sim 1/3$ switching period is usually assigned to the IGBT snubber, ensuring that 90% of the voltage-spike related energy stored in C_s could be dissipated in a timely manner. Thus, the snubber resistance is

$$R_s \leq \frac{1}{3C_s f} \qquad (2.35)$$

Here f is the IGBT switching frequency, C_s is determined by the stray inductance, loop current and the voltage-spike tolerance. An exemplary selection is $R_s = 25\Omega$ and $C_s = 0.1 \, \mu\text{F}$, which leads to a snubber time constant of $2.5 \, \mu\text{s}$.

4. Main-power loop

Take the three-phase two-level inverter-fed motor drive system as an example, as shown in Fig. 2.37. The motor could be modelled as an RL load. A typical 25 kVA induction motor has the stator winding resistance of 0.3 Ω and the stator leakage inductance of 1.2 mH. Therefore the time constant is 4 ms.

5. DSP control loop

The time constant of the control-loop transient is decided by the DSP calculation period and the time delay of the control feedback loop. Figure 2.38 shows the scheme of the DSP control loop for a three-phase two-level inverter. Voltages and currents of the main-power loop are sampled to generate analogue signals, which are converted to digital signals through ADC and used for the digital signal processing and control algorithms, e.g., d-q transformation, PI controller and PWM control. Ultimately DSP generates PWM pulses to drive power switches through gate-drive circuits.

Assume the switching frequency of the inverter is 6.4 kHz and the sampling frequency is the same or twice of the switching frequency. In general, the implementation of one DSP command only takes several ns. The conventional control strategy does not require much calculation, making it possible for DSP to finish the calculation within one switching period. Therefore it is applicable to equal the calculation

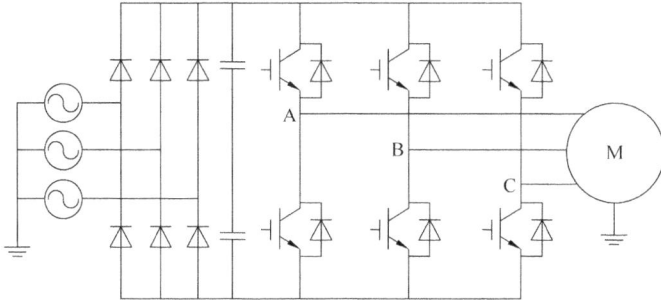

Fig. 2.37 The main circuit of a three-phase two-level inverter-fed motor drive system

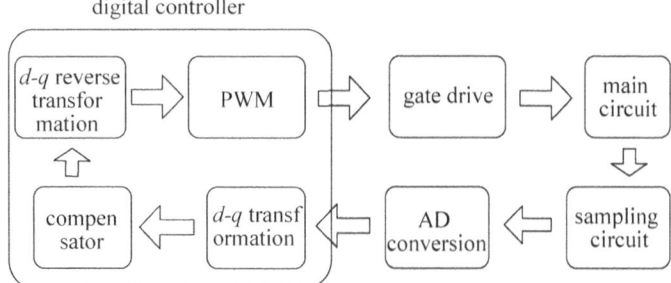

Fig. 2.38 The DSP control loop

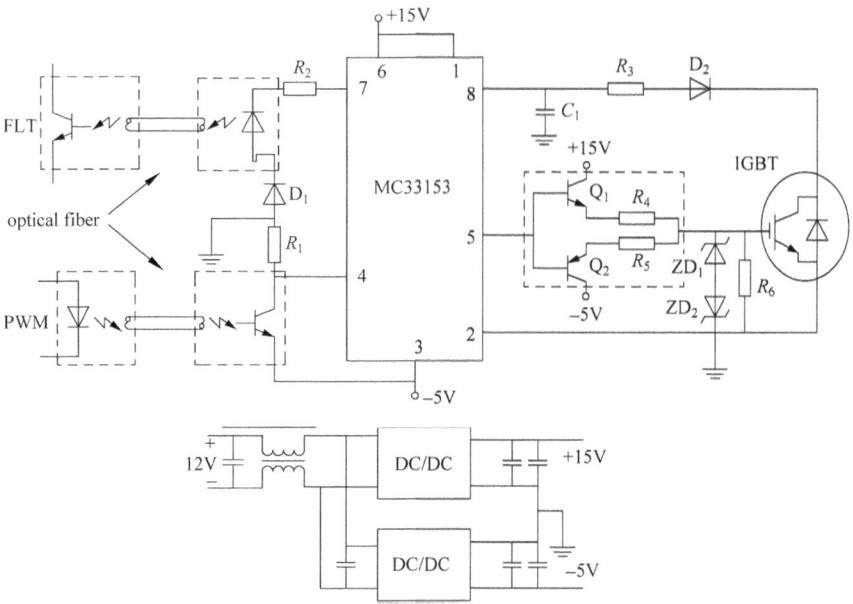

Fig. 2.39 IGBT gate-drive loop and the power supply

period or the control period to the switching/sampling period, i.e., 156.3 μs, which can be treated as the control-loop time constant.

6. Gate-drive loop

Shown in Fig. 2.39 is an exemplary IGBT gate-drive circuit with the power supply. PWM signals arrive at the gate-drive chip through the optical fiber and optical coupler, get amplified through the current amplifier, and reach the IGBT gate passing through the gate resistor. Meanwhile, the overcurrent protection signal is forwarded to the top-level control system through the gate-drive chip, optical coupler and optical fibers.

Table 2.1 Comparison of time constants in multiple transient loops

Case	Timescale level	Times ratio
Time constant of switching commutating loops without snubbers	100 ns	$1.0 * 10^0$
Turn-off time of some semiconductor	200 ns	$2.0 * 10^0$
Time delay by the gate driver	300 ns	$3.0 * 10^0$
Time constant of the snubber loop in high power converter	2.5 µs	$2.5 * 10^1$
Time delay by some control chip	150 µs	$1.56 * 10^3$
Switching period of the main circuit	4.0 ms	$4.0 * 10^4$

The signal propagation time is to be investigated by digging into the details of the gate-drive chip, MC33153, as shown in Fig. 2.39. The main specifications of such chip include: a typical time delay between the rising edges of the input and output is 120 ns, with the maximum value of 300 ns. The time delay between the overcurrent detection and the implementation of the protection is 300 ns. Therefore, the time constant of this gate-drive circuit is considered as 300 ns.

In summary, in a power electronic system, time constants of different loops and parts vary widely. Even though all these transients happen in the same system, each subsystem has its own transients and time constants. The time constants, switching period and time delays are summarized in Table 2.1. A clear difference is exhibited, where the shortest time constant is 100 ns while the longest is 4 ms, i.e., a difference of 40,000 times.

2.3.2 Correlations Among Different Time-Constant Loops

So many different time constants exist in one system. Comprehending their correlations is of importance, which will be detailed as follows, including the DSP control board, gate-drive board, switches, snubber circuit, connecting bus bars and feedback loops.

1. The DSP control board and the gate-drive board

The control loop is usually 500 times slower than the gate-drive loop. Therefore compared to the DSP control loop, the dynamic process of the gate-drive loop is negligible. However, when the connection in between uses the optical fiber or wires, various propagation delays is introduced. When using optical fibers, both sides need adopt optical couplers as well, which also bring the time delay and distortion. When cables are used, due to the existence of the stray capacitance, pulses generated by the

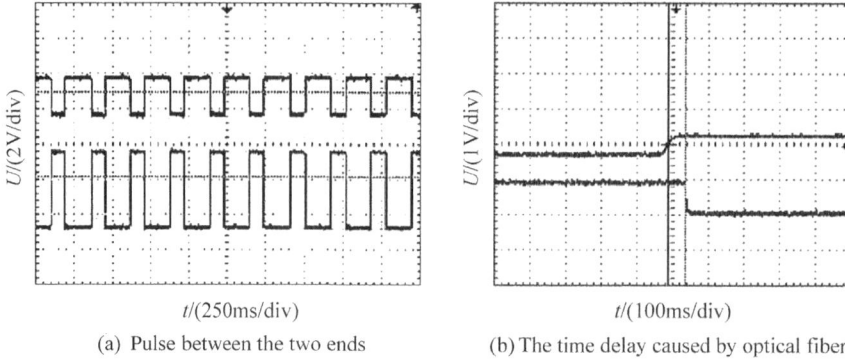

(a) Pulse between the two ends   (b) The time delay caused by optical fiber

Fig. 2.40 The measured pulses at transceiver and the receiver side of the optical fiber between the control board and the driver

DSP will be filtered thereby creating the distortion. Therefore optical fibers, couplers and cables will all create the difference between the control board output and the gate-drive board input. However, such distortion is minor in the real practice, not the main contributor to main-power-loop pulse distortions. Shown in Fig. 2.40 are pulses measured at the transceiver and receiver side of the optical fiber, respectively. From Fig. 2.40a, pulses at the two ends are out of phase. It is indicated in Fig. 2.40b that the time delay caused by this optical fiber is ~120 ns.

Besides, PWM signals generated by the DSP usually have no dead-band settings. To avoid zero or too small dead-band settings, before going into the gate-drive circuit a hardware dead-band using analogue circuits is inserted between falling and rising edges of two complimentary signals, even though the software dead-band might exist already. The dead time is usually longer than the sum of the switch turn-on and turn-off time, i.e., minimum pulse width. It is one of main contributors to the difference between DSP pulses and the gate-drive-board input, representing the dynamic relationship between the control system and gate-drive system.

2. **The power switch and the gate-drive board**

The pulse transfer between the control board and the gate-drive board is the information-signal-level behavior, with a relatively low-power level. After receiving control signals, the gate-drive system amplifies the power and delivers them to power switches, such as IGBTs. To charge the IGBT gate, gate-drive signals are tuned into the power level. In another word, the gate-drive circuit needs consume the power. Due to the existence of stray parameters, energy pulses are easy to be distorted and delayed. Therefore compared to pulses from the control board to the gate-drive board, gate-drive output pulses are more distorted.

Shown in Figs. 2.41 and 2.42 are the measured U_{CE} and U_{GE} of the IGBT during turn-on and turn-off processes, respectively. In addition, the smoothing effect caused by the stray capacitance will slow down the pulse rising edge, and even loses the pulse if the pulse width is too narrow thereby failing the pulse modulation. To avoid the potential pulse loss, a minimum pulse width could be set on DSP signals to bridge

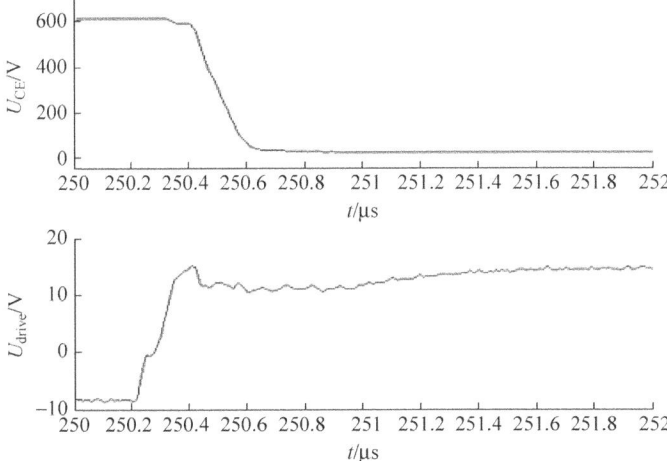

Fig. 2.41 U_{CE} and U_{GE} of the IGBT in the turn-on process

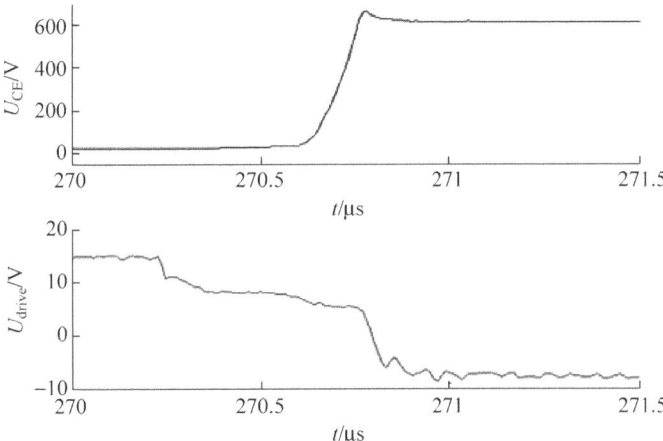

Fig. 2.42 U_{CE} and U_{GE} of the IGBT in the turn-off process

the gate-drive system with switches seamlessly. Therefore, the minimum pulse width is the dynamic connection between the gate-drive system and power switches.

3. The power switch and the snubber circuit

The snubber circuit is to prevent the switch from being damaged by the over voltage, which however affects the switching process. All typical snubber circuits shown in Sect. 2.3 have the resistor R, which will dissipate the energy and lower the converter

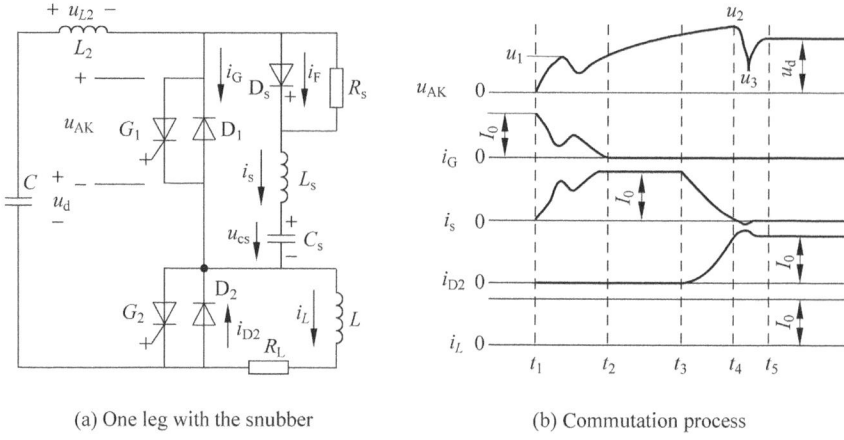

(a) One leg with the snubber   (b) Commutation process

Fig. 2.43 The snubber circuit and the commutation process of one GTO

efficiency. Such price is worthwhile given that the snubber will reduce the voltage spike thereby keeping the switch safe. The snubber circuit buffers the harmful impact brought by stray parameters and improve loop parameters.

Take one GTO leg as an example. Its snubber circuit with operational principles is shown in Fig. 2.43. The leg equivalent circuit is shown in Fig. 2.43a, where the snubber for the bottom switch is skipped. When G_1 turns off, the current commutates from G_1 to D_2, with the detailed commutating process shown in Fig. 2.43b.

Note that in the turn-off process three voltage spikes appear across the anode and cathode of the GTO, all of which are resulted from the interaction between the switch and the snubber circuit. Such waveform is also a reference for the snubber component selection and the switch safe operation. Take the first voltage spike u_1 as an example. When G_1 turns off, the load current is assumed constant. Both the turn-on snubber and the switch undertake a current variation of ~kA/μs, which induces voltage spikes on the loop stray inductance and the snubber capacitor. In addition, the snubber diode D_s has a forward turn-on voltage much higher than the steady forward voltage u_d, which is determined by the diode forward turn-on characteristics. Therefore the switch loop and the snubber circuit together induce the voltage spike u_1, which endangers the switch reliability even though its amplitude is much lower than the GTO steady-state voltage rating. This is because with the occurrence of u_1, quite an amount of the anode current, i.e., tailing current still exists, yielding the GTO full of current carriers with a low voltage-blocking capability. A large u_1 will create a dynamic avalanche thereby damaging the switch. Thus, in reality, parameter matching of the snubber circuit, such as an inductance-free capacitor and a low-forward-voltage diode is of importance. Settings of snubber parameters represent the dynamic linkage between the switch and the snubber circuit.

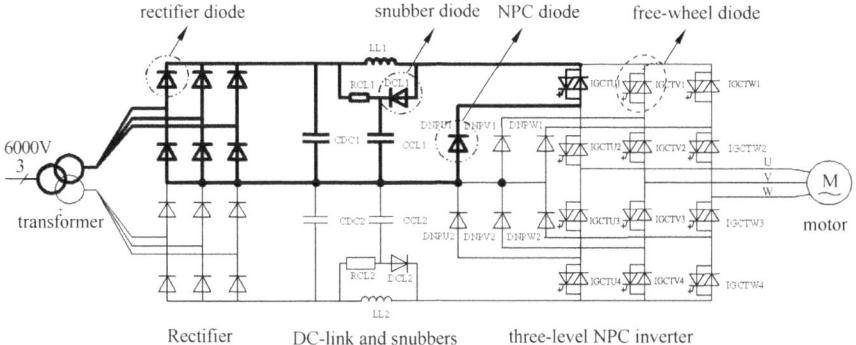

Fig. 2.44 The main-power-loop layout of an IGCT based three-phase NPC inverter

4. Power switches, snubber and bus bars

The main-power loop of an IGCT based three-level NPC inverter is shown in Fig. 2.44, where the DC-bus capacitor is an energy buffer. The input energy reaches the power switches through the DC-bus capacitor and bus bars. Even equipped with a snubber capacitor, most of the electromagnetic transient energy flows through the DC-bus capacitor. Therefore, the DC-bus capacitor always participates in the current commutation. Besides, stray parameters of DC-bus capacitor are critical, e.g., its stray inductance causes most of turn-off voltage spikes. Given that the switch current flows through DC-bus bars, reducing its resistance helps lower the power loss. Therefore the bus-bar optimization is critical for the switch safe operation.

The snubber circuit and bus bars determine the safe operation area of the power switch. Specifically, the stray inductance of the bus bar and the snubber capacitance counterpart each other thereby lowering the turn-off voltage spike. In the real applications, stray inductance on the bus bar is inevitable. The snubber circuit must match the bus bar to shrink the voltage spike thereby securing the switch's safety. Optimization of the snubber circuit needs incorporate with the bus-bar structure and the switch characteristics. The experimental waveform of the IGCT turn-off voltage and the diode forward recovery voltage is shown in Fig. 2.45.

It is indicated in Fig. 2.45 that the second turn-off voltage spike U_{p2} is because that the stray inductance of the bus bar transfers the stored energy to the snubber capacitor C_s, creating $U_{p2} = \sqrt{L_s/C_s} \bullet I_C$. Therefore, the snubber-circuit performance is influenced by bus bars. To limit U_{p1} and U_{p2} below the voltage rating, the upper limit of the bus-bar stray inductance L_s needs be determined based upon the switch selection. Therefore the snubber and bus bar parameters are vital to the transients of switches, main-power loop and snubber circuits.

Fig. 2.45 Experimental
waveform of the IGCT
turn-off voltage and the
diode forward recovery
voltage

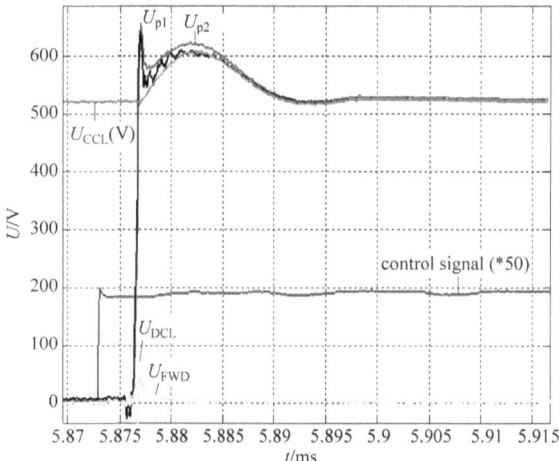

2.3.3 Impact of the Time-Constant Difference

The time-constant difference among various transients highly impacts the power
electronic system. Qualitative analysis of such impact will be carried out in this
section on the output pulses of the DSP, input pulses of the gate-drive circuit, switches,
snubber, connecting bus bars, feedback loop and main-power loop. In later sections,
a quantitative analysis of each subsystem will be done.

1. **Difference between the DSP output and gate-drive input pulses**

Such difference is mainly revealed by the dead-band, distortion, minimum pulse
width and time delay. Reasons of such difference have been discussed in Sects. 2.3.1
and 2.3.2. For the dead band, its main purpose is to prohibit the leg shoot-through
thereby protecting switches. However, it will deteriorate the control performance
and distort the voltage and current waveform, though dead-band compensations in
previous literatures could mitigate such impact to some extent. Distortions of the
gate-drive input slow down the switching speed, vary the pulse width and even eliminate pulses thereby worsening the control performance. The setting of the minimum
pulse width avoids the pulse loss, but it meanwhile changes the modulation performance especially in the open-loop operation. In the closed-loop control, effects of
the minimum pulse width limitation can be compensated by the feedback loop. The
time delay of pulses restrains the dynamic response of the control system and sets
upper limits to the switching frequency and control frequency.

2. **Difference between the gate-drive input and output**

Such difference includes the time delay, distortion and the inconsistency of the input
and output under faults. All the delay and distortion will be revealed through switching behaviors. Given gate-drive pulses are all energy-level, the impact of such delay

and distortion becomes more severe. For the inconsistency of the input and output at faults, a simple example is the gate-drive circuit detects the IGBT overcurrent, yielding the hardware protection is triggered to block pulses and turn off the switch. At this moment the input of the gate-drive circuit is still normal while the output signal is already blocked. Such inconsistency is positive to avoid the switch damage. Another example is that an inappropriate design of the gate-drive circuit will result in a high du/dt for the bottom switch when the top switch turns on, which introduces the reverse current to mis-trigger the bottom switch even though the actual command is to turn off. Such inconsistency is negative, which potentially leads to the bridge shoot-through and destroys switches.

3. **Impact of the snubber circuit on the switching process**

The snubber circuit reshapes the switching process of power semiconductors, protects the switch from the vital voltage spike, however, also consumes the energy thereby lowering the system efficiency. It meanwhile slows down the switching process.

4. **Impact of the bus-bar design on the switching process**

Stray parameters of bus bars are one of the main contributors for the switch turn-on voltage spike. Optimizations of the bus bar significantly reduce the stray inductance, improve switching processes and protect switches. Incorporating with the snubber circuit together will improve the switching performance. Such impact will be detailed in later sections.

5. **Impact of the feedback control on the switching process**

Feedback loops contain the hardware direct feedback and software indirect feedback. The former one has shorter timescales and much faster speed than the latter one, however, highly relies on the hardware with more complex structures. The hardware direct feedback is mainly to protect the switch from over-current for the safety purpose and dynamically balance the voltage distribution of series connected switches in switching processes, both of which do not participate in the top-level software control. The software indirect feedback samples the real-time data, controls the pulse width, and ultimately generates the voltage and current waveforms for the control purpose. It does not influence the switching process directly. Therefore, the system time constant is mainly influencing the software indirect feedback. For the converter shown in Fig. 2.31 with the closed-loop control, the simulated three-phase output currents with a sudden load change are shown in Fig. 2.46. Compared to Fig. 2.46a, the time delay of the control loop is considered in Fig. 2.46b.

A clear difference is shown between two transient processes, revealing the direct impact of the time delay of the feedback loop on the control performance.

6. **Difference of power switching processes**

Switching actions will form switching loops. Coupled with stray parameters in the loop, different switching waveforms appear. In an active NPC (ANPC) converter shown in Fig. 2.47, two switching loops are exhibited, the left of which is shorter with

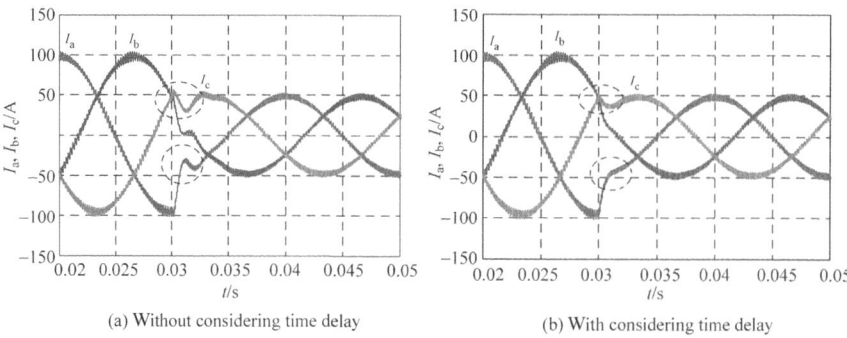

(a) Without considering time delay   (b) With considering time delay

Fig. 2.46 Comparison of the closed-loop control under the ideal operation and with time delay

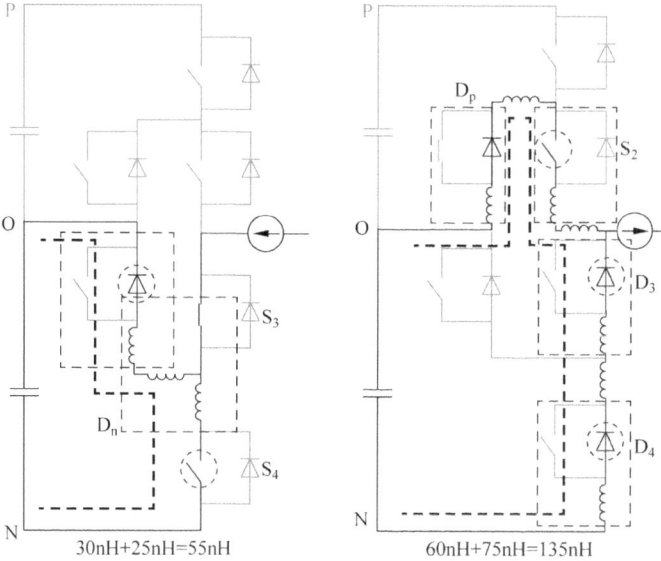

Fig. 2.47 Two commutation loops with different stray inductance of different power switches

55 nH stray inductance while the right of which is longer with 135 nH inductance. Variety of the inductance leads to the variation of the switching process thereby different switching loss and electric stress. Redundancies exist in the ANPC converter, which allows the selection of different switching actions and sequences based upon the diversity of the switching-loop inductance. Such effort results in the reduction of the power loss and electric stress.

2.3.4 Loop-Parameter Matching for Energy Balancing

To smoothly bridge various electromagnetic transients, parameter matching must be realized for different loops, aiming at the balance of the transient energy. Such matching includes: DSP and communication, feedback loop and the control loop, snubber and main-power loop, switches and main-power loop, filters and the main-power loop, various parts in the main-power loop.

1. **Matching between the DSP and the upper-computer communication**

Such matching is mainly focused on the operational frequency/period. In general, communication can only be implemented after the DSP finishes data sampling, processing, control and modulation. In reality, the communication frequency cannot be set too high in case that the computation resource of the DSP is occupied. It definitely cannot be too low to affect the command transmission speed of the upper-level computer.

2. **Matching between the feedback loop and control loop**

This can be detailed from the hardware and software feedback, respectively. The hardware feedback needs provide fast dynamic response for over-current protections and voltage balancing of series connected switches. The timescale needs be within one switching period so that the switch can be protected and the dynamic voltage balancing can be realized. It also cannot be too fast. Otherwise mis-protections might be triggered by disturbances or the dynamic balancing circuits will act frequently.

Frequency of the software feedback can be selected the same as the switching or control frequency. A too slow feedback will affect the control performance and slower the transients. A too fast feedback will be submerged by the control period thereby not directly benefiting the system dynamic response.

3. **Matching between the snubber loop and the main-power loop**

Stray parameters in the main-power loop, capacitors, cables and switches tend to induce turn-off voltage spikes. The design of the snubber circuit must match the main-power loop to further reduce the over voltage. While effectively reducing the voltage spike, the larger the snubber parameters the more energy consumption thereby the lower the converter efficiency. On the contrary, small snubber values will discount the snubber performance. Therefore the design of the snubber should be based upon main-power-circuit parameters.

4. **Matching between switches and the main-power loop**

Stray parameters in the switching loop contain the internal and external parasitics of the switch, all of which determine the switching behavior. The resonance between the stray inductance and the switch output capacitance is prone to creating the oscillation when switching off, which further creates EMI and slows down the switching speed. Such oscillation does not alter the switching loss much. Some moderate oscillation is

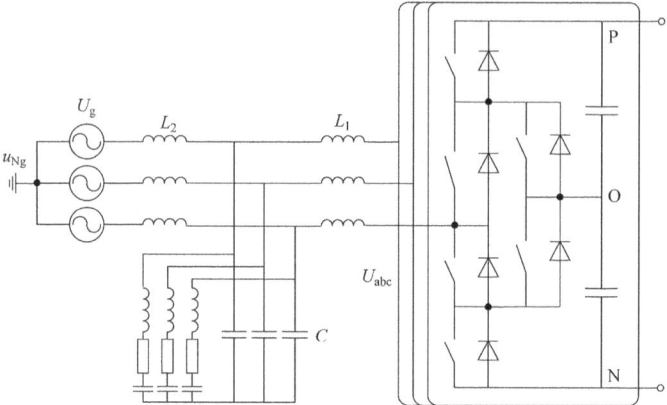

Fig. 2.48   Three-level grid-tied ANPC inverter with LCL filters

allowed if the switching speed is not strictly required unless the destructive voltage spike occurs.

Switching behaviors are coupled with switching loops in the power electronic converters. In general, a symmetric design of switching loops is recommended for the consistency of stray parameters, which will secure the same power loss for all switches. Thus the symmetry of the switching loop should be considered when designing the overall converter structure, such as component placement, bus-bar structures and switching actions.

5.   **Matching between the filters and the main-power loop**

In this section a grid-tied three-level ANPC inverter with an *LCL* filter is employed to detail filter design principles. As shown in Fig. 2.48, the damping circuit is equipped with the *LCL* filter, with the equivalent circuit of which shown in Fig. 2.49. Here L_1 restrains the output current harmonics of the inverter. When designing L_1, the cost and dimension needs be considered. C is used to provide the path for high-frequency current components. The reactive-power-compensation capability of the capacitor needs be addressed. L_2 is for the attenuation, the voltage drop of which needs be taken into account. In addition, the resonant frequency of the *LCL* circuit ω_r should meet $10f_{line} < \omega_r < 0.5f_{sw}$ for the demanded filtering performance. The damping circuit aims at an excellent damping effect thereby meeting the grid requirements of THD and harmonics, with a small power loss.

As shown above, the filter will reduce current harmonics generated by the main circuit thereby meeting specifications of the grid current harmonics. Oscillations might happen if the filter is not appropriately designed. An overdesigned filter will meet the filtering requirement, however, increase the cost and loss. Thus, an optimal filter design needs root from main-circuit parameters and the control strategy.

Fig. 2.49 Equivalent circuit of the LCL filter

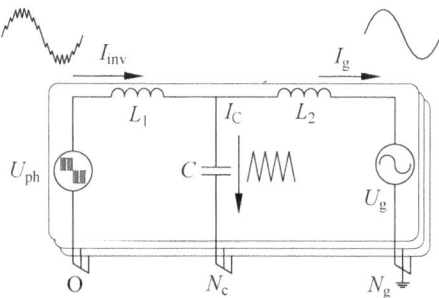

6. **Matching between different parts in the main-power loop**

In general, the symmetry of commutation loops should be maintained if possible so as to evenly distribute the power loss and electric stress across all switches. Same principles apply to the main-power loop per phase, given that main-power-loop parameters and control strategies are based on the precondition that all phases are identical. Some converters are not subject to such restraints if the symmetric operation is not a mandate.

As a summary, a power electronic converter can be treated as the mixture of electromagnetic transient loops with different time constants. Some exemplary time constants are those of the switching loop, DC-bus-bar loop, snubber circuit, main-power loop, DSP control loop and the gate-drive loop. A smooth conjunction of all such transients with different time constants is the premise of the system controllability and reliability. A timely parameter matching among subsystems is recommended, which essentially is the optimization and accurate control of the power electronic systems.

2.4 Electromagnetic Pulses and Pulse Sequences

Controllable electromagnetic pulses are employed in power electronic systems for energy transformation and transfer, which are the norms of transients. In this process, regardless of the shape and timescales of the pulse or pulse sequences, the energy conversation and no energy sudden change need be complied with, which is the fundamental of the transient analysis.

2.4.1 *Mathematical Expression of the Electromagnetic Pulses and Pulse Sequences*

Electromagnetic pulses are energy impulses. Mathematically, one pulse can be defined as the combination of two step functions, with one representing the rising edge and the other standing for the falling edge. The time interval between two edges is defined as the pulse duration time, or pulse width. If the rising edge is prior to the falling edge, the pulse is defined as the positive pulse, otherwise, the negative pulse. To detail the pulse characteristics, we take one voltage pulse as an example, as shown in Fig. 2.50. Here t_r is defined as the time interval when climbing from 10–90% of the pulse amplitude. t_f is the time interval for the pulse dropping from 90–10% of steady-state amplitude. The pulse width t_w is the time between rising and falling edges. The amplitude U is the steady-state value. Overshoot time t_s is the time for the pulse to rise from U to the maximum value U_{max} then drop back to U.

The definition of the pulse sequence is based upon the single pulse. In addition to the basic characteristics of a single pulse, following characteristic parameters need be considered for a pulse sequence.

1. Time property

If pulses among a pulse sequence are periodic, the pulse repetition frequency (PRF) is applied to represent the time interval between periodic pulses. Otherwise, the pulse repetition rate (PRR) is used to indicate pulse numbers per unit time, even though these pulses might not be periodic or have different shapes.

Shown in Fig. 2.51 are the voltage pulse sequences with induced current employed in the pre-excitation process of a variable speed motor drive system.

As shown above, two distinct stages exist in the voltage pulse sequence. The first is the current-rising stage, when a single voltage vector is adopted by the inverter yielding periodic pulses. Here $PRF_1 = PRR_1 = 1/(250 \ \mu s) = 4$ kHz. The second is the current-maintaining stage, when different voltage vectors are used to limit the current ripple within a narrow hysteresis. Here periodic pulses are replaced by

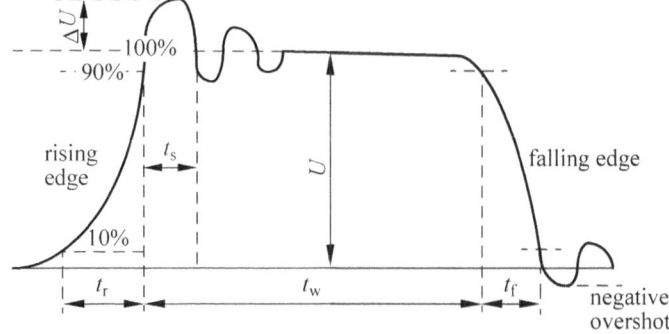

Fig. 2.50 Definitions about a pulse

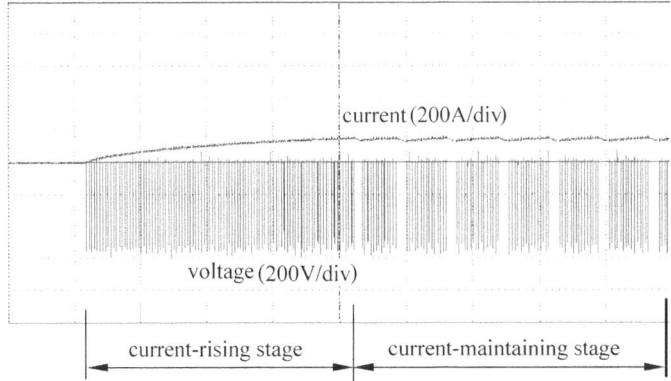

Fig. 2.51 Experimental voltage and current waveforms of the DC pre-excitation control

Fig. 2.52 Structural property of the pulse sequence

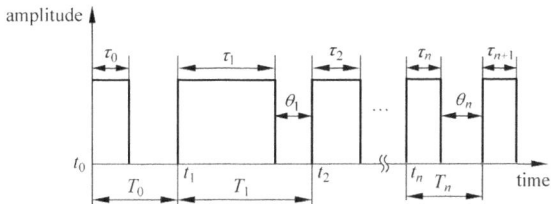

periodic pulse sequences. The pulse-sequence repetition frequency is $PRF_2 = 133\,Hz$ and the pulse $PRR_2 \approx 2\,kHz$. PRF_1, PRF_2, PRR_1 and PRR_2 are all related to the pulse duty cycle and amplitude, which are determined by the control strategy and the load time constant.

2. Structural property

Shown in Fig. 2.52 is a pulse sequence randomly distributed along the time axis. Here t_n is the starting time of the nth pulse, indicating its relative location. θ_n is the time interval between the falling edge of the nth pulse and the rising edge of the (n + 1)th pulse, representing the structure of the pulse flow. $\theta_n > 0$ means no overlap among pulses, e.g., the gate signals of the power semiconductors. $\theta_n < 0$ indicates the overlap among pulses, e.g., line-line voltage of the multi-level inverter output.

3. Internal energy distribution

Pulses are the form of the energy. So are pulse sequences made of multiple pulses. To evaluate the energy of the pulse sequence, not only the overall energy but also the energy distribution inside the pulse sequence needs be considered.

Assume one pulse sequence $x(t)$ ($t > 0$) has the energy definition as below

$$E = \int\limits_{t=0}^{+\infty} x(t)^2 dt$$

$$E(t) = x(t)^2 \tag{2.36}$$

Here E represents the overall energy. $E(t)$ is the instantaneous energy, a function of time.

All electromagnetic pulses can be characterized by the voltage and current, summarized as energy pulses. Their mathematical description should include the pulse generation and transfer, and particularly, differentiate the energy pulse from the information pulse. From the short-timescale perspective, e.g., ns ~ μs, energy pulses are not either 0 or 1, but a continuous energy form with the rising edge, falling edge, pulse width and amplitude, i.e., continuous in both the time and space domains. A voltage pulse can be expressed as

$$\begin{cases} u(x,t)_{|t=t_0^+} = u(x,t)_{|t=t_0^-} \\ u(x,t)_{|x=x_0^+} = u(x,t)_{|x=x_0^-} \end{cases} \tag{2.37}$$

Medium is needed to carry the energy pulse with the limited propagation speed, as shown in (2.38)

$$k\frac{\partial^2 u}{\partial x^2} = \tau\frac{\partial^2 u}{\partial t^2} + \frac{\partial u}{\partial t} \tag{2.38}$$

Here k the dielectric property and τ is the time constant. Different k and τ result in variety of the pulse propagation, even though the propagation Eq. (2.38) is always true.

2.4.2 Propagation and Deformation of the Pulse and Pulse Sequence

The propagation of pulses is a time-space function. The pulse shape will be distorted by nonlinear factors along transmission lines, which generates various unknown waveforms deviating from the conventional theoretical analysis. Some of distortions are harmless, even though their internal mechanism is to be comprehended. Others are destructive in need of detailed solutions.

The PWM voltage pulse under the ideal topology is the perfect square waveform. However, as shown before, the pulse is deformed by loop parameters at different locations. The actual voltage pulses generated by an inverter are shown in Fig. 2.53, where multiple abnormal pulses are exhibited.

Internal mechanisms of such pulses are diverse, which are attributed to the deadband, minimum pulse width, reverse recovery process of the snubber circuit, loop stray inductance, load characteristics, etc. All these pulses occur in transients with short timescales. Some pulses are due to the control strategy not considering microscopic transients, named as control abnormal pulses. Such pulses distort the output waveform and even affect commutation processes. Some pulses are caused by loop parameters, e.g., inappropriate settings of snubber parameters, which are named as loop narrow pulses. These pulses are usually harmless while only distorting the output waveform. Efforts are needed to clarify the reason of such pulses. Other pulses are caused by loop parameters and the control algorithms together, either harmful or not, which are subject to be differentiated.

1. Abnormal pulses in the dead-band

In a three-phase three-level SVPWM based inverter, a dead-band is inserted between complementary switches in each leg to avoid the shoot through, during which the leg is uncontrollable, i.e., dead-band effect. For an ideal operation, dead-band pulses are symmetric, as shown in Fig. 2.54.

In the real practice, various asymmetric dead-band pulses are observed in the actual inverter output, as shown in Fig. 2.55.

The zero current of the Phase B results in the instability of the phase output voltage, which generates the asymmetric dead-band effect.

Fig. 2.53 The actual voltage pulses generated by an inverter

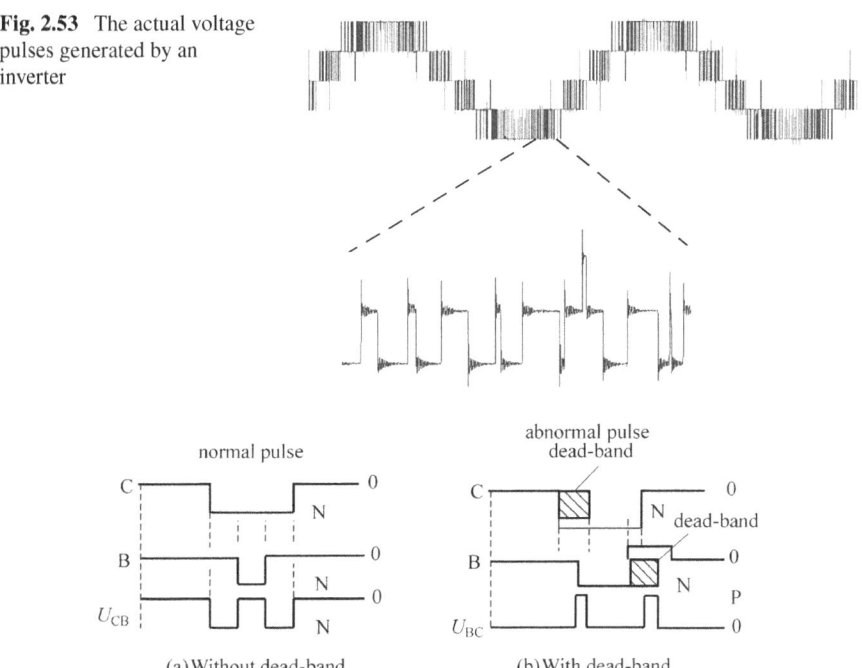

(a) Without dead-band   (b) With dead-band

Fig. 2.54 Symmetric dead-band effect (Phase-B current flows out and Phase-C current flows in the inverter)

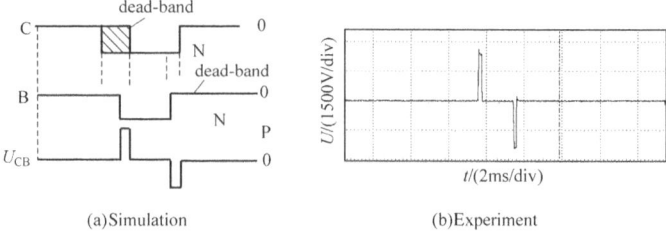

(a)Simulation   (b)Experiment

Fig. 2.55 Asymmetric dead-band effect (Phase-B current crosses zero and Phase-C current flows in the inverter. Pulse widths are 25 μs and 22 μs, respectively)

2. Abnormal pulses in the loop

The line-line voltage of the inverter output is measured as U_{AB} and U_{CB}, as shown in Fig. 2.56. The voltage leap of one phase generates a voltage spur on the other, i.e., the commutation process of one phase changes the voltage of the other two phases.

For the three-phase three-level NPC inverter using SVPWM control in Fig. 2.44, it's assumed that the phase-C current flows out of the leg. During the current commutation, the current flows through the current-limiting inductor, which reduces the Phase-A voltage nearly to zero. This explains why a short-duration pulse appears on U_{AB}, even though U_{AB} quickly recovers back to $1/2 U_{dc}$. Such short pulses are resulted from the shared snubber circuit of the three phases. The commutation at one phase will impact the other two phases.

3. Abnormal pulses caused by the control and the loop

In this section, we will still use the inverter topology shown in Fig. 2.44. The waveform shown in Fig. 2.16a is because that Phase-A voltage jumps from 0 to positive (P) within the dead band while Phase-B voltage remains at 0. Phase-A current flows into the leg, generating P as the leg output voltage within the dead band and further yielding $U_{AB} = 1/2 U_{dc}$. Within the dead band, the Phase-A current changes the polarity, flowing out of the leg thereby resulting in 0 output voltage. Given that

Fig. 2.56 The experimental waveforms of the line-line voltage

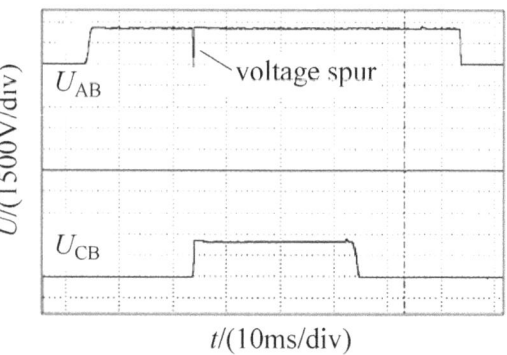

Phase-B remains 0, now $U_{AB} = 0$. Upon the completion of the dead band, Phase-A voltage becomes P, i.e., jumping from 0 to $1/2U_{dc}$. All the above together yield the abnormal pulse in Fig. 2.16a.

The combination of the control strategy and loop parameters generates another abnormal pulse. Shown as U_{CB} in Fig. 2.16b, Phase-C current flows into the neutral point while Phase-B current flows out of the neutral. The pulse width is narrow, and becomes even narrower when eaten up by a dead band. For a leg of an NPC inverter, the narrow pulses jumping from the low to high voltage are mainly due to the control strategy. When shorter than the recovery time of the snubber circuit, such abnormal pulses appear. In terms of the current commutation, such short pulses are not harmful as long as sufficient time intervals are given to turn on or off the switch. To eliminate such pulses, widening the minimum vector time to a value larger than the snubber recovery time is the solution.

2.4.3 Time and Logic Combination of Pulse Sequence

In a power electronics converter, to approach the reference waveform by utilizing the single-switch output, pulse sequences are employed by manipulating corresponding switching actions. To obtain the required frequency feature, we adopt the pulse-sequence time combination in power electronic systems. With the increasing demand of the power rating, output frequency and output performance, one single pulse sequence is difficult to meet all requirements. Here we introduce the pulse logic combination, i.e., at the same moment combining multiple pulses or pulse sequences in the space to realize the reference output.

1. **Pulse time combination and its limitation**

A power electronic converter essentially is a power amplifier. Theoretically it could realize any random-waveform transformation, as shown in Fig. 2.57 transforming the DC input to AC or any other waveform. When reflected to the frequency domain, it is supposed to output any component at any frequency within the bandwidth, as shown in Fig. 2.58.

The problem for the voltage-source converter is, the basic topology is made of a complementary pair of switches, which theoretically can only generate square pulses with specific amplitude and pulse width, with its mathematical expression in the related frequency domain shown as

$$p_t(t) = U_m[u(t + \frac{t_w}{2}) - u(t - \frac{t_w}{2})] = \int_{-\infty}^{\infty} U_m t_w \frac{\sin(\frac{\omega t_w}{2})}{\pi \omega t_w} e^{j\omega t} d\omega \qquad (2.39)$$

Here U_m and t_w are the pulse amplitude and pulse width, respectively. $u(t)$ is the step function. The spectrum in the frequency domain with different pulse width is shown in Fig. 2.59c. With all-frequency components being included, it is obviously impossible to tune each frequency component close to the target with only one pulse.

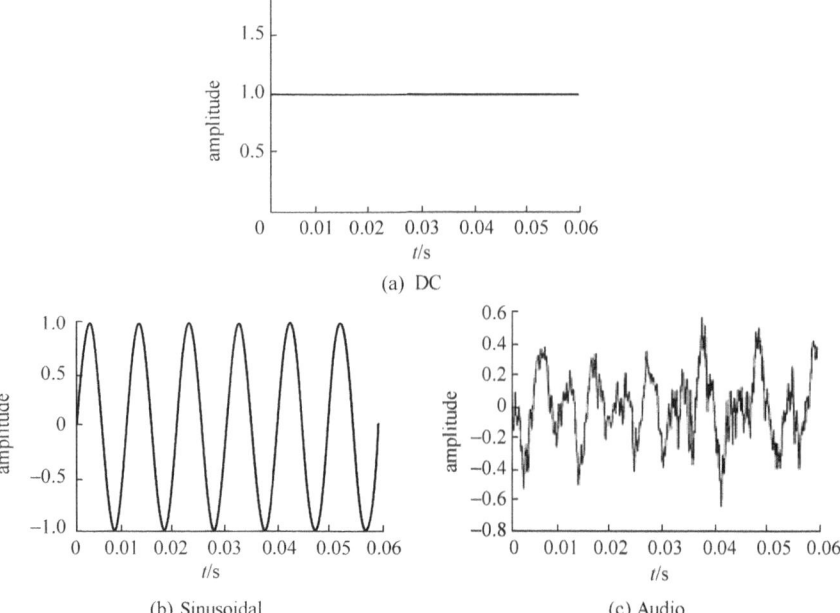

(a) DC

(b) Sinusoidal

(c) Audio

Fig. 2.57 Output waveforms of a power electronic amplifier

Fig. 2.58 Typical
frequency-amplitude
characteristics of a power
electronic amplifier

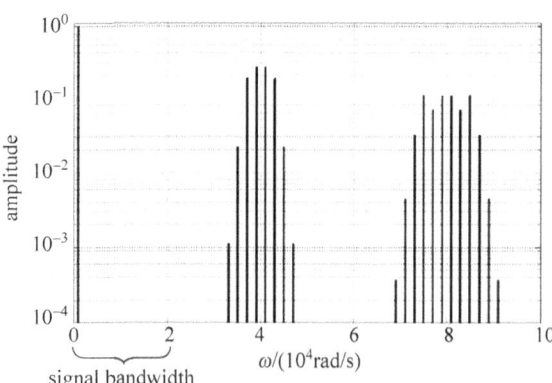

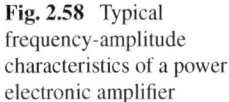

signal bandwidth

The solution is to generate the pulse sequence made of pulses with adjustable
pulse width, by utilizing interlocked switches at the same leg to approach the targeted
output, named as PWM control. The mathematical expression is shown as below

$$f(t) = p_{t_1}(t - t_1) + \arcsin \theta \sum_{i=1}^{\infty} p_{t_i}(t - t_i) \qquad (2.40)$$

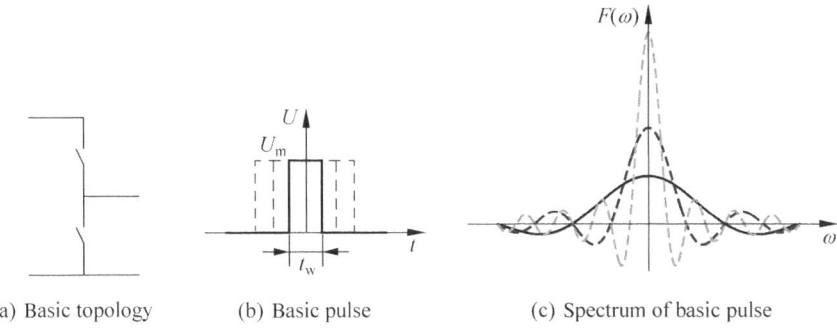

(a) Basic topology   (b) Basic pulse   (c) Spectrum of basic pulse

Fig. 2.59 Basic topology and pulse of a power electronic converter

A lot of applications assume the difference of t_i as a constant, even though it is not always true, e.g., in the hysteresis PWM control.

The above is an example of the pulse time combination, i.e., forming a pulse sequence by moving pulses to the appropriate time. Through combining and regulating the pulse sequence, the electrical component at the targeted frequency is obtained.

The pulse time combination can take care of most of the spectrum. For the SPWM using the symmetric sampling technique, the pulse-sequence spectrum without the DC component is

$$f(t) = \frac{4E}{\pi} \sum_{n=1}^{\infty} \frac{1}{q'} J_n\left(q'\frac{\pi}{2}M\right) \sin\left[(q'+n)\frac{\pi}{2}\right] \cos[n(\omega_o t + \theta_o)]$$

$$+ \frac{4E}{\pi} \sum_{m=1}^{\infty} \sum_{n=-\infty}^{\infty} \frac{1}{q} J_n\left(q\frac{\pi}{2}M\right) \sin\left[(q+n)\frac{\pi}{2}\right] \cos[m(\omega_c t + \theta_c) + n(\omega_o t + \theta_o)]$$

$$(2.41)$$

Here $q = m+n(\omega_o/\omega_c)$ and $q' = n(\omega_o/\omega_c)$. ω_o, θ_o and ω_c, θ_c are the angular velocity and the phase of the reference and carriers, respectively. M is the modulation index. $J_n(x)$ is the n-order Bessel function. Assume the reference frequency is 1 kHz, the carrier frequency is 20 kHz and the modulation index is 0.9. The normalized pulse sequence then is shown in Fig. 2.60a with the spectrum shown in Fig. 2.60b. Eliminating the carrier-frequency, integer-order and other side-band harmonics yields the required sinusoidal waveform. The sampling principle generates some side-band components, which due to the small amplitude do not affect the signal fidelity. With the natural sampling even such side-band harmonics around the fundamental could be further erased.

With more applications of power electronics, higher performance is required, which sees the limitation of the pulse time combination. For the wide-frequency-range high-power converters, power ratings, output frequency range and the waveform distortion are the three most important performance indexes, which usually

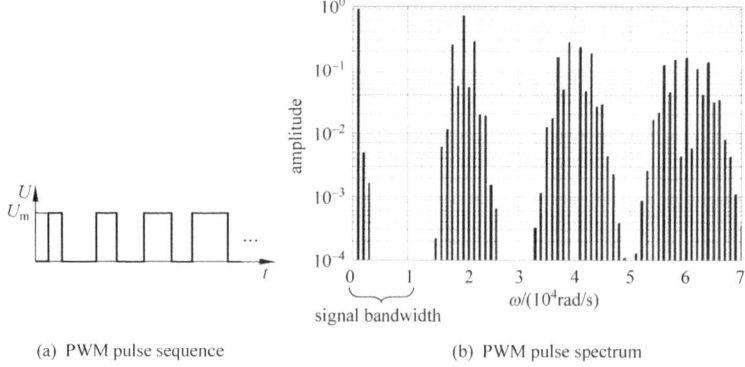

(a) PWM pulse sequence   (b) PWM pulse spectrum

Fig. 2.60 The typical PWM pulse sequence and its spectrum

confine each other. From Fig. 2.60b, to get a low THD needs secure the clean signal within the target frequency range, i.e., harmonics free. Otherwise, it is difficult to filter all harmonics. Theoretically when no switching transients or other nonlinear factors are involved, harmonics within the target frequency range mainly come from two sources:

(1) Side-band harmonics of the signal itself, mainly caused by the symmetric sampling. Such harmonics will not be eliminated unless the natural sampling is used. This type of harmonics usually has a low amplitude, which drops with the increment of the PWM frequency;

(2) Side-band harmonics caused by the low carrier frequency. If the PWM frequency is not high enough, side-band harmonics will appear at the left side of the carrier frequency, or double and triple of carrier frequency. Such harmonics will merge into the signal frequency band and create the intolerable distortion.

Therefore, to widen the output frequency band of the converter and keep the high signal fidelity, a high PWM frequency is a must. For a simple time combination of pulses, i.e., one single pulse sequence, the PWM pulse frequency is the switching frequency. In the medium or high power converters, the power switches will be restrained by the switching speed, power loss and EMI, which confines the switching frequency to a low level.

2. The logic combination of pulses and pulse sequences

To remove the bottleneck of the pulse time combination, the logic combination of pulse sequences is recommended, i.e., logically superposing pulses or pulse sequences generated by multiple switching units and synthesizing output components at the target frequency, which realizes the high-performance output with limited switch capability.

The logic combination of pulse sequences, which are highly related to the topology and modulation strategy, is diverse. Three main methods are summarized as Fig. 2.61.

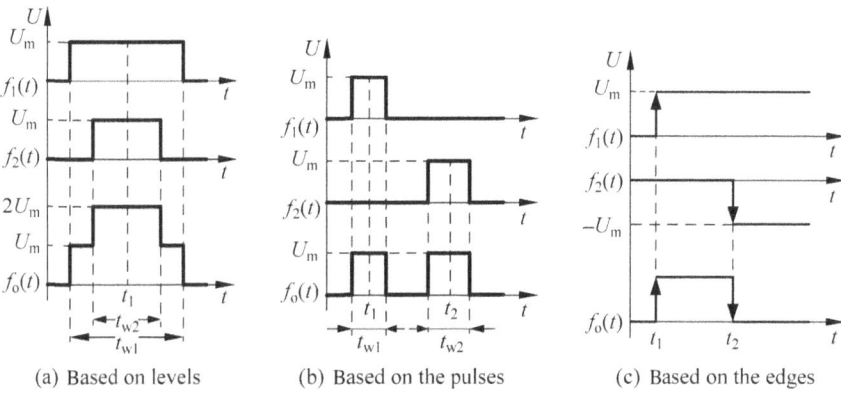

Fig. 2.61 Three logic combinations of pulse sequences

(1) Combination based on the level;

The goal of such combination is to maximize the number of voltage/current levels, which allows the output pulses to well approach the target waveform. While reducing the harmonics, it also reduces the voltage amplitude of every single pulse. One example is shown in Fig. 2.61a and formulated as Eqs. (2.42) and (2.43)

$$\begin{cases} f_1(t) = p_{t_1}(t - t_1) \\ f_2(t) = p_{t_2}(t - t_1) \end{cases} \tag{2.42}$$

$$f_0(t) = f_1(t) + f_2(t) = \begin{cases} 0, & t < t_1 - \frac{t_{w1}}{2} & \text{or } t > t_1 + \frac{t_{w1}}{2} \\ U_m, & t_1 - \frac{t_{w1}}{2} \le t \le t_1 - \frac{t_{w2}}{2} \text{ or } t_1 + \frac{t_{w2}}{2} \le t \le t_1 + \frac{t_{w1}}{2} \\ 2U_m, & t_1 - \frac{t_{w2}}{2} \le t \le t_1 + \frac{t_{w2}}{2} \end{cases}$$

$$\tag{2.43}$$

This method is widely used in multiplex and multilevel converters, exhibiting various features based upon the topology and modulation methods. The overall merit is that through introducing one extra control freedom, pulse width modulation and multiplex are implemented simultaneously, resulting in more precisely modulated output thereby enhancing the signal fidelity with the limited voltage rating and switching frequency.

(2) Combination based on pulses. Its goal is to increase the number of effective pulses within one unit time interval, relocate the pulse sequence and increase the output pulse frequency without increasing the switching frequency. For Fig. 2.61b, its mathematical expression is

$$\begin{cases} f_1(t) = p_{t_1}(t - t_1) \\ f_2(t) = p_{t_2}(t - t_2) \end{cases} \tag{2.44}$$

$$f_o(t) = f_1(t) + f_2(t) = p_{t_1}(t - t_1) + p_{t_2}(t - t_2) \tag{2.45}$$

It is worthwhile pointing out that such method is only applied to the pulse sequence, i.e., needs incorporate with the time combination of pulses. It is invalid to apply such method for a single pulse. The reason is that such method is an extension of the pulse time combination. All these two methods utilize the information contained in the time based pulse sequence to synthesize the waveform. Compared to the pure time combination, this method divides information of one pulse sequence into multiple sequences, each of which in reality corresponds to one pair of complementary switches. Such effort lowers the requirement of the switching frequency. On the other hand, it could be assumed that each pulse sequence independently samples the target waveform at the specific sampling pace, which ultimately superposes together to increase the equivalent sampling frequency, widens the frequency range and maintains the fidelity.

(3) Combination based on edges. The fundamental of two methods above is a single pulse. By dividing the pulse into the rising and falling edges, we can further use the edge as the fundamental and combine their edges reasonably to synthesize the reference output. Its scheme is shown in Fig. 2.61c with the mathematical expression shown as

$$\begin{cases} f_1(t) = U_m u(t - t_1) \\ f_2(t) = -U_m u(t - t_2) \end{cases} \tag{2.46}$$

$$f_o(t) = f_1(t) + f_2(t) = p_{t_2-t_1}(t - \frac{t_1 + t_2}{2}) \tag{2.47}$$

Similar to other pulse logic combinations, such method needs involve the pulse time combination as well, given that it does not contain any modulation information. The difference is that such method further divides the pulse sequence into the edge sequence, doubling the basic unit. For a power electronic converter, pulse edges correspond to switching actions of a pair of complementary switches. Therefore, such method can further optimize the switching sequence of switches. For instance, the rising and falling edges of one pulse could be assigned to two pairs of complementary switches, respectively to control commutating loops aiming at improved transients or balanced conduction loss. Same as other pulse logic combinations, this method multiples the pulse sampling frequency with the same switching frequency and assigns extra control freedoms for switching actions.

With higher requirements of the waveform quality, the time combination and logic combination of pulses need be coordinated, i.e., logic combination of pulse sequences. Different pulse synthesizing strategies are generated when integrating the three above with the time combination.

Time and logic combinations of pulses and pulse sequences significantly enhance the capability of power electronic systems, which however complicates transient processes. In-depth principles on multi-timescale and multi-dimension-coupled transients and their interactions deserve further research.

Chapter 3
Transient Characteristics of Power Switches

Power switches are building blocks of power electronics, with their characteristics fully exhibited only in power electronic converters. At the system perspective, power switches are one of main factors in power electronic systems and determine electromagnetic transients when coupled with other factors like the main-power circuit and control algorithms. As the critical bridge between the software and hardware, they are the places where information and energy mix together, displaying significant nonlinear behaviors. Therefore transient characteristics of power switches are the key to investigate power electronic transients, as determined by the internal physical mechanism and external influential factors.

3.1 Physical Mechanism and Characteristics of Power Switches

Power electronics is a technique of efficiently transforming and controlling energy through the effective usage of power semiconductor devices, applied circuits, design theory and analyzing tools. In the high-power power electronic systems, power switches are more prominent for the system reliability, cost and performance. State-of-the-art power devices are far from meeting system requirements. Therefore, comprehending the transient characteristics of power devices thereby maximizing their potential is one of research focuses. Shown in Fig. 3.1 are several typical high-power devices with various voltage and current ratings, indicating the existence of limitations of the power ratings.

So far, there is no accurate definition of the power switch characteristics, which in real applications are revealed as the electrical, thermal and mechanical performance, including on-state, off-state, switching-on, switching-off, reverse recovery, gate-drive, mechanical and thermal characteristics. All of those are externally measureable, though essentially determined by the switch internal mechanism and

© Tsinghua University Press and Springer Nature Singapore Pte Ltd. 2019   91
Z. Zhao et al., *Electromagnetic Transients of Power Electronics Systems*,
https://doi.org/10.1007/978-981-10-8812-4_3

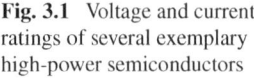

Fig. 3.1 Voltage and current ratings of several exemplary high-power semiconductors

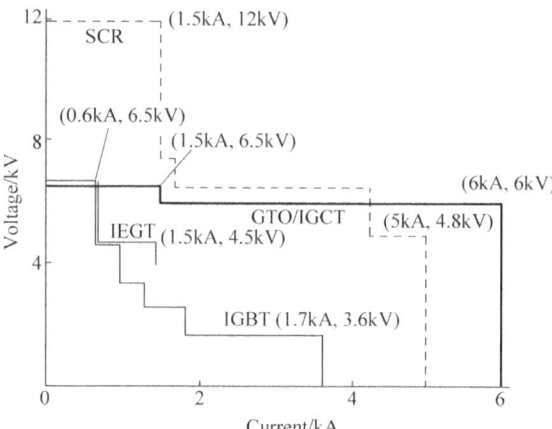

external constraints together. With various power devices existing on the market, different physical mechanisms divide them into several categories, even though the characteristics of each type of devices have certain similarities.

3.1.1 Physical Mechanism Versus the Switching Characteristics

The physical essence of power switches is very similar to integrated chips (ICs), made of PN junctions, BJTs and MOSFETs complying with the semiconductor fundamentals. However, differences exist in terms of manufacturing and applications. Specialties of high-power circuits need be considered for power switches, such as insulation and current capability. In the real practice, all power switches are working at the switching mode without staying in the active region. Energy transformation is carried out instead of the simple digital signal logics or on/off states. Therefore, investigating the specialty of power switches in power electronic systems is critical.

1. **Category of the physical mechanisms**

Physical mechanisms of power switches are highly related to the switch current carriers and internal structures. Based on the property of current carriers, power devices could be categorized as bipolar, unipolar and hybrid types. Based on structures they could be classified into two-layer one-junction diodes, three-layer two-junction transistors and four-layer three-junction thyristors.

(1) Bipolar devices, which contain two types of current carriers, holes and electrons participating in the current conduction inside the device. The fundamental of such devices is the PN junction. Thus, these devices are also named as junction devices. However, the junction field effect transistor (JFET) only employs

majorities for the current conducting, which belongs to the unipolar device. The bipolar devices have low on-state voltage drop, high blocking voltage, high current capability and relatively low switching frequency, making them suitable for high-power converters. The frequently used devices are BJTs, GTOs and GCTs (thyristor of the IGCT, excluding the integrated gate circuit).

BJTs are three-layer bipolar devices with merits of easy control, short switching time, low forward voltage drop and excellent high-frequency performance. The demerits such as the second breakdown and low-voltage withstanding, however, obstruct their applications in high-power converters.

GTOs are the four-layer bipolar devices, a fully-controlled type with high voltage and current ratings. Derivatives from original GTOs include reverse blocking, reverse conducting, non-reverse blocking, buried-gate, amplified-gate and MOS types, etc. The demerits of GTOs lie on the low turn-off gain and high turn-off gate-drive power. To limit dv/dt and switching-off loss, a particular snubber circuit is needed, which introduces the extra power loss. Even so, compared to conventional thyristors, GTOs are advantageous in terms of the size, weight, efficiency and reliability. Tradeoffs between the forward voltage drop and switching loss have been made for GTOs through controlling the carrier lifetime. It has replaced thyristors in majority of high-voltage high-current applications, but meanwhile is being replaced by other devices such as IGCTs in lower-power circumstances.

(2) Unipolar devices, which only have one type of current carriers inside to participate in the current conduction. Such switches usually have the high switching frequency, e.g., tens to hundreds of kilohertz under several hundred volts. The typical example is MOSFET.

The power MOSFET is a voltage controlled device, which has low gate-drive power, high switching speed, no second breakdown and wide safe operation area. It also has the negative current-thermal coefficient, outstanding current self-adjusting capability, stable thermal performance and high anti-disturbance capability. Due to its mechanism and structure, the current and voltage ratings are difficult to increase, making it a perfect candidate for medium-to-low power applications with the high switching frequency.

Schottky diodes and JFETs only have majorities participating in the current conducting. Therefore, they are unipolar devices too. Their switching frequency can go up to ~MHz, but their power ratings had been limited until wide-bandgap (WBG) devices emerged.

(3) Hybrid devices, the mixture of unipolar and bipolar devices together. Such devices utilize bipolar devices as the power channel (high-voltage withstanding, high current density and low forward voltage when conducting the current, such as SCRs, GTOs and BJTs) and unipolar devices as the control channel (high input impedance, fast response, such as MOSFETs) thereby inheriting merits of both devices. The representatives include IGBTs and IEGTs. In addition, IGCTs integrate the external control channel with the power device thereby forming another example of hybrid devices.

IGBTs are more and more widely used due to its excellent performance. It has a low forward voltage drop and high current density like BJTs, with the high switching frequency, low switching loss and easy control like MOSFETs. Therefore, merits of IGBTs include the low heat dissipation, low gate-drive power, small size, wide safe-operation area, low noise and easy protection. Its positive thermal-resistance coefficient facilitates its parallel connection.

IEGT is an electron-injected enhancement-mode (E-mode) insulated gate transistor. Its gate has an improved structure and its cathode is specifically designed, yielding the excellent switching-off performance like IGBTs and the low forward voltage at the high current.

IGCT is an integrated gate commutating thyristor, the performance of which is between GTO and IGBT. It has the large power rating, high voltage withstanding, high switching frequency and low gate-drive power, a great candidate for medium-to-high voltage converters.

With the development of power semiconductor switches and emerging of new materials, manufacturing and technologies, the categories above might not reveal the difference clearly. For instance, with the WBG materials, SiC and GaN based devices are becoming the new research focuses. At the same time, based on the assembly in the actual system, power devices could be classified according to the packaging techniques, e.g., crimping and power modules. Usually the crimping devices have high current ratings.

Here we select several characteristics as examples, such as junction-temperature limits and voltage/current withstanding capability to analyze internal physical mechanisms.

2. Junction-temperature limits versus intrinsic temperature

Nearly all characteristics of power semiconductor devices are related to the temperature. The limit of the junction temperature for all power switches is written in datasheets, which is highly related to semiconductor materials. In the analysis of semiconductors, the concentration of current carriers introduced by the extrinsic plays the major role. A general assumption is that all the extrinsic has been ionized with current carriers having much higher concentration than intrinsic semiconductors. This is the fundamental of semiconductor physics. All the above meanwhile is related to the temperature.

Assume the current carrier concentration inside the semiconductor is $|N_D - N_A|$. Here N_D is the concentration of the quintavalent extrinsic, named as the donor concentration. N_A is the concentration of the trivalent extrinsic, named as the acceptor concentration, much higher than that in intrinsic semiconductors. With the temperature rising, the energy received by electrons increases. When above some threshold, the electron concentration is not equal to $|N_D - N_A|$ any more. Shown in Fig. 3.2 is the concentration of free electrons at different temperature with different $|N_D - N_A|$.

At very low temperature, electrons are trapped by donor atoms and holes are restrained by the acceptor atom. No ionization happens, defined as carrier frozen significantly lowering the carrier concentration. Figure 3.3 shows the complete process

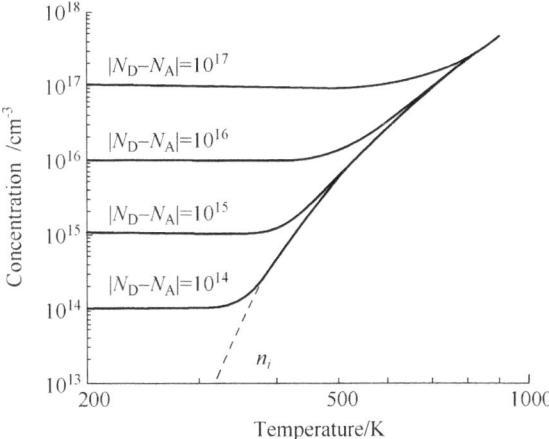

Fig. 3.2 Concentration of free electrons at different temperature

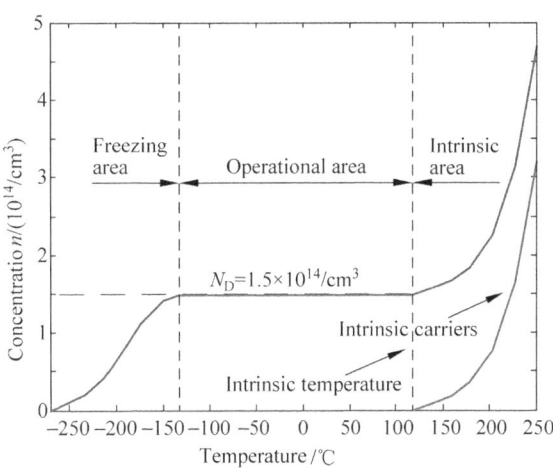

Fig. 3.3 Electron concentration versus temperature for one N-type semiconductor

of the electron concentration from low to high temperature for an N-type semiconductor. Within a wide temperature range, concentration of electrons is the same as that of extrinsic atoms, i.e., operational area. At the high temperature, the number of intrinsic carriers exceeds extrinsic atoms, defined as the intrinsic temperature. Above this point, extrinsic atoms affect less on current carriers. Therefore, this temperature is vital for semiconductor switches, given the intrinsic temperature is highly related to the highest operational point. Note these two temperature values are different. For example, in order to increase the voltage blocking capability, a lightly doped region is required. Its intrinsic temperature is not equal to the highest operational temperature, which in general is the rated junction temperature. As shown in Fig. 3.3, the rated junction temperature is 125–150 °C, determined by the silicon material. For the WBG device, this temperature can be greatly increased, which is one of the superior characteristics of WBG devices over Si devices.

Fig. 3.4 I-V curve of the PN
junction

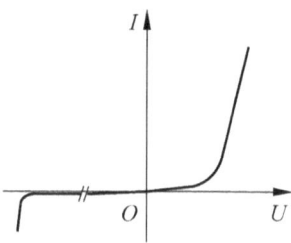

3. Breakdown of the PN junction

Nearly all the voltage of the semiconductor switches is undertaken by PN junctions, though the forward or reverse voltage is withstood by different PN junctions.

The I-V curve of the PN junction shows when reverse biased the reverse current remains as a small value and has little relationship with the reverse voltage, which is named as the reverse saturation current or leakage current. In reality, such current will increase with the reverse voltage and abruptly rises once the reverse voltage exceeds some threshold, as shown in the third quadrant of Fig. 3.4.

The breakdown of the PN junction is due to abundant extra electrons and holes generated by the atom ionization in the space-charge region. Two different types of the atom ionization exist. One is the avalanche breakdown, the other is the Zener breakdown.

The avalanche breakdown is the most frequently breakdown phenomenon in power electronics, which is related to the multiplier effect of carriers under the strong electric field. When the reverse voltage across the PN junction increases, the electric field in the space-charge region is enhanced, accelerating the drifting speed and increasing the kinetic energy of electrons and holes. Such high-speed and high-energy carriers keep colliding with crystal atoms thereby activating more electron-hole pairs, named as the collision ionization. Newly generated carriers are accelerated again in the strong electric field, gaining more energy and creating more collisions. The carriers are then multiplied, named as the avalanche multiplier effect. Once happening, the carrier concentration increases quickly, yielding the increased reverse current thereby breaking down the PN junction, i.e., the avalanche breakdown shown in Fig. 3.5.

The avalanche breakdown is not only dependent on the electric field in the space-charge area, but also related to the width of that region. This is because the increment of the kinetic energy needs the sufficient accelerating distance. On the other hand, in order to multiply the carrier concentration, each carrier needs enough chances to collide with atoms before entering the neutral regions. Therefore, the avalanche usually happens at the lightly doped side of the space-charge region. The avalanche breakdown voltage U_B drops as the doping concentration increasing. Equation (3.1) shows the relationship between U_B and the extrinsic concentration of N-type side for a P^+N junction. Usually, N represents the concentration of the lightly doped region.

Fig. 3.5 Avalanche
breakdown of the PN
junction

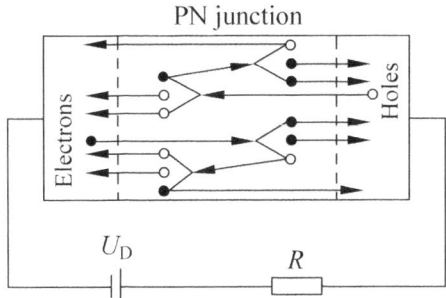

$$U_B \propto N^{-0.75} \tag{3.1}$$

Zener breakdown is also defined as the tunnel breakdown, which only occurs in the heavily doped PN junction. Different from the avalanche breakdown, Zener breakdown happens at the low voltage, when the crystal atoms in the space-charge region are directly ionized by the electric field, instead by accelerated high-energy carriers. When both sides of the PN junction are heavily doped, the space-charge region cannot be widened even at a very high reverse biased voltage. There will be no enough distance to expedite the carrier to ionize the crystal atoms. However, the short width of the space-charge region enhances the electric field, ionizing the atoms and freeing the trapped electrons into the N region directly. This could be further explained by twisted energy bands. As shown in Fig. 3.6, when the P-region valence band is lifted above the N-region conductance band, even though the barrier for N-region electrons to diffuse to the P-region conductance band is high, there is no any obstacle for electrons at the top valence band of the P-side depletion region to drift to the N-region conductance band. Only one tiny region in the forbidden energy band needs be crossed. Here E_c is the bottom of the conductance band, E_v is the top of the valance band, and the difference between is the forbidden bandwidth. Such crossing is possible if the space-charge region is not wide enough. The more heavily doped at the both sides of the PN junction, the narrower the space-charge region and the more twisted the energy band thereby the easier for electrons to travel from the P-region valance band to the N-region conductance band. This could be explained as that the reverse-biased electric field withdraws electrons from the P-region valence band instead of the conductance band, resulting in the rapid increment of the reverse current. In the real practice, the Zener breakdown could be utilized to fabricate the Zener diode for the voltage stabilizer.

As a summary, the avalanche and Zener breakdowns have three major differences:

(1) Impact of the doping concentration

As shown above, the Zener breakdown only happens inside the PN junction which has both sides heavily doped. The avalanche breakdown is due to the collision ionization, related to both the electric field strength and the width of the space-charge region. The wider the space-charge region the more possible the avalanche breakdown. When

Fig. 3.6 Zener breakdown
of the PN junction

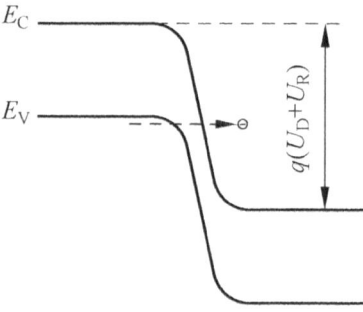

the PN-junction extrinsic concentration is not high, most of the breakdowns are the avalanche type.

(2)   External impact factors

Lighting and fast-ion collision can both activate electron-hole pairs thereby triggering the avalanche breakdown, which however will not affect the Zener breakdown.

(3)   Thermal impact on the breakdowns

The Zener-breakdown voltage has a negative thermal coefficient, i.e., the breakdown voltage drops with the temperature rising. For the avalanche breakdown, since the collision-ionization rate drops with the temperature rising, the breakdown voltage has positive correlation with the temperature, i.e., a positive thermal coefficient.

Both breakdowns are reversible and the PN junction could recover after the breakdown, as long as the external protection is timely to limit the reverse current and secure the multiplication of the reverse voltage and current not to exceed the maximum power dissipating capability of the PN junction.

The PN-junction reverse current has a positive correlation with the junction temperature T. This is because when the temperature increases both the minor concentration in the neutral region and the space-charge-region current will go up. Besides, the avalanche breakdown will be affected by the temperature as well. To avoid the severe thermal impact on the stability of junction devices, effective thermal dissipations are the must, which explains why all power devices are equipped with heatsinks. In extreme cases when the multiplication of the voltage and current exceeds the maximum power dissipating capability of the PN junction, the junction temperature will rise due to the ineffective power dissipation, which in return increases the reverse current and the power loss. Without further actions, a thermal runaway is formed until the overheating destroys the PN junction. The breakdown caused by such thermal instability is named as the thermal breakdown, or thermal-electric breakdown. Different from the avalanche and Zener breakdown, the thermal breakdown is irreversible, which must be avoided.

4. **Failure mechanism due to large di/dt in thyristors**

The large turn-on di/dt is prone to failing devices especially those four-layer three-junction ones, such as thyristors, GTOs and IGCTs. Compared to thyristors, GTOs and IGCTs employ multiple cells in parallel, distributing the current evenly thereby yielding larger di/dt capability.

The di/dt capability refers to the maximum current rising rate of the device at specific conditions without harming the device. The impact factors of the di/dt capability include initial conducting area, turn-on time, expanding time, switching frequency and thermal impedance. When triggered, the device needs time to switch from the off state to the on state. Initially inside the thyristor there is only limited cathode area close to the gate conducting the current. Then the conducting area is expanding to the rest of the cathode at the specific speed, e.g., 0.1 mm/µs. If the initial conducting area is not sufficient with a too high current rising rate, a high current density will emerge in the area, resulting in the rapidly increasing temperature and ultimately damaging the device. For GTOs and IGCTs, instead of expanding the conducting area to the whole cathode, the conducting process is focused on the finely defined cathode cells.

In addition to the initial conducting area and the expanding speed, the di/dt capability is also determined by the thermal capability of the device. Silicon devices have relatively low specific heat, thermal capacity and thermal conductance. Therefore, when current rises, most of the heat is congested in the conducting region with very little heat effectively dissipated, yielding a rapid temperature rise. When the temperature exceeds some threshold, the hot carrier injection is dominant over regular current carriers, which further creates the positive thermal feedback, i.e., the higher the current the higher the temperature, followed by the lower resistance then the higher current. In a short period, majority of the current will form the cluster at the hottest area, i.e., heat spot, disabling and even destroying the device. In some cases, the over-temperature inside the switch due to the large di/dt exceeds 1000 °C, far beyond the melting point of the interface between the metal and semiconductors. This tends to destroy the switch within a very short period.

For thyristors, the large di/dt tends to create heat spots around the gate, as shown in Fig. 3.7. For GTOs, heat spots usually emerge around the edge of cathode strips, which is also the edge of the gate, as shown in Fig. 3.8. For the reverse conducting IGCTs, damages appear on the anti-paralleled diode as well as the edge of cathode strips.

For the IGCT, to prevent the damage caused by the large di/dt, a turn-on snubber circuit is adopted, as shown in Fig. 3.9. Here the inductance L_s limits the large di/dt in the commutating loop when the IGCT turns on.

Fig. 3.7   Damage by large di/dt in the thyristor

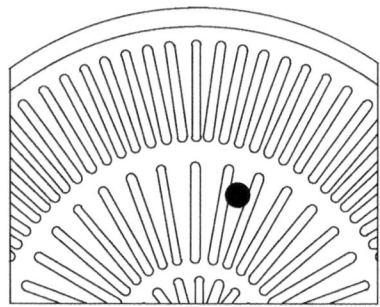

Fig. 3.8   Damage by large di/dt in the GTO

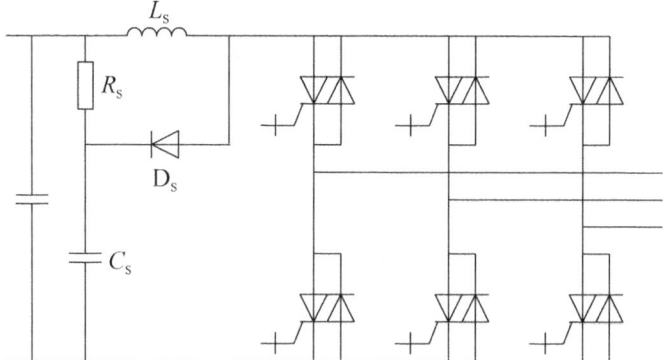

Fig. 3.9   An IGCT converter equipped with the turn-on snubber circuit

3.1.2 Different Characteristics of Different Semiconductor-Physics Based Power Devices

1. **Difference caused by semiconductor materials**

Schottky diodes have the similar I-V curve as PN junctions. The comparison between these two types of diodes is shown in Fig. 3.10. Regardless of the resemblance, some obvious difference is exposed.

Firstly for current carriers, when forward biased, the PN junction injects the holes from P-type to N-type semiconductors. The electrons move from the N-type to P-type semiconductor. All carriers are minors, named as the minor injection. When accumulated enough, minors diffuse to form the current. This limits the switching speed. For the Schottky diode, the current is formed by the majority without the minor accumulation. Therefore, the switching speed is not slowed, providing better high-frequency performance than the PN junction diode.

Secondly, with the same voltage barrier, the saturation current of the Schottky diode is much higher than that of the PN junction. In another word, at the same forward current, the voltage drop of the Schottky diode is lower than the PN junction. Without accumulated minors, no conductance modulation exists in the Schottky diode, which provides a softer I-V curve of the forward characteristics than the PN junction.

Thirdly, the reverse leakage current of the PN junction is caused by the minor withdrawal through the space-charge region, sensitive to the temperature. For the Schottky diode, when reverse biased, the electrons are drawn out of the metal side to the silicon. Given the electron density of the metal and the junction characteristic of the metal oxide are both insensitive to the temperature, the reverse current does not change much with the temperature. However, the space-charge region of the Schottky diode is narrower than the PN junction. When reverse biased, the current tunneling effect will contribute to the reverse current, leading to a softer I-V curve for the reverse characteristics.

As shown in Fig. 3.10, compared to the PN junction diode, the forward characteristics of the Schottky diode is not "stiff", its forward voltage is lower, and its reverse

Fig. 3.10 I-V curves of a Schottky diode and a PN junction diode

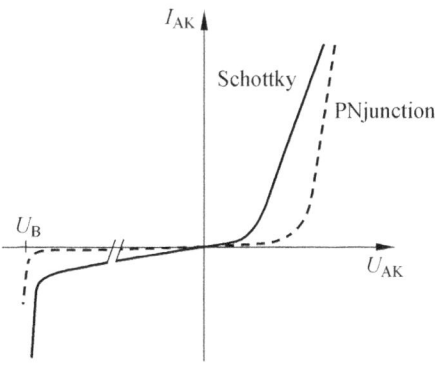

Fig. 3.11 Reverse recovery
processes for a silicon FRR
diode and SiC Schottky
diode

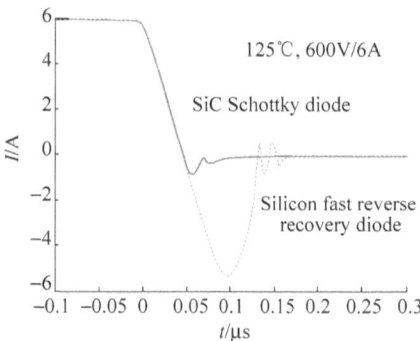

leakage current is higher. The semiconductor side of the Schottky diode uses silicon. With different materials at the metal side, the characteristics of both forward and reverse biased are varied, both of which are hard to be taken care of simultaneously, i.e., a low forward voltage drop results in a large reverse leakage current while a small reverse leakage current yields a large forward voltage drop.

Lastly, look at the reverse recovery characteristics of the Schottky diode, which is superior to the PN junction with faster and softer recover recovery process. This is due to the absence of the minor accumulation. Such an excellent reverse recovery, however, is hard to be used in the high-voltage applications, mainly due to its low voltage-blocking rating. The stake was changed once SiC Schottky diodes emerged. Compared to silicon fast reverse recovery (FRR) diode, the SiC Schottky diode is even faster and softer. Shown in Fig. 3.11 is the comparison of reverse recovery processes for a 600 V silicon FRR diode and a SiC Schottky diode.

2. Difference caused by semiconductor structure

Take a three-layer two-junction IGBT and a four-layer three-junction GTO as examples. For an IGBT based converter, as shown in Fig. 3.12, the top switch is kept on with the bottom switch off. When the load current changes the direction from "out" to "into" the leg, the current will shift from the main IGBT to its anti-paralleled diode, otherwise, from the diode to the main IGBT. A similar process exists for the bottom switch.

Shown in Fig. 3.13 is a GTO based converter. When we keep the top switch on and the bottom switch off and let the load current change the direction from out to into the leg, current will shift from the main GTO to its anti-paralleled diode. This is the same as the IGBT converter. Difference happens once the current direction changes from "into" to "out" of the leg. At the on-state, without sufficient current flowing through, the GTO will turn off naturally. Its gate adopts the pulse current to trigger. Therefore, no sustaining current is provided after the triggering moment. When the leg current flows in, the top GTO has no current going through. Neither does it withstand any voltage. Even the control unit still treats it on, the top GTO is not fully on. When the load current changes the polarity with a high di/dt, no path is

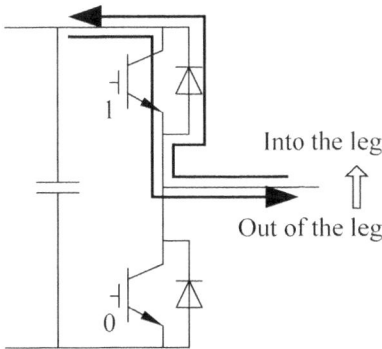

Fig. 3.12 Current changes the polarity in an IGBT converter

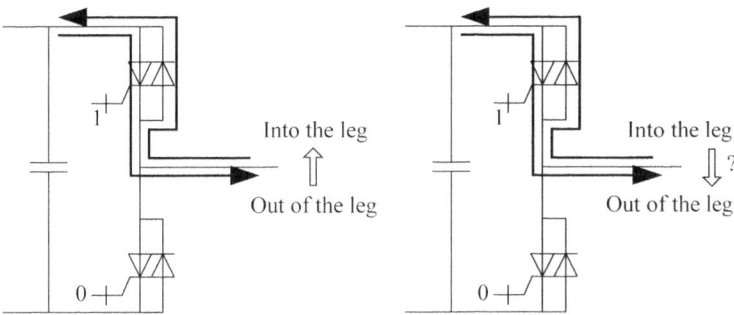

Fig. 3.13 Current changes the polarity in a GTO converter

provided for the current to commutate from the anti-paralleled diode to the main GTO, which obstructs the current commutation and even creates the converter fault. One feasible solution is to retrigger the GTO once the current polarity changes, providing an effective current path. For the IGCT, such retriggering function is realized by the integrated gate circuit. Note such feature might be required not only when the current changes the polarity but also at other circumstances, which will be detailed in the following chapter. On the contrary, IGBTs do not need retriggering because its gate is triggered by a voltage level. As long as the switch is on with the gate voltage high enough, its main body is always on, in spite of the current direction.

3.2　Transient Performance Testing of the Power Switch in the Converter

The switch transient characteristics, converter topology and control algorithms will directly affect the converter performance. Understanding transient behaviors of power switches is critical for the design, analysis and control of power electronic converters. In this section, an IGCT based multi-level inverter will be used to demonstrate the measurement and analysis of switching transients. Tightly coupled with other influential factors in the converter, at different stages of design, development and operations, switching transient behaviors might diversify. Therefore, the study of the switch transient performance is closely coupled with the converter design. In this section, we will analyze it in a single-switch test bench and the whole converter, respectively.

To test the switching performance on a single-switch test bench, we need consider the topology, parameters and mechanical assembly, process the simulation along with the actual test, and analyze the IGCT characteristics. Such single-switch test is an important step prior to the system operation. The IGCT performance and quality then can be examined. Note majority of parameters given in the IGCT datasheet were only obtained through the test circuit with specific peripherals, such as the turn-off delay and tailing time. The performance will change once the test circuit changes. Therefore, through carrying out the single-switch test we could obtain the first-hand data of the IGCT performance. Additionally, some short-timescale transients can be fully investigated, such as snubber parameters and performance. The operational modes of the IGCT in the single-switch tester are related to some in the two-level and three-level converters. Therefore, the design principle of the converter need be taken into account when designing the single-switch test bench. The last but not the least, some specific experiments are not easy to implement in the ultimate converter system, such as the turn-off voltage spike induced by the leakage inductance, which however could be carried out in the single-switch tester.

We need understand that the single-switch test cannot fully represent the whole system. Some flaws might exist. Therefore, in addition to the single-switch test, IGCT test in the overall converter is also recommended. Such test as an online operation will reveal not only the actual characteristics of the IGCT in the system but also the impact of peripherals, such as the stray inductance.

3.2.1　Topology and Control of the Single-Switch Tester

The IGCT single-switch tester could have various circuits, though the core part remains the same, i.e., a buck converter made of one IGCT and one diode. The device under test (DUT) could be the reverse conducting IGCT, asymmetric IGCT or the IGCT body diode. The snubber circuits for the di/dt and du/dt are diverse. The overall test bench is shown as Fig. 3.14, with the setup able to fully reveal the current

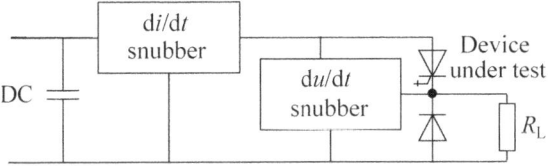

Fig. 3.14 Scheme of the single-switch test bench

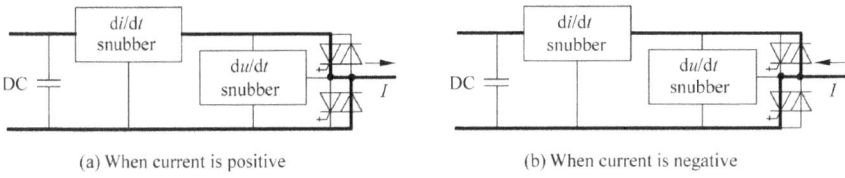

(a) When current is positive   (b) When current is negative

Fig. 3.15 The current commutation for the two-level converter

Table 3.1 Three-level switching logic (1: on, 0: off)

	T_1	T_2	T_3	T_4	States
$U_{dc}/2$	1	1	0	0	Positive level
0	0	1	0	0	Dead band
0	0	1	1	0	Zero level
0	0	0	1	0	Dead band
$-U_{dc}/2$	0	0	1	1	Negative level

commutation in the main-power circuit of the two-level or three-level converter. In the ultimate converter, the di/dt snubber circuit could be one set per phase or per three phases. The du/dt snubber can be one set per phase or without.

For a two-level converter, when the phase current is positive, i.e., flowing from the bridge to the load, or negative, the bridge commutation loops are shown in Fig. 3.15a, b, respectively, the same as the single-switch tester.

The current commutation for the three-level NPC inverter is quite complex, which needs consider the logic among four switches in one leg simultaneously, as shown in Table 3.1. The current commutation among four IGCTs, T_1–T_4, is shown in Fig. 3.16a–d, respectively. Exhaustive attacks of switching actions and phase current polarities indicate that the commutation process of T_1 is 1100 ⇔ 0100 ⇔ 0110 when the current is positive. In this process, T_3 has no current. Therefore, the action of 0100 ⇔ 0110 does not affect the overall commutation, as shown in Fig. 3.16a. Similarly, the commutation of T_2–T_4 is shown in Fig. 3.16b–d, respectively. All four switches have the same commutation circuitry structure as Fig. 3.14.

On the single-switch tester, a close correlation exists among the DC-bus voltage, load current, load impedance and control parameters. The parameter mismatch leads to the poor test data and even destroyed components. Presently, multiple-pulse testing is widely adopted, where the pulse duty cycles and numbers are critical. Prior to the

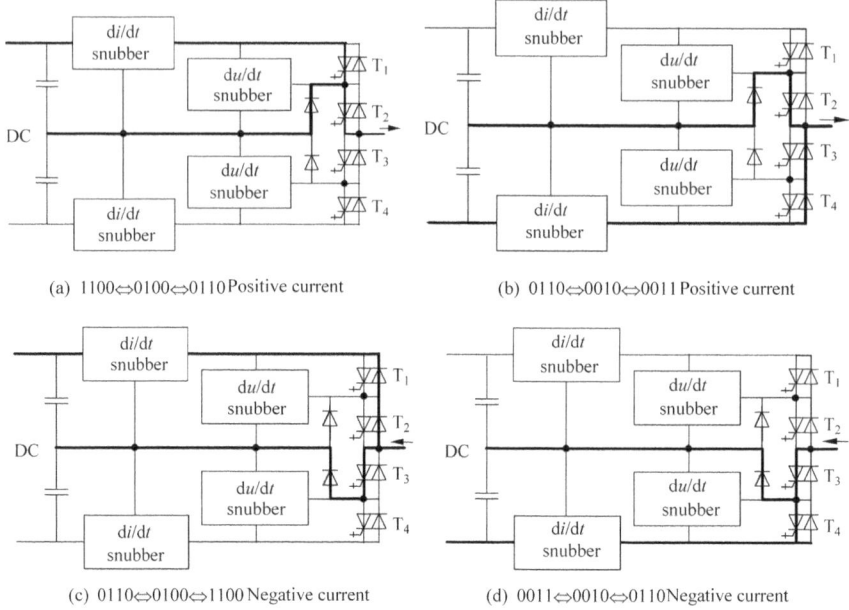

(a) 1100⇔0100⇔0110 Positive current

(b) 0110⇔0010⇔0011 Positive current

(c) 0110⇔0100⇔1100 Negative current

(d) 0011⇔0010⇔0110 Negative current

Fig. 3.16 Commutation processes of $T_1 \sim T_4$

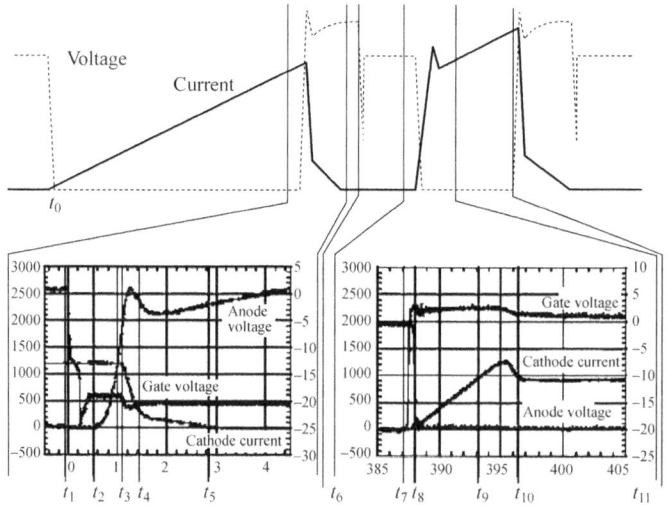

Fig. 3.17 A typical double-pulse-test waveform of an IGCT

actual test, simulation analysis needs be implemented to predict circuit behaviors under multiple pulses. A typical double-pulse-test waveform is shown in Fig. 3.17.

Referring to the test waveform, switching modes of the IGCT and related circuits in the single-switch tester are detailed as follows.

(1) t_0–t_1. At t_0, the control system triggers the IGCT gate and turns it on. The DC-bus capacitor gets discharged through the load. The load current then reaches the reference at $t=t_1$, when the control system imposes the turn-off signal on the IGCT gate;

(2) t_1–t_2. The external behavior of the IGCT maintains the same. This period is named as the storage time, corresponding to the turn-off delay. The IGCT is about to turn off;

(3) t_2–t_3. At $t=t_2$, the IGCT anode voltage begins to rise, initiating the actual turn-off process. The diode is however still reverse biased, with the current flowing through the switch unchanged;

(4) t_3–t_4. At $t=t_3$, the IGCT anode voltage reaches the DC-bus voltage, resulting in the drop of the IGCT current. The diode begins to be forward biased. In this time interval, the overlap between the switch voltage and current emerges, yielding the largest electric stress;

(5) t_4–t_5. At $t=t_4$, the IGCT current drops to nearly zero. The tailing process begins. At $t=t_5$, the switch is close to be fully turned off;

(6) $t_5 \sim t_6$. In this time interval, the resistor of the di/dt snubber undertakes the current, generating an extra voltage spike across the switch. The snubber inductor keeps discharging until $t=t_6$, when the snubber diode is reverse biased;

(7) t_6–t_7. The IGCT is reverse biased, undertaking the full DC-bus voltage. Both the load and the diode have the current flowing through. Such current will fade due to the power loss in the loop, however, slowly;

(8) t_7–t_8. At $t=t_7$, the IGCT is triggered again to fully test its actual turn-on performance. Externally the switch status does not change. We name this period as the turn-on delay;

(9) t_8–t_9. The load current begins to switch from the freewheeling diode to the IGCT. Meanwhile the di/dt snubber kicks in, limiting the current rising slope;

(10) t_9–t_{10}. The load current begins to switch from the diode to the IGCT. The diode enters the reverse recovery stage. A large extra current is then overlapped to the IGCT turn-on current. Now the switch needs be fully turned on to shut down the diode, otherwise a significant loss will appear in the diode;

(11) t_{10}–t_{11}. The switch remains on with the current varying due to the DC-bus capacitance and the load. If the load is the inductor, the current will increase. At $t=t_{11}$, the IGCT is turned off again.

At this point, all the electric performance of the switch is fully investigated. At the end, we need discharge the DC-bus capacitor and the load then wrap up the whole test.

3.2.2 Stand-Alone Tester for Single-Switch Dynamics

Various circuitry topologies are available for the single switch test. Majority of the differences lie in the snubber circuit. For the single IGCT tester shown in Fig. 3.18, no du/dt snubber is equipped. Here L_{s1}–L_{s3} are the equivalent stray inductance in the loop to test the switching-off performance. All such inductance might be introduced by long bus bars or cables, not the physical inductors.

From the layout of the stand-alone tester, it can be seen that complementary switch pairs are not only made of top IGCT and the bottom diode, but also the neutral point clamping (NPC) diode. The stack of the NPC tester is shown in Fig. 3.19, devices of which from top to bottom are snubber FRR diode, IGCT, NPC diode, insulator and IGCT, respectively. Multiple heatsinks are employed, not to dissipate the heat since devices in the tester usually do not run for a long term, but to fully investigate the switch mechanical and electrical performance. In this way, we can obtain the switching characteristics similar to the actual system.

The stand-alone tester can be divided into two parts. One is to test the IGCT with its complementary diode, as shown in Fig. 3.20. The other is to test the IGCT with the NPC diode, as shown in Fig. 3.21.

During the test, the assembly of the peripheral circuit is critical. The snubber capacitor usually is assembled by using the bus bar to stay close to the DUT thereby avoiding the stray inductance.

Experimental waveforms on the IGCT stand-alone tester are shown in Fig. 3.22. Here U_T is the IGCT terminal voltage, I_T is the IGCT anode current, U_G is the IGCT gate signal with the high-level signal triggering the switch and the duty cycle of 75%. The detailed turn-off waveform is shown in Fig. 3.23. Experimental results of the turn-off loss are shown in Fig. 3.24 with the turn-off time shown in Fig. 3.25. Here U_{T0} is the IGCT steady-state voltage after fully turned off, I_{T0} is the IGCT current

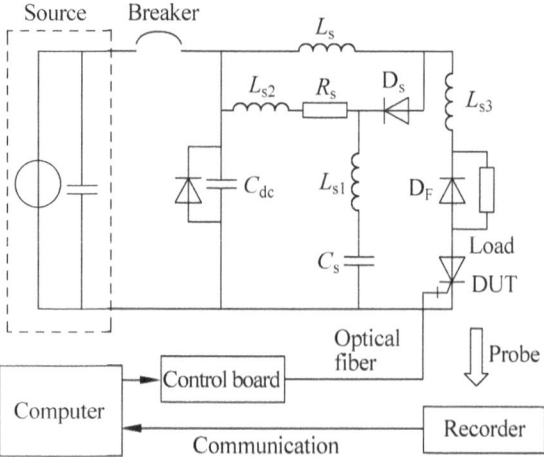

Fig. 3.18 Scheme of the stand-alone tester

Fig. 3.19 Stack of the
stand-alone tester

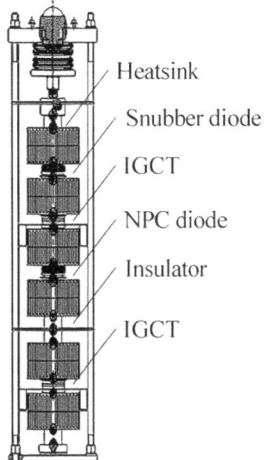

Heatsink

Snubber diode

IGCT

NPC diode

Insulator

IGCT

Fig. 3.20 The tester circuit
(1)

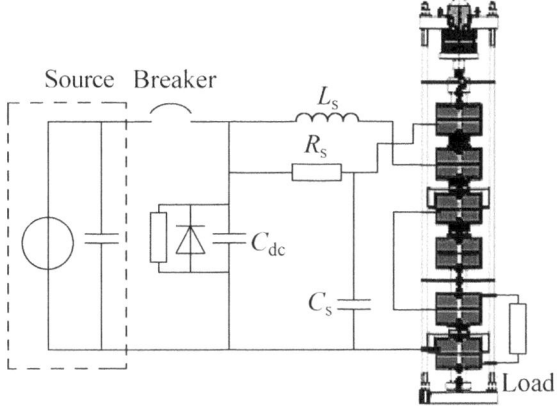

prior to the turn-off, E_{off} is the switching-off energy, t_{doff} is the switching-off delay time, t_f is the falling time, t_t is the tailing time beginning with 5% of the turn-off current, and t_{total} is the total turn-off time. Note with the small turn-off current, the error of the measured turn-off time increases. The experimental results above are of importance for the converter design and transient analysis.

Fig. 3.21 The tester circuit
(2)

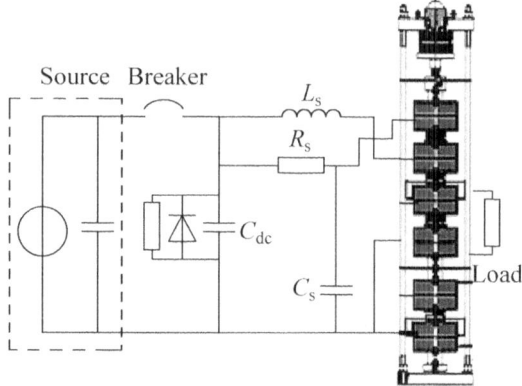

Fig. 3.22 Experimental
waveform of the IGCT

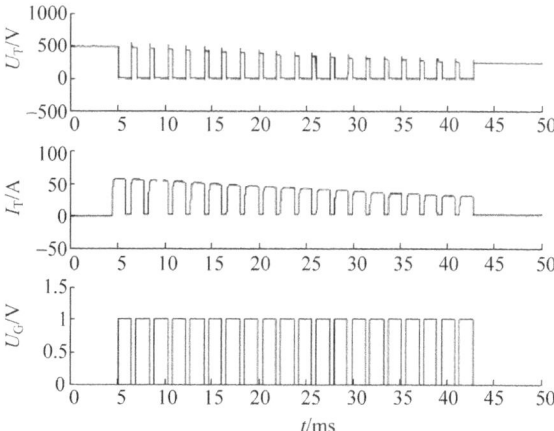

Fig. 3.23 The detailed
turn-off waveform

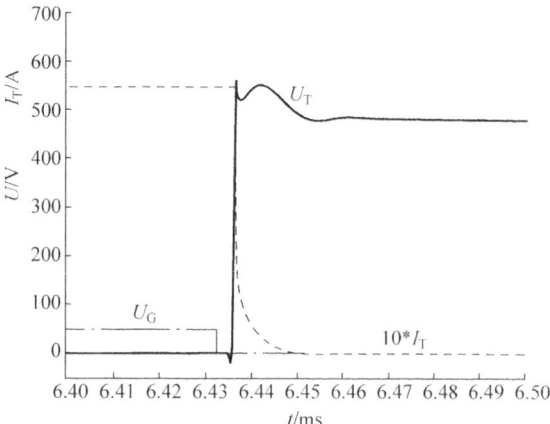

Fig. 3.24 Experimental
turn-off energy

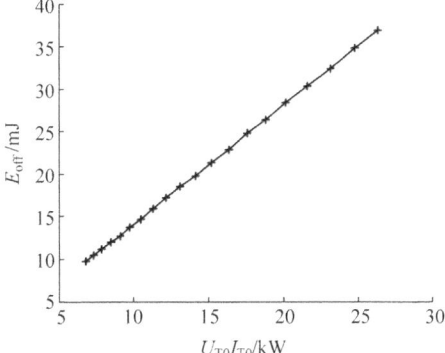

Fig. 3.25 Experimental
turn-off time

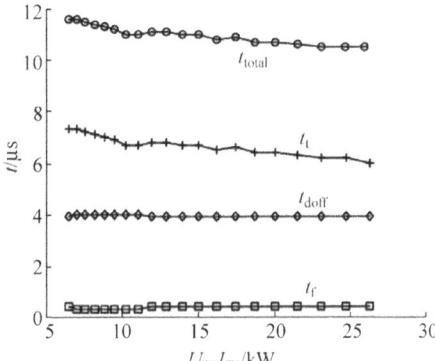

3.2.3 Transient Characteristics of a Single Switch in the Converter

The example used in this section to analyze the IGCT transient performance is a 550 kW three-level inverter. The IGCT ID # is 5SHX08F4502, the clamping and snubber diodes are 5SDF03D4502, and the maximum DC-bus voltage is 5 kV.

Each leg of the three-level inverter contains four reverse-conducting IGCTs, i.e., equipped with anti-paralleled diodes, and 2 NPC diodes, which can form 20 buck converters without considering the DC-bus capacitors and the turn-on snubber. Each buck converter consists of one IGCT and one diode. With some modifications each buck converter could be the tester of the IGCT performance, even though not all of them are needed. Similar to the previous section, four loops in each leg are selected to test commutation characteristics of each IGCT in the inverter, which are totally aligned with the loops driven by the actual control system. To facilitate the analysis, all the commutation circuits are illustrated in Fig. 3.26. Note the DC-bus capacitor and the turn-on snubber are shared by the three-phase converter.

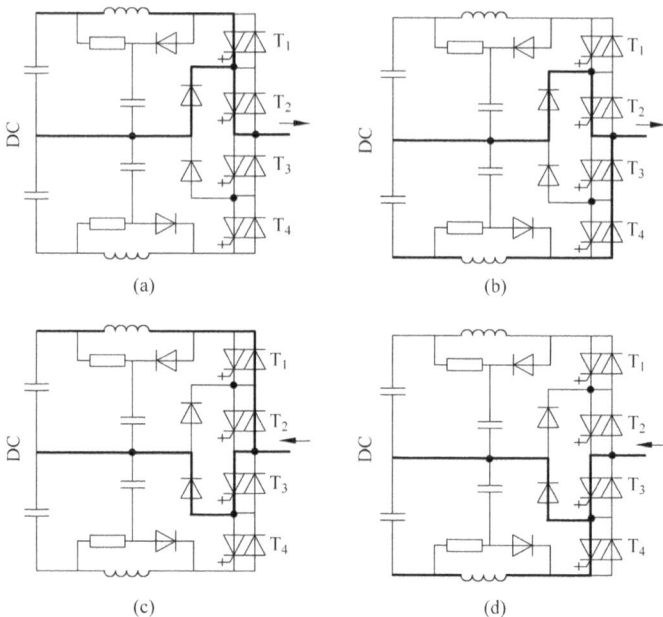

Fig. 3.26 Commutation loops of one IGCT in the leg

Same as the single-switch tester, the IGCT tested in one leg is connected to the inductive load. When testing different loops, we connect the load with the converter in different ways:

(1) To test T_1, as shown in Fig. 3.26a, the load is in parallel to the top NPC diode. The top half DC-bus capacitor is charged. The control board triggers T_1;

(2) To test T_2, as shown in Fig. 3.26b, the load is connected to the T_3 anode and T_4 cathode. The bottom half DC-bus capacitor is being charged. The control board triggers T_2;

(3) To test T_3, as shown in Fig. 3.26c, the load is connected to the T_1 anode and T_2 cathode. The top half DC-bus capacitor is being charged. The control board triggers T_3;

(4) To test T_4, as shown in Fig. 3.26d, the load is in parallel to the bottom NPC diode. The bottom half DC-bus capacitor is being charged. The control board triggers T_4.

Since three phases share same DC-bus capacitors, all input terminals of other IGCT optical fibers should be disabled except that of the DUT. All the gate-drive circuits of other IGCTs should be powered to secure the voltage-withstanding capability. The power level won't be too high given the test period is relatively short.

1. **Procedures of testing one single IGCT in the converter**

With locating the target commutation loops, the detailed test procedure is shown as follows.

(1) Complete preparations prior to the IGCT test, such as the insulation test, voltage withstanding test, examinations of DC-bus capacitors and gate-drive circuits.
(2) Select appropriate control parameters and estimate the test voltage and current. All control systems and test equipment are the same in the stand-alone tester as those in the converter system.
(3) Connect the control board with the DUT through optical fibers, effectively disable other IGCTs and connect the load in the appropriate way.
(4) Charge DC-bus capacitors. The pre-charge circuit could be either the external power supply or the rectifier circuit in the original system. Disconnect the charging circuit once the voltage reaches the target.
(5) Trigger the IGCT with the preset switching frequency, duty cycles and pulse numbers through the control system. Record triggering signals, IGCT voltage and related current. Other waveforms like the snubber voltage and current could be recorded as well if needed. One test example is shown in Fig. 3.27.
(6) Discharge the DC-bus voltage below the safe value, which could be accomplished by the discharging resistor.
(7) Test other IGCTs through repeating steps (3)–(6) until all IGCTs are tested.
(8) Download the data from the waveform recorder, process the signal analysis and compare the IGCT characteristics in details. Use the load current as x-axis and compare the first and second spikes of the IGCT turn-off voltage. The internal switches (T_2 and T_3) and external switches (T_1 and T_4) in each phase of the three-level NPC inverter are comprehensively compared, as shown in Fig. 3.28. Here U_2 corresponds to T_2 in Phase U. Investigating the IGCT turn-off voltage through such curves is able to reveal interactions among various elements in the actual circuit.

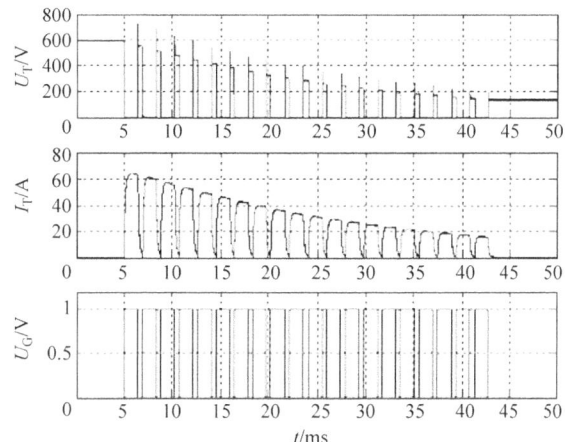

Fig. 3.27 The test waveform of one IGCT

(9) Lastly, select internal and external switches, e.g., W_2 and V_4, respectively, to repeat the test based on inverter ratings. Here W_2 has the highest turn-off voltage spike while V_4 has the lowest. The turn-off waveforms of W_2 and V_4 in one switching process are shown in Figs. 3.29 and 3.30, respectively. Then, the first and second voltage spikes in each turn-off process of W_2 and V_4 are shown in Figs. 3.31 and 3.32, respectively. Here I_{T0} is the IGCT current prior to the turn-off, which is related to the load current and the gate signal. It is indicated that though all IGCTs are from the same patch with same parameters of circuits, control and load, some differences are still exhibited between two IGCTs, which is the further focus of the IGCT test in the converter.

The analysis above uses the tested turn-off characteristics as an example. It is worthwhile pointing out that the same procedure applies to other tests, such as the switching time intervals.

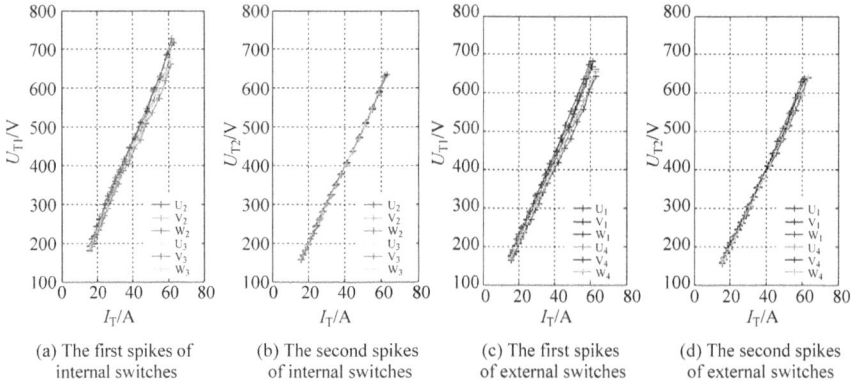

(a) The first spikes of internal switches

(b) The second spikes of internal switches

(c) The first spikes of external switches

(d) The second spikes of external switches

Fig. 3.28 Comparison of IGCT turn-off characteristics in the converter

Fig. 3.29 Turn-off waveform of W_2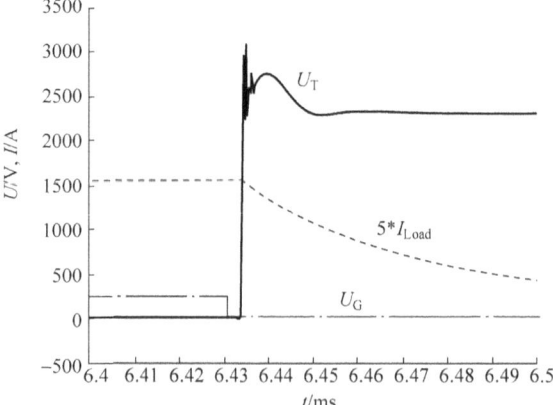

Fig. 3.30 Turn-off waveform of V₄

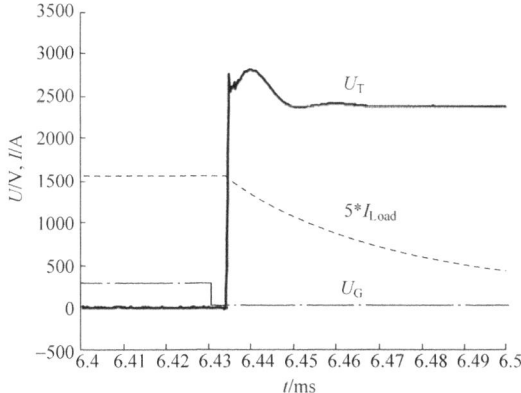

Fig. 3.31 Turn-off voltage spike of W₂

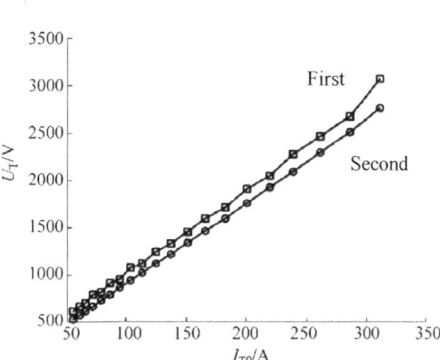

Fig. 3.32 Turn-off voltage spike of V₄

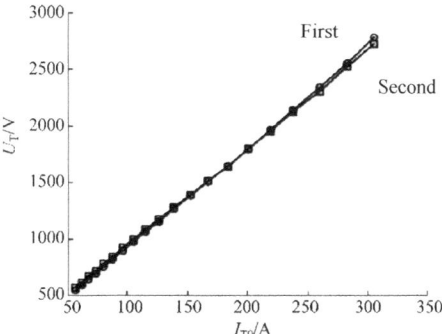

2. **Analysis of the test result**

The test of single switch in the converter exhibits differences of switching transients for different IGCTs in the same system. Such differences are resulted from influential factors other than switches, as shown below.

(1) Switch location. As shown in Fig. 3.28, voltage spikes of all IGCTs are symmetrical, i.e., all inner switches have similar performance. So are outer switches, though inner switches have higher voltage spikes. Referring to Fig. 3.26, we can tell the inner switch has the commutation loop enclosing more devices and larger loop areas, which yield the worse switching-off performance. Therefore, when designing and optimizing the system, the characteristics of inner switches should be referred.

(2) Mechanical structure. Each phase of the inverter has the same topology and devices and shares the same turn-on snubber. However, there is still some difference among switches at the same topological location. When assembled, the snubber diode and capacitor are closest to Phase V, which causes relatively low turn-off voltage spike observed in experiments. Such finding is beneficial for the mechanical design, reminding engineers of locating the snubber close to all three phases simultaneously. Comparably, the leg with the most compact bus-bar structure has the best IGCT switching transients, even though bus bars might be slightly twisted. On the other hand, testing each IGCT in the converter helps double check the busbar design and assembly of each phase.

In the real practice, the stray inductance of the commutating loop influences the most on the IGCT switching-off performance. One effective way to reduce the stray inductance is to compact the mechanical design, which however challenges the thermal and insulation design. Overall, the design and optimization of the high-voltage high-power power electronics converter is to find a reasonable and effective solution out of all contradictions.

3. **Pros and Cons of different testers**

Setting up the standalone tester for the device transient analysis is of importance after selecting the switch and before the switch is equipped in the final converter. It provides first-hand test data of switching transients instead of solely relying on the datasheet, highly beneficial to recognize switching characteristics, reliability, protections and maximize the switch potential. As a summary, testing and analyzing the switch on the standalone tester helps:

(1) Examine and test if the switch performance is aligned with specifications, and if each switch is qualified;

(2) Test the switching transients and snubber performance, which provides the reference for the design of the main circuit and control system;

(3) Correlate the switching transients with the mechanical assembly, guiding the mechanical design of each phase;

(4) Investigate the switching transients under any extreme cases, which cannot be realized by the actual converter, thanks for the flexibility that the single-switch tester offers;

(5) Comprehend switching characteristics accurately with relatively simple tests;

(6) Provide safe testing when discharging the DC-bus capacitors during the test. It can also test the device at any voltage and current needed.

Obviously, the switch tester is not equal to the converter. The cons include:

(1) The switch tester has different mechanical designs, cooling methods and circuit parameters from the final converter. It is hard to rely on the outcome of the single-switch tester to estimate the switching performance in the final converter, given the ultimate converter operation is a highly complex process involving the snubber, topology, control method and mechanical structure;

(2) The switch tester is equivalent to only part of the voltage-source converter. Though we can use it to predict switching behaviors in the converter, the inter-actions between switches and other peripherals are missing.

3.3 Transient Performance Analysis of Power Devices in the Converter

3.3.1 Switch Performance During the Operation

Take one IGCT based converter as an example. To fully comprehend its character-istics requires the test and analysis in the actual converter, instead of only relying on the test-bench based data or the single-switch test results in the converter. Note when testing the switch during the converter running, the converter is triggering the IGCTs based upon the pre-set control strategy. In this way some switching char-acteristics other than those observed in the single-switch tester could be exposed, which is helpful to study the relationship between the system parameters and the IGCT dynamic behaviors. Of course testing IGCTs during the converter operation is relatively inconvenient. Firstly each IGCT is running in different modes, which does not provide a direct head-to-head comparison. Secondly the cooling system needs be added during the test, which complicates the test procedure. Thirdly due to the high-voltage high-power operation, risks must be undertaken and a much stricter test procedure must be complied with.

Despite the risk and complication, testing the IGCT online while the converter is running is critical. The test content includes: with the specific DC-bus voltage, when the converter is in the steady-state operation, measuring each IGCT voltage and the phase output current. To save the effort, complimentary switches could be tested together along with the phase current. In order to reduce the failure risk, usually the DC-bus voltage is settled lower than the ratings. After the test, select some exemplary test data to analyze IGCT dynamic behaviors, look through and statistically compare all waveforms, and pick the typical waveform for the final analysis.

Comparisons of IGCT turn-off actions at different moments provide us the turn-off waveforms at different currents with a relatively stable DC-bus voltage, which can be done through reconstructing the time axis. Note from previous sections not all triggering actions will result in the change of the IGCT voltage, which is related to the phase current polarity. Shown in Fig. 3.33 is the voltage across T_2 in Phase

Fig. 3.33 Turn-off voltage of V_2 when the converter is running

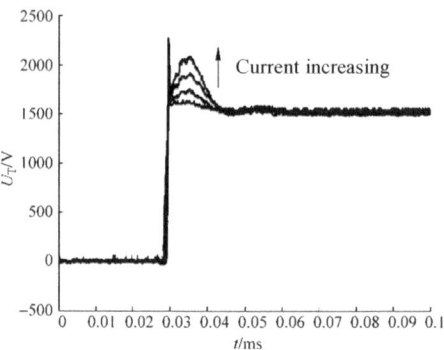

V, i.e., V_2 at different currents. Curves from the bottom to top represent the turn-off voltage with current increasing.

The same approach is applied to other IGCTs for the quantitative analysis and comparison of switching transients and commutation loops. Experimental results indicate that tested switching behaviors are aligned with the outcome of the single-switch testing in the converter, further verifying the effectiveness of the single-switch tester.

3.3.2 Interactions Among Switches

Some special phenomena are to be observed when testing the switch during the converter operation, which cannot be measured in the single-switch tester. Mostly such phenomena are related to interactions among switches.

1. **Dead-band plateau**

When one IGCT is turned off in Phase U, the waveform of its complementary switch is shown in Fig. 3.34.

Particularly when outer switches U_1 and U_4 are switched off, their related complementary inner switches exhibit a voltage plateau with some level and lasting time, which cannot be observed in the line-line voltage. The width of such plateau is approximately equal to the dead-band setting in the control unit, thus named as the dead-band plateau. Wheninner switches are turned off, no such plateau is observed across related outer switches.

Two dead-band states exist in one phase, i.e., 0100 and 0010, as shown in Fig. 3.35. Resistors in leg are balancing the steady state voltage.

Take Fig. 3.35a as an example, where T_3, T_4 and bottom NPC diode together undertake the lower half DC-bus voltage. We can roughly calculate the off-state equivalent resistance of the IGCT and diode as U_{DRM}/I_{DRM}. Based on the datasheet, the calculated voltage undertaken by T_4 is 23% of the lower half DC-bus voltage,

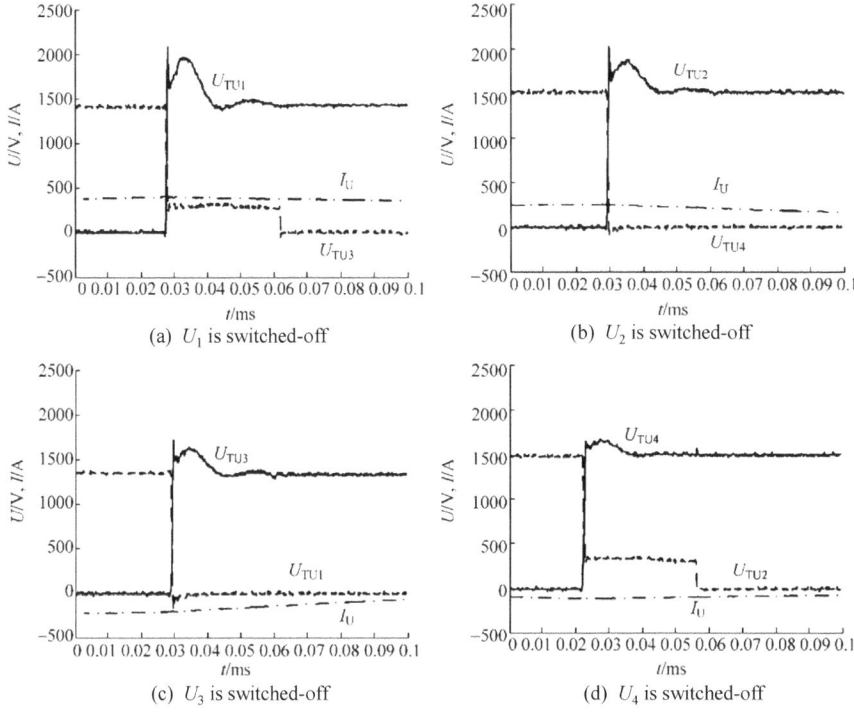

Fig. 3.34 The dead-band plateau in Phase U

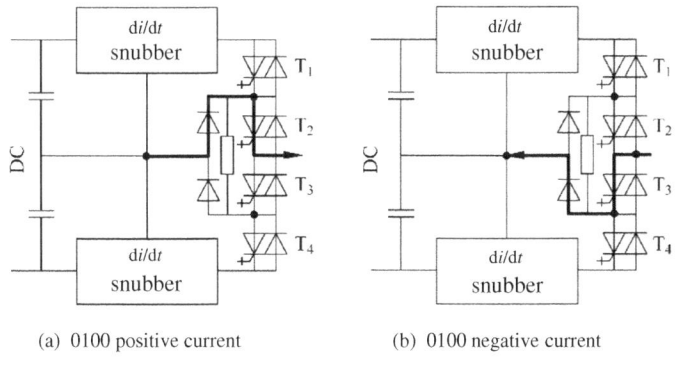

Fig. 3.35 The state of the leg during the dead-band plateau

i.e., the left 77% for T_3, which is aligned with experimental results. When the inner switch exhibits the dead-band plateau, a voltage drop across the other switch, which undertakes the half DC-bus voltage together with the inner switch, emerges as shown in Fig. 3.36.

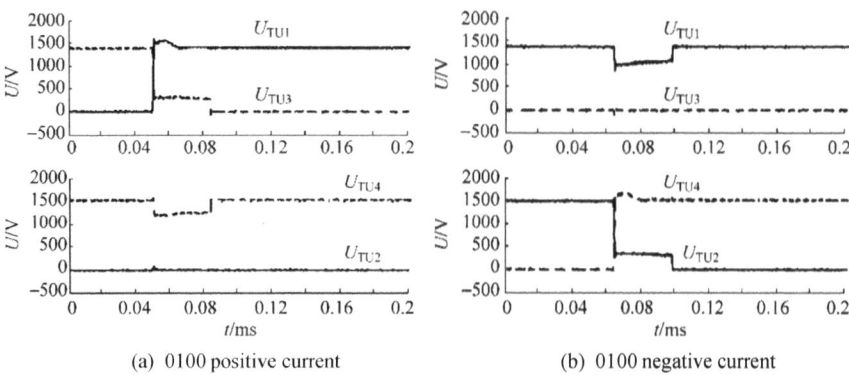

(a) 0100 positive current   (b) 0100 negative current

Fig. 3.36 All dead-band plateaus in Phase U

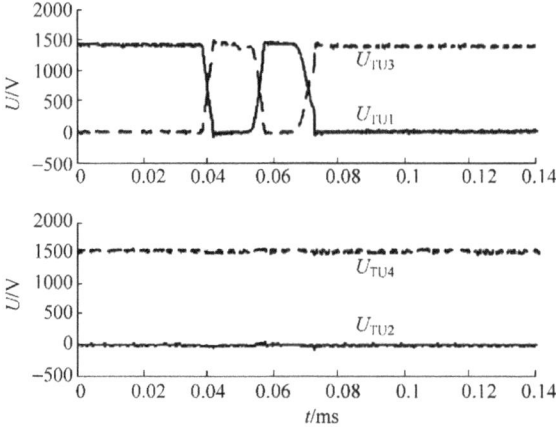

Fig. 3.37 Load current changes the polarity during the dead band

In addition to the dead-band plateau above, a special dead-band plateau also exists, though its occurrence probability is low. When the current changes the polarity during the dead band (0100) in Phase U, the measured Phase-U voltage is shown in Fig. 3.37. Such phenomenon usually happens when the current is small, having little impact on the reverse recovery process of NPC diodes and IGCT anti-paralleled diodes. If the load-current changing rate is high when crossing zero, it might create the device failure. Such harmful current changing rate could be limited by the output filter and load impedance.

2. **Off-state voltage spikes**

At the off state, IGCT voltage has spikes upwards and downwards, as shown in Fig. 3.38. The reason is that the snubber capacitor voltage is altered by other IGCTs' switching actions. The spikes downwards are nearly half of the DC-bus voltage, due

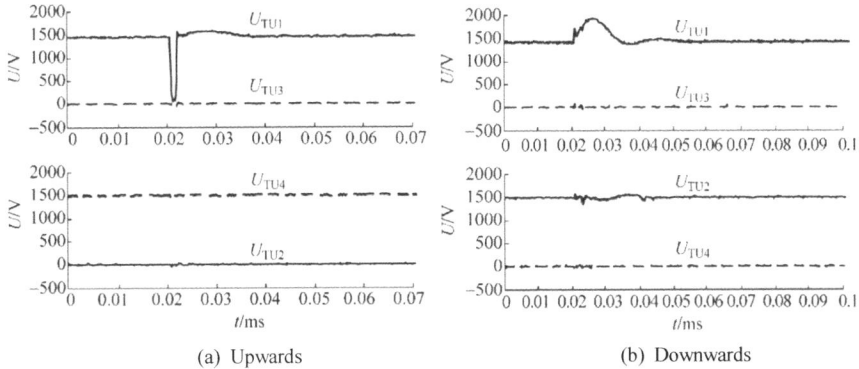

(a) Upwards   (b) Downwards

Fig. 3.38 Off-state voltage spikes

to the voltage drop of the snubber capacitor caused by other IGCTs' turning on. Those spikes upwards are caused by the snubber-capacitor voltage increment during other IGCTs' turning off. Note such spikes are different from IGCT turn-off voltage spikes. Their first spike is relatively low, which is caused by loop stray parameters. The second spike is consistent, which is introduced by the clamping capacitor of the turn-on snubber.

Other very tiny off-state spikes also exist, which will be analyzed together with on-state spikes later.

3. **On-state voltage spikes**

Theoretically no voltage spike should emerge during the on-state of the IGCT, though experimental waveform shows the existence of such spike, which is usually in oscillation with very small time constants, e.g., <0.5 μs. Similar voltage spikes also exist across the interlocked IGCT during the off-state, as shown in Fig. 3.39. This is due to switching actions of IGCTs in the same leg or different legs, which can be explained from two aspects, (1) some disturbance caused by switching actions of other IGCTs is coupled through the circuit or spatially with measured IGCTs; (2) some disturbance caused by switching actions of other IGCTs is coupled with probes through the circuit or spatially then gets recorded.

4. **Fake shoot-through**

It looks like the first and third IGCTs, the complementary switch pair are turned on simultaneously, as shown in Fig. 3.40 given both IGCT voltage drops to zero. Such behavior is called fake shoot-through. It happens when the leg is transitioning from 0110 to 1100, with the phase current being positive. The third IGCT is first zero-current switched off, resulting in that T_1 undertakes the top-half DC-bus voltage, T_4 undertakes the bottom-half DC-bus voltage and T_3 undertakes zero voltage. When T_1 turns on, the turn-on snubber circuit makes the snubber inductor undertake the

Fig. 3.39 On-state voltage
spikes

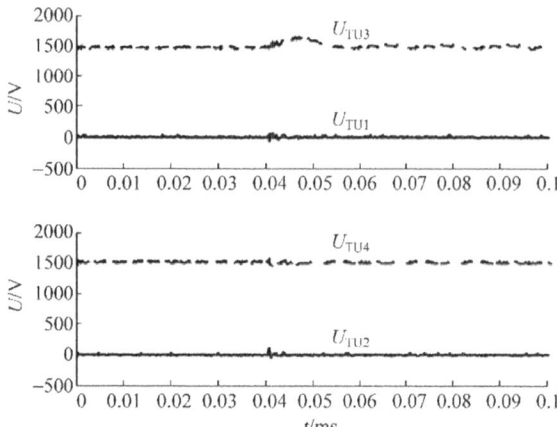

top half DC-bus voltage, which further results in interlocked switches undertake zero
voltage. Upon the completion of the current commutation, the upper DC-bus voltage
is undertaken by T_3. Such phenomenon happens when two interlocked switches turn
off with zero current or turn on with non-zero current. As a matter of fact, the voltage
spike downwards during the off state is one kind of the fake shoot-through, though
it is caused by other legs. The time period of the fake shoot-through along with the
leg output current could be used to evaluate the current limiting effect of the turn-on
snubber inductor.

Fig. 3.40 Fake
shoot-through

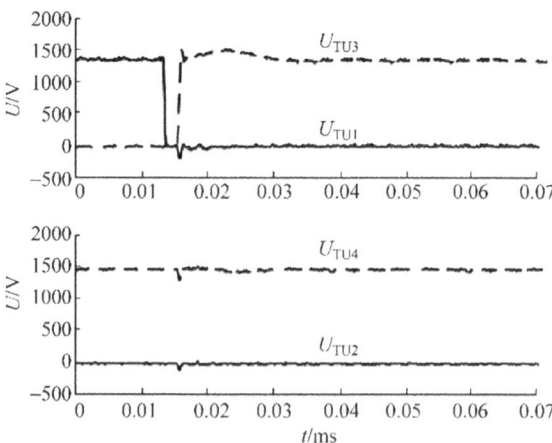

3.4 Power Devices in Parallel Connection

Due to the cost limitation and technical needs, we need frequently parallel power devices in power electronics systems. While paralleling devices becomes one of the focuses to secure the high current with high switching speed, the dynamic current balancing is one of major transient processes to investigate. The key is to guarantee in any operational modes all paralleled switches will undertake current evenly without any obvious imbalance. Therefore, the switch selection, circuit design, control implementation and mechanical design are to be analyzed and compared.

The hybrid switch like the IGBT inherits merits of both MOSFET and BJT. Some types of IGBTs exhibit a negative thermal-resistance coefficient at the small current while a positive one at the high current. Such IGBT is prone to being paralleled given its auto current balancing capability in the steady state. In fact high-power IGBT modules have already adopted the parallel connection inside, e.g., 3–4 individual IGBT dies in parallel connection, which though is not the focus of this section. We will emphasize more transients during the switch parallel.

IGBT and IGCT are two exemplary devices in high-power applications. The core of the IGCT is a four-layer three-junction thyristor with the high current capability up to 6000 A. Therefore, we seldom see IGCTs in parallel connection. The IGBT parallel is widely used even in the low-voltage converter, e.g., 380 V/300 kW AC motor drive systems, which is in need of the current rating of ~700 A. At such voltage level, the IGBT current rating is ~450 A, which makes the IGBT parallel necessary. In addition to the power and thermal capability, using low-current IGBTs in parallel to replace expensive high-current modules is also an effective way to reduce the cost.

3.4.1 Key Influential Factors on the Switch Parallel

The switching characteristics are of importance for the switch parallel, including the gate-drive resistance, gate voltage, junction temperature and loop inductance. Through changing such parameters to further investigate their impacts on the switching performance, we could better picture the current distribution among paralleled devices.

1. **Impact of the gate turn-on voltage**

For the IGBT, the gate turn-on voltage nearly determines its whole switching-on behaviors. Within the maximum ratings, the gate turn-on voltage directly affects the device channel resistance. A too high gate voltage will result in the gate damage. Therefore, the gate voltage should not exceed 20 V. Meanwhile a gate voltage lower than the threshold yields the switch operated in the active region thereby causing the device damaged. A recommended gate voltage is 10–18 V.

The transfer characteristics and output characteristics are able to reveal the relationship between the gate voltage and static parameters of the device. The output characteristics of each IGBT play an important role for the current balancing, as measured in Fig. 3.41 for one type of IGBTs.

At the same current level, the increment of the gate voltage will lower the saturation voltage, i.e., a negative correlation between the channel resistance and gate voltage. This provides the possibility of actively balancing the current distribution among paralleled switches through adjusting the gate voltage thereby the trans-conductance of each switch.

2. Impact of the gate turn-on resistance

Such resistance does not affect the saturation voltage, but switching transients significantly. The impact of the gate turn-on resistance on the switching performance has been validated in Fig. 2.12. According to experimental waveforms, the influence of the gate resistance on the device dynamic parameters is summarized as Table 3.2.

It is indicated in Table 3.2 that the device turn-off delay t_{d_off} is a monotone increasing function of the gate resistance. Therefore, when in parallel, switches could adopt different gate resistances to actively balance the current distribution.

3. Impact of the junction temperature

On the heatsink, the thermal gradient exists not only along the air flow but also in the vertical direction. Experimental results show that the worst case could result in a 25 °C temperature difference. Under such circumstance, the performance of paralleled devices is highly affected by the junction temperature. Neglecting the thermal impact will eventually endanger the system safety and worsen the reliability.

Most characteristic parameters of the device are highly sensitive to the temperature. So are those for paralleled switches. Increasing the junction temperature will increase the switch on-state resistance as well as the turn-on/off delay thereby affecting the static and dynamic current balancing. Five switching curves under different temperatures are illustrated in Fig. 3.42, with the junction temperature rising from

Fig. 3.41 Measured IGBT output characteristics

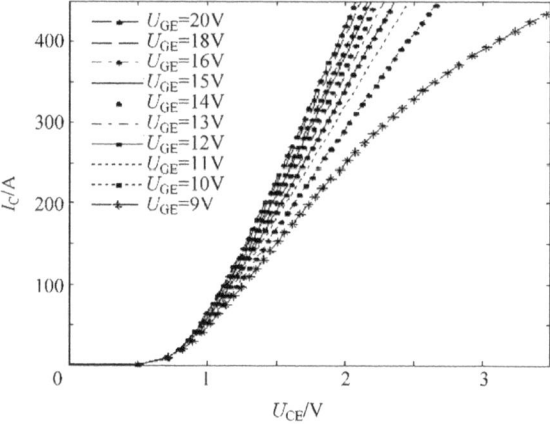

Table 3.2 Experimental results of gate resistance versus switching parameters

R_G/Ω	t_{d_on}/ns	t_r/ns	t_{d_off}/ns	t_f/ns
1.0	300.8	82.4	336.8	99.6
1.6	321.6	95.6	371.2	100.8
2.0	331.2	101.6	391.6	104.8
2.4	329.2	117.2	433.2	97.6
3.0	313.6	126.0	480.4	98.4
3.9	313.6	150.8	542.0	92.8
4.7	303.6	154.0	608.8	85.6
5.1	306.8	169.2	634.4	85.2
6.2	315.2	177.2	712.4	83.2
7.5	339.6	184.4	812.0	83.6
8.2	364.4	194.8	864.8	84.0

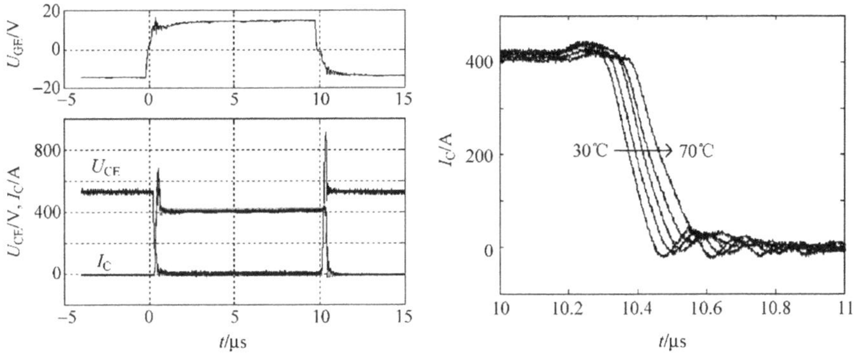

(a) Experimental switching process   (b) Detailed turn-off processes

Fig. 3.42 Experimental turn-off current of IGBTs under different temperature

30 to 70 °C. Each adjacent curve has a 10 °C gap. The increment of the t_{d_on} and t_{d_off} resulting from the temperature rise leads to the different current stress among paralleled IGBTs, if no further adjustment is done. When switched off, the higher-junction-temperature devices will have to turn off a higher current. When turned on, the lower-junction-temperature devices will undertake a higher turn-on current.

Shown in Table 3.3 are the experimental switching characteristics when the junction temperature rises from 30 to 75 °C. The DC-bus voltage is kept at 550 V and the load current is 400 A. The variation of t_{d_on} and t_{d_off} with the junction temperature is obvious.

The current imbalance among paralleled switches due to the different junction temperature could be controlled by adjusting the gate-drive voltage and resistance.

Table 3.3 Experimental
results of t_{d_on} and t_{d_off} with
the junction temperature

$T_j/°C$	t_{d_on}/ns	t_r/ns	t_{d_off}/ns	t_f/ns
30	310.8	94.4	402.0	114.4
35	313.6	94.4	406.0	119.2
40	314.0	94.0	408.4	122.4
45	316.0	96.4	412.0	126.8
50	319.2	96.4	414.0	129.6
55	320.8	96.8	416.4	142.0
60	322.4	98.4	423.2	146.0
65	324.4	95.2	424.8	146.0
70	326.4	96.0	434.0	160.8
75	328.0	96.0	434.4	160.0

3.4.2   Performance Analysis of Paralleled IGBTs

1.   Analysis of paralleled-switch steady states

The main IGBT static parameters include: on-state saturation voltage drop U_{CE_sat}, off-state blocking voltage U_{CES}, maximum repetitive turn-off current $I_{C(nom)}$, and the gate turn-on threshold U_{GE_th}. Here U_{CE_sat} and U_{GE_th} are two parameters determining the current balance. U_{GE_th} can be mathematically expressed as

$$U_{GE\_th} = U_{FB} + 2\Phi_{FB} + \frac{\sqrt{2\varepsilon_0\varepsilon_{si}qN_{Amax}(2\Phi_{FB})}}{C_{ox}} \tag{3.2}$$

Here U_{FB} is the flat band voltage, ε_{si} is the Si relative dielectrics, N_{Amax} is the p-well surface concentration, C_{ox} is the capacitance per unit area of the oxide layer, and Φ_{FB} represents the Femi energy level, which is a function of the junction temperature T_j shown in Eq. (3.3).

$$\Phi_{FB} = \frac{kT_j}{q} \ln \frac{N_{Amax}}{n_i} \tag{3.3}$$

Differential of U_{GE_th} with respect to T_j is

$$\frac{dU_{GE\_th}}{dT_j} - \left[\frac{\Phi_{FB}}{T_j} - \frac{k}{q}\left(\frac{E_g}{2kT_j} + 1.5 \right) \right]\left(2 + \frac{\sqrt{2\varepsilon_0\varepsilon_{si}qN_{Amax}(2\Phi_{FB})}}{2\Phi_{FB}C_{ox}} \right) < 0 \tag{3.4}$$

Therefore, U_{GE_th} will slightly decrease when the junction temperature T_j increases.

The device on-state saturation voltage is the voltage drop across its internal resistance once the current flows through. The output characteristics of the switch at different temperature are shown in Fig. 3.43.

With the large output current, a strong linear relationship exists between U_{CE_sat} and the load current, which could be approximated as

$$U_{\text{CE\_sat}} = I_C R_m = I_C(R_{ch} + R_a + R_j + R_{epi}) \tag{3.5}$$

For the third-generation trench-stop IGBTs, the on-state resistance R_m is made of four parts, i.e., channel resistance R_{ch}, accumulation-layer resistance R_a, JFET resistance R_j and epitaxial resistance R_{epi}. Among all, R_a and R_j are related to the manufacturing technology, and R_{epi} will increase with T_j-increasing. R_{ch} is the majority of the on-state resistance, which is influenced by the gate turn-on voltage and T_j, i.e.,

$$R_{ch} = \frac{L}{Z\mu_{ns}C_{ox}(U_{\text{G\_on}} - U_{\text{GE\_th}})} \tag{3.6}$$

Here L is the channel length, Z is the channel width per unit area and μ_{ns} is the electron mobility of the inversion layer, which is a monotone decreasing function of T_j, i.e.,

$$\mu_{ns}(T_j) = \mu_{ns}(T_0)(T_j)^{-m} \tag{3.7}$$

Therefore, the channel resistance will increase with the temperature rising, which is aligned with the device characteristics shown in Fig. 3.43.

The steady-state current distribution of paralleled devices is determined by the on-state resistance. The smaller-resistance device will undertake the larger portion of the current, creating the difference of the conduction loss. We can increase the gate turn-on voltage $U_{\text{G\_on}}$ to lower the on-state resistance thereby implementing an active current-balancing control.

2. **Analysis of the paralleled-switch dynamic performance**

Asynchronous turn-on current rising edges or turn-off current falling edges will result in the dynamic current imbalance among paralleled switches. Compared to on

Fig. 3.43 The device output characteristics at different temperature

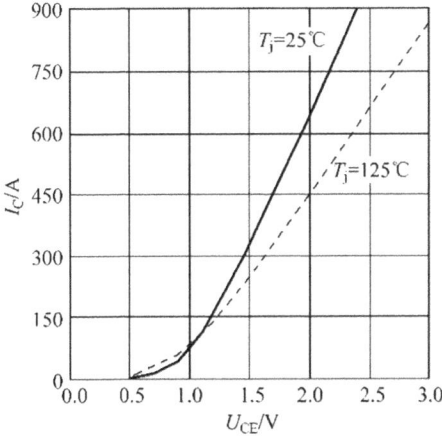

or off steady states, the dynamic current imbalance imposes more electric stress on switches. Therefore, the dynamic current balancing is even more important for the sake of the device safety.

The dynamic characteristic parameters include the turn-on delay t_{d_on}, rise time t_r, turn-off delay t_{d_off} and the fall time t_f, all of which determine the synchronization of switching behaviors of paralleled switches, and furthermore, the dynamic current balancing. Particularly t_{d_on} and t_{d_off} are two decisive parameters on the dynamic current balancing of parallel switches employing the synchronization trigger. These two parameters are both related to U_{GE_th} and U_{G_on}.

At the first stage of turning on the IGBT, before u_{GE} reaches U_{GE_th}, the gate turn-on voltage U_{G_on} charges the device junction capacitance C_{GE} and C_{GC} through R_G, i.e.,

$$u_{GE}(t) = (U_{G\_on} - U_{G\_off})(1 - e^{-\frac{t}{\tau_{on}}}) + U_{G\_off} \tag{3.8}$$

Here the turn-on time constant is

$$\tau_{on} = R_G(C_{GE} + C_{GC}) \tag{3.9}$$

C_{GC} is dependent on the voltage across it, which increases with the voltage dropping. Therefore, an observable difference exists for C_{GC} in turn-on and turn-off processes, which further results in the difference between t_{d_on} and t_{d_off}.

At the second stage of the turn-on process, u_{GE} exceeds U_{GE_th}, yielding the emergence of the collector current i_C.

The whole time of the first stage is defined as t_{d_on}. Similar to MOSFETs, we have

$$t_{d\_on} = -\tau_{on} \cdot \ln\left(\frac{U_{G\_on} - U_{GE\_th}}{U_{G\_on} - U_{G\_off}}\right) \tag{3.10}$$

Differential of t_{d_on} with the respect to T_j results in

$$\frac{dt_{d\_on}}{dT_j} = -\tau_{on}\left(\frac{U_{G\_on} - U_{G\_off}}{U_{G\_on} - U_{GE\_th}}\right) \cdot \frac{dU_{GE\_th}}{dT_j} > 0 \tag{3.11}$$

Therefore, t_{d_on} is a monotone increasing function of T_j. Meanwhile, increasing R_G will enlarge t_{d_on} while increasing U_{G_on} will slightly reduce t_{d_on}.

Similarly, the first stage of the turn-off process for the IGBT is when the gate discharges through R_G, i.e.,

$$u_{GE}(t) = (U_{G\_on} - U_{G\_off})e^{-\frac{t}{\tau_{off}}} + U_{G\_off} \tag{3.12}$$

Here the turn-off time constant is

$$\tau_{off} = R_G(C_{GE} + C_{GC}) \tag{3.13}$$

Due to the dependence of the junction capacitance C_{GC} on the voltage, τ_{off} is far longer than τ_{on}. The turn-off delay t_{d_off} has similar characteristics to the turn-on delay, in terms of being influenced by T_j, R_G and U_{G_on}. All the analysis above is of importance given the current dynamic imbalance results from asynchronous switching actions of paralleled switches. When two switches are paralleled, the earlier turned-on device turns on a higher current and the later turned-off device turns off a higher current. Appropriate adjustment of gate-drive parameters could further synchronize switching actions of paralleled switches thereby realizing the dynamic balancing.

3. **Analysis of the peripherals' impact**

The load-loop asymmetry among paralleled switches roots in the conflict between the component placement and mechanical design, which results in the current imbalance. By optimizing the bus-bar connection for the load, symmetric load loops will be realized.

Neglecting the snubber circuit, we find in a two-level inverter the largest impact factor on the commutation loops of paralleled switches is the bus-bar stray inductance. For instance, when an IGBT turns on, the current rising rate is

$$\frac{di_C(t)}{dt} = \frac{U_{G\_on} - U_{GE\_th}}{R_G C_{GE}/g_m + L_s} \tag{3.14}$$

Here g_m is the device trans-conductance and L_s is the stray inductance of the main circuit of paralleled IGBTs. Due to different bus-bar structures, the stray inductance of paralleled switches is diverse, which results in different di/dt and current dynamic distribution for paralleled switches. The device with a higher di/dt will undertake a higher electric stress.

The constraints of the component placement make it impossible to fully symmetrize all paralleled devices, creating the difference among the loop inductance of paralleled switches. Even though the multi-layer bus bars are widely used, which could further reduce the stray inductance and improve the symmetry of the commutation loops thereby minimizing the parameter difference, it is still hard to ignore the impact of the inductance difference on the dynamic current balancing.

Despite the negative impact brought by the loop inductance difference, we still can compensate the current imbalance through adjusting gate-drive parameters of paralleled switches thereby actively controlling the current distribution.

3.4.3 Experimental Study of IGBT Parallel

Majority of voltage-source power electronics converters share the same basic topology unit for the power conversion, i.e., commutating the current through two interlocked switches. Therefore, studying such basic topology unit is the precondition to understand the working mode of power electronic converters. Shown in Figs. 3.44

and 3.45 are the test bench and circuit measuring current distribution of paralleled switches, respectively. The benches are able to emulate the different junction temperatures among paralleled switches in actual converters. Here we investigate two switches in parallel, as the fundamental of multiple switches in parallel.

The test bench includes four parts, i.e., mechanical assembly, temperature control, pulse control and the voltage and current measurement. The mechanical part secures the symmetry of commutation loops of paralleled switches. A ~mm gap is left on the device cold plate to separate the thermal conductivity between two modules. This facilitates the independent temperature control of paralleled switches, which is realized by a dual-channel temperature controller and four heating plates. When the temperature reaches the steady state, the junction temperature of idle switches is assumed the same as the plate temperature. A double-pulse test is carried out to monitor the current distribution. The first pulse is to let the current reach the target while the second pulse with 10 μs width is to test paralleled switches. The switching loss during the test is assumed to have negligible impact on the switch junction temperature. Therefore, the switch junction temperature could be assumed the same as the plate temperature. High-bandwidth Rogowski coils and oscilloscopes are used for the current measurement.

Under the same junction temperature, gate resistance and gate turn-on voltage, the automatic current balancing could be realized among paralleled switches, as shown in Fig. 3.46.

1. **Current balancing among two paralleled switches**

To emulate the actual working modes of paralleled switches in the converter, two sets of experiments were carried out under different ambient and ambient-to-junction temperatures, respectively. The first set of experiments fixes the temperature difference between two paralleled IGBTs. By setting the dual-channel temperature controllers to regulate the ambient temperature, we could test the current distribution among two switches. Note the temperature difference between two switches is always kept the same. The test starts once the temperature reaches the steady state. Such test is equivalent to scenarios where the converter works in different ambients. The second set of experiments keeps T_1 temperature constant and alters T_2 temperature thereby vary-

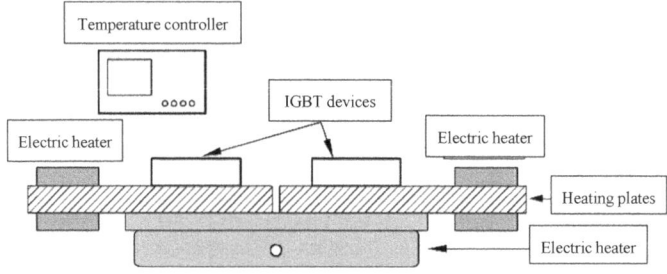

Fig. 3.44 Test bench of multiple switches in parallel

Fig. 3.45 The test circuit of paralleled switches

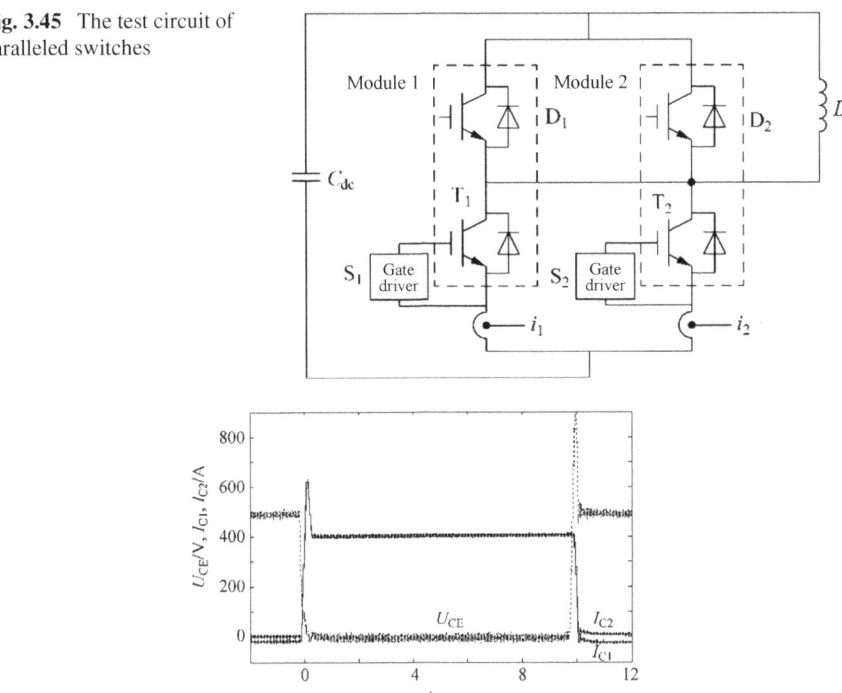

Fig. 3.46 Experimental waveforms of automatic current balancing among paralleled switches

ing the temperature difference between two switches. This is equivalent to scenarios where the converter works under different loads.

Specifically, the first experiment keeps the temperature difference between T_1 and T_2 as 10 °C. The ambient temperature rises from 40 to 70 °C. Experimental waveforms are shown in Fig. 3.47.

The second experiment keeps T_1 temperature ~50 °C while gradually increasing T_2 temperature, which widens the temperature gap. Experimental results are shown in Fig. 3.48a–d, where the temperature gap is set as 12.0, 21.2, 25.1 and 29.8 °C, respectively.

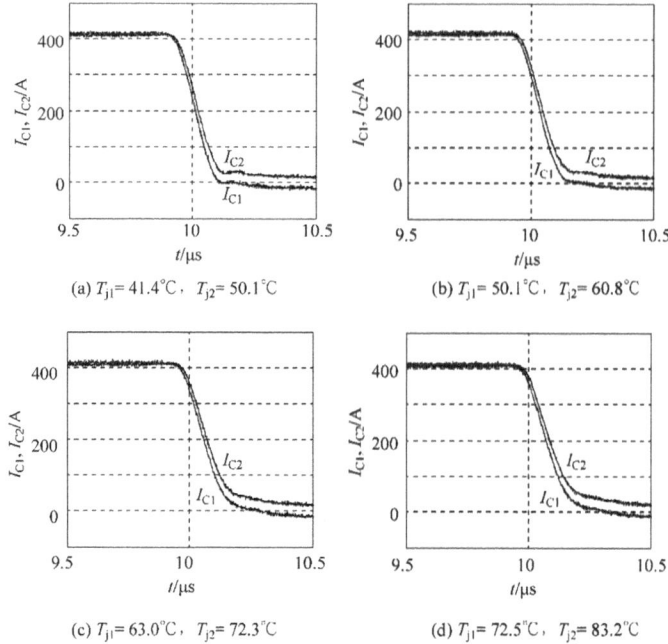

Fig. 3.47 Turn-off current of paralleled switches under the same temperature difference with different ambient temperatures

As shown in Fig. 3.47, the current gap between paralleled switches barely changes with the ambient, when the temperature gap is a constant. T_2 with a higher junction temperature has lower on-state current, however, a larger turn-off current due to the impact of the turn-off delay. The IGBT current capability has a negative correlation with the junction temperature. This yields the high-temperature switch undertakes the higher turn-off current, which is certainly an unreasonable design.

Figure 3.48 indicates that the difference between on-state current of paralleled switches has a positive correlation with the junction temperature. This number is 11 A when the temperature difference is 12.0 °C, and 26 A when the temperature difference is 29.8 °C. Meanwhile with the temperature difference increasing, the turn-off current difference becomes more obvious. In a higher-power converter with more switches in parallel, a large turn-off current difference confines the output-current capability of paralleled switches.

To verify the feasibility of the active gate control coping with the current imbalance under different temperature, we modified the gate-drive circuit for current balancing among paralleled switches. As shown in Fig. 3.49a, the gate resistance for each switch is 1.7 Ω, the DC-bus voltage is 600 V, the device turn-off current is 420 A, and junction temperature of T_1 and T_2 are 38.1 and 50.4 °C, respectively. The turn-off current gap is ~15 A. After revising the gate resistance for the higher-temperature

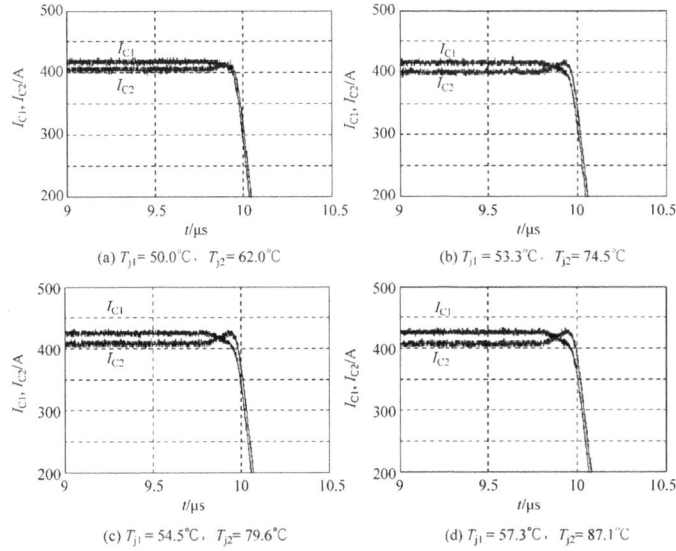

Fig. 3.48 Turn-off current of paralleled switches under different temperature gaps

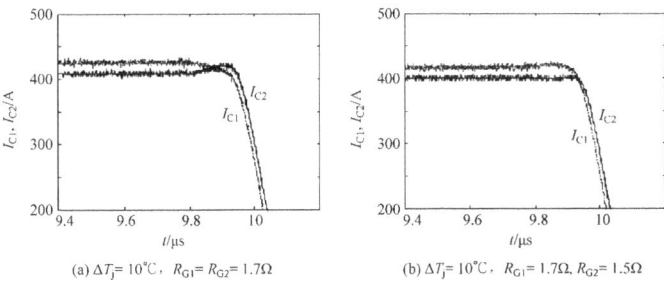

Fig. 3.49 Turn-off current of paralleled switches under different gate resistances

device to 1.5 Ω, the turn-off current peak becomes identical for different-temperature devices. Here the junction temperature of T_1 and T_2 is 38.9 and 51.5 °C, respectively. Experimental waveform is shown in Fig. 3.49b.

2. **Current capability of paralleled switches**

When testing a 160 kW/380 VAC two-level inverter, we use two IGBTs in parallel to form one switching module. Experiments were carried out to measure the turn-off current capability. Here we set up the test bench as a typical buck converter for DUTs. The main control board generates control pulses, with appropriate pulse parameters and protections. Pulses are continually generated until the protection is

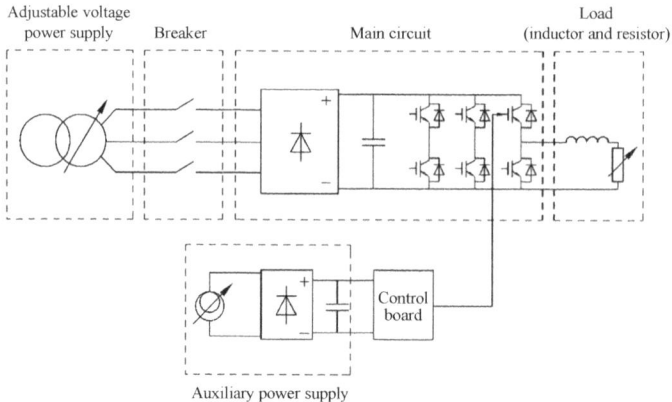

Fig. 3.50 Scheme of the test bench for devices in parallel

Fig. 3.51 Experimental
waveform of the current
capability of two switches in
parallel

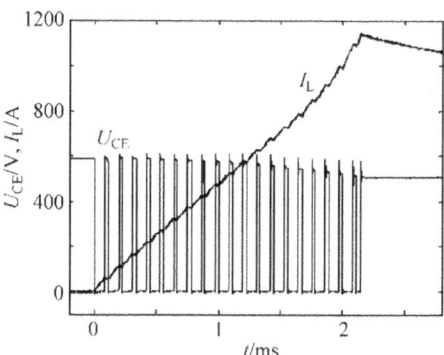

triggered. The scheme of the test bench is shown in Fig. 3.50. The load is made of
inductors and resistor tanks. The input voltage of the buck circuit is generated by
the adjustable voltage power supply and maintained by DC-bus capacitors. During
the test, the front-end circuit breaker is open and the current capability of paralleled
switches in such system could be tested.

The ambient is 30 °C and the DC-bus voltage is 600V. Experimental results shown
in Fig. 3.51 indicate the switch could reliably turn off 1146 A load current.

In an actual 315 kW/380 VAC two-level inverter, the main switch is made of
four IGBTs in parallel. Based upon the test bench shown in Fig. 3.50, experimental
results are shown in Fig. 3.52. The ambient is 25 °C and the DC-bus voltage is 600 V.
Paralleled switches could reliably turn off 2300A load current. Such test provides
the confidence on the reliable high-power operation.

Fig. 3.52 Experimental
waveform of the current
capability of four switches in
parallel

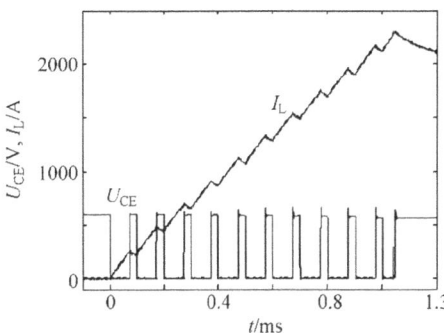

3.5 Power Devices in Series Connection

Series connecting power switches represents another working mode of high-power
converters. IGBTs and IGCTs are two exemplary devices which could be used in
series connection to withstand higher voltage thereby enhancing the power capability.
Due to the device difference, the series-connecting technology of IGBTs is different
from that of IGCTs. Some later chapters will discuss transients and active control of
IGBTs in series connection.

3.5.1 Fundamentals of the Switches in Series Connection

Series connecting devices is one of the approaches enhancing the voltage withstanding capability. Presently it is witnessed that industries widely use IGBTs and
IGCTs in series connection. With the increasing demand of high-voltage and high-
power converters, such series-connection technology becomes more and more popular and diverse. The critical point is to balance the voltage distribution among
series-connected switches, especially in the dynamic process.

Two balancing methods are widely adopted. One is the open-loop balancing, as
shown in Fig. 3.53. With synchronized gate signals, the selected power devices need
have the switching performance as close as possible. At some point the dynamic balancing circuit is needed. With various balancing approaches available, their common
point is the absence of feedback on the voltage distribution in the balancing process.

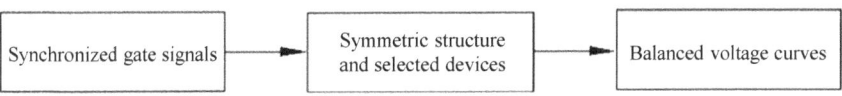

Fig. 3.53 Open-loop voltage balancing for switches in series connection

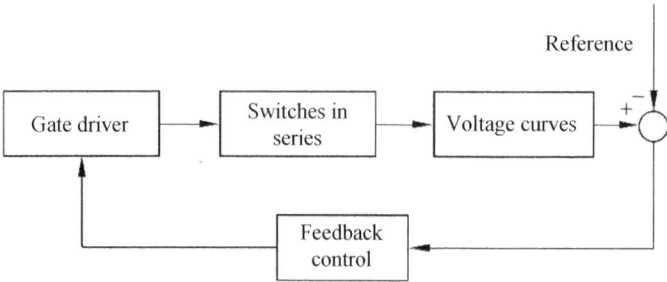

Fig. 3.54 Closed-loop voltage balancing for switches in series connection

The other approach is to adjust the gate-signal timing using the feedback circuit thereby realizing the voltage balancing, named as the closed-loop voltage balancing. Firstly the threshold of the voltage imbalance needs be set, i.e., the reference value of the voltage across the switch. Secondly the actual voltage drop across the switch is introduced as the feedback. Comparing the actual with the reference value allows the adjustment of the control thereby alleviating the voltage imbalance, as shown in Fig. 3.54. For such kind of approaches, the ultimate goal is not to synchronize gate signals of switches, but to realize the consistency of the voltage across each switches, including rising and falling edges.

In the real practice, asynchronous gate signals might emerge due to the driver signal generation and propagation delay, yielding to the voltage imbalance. Even though gate signals are perfectly aligned with each other, the voltage imbalance of switches in series connection is still possible due to the diversity of the switching performance, temperature and circuit parameters. The open-loop method is to minimize the diversity, while the closed-loop method uses the feedback information. They both have pros and cons. It is worthwhile to point out that two methods could be combined, especially for series connected IGBTs.

State-of-the-art balancing technology for series IGBTs and IGCTs is mostly based upon Figs. 3.53 and 3.54, which could be further extended to other high-power devices. Below are several typical techniques.

(1) Passive balancing

To avoid any destructive voltage imbalance among series connected switches, balancing circuits are in parallel to switches, including the dynamic balancing and steady balancing circuits. This method adopts passive components to cover the diversity of the switching performance and main-circuit parameters, yielding more robust switches through enhancing output characteristics of switches and balancing circuits (Fig. 3.55).

(2) Active balancing

The voltage across each switch in series connection will be detected. Once exceeding the threshold, a time delay will be added to its gate in the next switching cycle,

Fig. 3.55 Passive balancing circuit

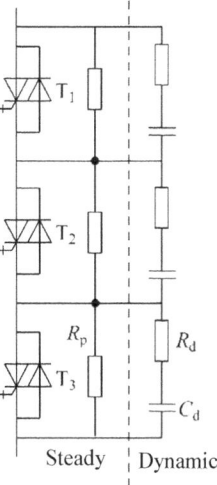

Fig. 3.56 Active gate-signal balancing

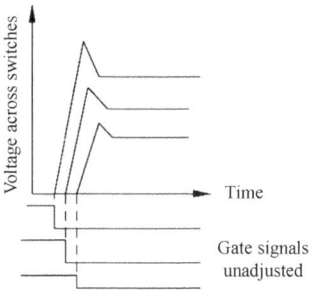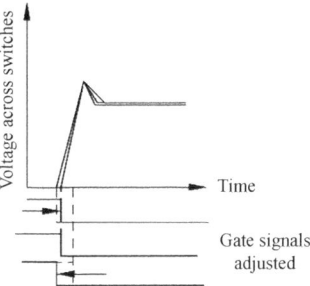

otherwise a time advance. Essentially such method adjusts the timing of gate signals to realize the voltage balancing, as shown in Fig. 3.56.

(3) Over-voltage clamping

Such approach is only suitable for IGBTs in series connection or its over-voltage protection, as shown in Fig. 3.57. Taking the advantage of the close relationship between gate signals and voltage drops of IGBTs, we could realize the voltage balancing by using the voltage feedback to adjust its gate signal.

Among all three approaches above, method (1) is suitable for all switches in series connection. Method (2) theoretically can be used for both IGBTs and IGCTs though it becomes quite complex when used in IGBTs due to sampling and control units. Method (3) is only suitable for IGBTs, which is subject to affecting the switching performance and increasing the switching loss when design parameters are mismatched, even though the extra circuit is not bulky.

Fig. 3.57 Active balancing
with over-voltage clamping

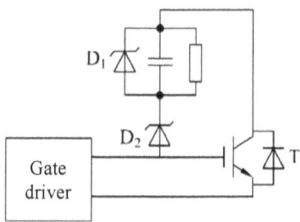

3.5.2   IGCTs in Series Connection

In this section we focus on IGCTs in series connections and related voltage balancing
technologies. Both the static and dynamic balancing need be involved. The voltage
imbalance in the steady state is mainly caused by different off-state leakage current
of switches. Such leakage current is related to the junction temperature and inherent
parameters. The dynamic voltage imbalance is mainly due to the diversity of the
switching performance and the propagation delay along the gate-drive loop. All
impact factors of the voltage imbalance are listed in Table 3.4.

When using the passive balancing circuit, one simple approach is to parallel one
resistor across each switch, as shown in Fig. 3.58a, namely the static balancing circuit.
The goal is to let the imbalanced leakage current go through the paralleled resistor
thereby yielding a smaller voltage imbalance. The resistance could be calculated by
the following empirical equation.

Table 3.4 Influential factors
of the voltage imbalance

Parameter differences		Static voltage balancing	Dynamic voltage balance
Semiconductor	Off-state leakage current	Significant	
	Junction temperature	Significant	Significant
	Turn-on/-off delay		Significant
	Storage time		Significant
	Reverse recovery charge		Significant
Gate driver	Turn-on delay		Significant
	Turn-off delay		Significant

Fig. 3.58 Passive balancing circuits

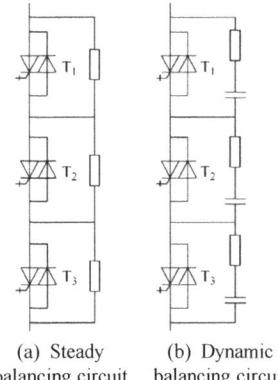

(a) Steady balancing circuit　(b) Dynamic balancing circuit

Fig. 3.59 Voltage imbalance versus the paralleled resistor

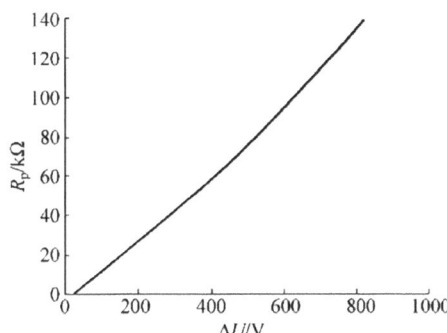

$$R_p = \frac{\Delta U_T}{\Delta I_{lk}} \qquad (3.15)$$

Here ΔU_T is the tolerance of the voltage imbalance and ΔI_{lk} is the tolerance of the off-state leakage current.

In real applications, R_p needs consider the steady-state balancing, i.e., the voltage difference among series connected switches, and the loss. Take the 4500 V/630 A IGCT as an example. Its leakage current is less than 20 mA under the repetitive turn-off voltage of 4500 V. When two are in series connection to undertake 4.6 kV DC bus voltage, the maximum voltage difference ($\Delta U_T = |U_{T1} - U_{T2}|$) of two switches ($T_1$, T_2) under different R_p is shown in Fig. 3.59, with the power loss of R_p shown in Fig. 3.60. Note the tolerance of the resistor, $\Delta R_p/R_p$ will impact the voltage imbalance, as shown in Fig. 3.61.

Obviously the smaller the balancing resistance, the better the balancing performance, however, the larger the current going through the resistor thereby the higher power loss. Therefore, the selection of R_p is the tradeoff between the voltage balancing and the power loss. The usual range for R_p is 20–100 kΩ. In addition, the tolerance of the balancing resistance should be minimal.

For the IGCT dynamic balancing, the procedures below need be followed:

(1) Adopt IGCTs with the same ID number from the same patch, to minimize the diversity of the switching performance such as the propagation delay, storage time and reverse recovery time;
(2) Double check the IGCT switching performance before equipped to the system. Make sure the series connected switches match;
(3) Design symmetric heatsink and thermal loop to reduce the temperature difference;
(4) Design identical gate-signal generation and propagation loops to reduce the difference of the turn-on/off delays.

Following the above will help minimize the voltage imbalance of series connected IGCTs. However, the ultimate balancing might not be ensured unless the dynamic balancing circuit is adopted, which usually is an RC snubber circuit shown in Fig. 3.58b.

The capacitance and resistance of the dynamic balancing circuit need be comprehensively optimized. The delay-time variations highly impact the dynamic balancing of IGCTs in series connection, among which the turn-on delay usually is much shorter than the turn-off storage time. Therefore, the IGCT turn-off balancing should be put on the top priority. Here C_d could be calculated through follows

$$C_d = \frac{\Delta t_{\text{delay}} I_{\text{T\_max}}}{\Delta U_{\text{T\_max}}} \tag{3.16}$$

where Δt_{delay} is the maximum tolerance of the whole turn-off delay time, including the turn-off storage time and the gate-signal turn-off delay. $I_{\text{T\_max}}$ is the maximum current going through the IGCT. $\Delta U_{\text{T\_max}}$ is the maximum allowed tolerance of the IGCT voltage.

Note in the current-source converter with series connected IGCTs, the switch turn-off due to the natural phase commutation exists, which results in the dynamic voltage imbalance due to the IGCT reverse recovery charge, Q_{rr}. Therefore, another criteria is needed for C_d selection, i.e.,

Fig. 3.60 Resistor loss versus the resistance

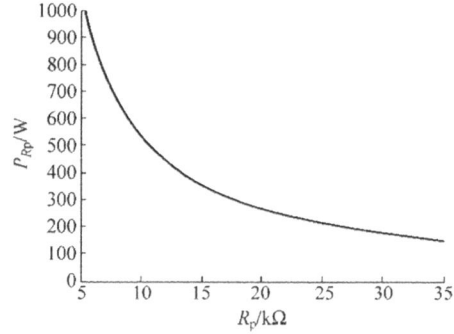

Fig. 3.61 Resistance
tolerance versus the voltage
imbalance

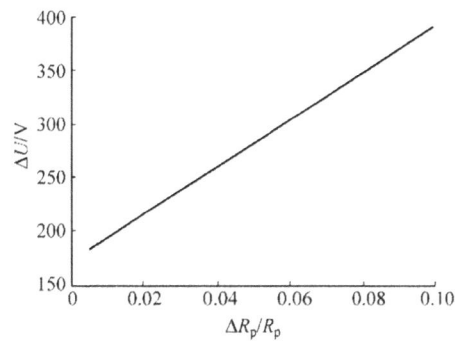

$$C_d = \frac{\Delta Q_{rr}}{\Delta U_{T\_max}} \tag{3.17}$$

For series connected IGCTs, $C_d = 0.1$–1 μF, much smaller than those for GTOs. For the R_d selection, it needs follow the specs of the control and IGCT characteristics, i.e.,

(1) R_d should be a small value so that the C_d discharging could be finished within the minimum on-state and off-state pulse width;
(2) R_d cannot be too small, otherwise once IGCT turns on the C_d discharging through R_d will add a large current spike on the IGCT;

In the real practice, the impact of the dynamic balancing circuit on the voltage balancing and power losses need be considered, as shown in Table 3.5.

To further detail the function of the static and dynamic balancing circuit, experimental results are provided here. Two 4500 V/630 A IGCTs are in series connection. A significant voltage imbalance is highlighted. Therefore, to secure the switch safety, all experiments were carried out under a lower DC-bus voltage.

Assume the off-state voltage values of two IGCTs are U_1 and U_2, respectively, and the voltage difference after the first voltage spike when switching off is ΔU. Here we define a voltage balancing performance index as

$$k = \frac{\Delta U}{(U_1 + U_2)/2} \tag{3.18}$$

Table 3.5 The impact of R_d and C_d

	Voltage difference ΔU_T	Turn-on loss	Turn-off loss	Capacitor and resistor loss
$C_d\uparrow$	↓	↑	↓	↑
$R_d\uparrow$	↑	↓	↑	–

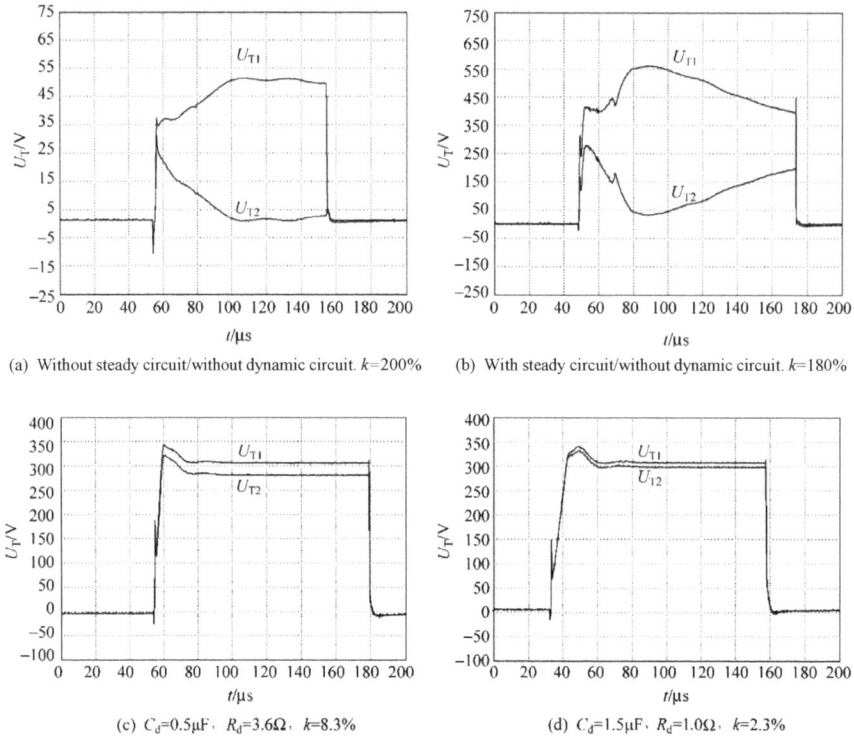

Fig. 3.62   Comparison of the voltage balancing of series connected IGCTs

The larger k, the worse the balancing effect. With a time delay of 100 ns between two IGCT gate signals, the experimental result of voltage balancing is shown in Fig. 3.62, where scenarios with or without balancing circuits under different balancing parameters are illustrated. It indicates significantly different k under different balancing parameters.

Without the dynamic balancing circuit, one switch nearly undertakes the whole DC-bus voltage while the other undertakes nearly zero, which makes the switch in series connection useless. The static balancing circuit improves the off-state voltage distribution, however, the dynamic performance is still not satisfactory. The introduction of the dynamic balancing circuit greatly improves the IGCT voltage balancing performance, though different parameters yield different results.

As shown above, the transient characteristics of the power switches are related to multiple aspects.

Firstly, the internal physical mechanism. The material, structure and conductivity will all influence the switch performance. The influence of the temperature remains always true.

Secondly, the external operational conditions. The quantification of switch performance is all related to the test circuit. Performance based on the stand-alone test bench and the converter has different features, revealing the transient performance from different aspects. Some critical factors such as stray parameters highly impact switching transient characteristics.

Thirdly, interactions among switches especially in switching processes when some specific transients could be observed and classified.

Lastly, device in parallel/series connection, which is an effective way to increase the system power capability. The voltage and current balancing represents an exemplary electromagnetic transient of power switches. Recognizing principles of such transients, establishing transient models and actively controlling transients are the preconditions of the active voltage and current balancing.

Chapter 4
Transient Commutation Topology and Its Stray Parameters

The transient commutation topology (TCT) is one of the fundamentals of power electronics transient analysis. Different from the conventional topology based on lumped parameters, the TCT is based on short-timescale transient commutations along with stray parameters. The core is to take stray parameters into account, including the impact analysis, extraction and allocation. In the synthesis of power electronics, the TCT and stray parameters involve multiple factors, such as power semiconductor devices, circuit and control, and emphasize the interaction between switches and TCT. In the decomposition of power electronics, it analytically reveals the relationship between the mechanical structure and circuit parameters, which has a strong non-linear performance. Such TCT and related stray parameters play an important role in the transient analysis and control.

4.1 Definition of the TCT

4.1.1 Definition of the Converter Topology

Topology is a concept of the geometry. It describes the relationship among various components by abstracting their locations. Specifically to the converter topology, it abstracts locations and relationships among electric components in the circuit. Therefore, the topology is not equal to the circuit. Different converter circuits could have the same topology. Given the topology only describes circuitry connections without focusing on components, no auxiliary circuits are considered, e.g., the topology of a three-phase two-level inverter does not care if the switching device is IGBT or IGCT.

For the converter topology, an exemplary one is a multi-level topology, a typical and feasible solution to resolve the conflict between the device ratings and the converter power ratings. It is widely acknowledged that the multi-level topology was firstly proposed in 1980, when an article was published using one pair of clamping

© Tsinghua University Press and Springer Nature Singapore Pte Ltd. 2019   145
Z. Zhao et al., *Electromagnetic Transients of Power Electronics Systems*,
https://doi.org/10.1007/978-981-10-8812-4_4

diodes to connect the upper and lower part of the leg together. Such approach clamped the neutral point thereby being called as a neutral-point-clamped (NPC) converter, the emergence of which represents the birth of the multi-level converter. Generally the multi-level converter is the extension of the three-level topology. The more levels, the more steps thereby the closer to the ideal sinusoidal with less harmonics. However, even though theoretically any number of voltage/current levels could be realized by the multi-level converter, too many levels are not the actual pursuit given the limitation of the hardware and the control complexity. In the real practice, as long as the performance specs are met, usually three-level or five-level converters are most feasible solutions.

Besides the multi-level topology, according to the device characteristics and applications, some other approaches are available, e.g., switches in series/parallel connections, multiplication of power modules and adoption of the transformer. Those approaches vary in terms of the performance and cost, mainly due to different main-power circuits. Take the adjustable speed system as an example. The typical solutions include the three-level NPC topology, hybrid five-level topology, cascaded H-bridge topology, two-level topology using series-connected switches and modular multilevel topology.

The three-level NPC topology adopts high-voltage (HV) IGBT devices in the main circuit, as shown in Fig. 4.1. Due to the controllable turn-on and turn-off, the current limiting is easy to be realized without any auxiliary snubber circuits. The cooling method could be either forced air or liquid cooling. No fuses are needed, which reduces failure points and enhances the system reliability. In such inverter, even though the three-level topology is adopted, high harmonics are inevitable. Therefore, a filter must be installed between the inverter output and motor terminals. Otherwise, a specially designed motor is needed. The motor voltage and current fed by such an inverter is shown in Fig. 4.2. In addition, a back-to-back three-level topology could be used to realize the four-quadrant operation.

The IGCT based three-level NPC inverter is usually able to drive a 6–6.9 kV motor. The circuit scheme is shown in Fig. 4.3. Such topology optimizes the output levels and device amount, which generates sufficient levels to be connected with the standard motors while keeping the device amount low. The output voltage and current waveform of such an inverter-fed motor is shown in Fig. 4.4.

The cascaded H-bridge topology adopts two-level inverter modules to form a high-power inverter. It does not need output transformers and directly realizes the 3.3 or 6 kV output. For the inverter part, it uses the single-phase H-bridge based inverter employing the SPWM control and overlapping their outputs. IGBTs are main switches. The input part is a multi-phase multiplication rectifier, realizing the input power factor >0.95 and THD < 1%. The overall structure is a modular design, as shown in Fig. 4.5 where five modules are used to output 6 kV. The merits of such topology include: (1) using affordable conventional IGBTs to be cascaded, which is able to reach any output voltage through adjusting the cascade numbers; (2) excellent input and output waveforms for standard-designed motors; (3) modular design to provide the redundancy, i.e., the bypass circuit could be added to short the faulty module thereby running the motor at the rated or de-rated power. The demerits

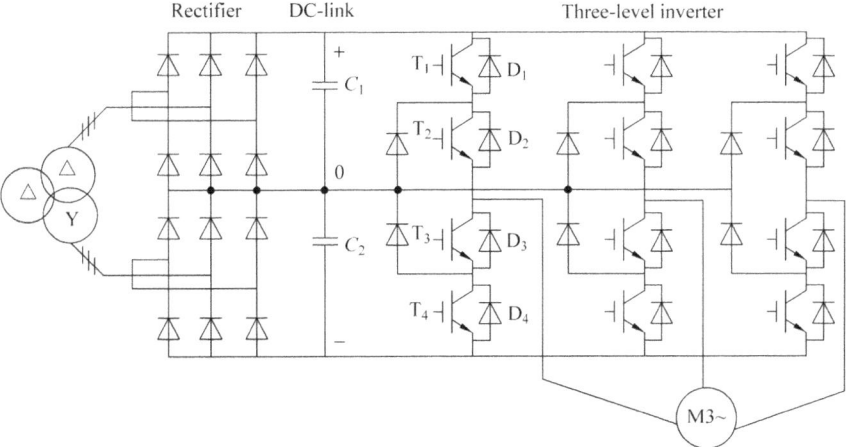

Fig. 4.1 The scheme of a three-level NPC inverter

Fig. 4.2 Filtered voltage and current of the motor driven by the three-level NPC inverter

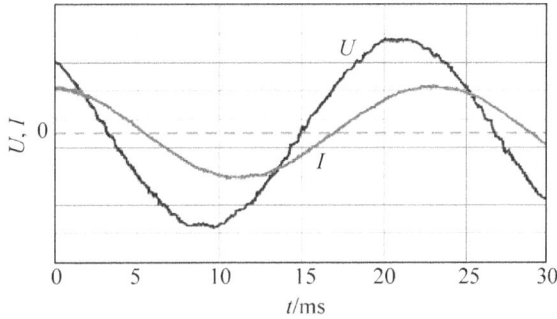

Fig. 4.3 Scheme of the three-level cascaded NPC inverter

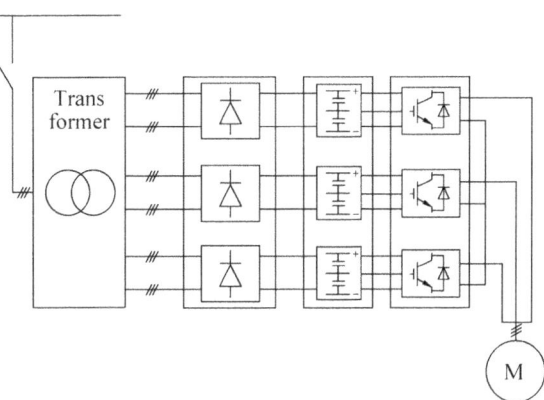

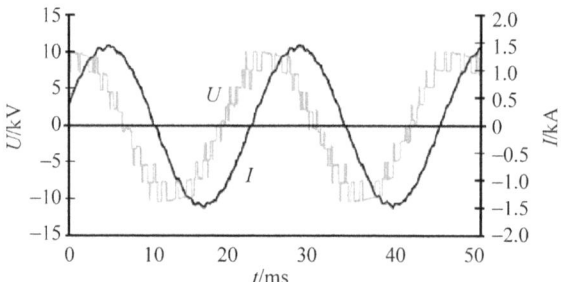

Fig. 4.4 Voltage and current of the motor driven by the three-level cascaded NPC inverter

include too many switches and power units, low power density, reduced reliability, and difficulty of realizing the regenerative and four-quadrant operations.

For adjustable speed drive systems and FACTS in power systems, some higher-voltage designs with simple cells become the interesting concern. The typical example is the modular multilevel converter (MMC), with its system scheme and sub-

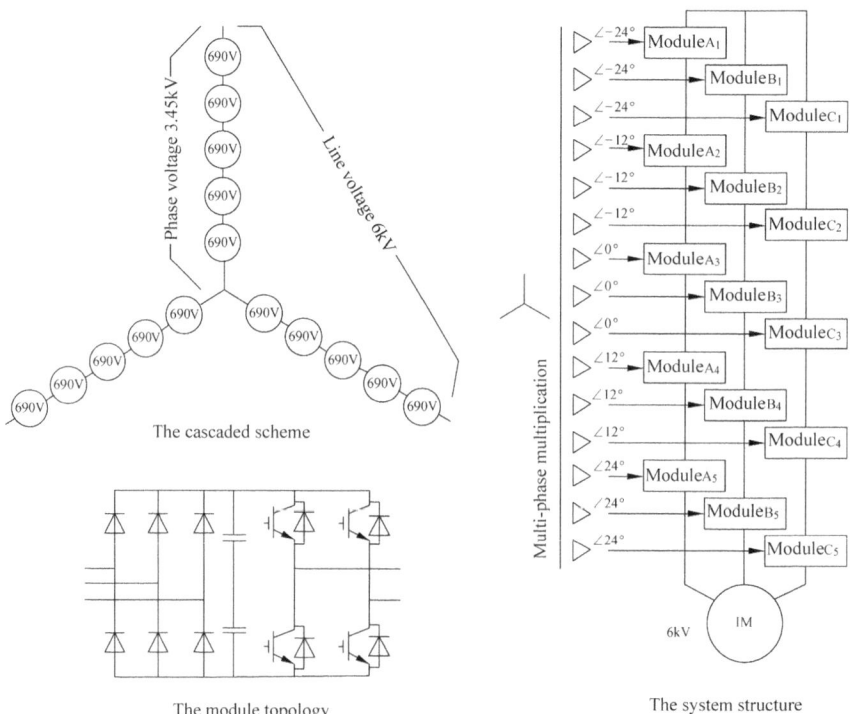

Fig. 4.5 The cascaded inverter system

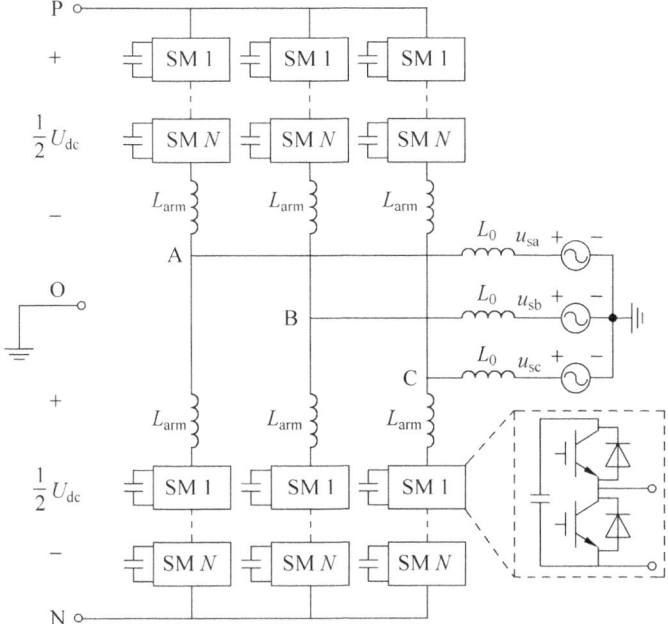

Fig. 4.6 MMC main structure and its sub-modules

modules shown in Fig. 4.6. Its each phase contains the upper and lower part, each of which is made of N half-bridge modules. Arm inductors are usually placed between the upper and lower parts in case the shoot-through during the switching transients. Two switches in each module are complementary, generating the effective voltage through connecting or bypassing its DC-bus capacitor to the main circuit. In this way a multilevel voltage output is created. Such MMC has simple structure and high efficiency. The demerits, however, are lacking of the DC-bus short-circuit protection. Meanwhile given each sub-module can only undertake quite a low voltage, atrocious amount of modules might be needed for the very high-voltage output. Feasible solutions include: (1) using the H-bridge module, half-voltage clamping module, clamping dual-half-bridge module, and interleaved dual-half-bridge module to realize the DC-side short-circuit protection; (2) using fly-capacitor based three-level modules, NPC three-level modules or dual-half-bridge three-level modules to increase the module voltage level, which helps reduce the module and switch numbers.

Whatever the multilevel topology is, particular modulation and control strategies are needed, which eventually differentiate those topologies accordingly. More importantly, we should focus on their common points and bottom-level characteristics, i.e., short-timescale transients.

4.1.2   Converter Transient Commutation Topology

Even though the equivalent circuit of the multi-level converter is quite straight forward, it is still incapable of the transient analysis, which is highly coupled with the switch commutation process, the fundamental of the converter analysis. For example, in the buck-type DC-DC converter, turn-on/off actions of the controllable switch correspond to the turn-off/on of the diode, which is a typical commutation behavior of the converter.

In the high-power converters, many switches are employed with complex commutations. Controllable switches are tightly coupled with diodes or other switches. Meanwhile, the VSC and CSC have different commutation processes. Except as otherwise defined in this book, VSCs are employed to explain the typical transient commutation behaviors, e.g., an IGCT based three-level inverter shown in Fig. 4.7.

For simplicity, the scheme of one leg is shown in Fig. 4.8, with switches from top to bottom defined as T_1–T_4, respectively. The leg output terminal is A and the DC-bus middle point is Z. Three voltage levels exist in the phase output voltage U_{AZ}.

For such a three-level NPC inverter, the commutation processes are complex mainly due to the complicated control logic. Four switches need be controlled in one leg, with their switching logic shown in Table 4.1. Only transitions between adjacent modes are allowed. Here we use 1 and 0 to represent the on and off states of T_1–T_4, respectively. Therefore, only transitions between 1100 and 0110, or 0110 and 0011 are allowed, with the dead band included. No transitions between 1100 and 0011 are permitted, given such transitions from the highest to lowest voltage levels lose the merit of the three-level topology and yield high du/dt and overvoltage during the current commutation.

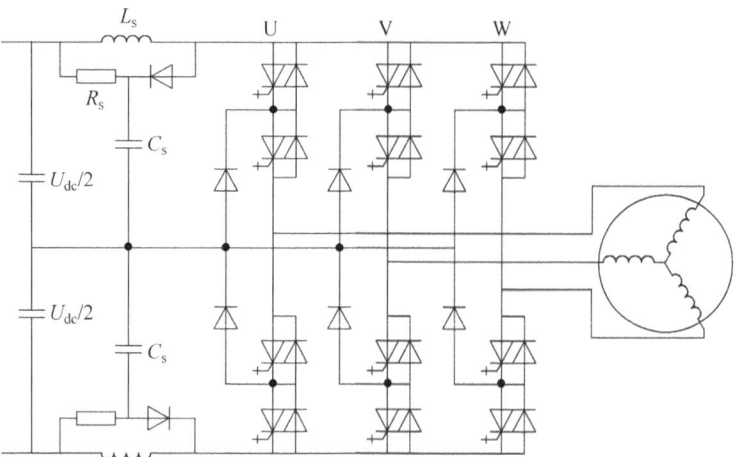

Fig. 4.7   The main circuit of an IGCT based three-level NPC inverter

All commutations of the three-level NPC inverter can be listed by enumerating switching actions of each leg with the consideration of the leg-current polarity. Four basic commutations are listed, corresponding to the switching on and off of four switches, respectively. The commutation process that T_1 participates in is $1100 \leftrightarrow 0100 \leftrightarrow 0110$. With the positive leg current, T_3 has no current. Therefore, $0100 \leftrightarrow 0110$ does little to the commutation, as shown in Fig. 4.9a. Here "increase" and "decrease" in the figures indicate the switch current trend when transitioning from the left states to the right states. If the transition is reversed, the trend is also reversed. Commutations of T_2–T_4 are shown in Fig. 4.9b–d, respectively. To illustrate such processes, we can bridge the control command with the circuit status, even though part of the control command does not necessarily change the circuit state.

In fact, whatever power electronic converter is employed, commutation behaviors are highly related to switches selected, which might vary commutations even under the same topology and controls. In addition to switch characteristics, other elements in the converter are equally critical, such as control parameters, mechanical structure, temperature, stray parameters, snubber circuit and load. Therefore, using ideal commutations above to analyze converter transients has major drawbacks.

To accurately analyze the transient behavior of the converter, non-ideal topology is of importance. From the aspect of the electromagnetic energy conversion, under the impact of transient processes and non-linear factors, such as transition time, stray parameters and magnetic saturation, the TCT of each system differs from others. To precisely describe transient processes of a diode NPC three-level inverter, not only

Fig. 4.8 The scheme of one leg in a three-level inverter

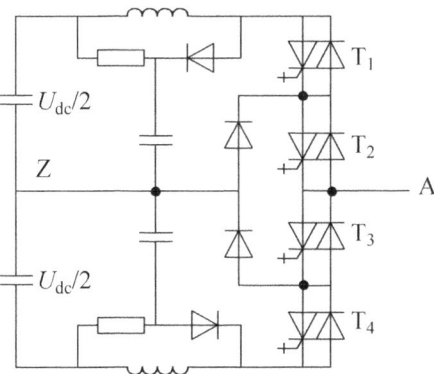

Table 4.1 Three-level converter switching logics

U_{AZ}	T_1	T_2	T_3	T_4	
$U_{dc}/2$	1	1	0	0	Positive level
0	0	1	0	0	Dead band
0	0	1	1	0	Zero level
0	0	0	1	0	Dead band
$-U_{dc}/2$	0	0	1	1	Negative level

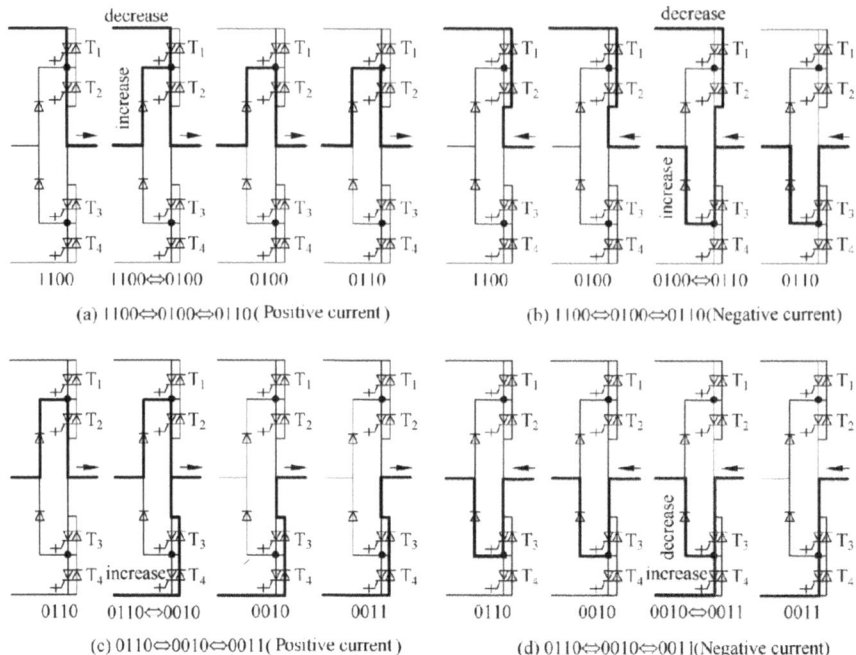

Fig. 4.9 Four basic commutations in the three-level inverter

non-ideal switching processes but also non-linear stray parameters need be included. This is the specialty of the converter transient commutation topology.

The commutation behaviors in Fig. 4.9 did not point out their specific locations in Fig. 4.7. For the case analysis of Fig. 3.28, stray parameters in each leg are diverse, creating different commutation loops with different stray parameters for switches at the same position but different legs, therefore, different transient behaviors. Due to the existence of such stray parameters, the transient voltage and current of switches are varied. Shown in Fig. 4.10 is the simulated turn-off process of one switch without considering stray parameters, in contrast to Fig. 4.11 where the turn-off voltage is measured in the actual inverter under the same DC-bus voltage and load current. A major difference is displayed. Hence getting rid of the constraints of ideal switch models and lumped parameters, introducing stray parameters and transient loops, and describing the commutation behaviors in the short-timescale is necessary, which is the essence of the transient commutation topology.

Different from the conventional topology based on lumped parameters, the TCT is defined as a topology considering stray parameters based on the short-timescale electromagnetic pulses, as shown in Fig. 2.8. Stray parameters play an important role in the transient commutation topology.

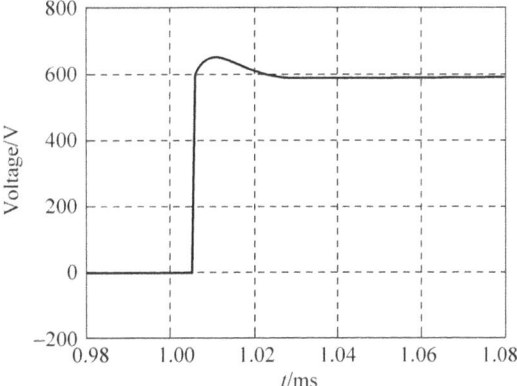

Fig. 4.10 IGCT turn-off voltage (simulation, without stray inductance)

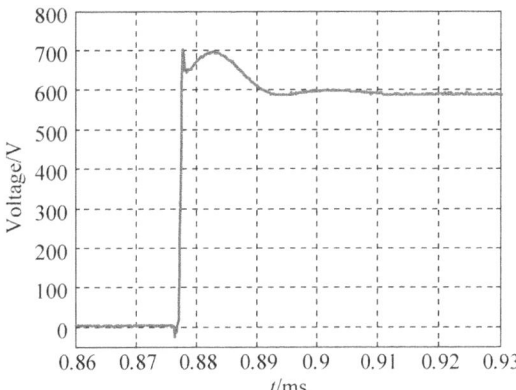

Fig. 4.11 IGCT turn-off voltage (experiments)

4.2 Extractions of Stray Parameters in Complex Main Circuits

Extractions of stray parameters are critical for the transient commutation topology. Particularly we need detail the relationship between stray parameters and current commutations. With computer aids, the technique for the stray-parameter extraction is rapidly progressing and diversifying. Various approaches will be compared in this section in terms of the accuracy and complexity, which is found particularly useful in the analysis of transient commutations.

4.2.1 Comparison of Extraction Approaches

Usually it is believed stray parameters include two categories, i.e., parasitics of passive components and those along the connecting cables. Conventional circuit design and theoretical analysis treat passive components as ideal and cables as non-lossy conductors. Such assumption remains effective when the working frequency is low. For power electronic converters, in the commutation process, the voltage and current change drastically. Plus the circuit size is large with more connector quantity. The impact of stray parameters is not negligible. Modelling of the resistor, capacitor and inductor with stray parameters is prevalent, however, modelling of the connectors such as bus bars is rather complex, given stray parameters are related to the material, relative locations, shape, size and layout. In summary, characteristics of stray parameters in power electronic converters are

(1) Large energy storage or delivery at short timescales within stray parameters. During the energy exchange in switching actions, large voltage or current spikes emerge with potential oscillation, causing significant EMI, switching stress and switching loss. In the worst case, it could damage switches;

(2) Relatively large value of stray parameters, which alter the main-circuit topology and parameters by changing the circuit impedance thereby causing the unexpected deviation;

(3) Differentiating from the ideal topology when considering stray parameters. The converter current and voltage will be distorted, creating large amount of high-order harmonics thereby affecting the power quality;

(4) Forming the EMI path through coupling main-circuit bus bars, semiconductors, PCBs and auxiliary circuits;

(5) The larger the power, the larger the mechanical size, which increases the length of cables thereby creating more stray inductance;

(6) Requirements on voltage withstanding, over-current capability, insulation and thermal dissipation are high. However, the electrical allowance of the switch is low, which adds stricter constraints on stray parameters.

Among numerous methods of extracting stray parameters, surveying calculation, analytical calculation, numerical calculation and partial element equivalent circuits (PEEC) are the exemplarities.

(1) Surveying method. For a single passive component, its circuitry model will be first built based on the theory. Then the impendence analyzer will be used for the parameter fitting in order to get the decent accuracy. For main-circuit connections, due to small stray parameters, the conventional method is not precise enough. Presently the time domain reflectometer (TDR) is the feasible choice. Such method, based on the microwave transmission theory, injects steep pulses (rising edge < 5 ps) through the TDR and measures reflected signals, which could be further used to calculate the inductance, capacitance and resistance. It has a high precision, but is very costly in terms of the measurement equipment and software;

(2) Analytical method. Such method is prevalent for the electromagnetic analysis given it provides analytical solutions and becomes very direct. Parameters could be approximated by using analytical equations. The application of such method is easy as well, especially for regularly shaped circuits. For majority of the circuits, the conductor shape is irregular, which results in the low precision of the calculation result, not aligned with the actual performance;

(3) Numerical method. The commonly used include method of moments (MoM), finite element method (FEM), finite difference method (FDM) and finite-difference time-domain method (FDTD). They all have their own pros and cons. MoM considers the interaction of sources and fields, which has high calculation precision, however, occupies large computation resource given it needs involve the fields when calculating the current. FEM and FDM have large amount of meshes, yielding heavy computation load as well.

(4) PEEC method, which was proposed back to 1970s involving the peripheral impact when modelling the circuit. Therefore, it has been popular since then and keeps improving. When using PEEC, we will first mesh the circuit, conductors and sometimes dielectrics into multiple fine elements. By assuming each element has the constant current density and surface charge density within, we could simplify Maxwell equations into multiple equation sets on those fine elements. Furthermore, we can get the partial inductance and capacitance of each element. Then the equivalent circuit model is built by incorporating all these partial parameters. At last, the general-purpose simulation software is applied for the calculation and analysis.

In general, the analytical method is simple and direct with the low precision. Numerical methods are more accurate with a high computation load, plus the complex source and boundary conditions. Therefore, numerical methods are only suitable for simple circuits and components. The surveying method has high accuracy, however, the measurement equipment and analysis software are expensive. So far, all these methods are applied in the design of integrated circuits, such as extraction of stray parameters along PCB traces given the size is small (μm) and the frequency is high (GHz). Therefore, such methods are frequently seen in the microwave and microelectronics. For power electronics especially the high-power converters, the circuit has relatively larger size (m) and lower frequency (kHz), with complex physical structures, irregular conductor shapes and various dielectrics. All these features obstruct the parameter extraction. Therefore, using PEEC is the more feasible choice for extractions of stray parameters.

4.2.2 Accuracy Analysis of PEEC

The accuracy of stray-parameter extractions is very critical for the quantitative analysis of the transient commutation process. Here we use one rectangular copper bar as an example to illustrate the accuracy of the PEEC method, as compared with the measured results.

The copper-bar loop has the dimension shown in Fig. 4.12a. The cross-section area is 12.5 mm * 3 mm. To shield any peripheral EMI and prohibit the energy within the model from being emitted to the far field, a standard ground is given for the calculation with a shield outside the circuit, as shown in Fig. 4.12b. When using the impedance analyzer, we connect its ground with the shield.

The shield is made of copper sheets with the thickness of 0.1 mm. The 3D dimension is 440 mm * 380 mm * 140 mm. For the measurement equipment, the shield and the objective are combined together. Four corners of the rectangular have been tied to the shield through thin wires.

The procedure of extracting stray parameters and verifying its effectiveness is shown in Fig. 4.13. Six main steps are included:

Step 1. Set up the maximum working frequency and divide elements

PEEC only requires the conductor geometry and materials. The maximum frequency (f_{max}) and element division are two critical factors. Influenced by the skin effect and proximity effect, stray parameters under different frequencies are different, with some non-linear relationship. In general, the frequency selection should refer to the interested frequency. In terms of element divisions, this is the essence of the PEEC method. It decides if some part should be modelled in piecewise or as a whole. This will eventually decide the equivalent circuit. When dividing elements, three basic criterions need be followed:

(1) All points along the conductor element have the same voltage potential, which requires the maximum size of the element far less than the wavelength of the studied electromagnetic field. The conductor size decides the modelling precision. The higher the precision required, the smaller the element thereby the

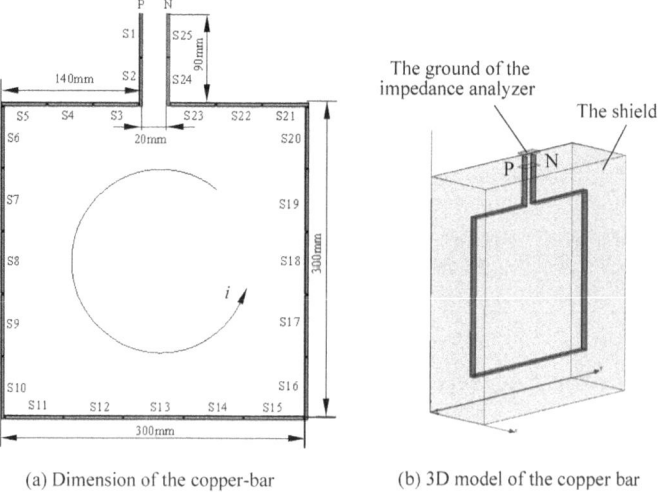

(a) Dimension of the copper-bar   (b) 3D model of the copper bar

Fig. 4.12 A simple copper-bar loop

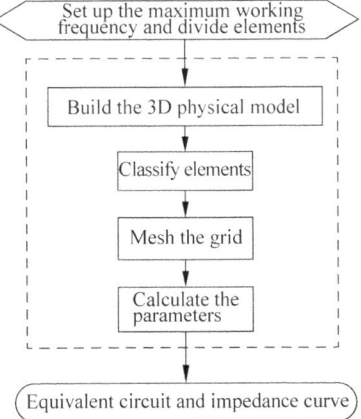

Fig. 4.13 The procedure of extracting stray parameters

more elements, computation cost and the complexity. Therefore, after meeting the precision requirement, the element numbers should be minimal. In general, the element length is defined as

$$d < 0.01\lambda \tag{4.1}$$

Here $\lambda = c/f_{max}$, the electromagnetic wavelength at the maximum frequency.

(2) Separate different circuits in the same topology as independent elements.
(3) Element divisions need facilitate the calculation of the stray inductance and capacitance, making the equivalent circuit concise and direct.

For the simple copper-bar loop, measurements of the impedance characteristics indicate the first resonant point happens ~50 MHz. Therefore, when employing such frequency to model the copper loop, the maximum element length is 60 mm. The whole copper bar is divided into 25 elements (S_1–S_{25}), as shown in Fig. 4.12a. The element separation is 0.01 mm.

Step 2. Build the 3D physical model

Based on the element divisions, copper-bar shape, size and material, the physical model could be built in three dimensions, as shown in Fig. 4.12b.

Step 3. Classify elements

The elements to be calculated and grounded are determined in this step. Since only stray parameters of the copper bar are needed, other elements like the shield and the impedance analyzer's ground are all modeled as the ground element. The copper bar is the unit to be calculated. According to the current direction, the current source is

Fig. 4.14 Grid meshing

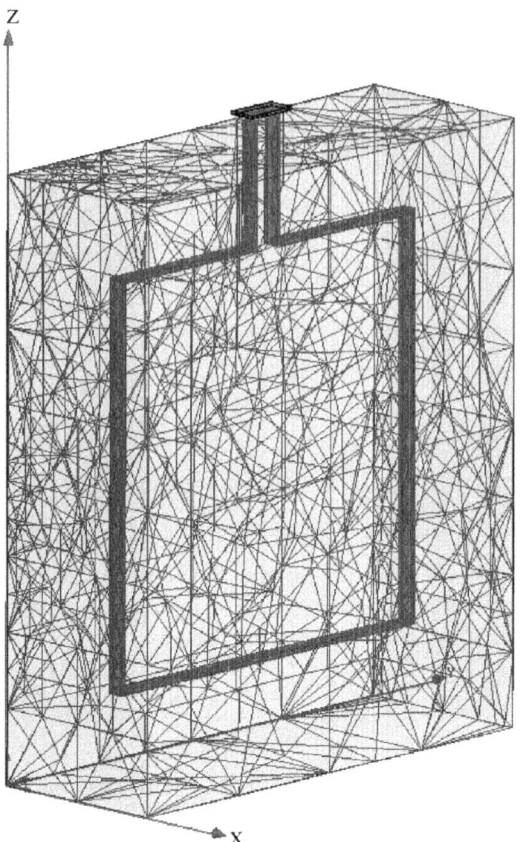

imposed to the element as the excitation for electromagnetic calculation in following steps.

Step 4. Mesh the grid

To consider the skin effect and proximity effect, the skin depth δ could be applied to mesh different layers of the copper bar. The meshing result is shown in Fig. 4.14. Here δ is calculated by the conductivity σ and permeability μ.

$$\delta = 1/\sqrt{\pi \ f_{max}\mu\sigma} \tag{4.2}$$

For the copper, $\sigma = 5.8 \times 10^7$ S/m and $\mu = \mu_0 = 4\pi \times 10^{-7}$ H/m.

Step 5. Calculate parameters

The stray resistance, inductance and capacitance will be calculated and generated as the matrix in the form of RLC. The calculated charge density is shown in Fig. 4.15, which indicates that the closer to the excitation the more concentrated the charge. Due to the skin effect, the closer to the corner the higher the charge concentration.

Step 6. Draw equivalent circuit and impedance curve

To simulate the impedance curve, the RLC matrix needs be exported to the circuit simulation software, such as PSpice to create the equivalent circuit. Figure 4.16 shows the equivalent circuit for element S_1 and S_2, each of which is a T-type circuit. For the element S_1, the T circuit is symmetric, with 1/2 of the self-stray inductance and resistance, $L_{1/2}$ and $R_{1/2}$ evenly distributed on the left and right side, respectively. The parasitic capacitance to the ground, C_{11}, is connected to the circuit middle point. The stray capacitance between S_1 and S_2, C_{12}, is located between middle points of the two circuits. M_{12} is the stray mutual inductance of two circuits.

Based on the equivalent circuit, PSpice could scan the frequency and generate the impedance curve of the equivalent circuit, as shown in Fig. 4.17 when the frequency varies from 100 kHz to 180 MHz. Some errors exist between the simulation and the

Fig. 4.15 Distribution of the charge concentrations

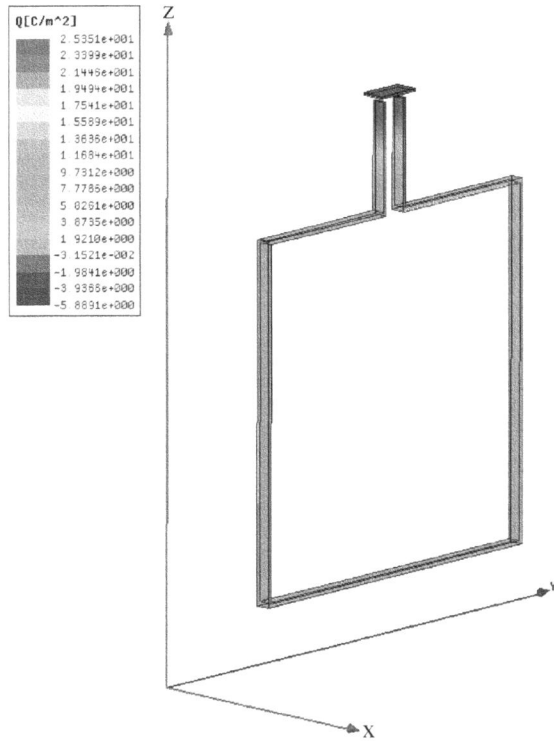

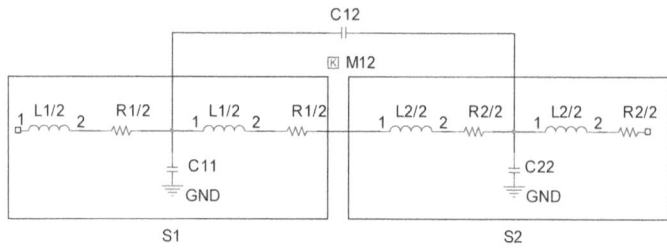

Fig. 4.16 The element equivalent circuit in PSpice

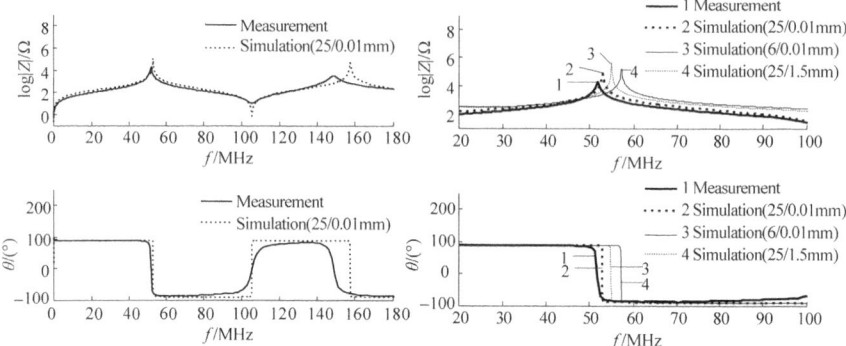

Fig. 4.17 Impedance curve of the rectangular copper-bar loop

experiments, which are mainly due to (1) precision of the element division, since the larger the element size the bigger the error, and (2) distance between elements, given the larger the distance the larger the error. When dividing the copper bar into 25 units with 0.01 mm distance in between, the simulation results and measurement have aligned amplitude and phase. Around the first resonance point the relative error of the amplitude is <10% and that of the phase angle is <6%. Therefore, it can be seen the PEEC method is suitable for the extraction of stray parameters in large-scale conductors, which is a perfect candidate for modelling bus bars in high-power converters.

4.2.3 Simplification of Stray-Parameter Extractions in the Complex Structures

In the actual converter analysis, the modelling and extraction of the bus-bar parameters are more difficult than simple-structured ones. The challenges remain at

(1) The actual bus bars in the high power converter have large dimension, irregular shapes, and complex working environments, all of which make the 3D modelling difficult. Therefore, the simplification of the modeling is the must;

(2) The TCT is highly related to time constants of the objective transients. Different time constants determine different topologies. In the commutation process, the frequency of all bus bars might vary. However, if stray parameters of all bus bars are calculated based on the same frequency, not totally aligned with the actual working condition, the calculation error will exist, which is subject to be analyzed and amended.

(3) Stray parameters of the bus bars are abundant and the equivalent circuit is complex, obstructing the system simulation and analysis. Analysis on the sensitivity of those parameters must be given to simplify the equivalent circuit.

Therefore, without sacrificing the precision, we could simplify some minor elements in bus bars of actual converters, for the simplicity of system modelling. In the large-scale bus-bar modelling, some small elements will complicate the grid meshing without altering the final analysis result. Such small parts include bolts and nuts, assembly holes of multi-layer bus bars, gaps and interconnections. All of these minor parts are not negligible in actual bus bars, however, could be simplified to enhance the efficiency of parameter extractions using the PEEC method, which is one of the most critical problems in the complex bus-bar modelling of power electronic converters.

Assembly holes of multi-layer bus bars can be taken as an example to illustrate the simplification of modelling and related impact on the accuracy. Assume the radius of the hole is r. The related resistance, capacitance and inductance could be swept with different valued r. Set the initial radius $r = 10$ mm, the end value of 30 mm, the step of 5 mm, and the working frequency of 30 MHz. Theoretical analysis shows that the variable hole radius results in variable parameters, especially the resistance and the mutual capacitance. The variation of absolute values of the self-inductance and the mutual inductance is small. However, the relative difference varies drastically, which is critical to the characteristics of the equivalent circuit.

Equivalent-area method is applied to simplify the modeling of assembly holes. The equivalent area could be calculated by the empirical Eq. (4.3), which results in a relatively low error between the simplified model and the actual model. The essence of such method is to add some aligned areas based on the original area. Figures 4.18 and 4.19 show the comparison between calculated C_{12} and L_1–M_{12} with actual values.

$$\Delta S = \pi (r - 5.5 \, \text{mm})^2, r > 10 \, \text{mm} \tag{4.3}$$

It can be seen with the equivalent-area method, the error between the simplified model and the original model drops significantly, for both the mutual capacitance and the equivalent inductance. As a summary, a small hole on the bus bars could be neglected. A large hole, if needed, could be modelled and simplified by the equivalent-area method to facilitate the meshing. The similar approach could be adopted when

Fig. 4.18 Capacitance based on the improved modelling

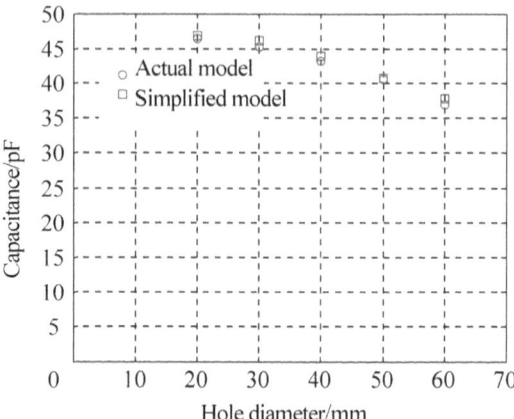

Fig. 4.19 Inductance based on the improved modelling

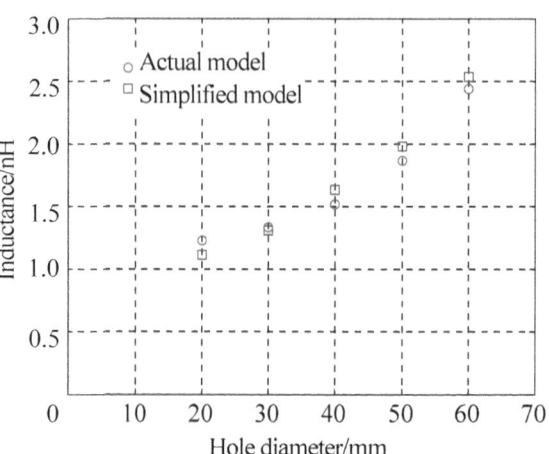

modelling the gap and layer-to-layer structure, yielding the high accuracy when simplifying complex bus bars.

The simplification of the stray-parameter extraction is essentially a tradeoff between the computation load and accuracy, a comprehensive problem involving multiple factors, such as the mechanical structure, short-timescale transients, charge and current distribution in circuits.

4.3 Analysis of Stray Parameters in IGBT Based Converters

4.3.1 Impact of Stray Parameters on the IGBTs in the Power Converter

In the power electronic converter, stray parameters in the main-power loop highly affect the reliability and performance in the transient commutation process. The switch will be the first to be affected. With the increment of power ratings, requirements on the voltage ratings, current ratings, insulation and thermal are increasing as well, which usually require large-scale switches, power modules and cooling systems. It then enlarges the size of interconnections and the distance between switches, which in return increases the stray inductance along the loop and highly impacts transient characteristics of power switches.

1. **Restraints of stray inductance on the IGBT applications.**

Such restraints can be experimentally analyzed. The test bench is shown in Fig. 4.20, where the switching processes could be measured with the bus-bar structure altered. Experimental comparisons could be used to qualify the impact of the stray inductance on the switches in the power electronics converter. The IGBT module used in Fig. 4.20 has the rating of 1200 V/450 A.

The stray inductance of the commutation loop is related to the bus-bar loop area. The longer the bus bars or the bigger the loop area, the larger the equivalent inductance. Hence the length and area of the bus-bar loop could be adjusted to change the stray inductance.

Three types of bus bars are experimentally compared through testing their inductance impact on the switching performance. These three designs are: compact multi-layer bus bars, small-area long bus bars, and large-area long bus bars. Their stray inductances increase progressively. To fully examine the leakage inductance,

Fig. 4.20 The test bench for the influence of stray inductance on IGBT

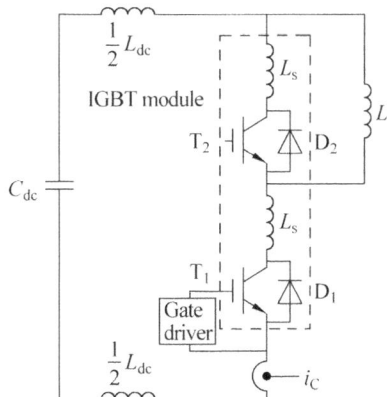

no snubber circuit is added in the test bench shown as Fig. 4.20. To further verify the effectiveness of the snubber circuit, the small-area long bus bars are equipped with the snubber and get tested at the same condition, which qualitatively analyzes the suppression on the stray inductance by the snubber circuit. Experimental waveforms on such three types of bus bars are shown in Fig. 4.21.

Figure 4.21a shows the switching waveform with compact multi-layer bus bars and the DC-bus voltage of 600 V. Figure 4.21c shows the switching waveform with the large-area long bus bars and the DC-bus voltage of 400 V. Figure 4.21b and c show the tests under the small-area long bus bars, the DC-bus voltage of 500 V, with and without snubber circuits, respectively.

The stray inductance will impose the electric stress on the switch in the turn-off process, i.e., creating a voltage spike across the switch. At the same turn-off current, the larger the stray inductance, the higher the voltage spike thereby the more potential to damage the device. As shown in Fig. 4.21b and c, the maximum voltage values are approaching to 1200 V, the maximum voltage rating of the switch. Increasing either the DC-bus voltage or output current is prone to destroying the device. From

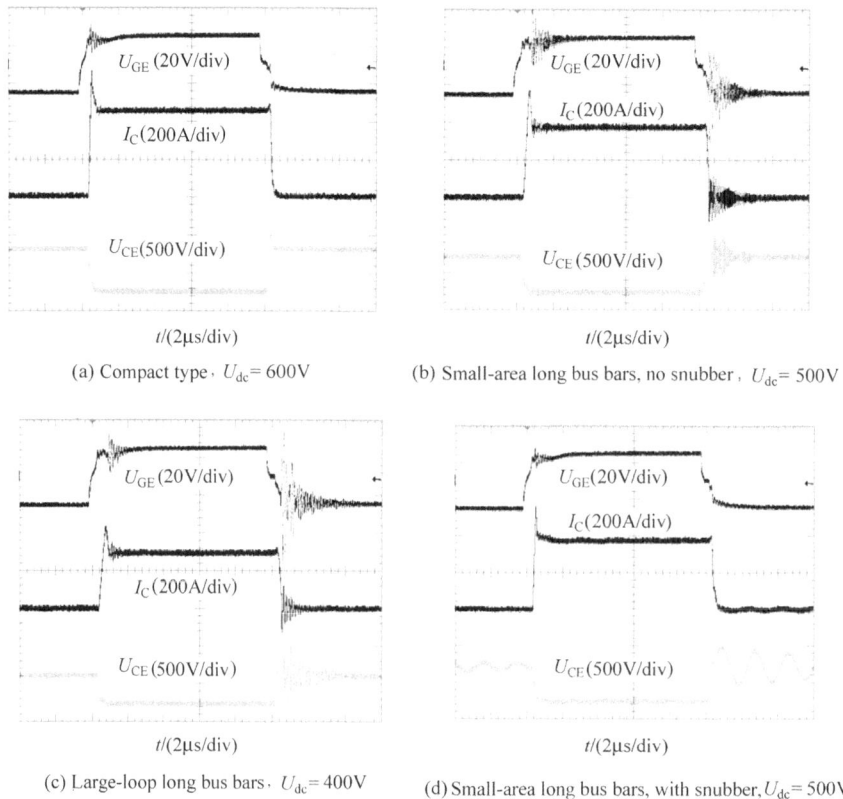

(a) Compact type, $U_{dc}= 600$V

(b) Small-area long bus bars, no snubber, $U_{dc}= 500$V

(c) Large-loop long bus bars, $U_{dc}= 400$V

(d) Small-area long bus bars, with snubber, $U_{dc}= 500$V

Fig. 4.21 Experimental waveforms of the stray-inductance impacts on switching performance

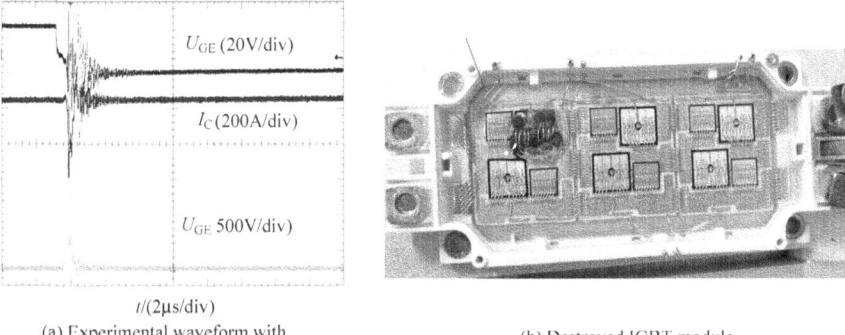

(a) Experimental waveform with long bus bars and large loop

(b) Destroyed IGBT module

Fig. 4.22 The overvoltage failure caused by the stray inductance

the aspect of the switch safety, the stray inductance of the DC-bus loop restrains the converter capability of the DC-bus voltage and output current.

At the same time, a large stray inductance on the DC-bus loop will create the gate-voltage oscillation during the switching actions, which has the positive correlation with the stray inductance. Such oscillation on one hand will create and spread EMI through the coupling circuit between the gate-drive power supplies and other peripherals, which tends to weaken the system EMC. On the other hand, it increases the failure risk of devices, given such mode is an uncontrollable abnormity. To secure the switch reliability, the reduction of the DC-bus stray inductance in the converter is rewarding.

Adding a snubber circuit will suppress the voltage spike induced by the stray inductance, which further reduces the gate-voltage oscillation, as shown in Fig. 4.21d. Compared to Fig. 4.21b, at the same DC-bus voltage and turn-off current, the equipment of the snubber circuit reduces the voltage spike to 1/3 of the original even though the DC-bus voltage shows some fluctuation.

2. IGBT failure caused by the stray inductance

Some destructive tests are carried out to investigate the impact of the stray inductance. With the test bench shown in Fig. 4.21c, the large-loop-area long-bus-bar system is given a 600 V DC-bus voltage. In the switching process, the turn-off voltage peak reaches 1700 V, higher than the claimed breakdown voltage thereby creating the device failure. The failure process is shown in Fig. 4.22a. The destroyed IGBT module is shown in Fig. 4.22b.

Three chips are paralleled to form one IGBT switch. Two such switches are in series connection to form one leg. The stray inductance for the chip further away from the DC-bus capacitor is larger than the other two, which explains why it undertakes a higher voltage stress and gets destroyed first.

3. **IGBT failure indirectly caused by the stray inductance**

Another destructive mode is indirectly caused by the stray inductance. Oscillations occur among the DC-bus inductance, snubber circuits and switches during switching actions. The switch is the source of the excitation, which determines the oscillating frequency is no higher than the switch equivalent upper-limit frequency. All the EMI related is conductive. Shown in Fig. 4.23 are two bus-bar structures with their equivalent circuits, both of which adopt the multilayer design to reduce the stray inductance.

The advantage of the structure A lies on the ease of assembly and test, a compact design of the overall converter, and a relatively small enclosure. The disadvantage is a long bus bar creating different commutation loops for different IGBT modules. The test results indicated a poor EMC performance. Structure B reduces the bus-bar length, provides an identical commutation loop for each module, and further embraces the better EMC. The disadvantage is the lack of the compactness thereby a larger enclosure.

Stray capacitance exists among the controller, the gate of the switch, the ground of the auxiliary circuits and the bus bar. Such capacitance provides the channel for the common-mode (CM) EMI, which can be converted to the differential-mode (DM) disturbance, influences the system reliability and worsens the EMC. Shown in Fig. 4.24a is the experimental line-line voltage of a motor fed by an inverter, with the bus-bar structure shown in Fig. 4.23a. Drastic oscillations emerge under the high-current operation, which further interferes with the system communication. Such discounted system reliability most likely results from the EMI disturbance caused by the inappropriately designed bus bar. In the long-term operation of the inverter, switch damages occurred, as shown in Fig. 4.24b. Among three paralleled IGBT

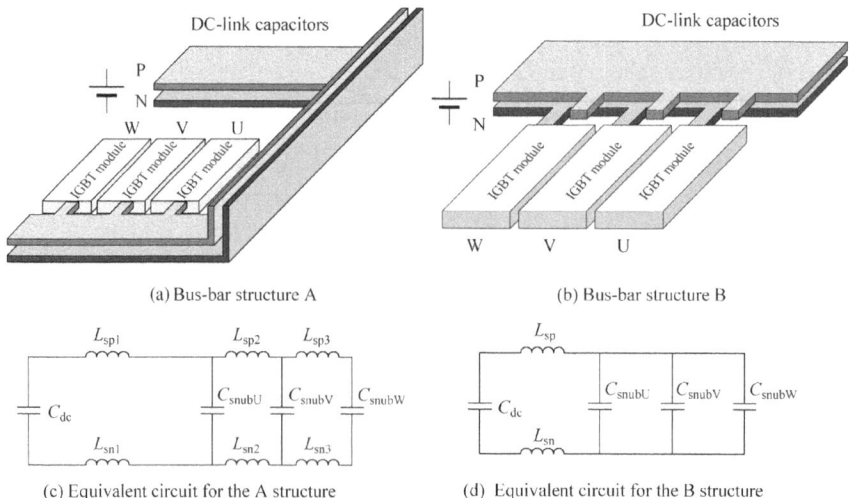

(a) Bus-bar structure A   (b) Bus-bar structure B

(c) Equivalent circuit for the A structure   (d) Equivalent circuit for the B structure

Fig. 4.23 Two bus-bar structures and their equivalent circuits

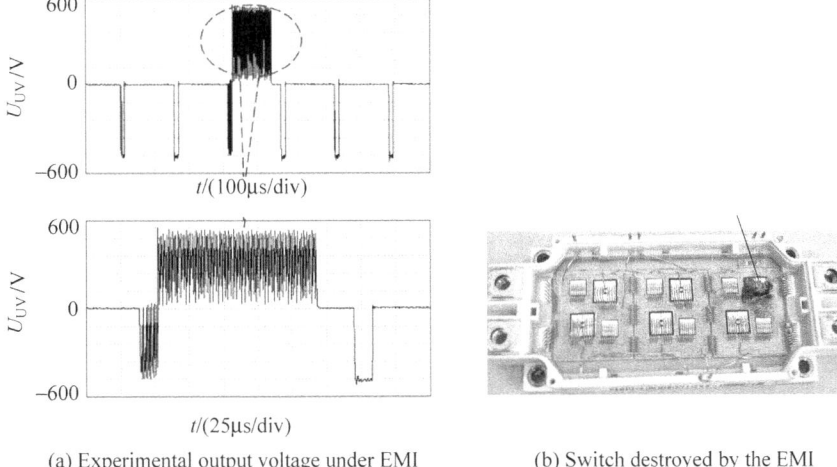

(a) Experimental output voltage under EMI   (b) Switch destroyed by the EMI

Fig. 4.24 EMI caused by the bus-bar layout

switches, the destroyed chip is located furthest away from the DC-bus capacitor. This chip is closest to the EMI source and bears the majority of the disturbance when the DC-bus voltage oscillates, which further induces the gate-voltage oscillation and creates the failure.

With the bus-bar structure B, EMI source is suppressed and the disturbance coupling is weakened with the enhanced EMC and reliability. Therefore, for the bus-bar design the less the stray inductance the better. Meanwhile the component placement needs be optimized to reduce the coupling between the bus bar and other peripherals vulnerable to noises. All of these help enhance the system EMC.

The DC-bus-bar stray inductance is a critical influential factor of switching transients, which makes the main-power circuit of the converter non-ideal. It changes the switch electric stress and amplifies the impact of di/dt and du/dt in commutation processes. For the high-power VSCs, DC bus bars are the must for the system main-power loop. Investigating the impact of the bus-bar inductance in the transient process is of importance for the converter reliability and performance.

4.3.2 Modelling of DC Bus Bars in the IGBT Based Converter

Multilayer bus bars are widely used in the IGBT based converter. Merits of such design include the significant reduction of the bus-bar stray inductance, compact design and ease of thermal dissipation. Therefore, modelling such bar bus remains as a focus.

The conventional extraction of stray parameters treats all conductors of bus bars as one solid piece. The PEEC method will be used to build its equivalent circuit for the further circuit simulation and performance analysis. However, for the IGBT converter, the large-scale multilayer bus bars have their own specialties, in terms of extracting and modelling stray parameters:

(1) Layers of bus bars are very close to each other, approaching to the limitation of the mechanical fabrication, assembly, and insulation;
(2) The bus-bar structure is highly related to the DC-bus capacitor;
(3) The bus-bar area is relatively large and the current flow inside are complex.

All the above characteristics impose some constraints on the extraction and modelling of the multilayer-bus-bar stray parameters in the IGBT based converter. Here we use one 315 kW converter to illustrate the design, modelling and analysis process of a large-scale low-inductance multilayer bus bar. For such a two-level 315 kW/380 VAC inverter, four 1200 V/450 A IGBT power modules are paralleled to form one main switch, as shown in Fig. 4.25. Instead of considering the current balance among IGBTs, here we focus on the impact of the bus bar on each IGBT module.

From the circuit topology point of view, the DC bus bar connects the DC-bus capacitor with IGBTs. With flat IGBT modules, the mechanical design of the bus-bar interface with IGBTs is relatively simple. For the DC-bus capacitor, given the maximum operational voltage of this two-level inverter is above 700 V while the rated voltage of conventional aluminum capacitors is 400–450 VDC, series connected

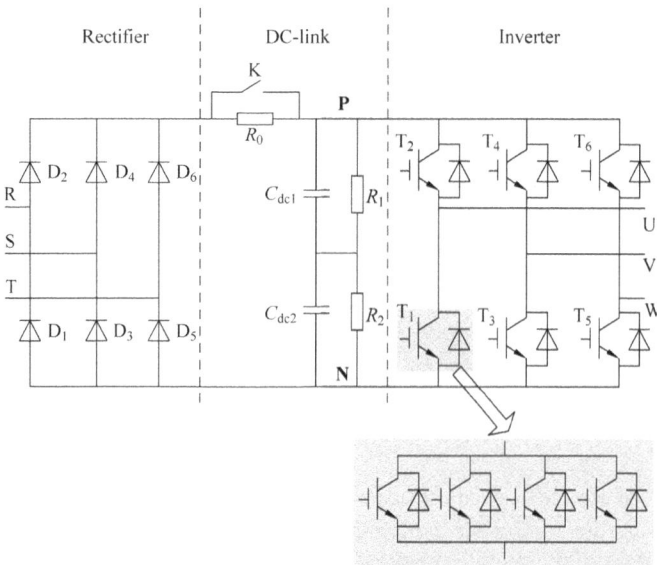

Fig. 4.25 The main circuit of one 315 kW/380 VAC two-level inverter

capacitors are needed. Furthermore, three voltage-potential bus bars are needed, i.e., DC+, DC− and 0. The two-layer DC bus bars are shown in Fig. 4.26a, where the top layer is the DC+ and the bottom layer is DC− and 0. The three-layer design is shown in Fig. 4.26b, where the top layer is DC+, middle layer is DC− and the bottom layer is 0. Both types of bus bars adopt the insulation pad between different layers for the electrical isolation. Assembly holes are used with bolts tightening DC-bus aluminum capacitors.

Prior to the fabrication of DC bus bars, the PEEC method can be employed to calculate related stray parameters based on the design dimension. The bus bars then can be further modelled with software. Meshing and simplification of such modelling is skipped here. Both types of DC bus bars have the similar placement of aluminum capacitors. To bridge with latter analysis, we divide both bus bars into four segments, with each segment illustrated in Fig. 4.26. A comparison of the stray inductance for each segment in each bus-bar structure is shown in Table 4.2. All these parameters are extracted by the PEEC method.

Both bus bars have the same equivalent circuits, as shown in Fig. 4.27, though parameter values might vary due to the structural difference. Here L_{bar1} is the stray parameter between the first segment of the bus bar and DC-bus capacitors, i.e., $L_{bar1\_P} + L_{bar1\_N}$. L_{bar2_*}, L_{bar3_*} and L_{bar4_*} are the stray inductance of corresponding segments, respectively. L_{IGBT} is the stray inductance of the IGBT internal. With such

Fig. 4.26 Two types of DC bus bars

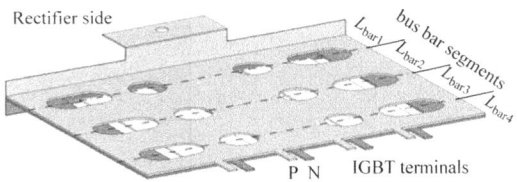

(a) Two-layer structure

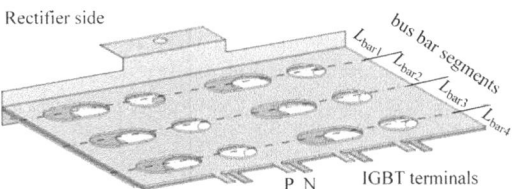

(b) Three-layer structure

Table 4.2 Parameter comparison of two DC bus bars

	L_{bar1} (nH)	L_{bar2} (nH)	L_{bar3} (nH)	L_{bar4} (nH)	Total (nH)
Two-layer	21.82	13.12	8.52	28.49	71.95
Three-layer	17.22	5.42	5.56	26.58	54.78

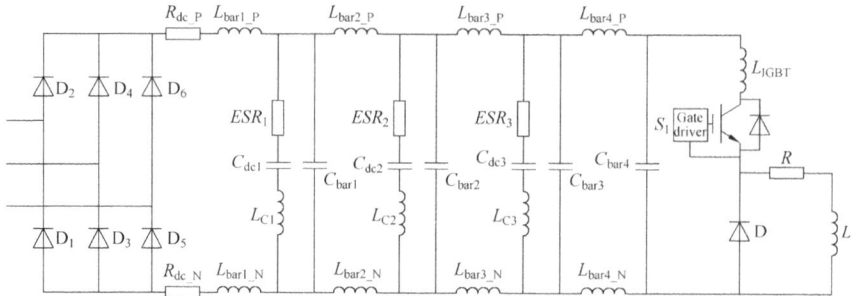

Fig. 4.27 The DC-bus-bar model of one IGBT converter

model of DC bus bars, both simulation and experiments could be used to analyze the impact of stray parameters.

With two different bus-bar structures, the IGBT converter of Fig. 4.25 is tested under a DC-bus voltage of 600 V and a turn-off current of 400 A. Both scenarios of with and without snubber circuit are studied with experimental waveforms shown in Fig. 4.28. Without the snubber circuit, turn-off voltage spikes of the IGBT vary with bus-bar structures. The turn-off voltage peak reaches 1000 V when the two-layer bus bar is used. In contrast, that with the three-layer structure is slightly lower, i.e., 950 V. With the snubber circuit equipped, the turn-off voltage peak values for both structures are significantly reduced to 720 V.

As shown in Fig. 4.26, the common point of both types of bus bars is that component terminals are not totally overlapped with bus bar. Such region is located at the negative DC-bus. Their difference is the different number of layers, with lower inductance for the three-layer structure even though more copper is consumed.

Experimental comparison shown in Fig. 4.28a and c indicates that the voltage spike with the three-layer structure is smaller. The larger overlapping area between positive and negative DC bus of such a three-layer design yields a lager mutual inductance thereby less leakage inductance internal of the DC capacitor array, i.e., lower L_{bar2} and L_{bar3}, which explains why its performance is better than the two-layer structure. However, at the most critical zones, i.e., connectors between the bus bar and the switch, no perfect alignment exists for either of bus bars, yielding a relatively large L_{bar4} in Fig. 4.27, i.e., stray inductance of the segment 4. Therefore, both types of bus bars introduce a quite large turn-off voltage spike of IGBTs. Given L_{bar4} is more critical than L_{bar2} and L_{bar3}, the merit of the three-layer design is not fully revealed, as validated in experiments. Even though the three-layer design reduces the overall inductance compared to the two-layer design, one noticeable issue is that the difference of IGBT switching transients with two types of bus bars is not obvious. Such analysis will be further extended in Sect. 4.5.2.

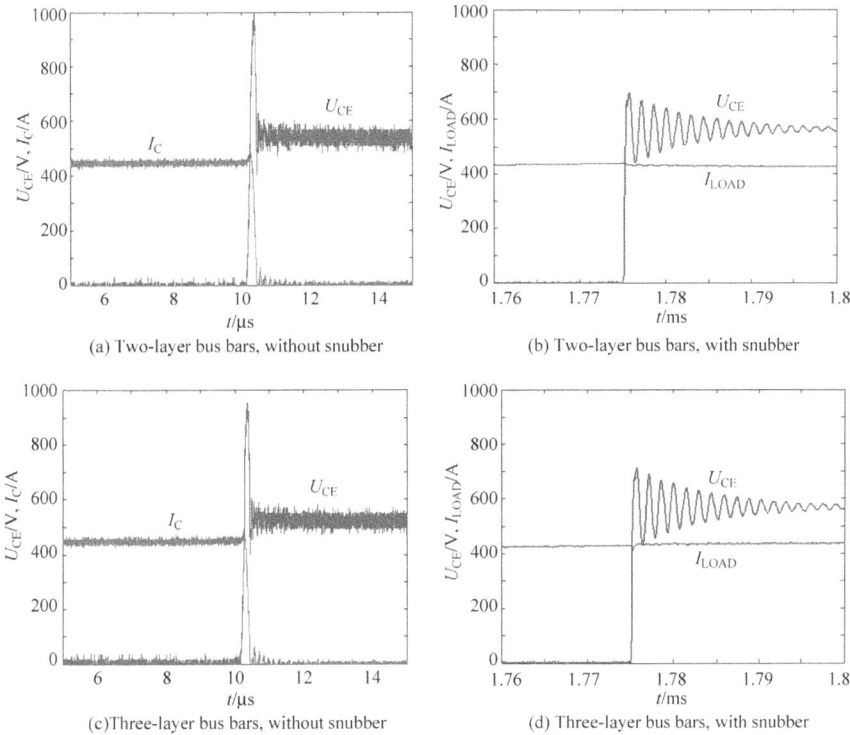

(a) Two-layer bus bars, without snubber

(b) Two-layer bus bars, with snubber

(c)Three-layer bus bars, without snubber

(d) Three-layer bus bars, with snubber

Fig. 4.28 Experiments of switching transients with two different bus bars

4.4   Analysis of Stray Parameters in IGCT Based Converters

4.4.1   Modelling of DC Bus Bars in a Three-Level IGCT Converter

Stray parameters in an IGCT based converter are different from those in IGBT based converters, mainly due to the device package. The mechanical assembly of power electronic converter is closely related to the switch package. In general, two types of DC bus bars are prevalent in the high-power converter. One is for the stack-clamping devices, the other is for modular devices. The former one is usually used in the high-current applications, where IGCTs and IGBTs especially IGCTs are preferred. In this section we take one 4500 V/600 A IGCT based three-level converter as an example to analyze and model stray parameters of its bus bar.

The actual system adopts the forced-air cooled clamping heatsink for both IGCTs and FRR diodes. By imposing some pressure, the anode and cathode of power devices are attached to the heatsink through the fixture, forming a switch stack. The heatsink is used for the thermal dissipation as well as the electricity conducting thereby acting as part of bus bars. All connectors between switches and external circuits are through the heatsink. In general, with the same topology for all legs, their stacking structure is the same. Some inverters apply one commonly used snubber for three legs. For the sake of the symmetry, the snubber diodes D_{CL1} and D_{LC2} are placed on the top and bottom of the middle phase V, respectively. As shown in Fig. 4.29, devices in the Phase-V stack from the top to the bottom are D_{CL1}, T_{V1}, D_{V1}, T_{V2}, T_{V3}, D_{V2}, T_{V4} and D_{LC2}, respectively. Here T_{Vi} (i = 1, 2, 3, 4) are IGCTs. D_{V1} and D_{V2} are neutral clamping diodes.

Bus bars in the inverter must comply with requirements of voltage and current capability, insulation and heat dissipation. They are meanwhile restrained by the device shape, dimension and locations. Due to the symmetry of the topology and switch stacks, bus bars are bilaterally symmetric and top-to-bottom symmetric. Based upon the location, bus bars could be categorized as snubber bus bars, connecting bus bars, and phase bus bars, as shown in Fig. 4.30. Here snubber bus bars connect the snubber capacitor C_{CL} with the diode D_{CL}, connecting bus bars are used to link the snubber with three legs, and phase bus bars are connections and heatsinks of four IGCTs and two clamping diodes.

Usually bus bars adopt the three-layer structure, placed in the middle of the converter. In addition, cables are important media as well to mainly connect passive components, for instance, the DC-bus capacitor and snubber components. To involve cable stray parameters in the equivalent circuit, when measuring the impedance of passive components we need include cables as well.

Fig. 4.29 Phase stack of the three-level IGCT inverter

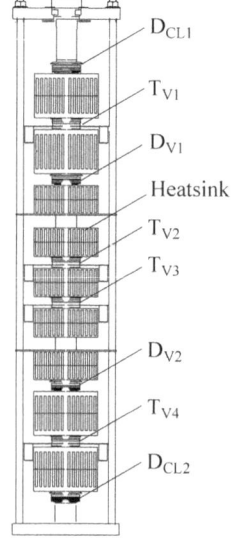

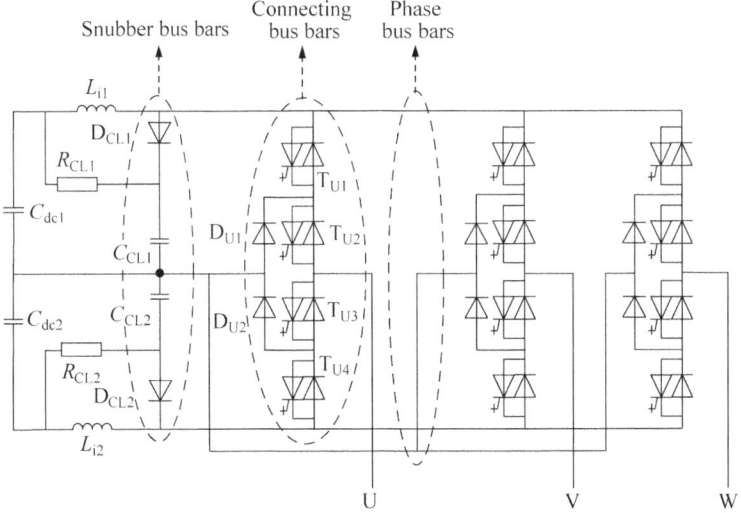

Fig. 4.30 Categories of bus bars for an IGCT based inverter

Compared to those bus bars used in the IGBT module based converters, the IGCT converters have following specialties:

(1) More conducting bus bars with the high independence from each other;
(2) Heatsinks conduct the current as well, which bring extra stray parameters;
(3) The DC-bus bars of three-level converters have three layers, each of which has large current flowing through;
(4) The snubber circuit is shared by three phases, the location of which determines the effectiveness of the stray-parameter suppression.

In summary, the IGCT based high-power converter has features of large size, high complexity, irregular shapes and strong diversity, etc.

4.4.2   Transient Commutation Topology

The highest frequency of the TCT is determined by the fall time t_f of the switch turn-off current, i.e.,

$$f_{max} = 1/(2\pi t_f) \tag{4.4}$$

The typical IGCT turn-off characteristics with related parameters are shown in Fig. 4.31. Here CS is the IGCT gate-control signal, I_T is the IGCT current, U_G is the IGCT gate voltage, I_{TGQ} is the maximum turn-off current, U_{DSP} is the first turn-off voltage peak, U_{DM} is the second turn-off voltage peak, U_{dc} is the DC-bus voltage and t_{d_off} is the IGCT turn-off time delay. t_f is the time interval during which the current drops from $0.8I_{TGQ}$ to $0.3I_{TGQ}$, around 300 ns. Therefore, the maximum frequency for the bus-bar modelling is 530 kHz, yielding the maximum bus-bar unit length of 500 mm.

The 3D model is shown in Fig. 4.32, with simplifications made below.

(1) Neglecting assembly holes and bolts on bus bars;
(2) Changing irregular bus bars into regular bus bars;
(3) Ignoring dielectrics, thermal vias on the heatsink and interactions of surrounding cables.

The whole bus-bar model contains 50 units, including 18 connecting bus bars, 25 phase bus bars and 7 snubber units. The inverter enclosure uses Aluminum coated zinc plates, with the dimension of 900 mm × 560 mm × 1800 mm. It acts as the shielding cabinet and the ground. The equivalent circuit containing bus-bar stray parameters is shown in Fig. 4.33.

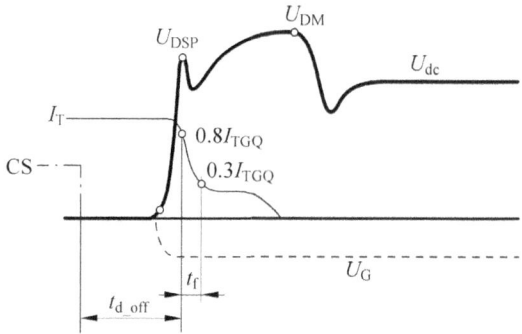

Fig. 4.31 IGCT turn-off characteristics and parameters

Fig. 4.32 3D model of the IGCT inverter

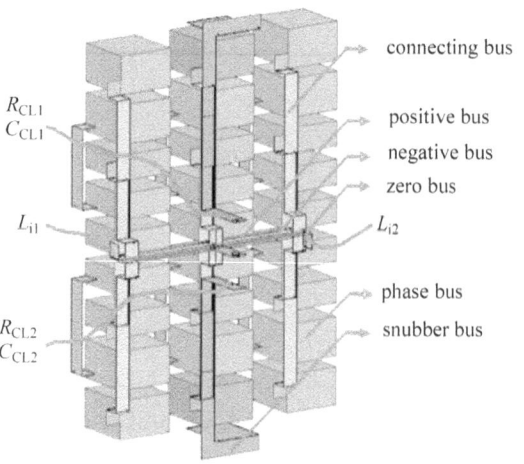

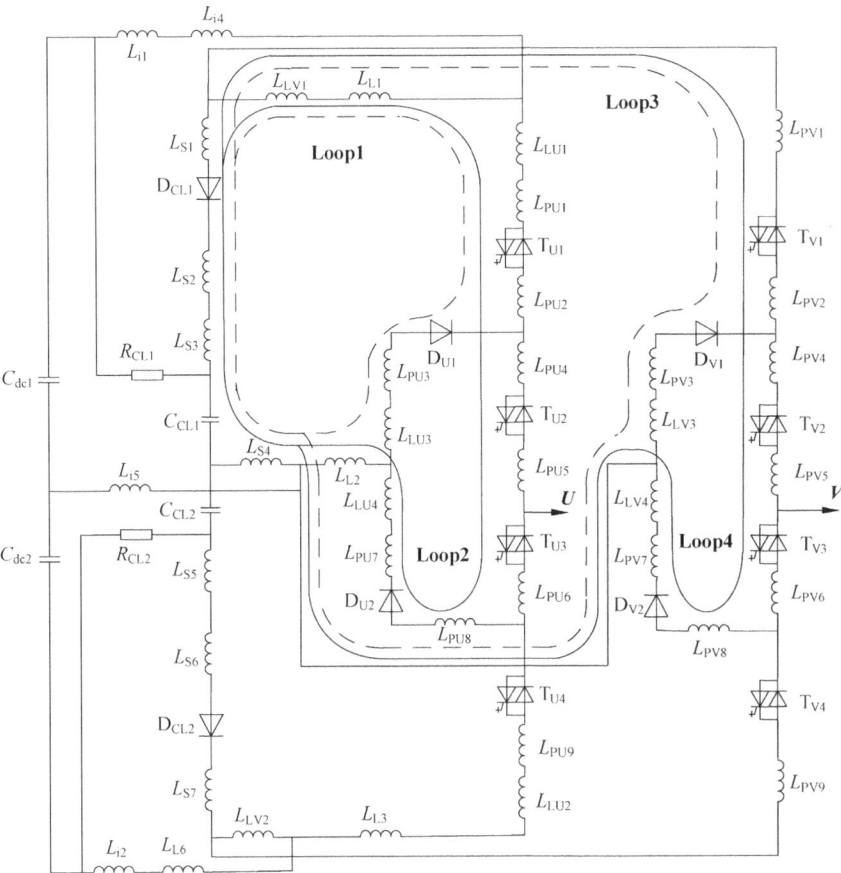

Fig. 4.33 Equivalent circuit including stray parameters

Due to the symmetry of the topology and circuit, only equivalent circuits of Phase U and V are given. For the simplicity, only the self-inductance of each bus-bar element is provided in Fig. 4.33. Here L_{L1}–L_{L6} are the inductance values of connecting bus bars, L_{S1}–L_{S7} are the inductance values of snubber bus bars, L_{PU1}–L_{PU9} are the inductance values of Phase-U bus bars, and L_{PV1}–L_{PV9} are the values of the Phase-V bus bars. Loop1, Loop2, Loop3 and Loop4 are the inductance values of the commuting loops for T_{U1}, T_{U3}, T_{V1} and T_{V3}, respectively. Obviously after stray parameters are taken into account, commutation loops look different from the original ideal topology, i.e.,

(1) Each IGCT in the same leg has different commutation loops with different stray parameters. The inner IGCT faces more stray inductance in the commutation loop than the outer IGCT, i.e., $L_{LOOP2} > L_{LOOP1}$ and $L_{LOOP4} > L_{LOOP3}$;

(2) The same-positioned switches in different legs have different loops with differ-
    ent stray parameters. For instance, the Phase-U and -W IGCTs have larger stray
    inductance than Phase-V IGCTs, i.e., $L_{LOOP1} > L_{LOOP3}$ and $L_{LOOP2} > L_{LOOP4}$.

In transient commutations of T_{U1}, three scenarios of the inverter-bus-bar current
need be considered:

(1) In the Loop1 (T_{U1}-D_{U1}-D_{CL1}-C_{CL1}), T_{U1} turns off with the current rapidly chang-
    ing, shown as the dashed line;
(2) The load current does not change with the Phase-V and Phase-W bus-bar current
    remaining stable, shown as the solid line;
(3) Other bus bars, such as the bottom half of the Phase-U and the top part of the
    Phase-V/W remain open-circuit. Namely the bus bar current is zero.

While all parameters of bus bars are calculated based on the same frequency,
the specialty of the bus-bar current distribution complicates the extraction of stray
parameters. Therefore, when modeling bus bars, we could consider the whole-bus-
bar modeling or commutation-bus-bar method.

The whole-bus-bar modeling assumes all bus bars work at the highest frequency.
All bus bars will be meshed and calculated with only the shielding cabinet as the
grounding unit. Such method allows all bus-bar units to be calculated and analyzed,
however, its demerits are obvious, e.g., heavy computation load and slow calculation
speed.

The commutation-bus-bar modeling only calculates working bus bars at the highest switching frequency. All other bus bars staying idle are modeled the same as the
shielding cabinet, i.e., the grounding unit. The merit is its fast calculation speed. The
demerit is that each commutation loop has to be modelled independently.

The stray inductance and capacitance of Loop1 has been calculated with the above
two methods, respectively (Tables 4.3 and 4.4). Here we only consider the inductance
of >30 nH and capacitance of >6 pF. The calculation results indicate that

(1) Mutual inductance and mutual capacitance exist among bus bars, due to the
    multi-layer bus-bar structure. For example, the inductive and capacitive coupling
    coefficients between bus bar elements L_{L1} and L_{L2} are 86 and 58%, respectively;
(2) The inductance relative error between the commutation-bus-bar method and
    the whole-bus-bar method is <10%. The capacitive relative error, however,
    is <1.3%. Therefore, other bus bars have some impact on the stray inductance
    of the commutation loop with little influence on the stray capacitance.

Table 4.3 Stray inductance of Loop1

Inductance	Whole-bus-bar	Commutation-bus-bar	Inductance	Whole-bus-bar	Commutation-bus-bar
L_{L1}	138.28	134.41	L_{S4}	44.03	43.82
L_{L2}	125.05	117.06	M_{L1L2}	−113.93	−107.81
L_{LU1}	223.11	235.93	M_{LU1LU3}	−86.28	−93.13
L_{LU3}	95.16	101.14	M_{LU1LV1}	−43.65	−46.92
L_{LV1}	217.18	224.03	M_{LU1S3}	31.94	34.88
L_{S2}	98.5	99.33	M_{LV1S2}	−54.02	−55.89
L_{S3}	172.79	176.66	M_{LV1S3}	−134.17	−138.93

Table 4.4 Stray inductance of Loop1

Capacitance	Whole-bus-bar	Commutation-bus-bar	Capacitance	Whole-bus-bar	Commutation-bus-bar
C_{L1}	7.44	7.55	C_{L1L2}	34.6	34.6
C_{L2}	38.69	39.01	C_{LU1LU3}	13.06	13.2
C_{LV1}	16.16	16.31	C_{PU1PU2}	10.89	11.02
C_{PU1}	7.17	7.24	C_{PU2PU3}	10.18	10.29
C_{PU2}	7.36	7.4	C_{LV1S2}	6.59	6.67
C_{PU3}	13.9	14	C_{LV1S3}	19.8	19.83
C_{S1}	16.52	16.6	C_{S1S2}	8.93	9.01

Furthermore, we could merge some bus bars to reduce the bus-bar numbers and facilitate the circuitry simulation. Three criterions are to be followed:

(1) Merge small-stray-parameter bus bars into large-valued ones;
(2) Merged elements need be in the same loop;
(3) Merged elements are adjacent in terms of the physical location and circuit topology.

To compare the results before and after merging bus bars, we assume IGCTs and diodes in the loop are all short-circuited and replace the snubber capacitor with a current source. In this way we could calculate the impedance characteristics of bus bars in the commutation loop when looking into two terminals of $C_{CL}+$ and $C_{CL}-$. Shown in Fig. 4.34 is the simplified Loop1 equivalent circuit in PSpice.

Since both L_1 and L_2 are geometrically vertical to other bus-bar units, little as 2% of the inductive and capacitive coupling is to be considered. The coupling between L_1, L_2 and other bus bars then could be ignored.

The simulated impedance characteristics of Loop1 with and without simplifications are shown in Fig. 4.35. As seen, the simplified model has the similar impedance characteristics to the original model, with relative errors of the amplitude and angle both less than 5%. This proves the feasibility of the simplification method and makes the established equivalent circuit more concise.

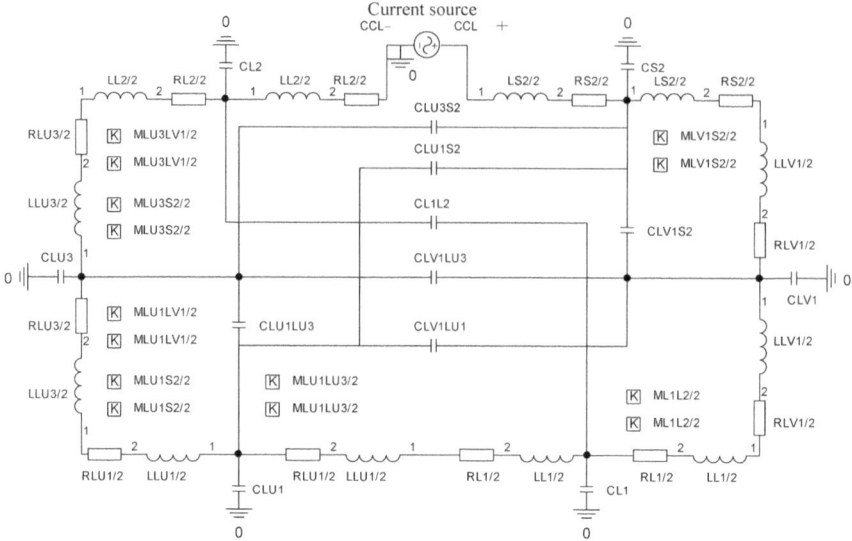

Fig. 4.34 The simplified Loop1 equivalent circuit in PSpice

In such analysis, we can further compare simulation results of the IGCT model with stray parameters to experimental waveforms, in order to verify the fidelity of stray parameters, as shown in Fig. 4.36. Experimental results could be used to reversely calculate stray parameters and compare with the 3D modelling, revealing 10% error. It is worth pointing out that using the IGCT turn-off voltage to reversely calculate the stray inductance lacks the accuracy, due to the influence of the diode

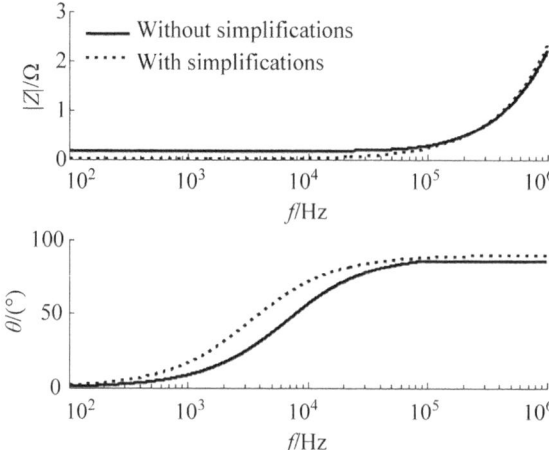

Fig. 4.35 The simulated impedance characteristics of Loop1

Fig. 4.36 Comparison of the simulation and experiments

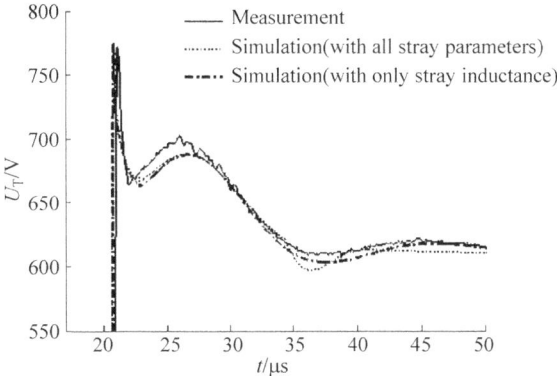

forward voltage. However, with the modeling of stray parameters in IGCT high-power converters, the transient model with only stray inductance could effectively emulate and predict the transient commutation process.

4.5 Quantitative Analysis and Optimization of Stray Parameters

4.5.1 Evaluation of Stray Parameters in the IGBT Module Based Converter

According to the bus-bar characteristics of the IGBT converter, optimization of its bus bars is restrained by

(1) Mechanical fabrication, assembly and insulation. The relative position and clearance are difficult to change with little optimization room left;

(2) Large area of any single conductor. Therefore, only some local zones of the bus bar are subject to optimizations.

Based on the above, a segment-based evaluation method is applied to evaluate stray parameters and optimize multilayer bus bars in IGBT converters. For the example in Sect. 4.3.2, the three-layer bus bars have 24% lower inductance than the two-layer bus bars. However, its suppression of the IGBT turn-off voltage spike is very limited. Hence we need analyze the influence of bus-bar stray parameters on the IGBT performance per segment, to seek the best location for the further optimization.

A simulation needs be firstly carried out based on the DC bus-bar model of the IGBT converter, as shown in Fig. 4.27, where stray inductance in each segment is adjusted thereby changing the overall stray inductance of positive and negative bus bars. Here we give five simulation waveforms, with the inductance of 5, 10, 20,

40 and 80 nH, respectively. Each simulation only changes stray parameters in one segment while keeping other segments by the default values calculated by PEEC. The simulation results are shown in Fig. 4.37.

The inductance L_{bar1} of the first segment bus bar, connecting the rectifier output with the first array of the DC-bus capacitors, has nearly no effect on the switch turn-off voltage. Being closer to switches, other stray inductance L_{bar2}, L_{bar3} and L_{bar4} are becoming more and more critical, as shown in Fig. 4.37b, c and d, respectively. The inductance of the segment closest to the switch terminals, L_{bar4}, plays the most important role on the turn-off process.

Simulation results also show that filtering capacitors suppress the stray inductance of upper-level bus bars. The stray inductance between the last piece of bus bars and switch module terminals, as the most sensitive parameter of DC bus bars, affects the turn-off voltage most. Therefore, when designing multilayer bus bars, the optimization design should be carried out particularly at the area close to switch terminals. When placing power components of the converter, shortening the distance between switch terminals and filter capacitors should be put on the top priority to minimize the stray inductance in between.

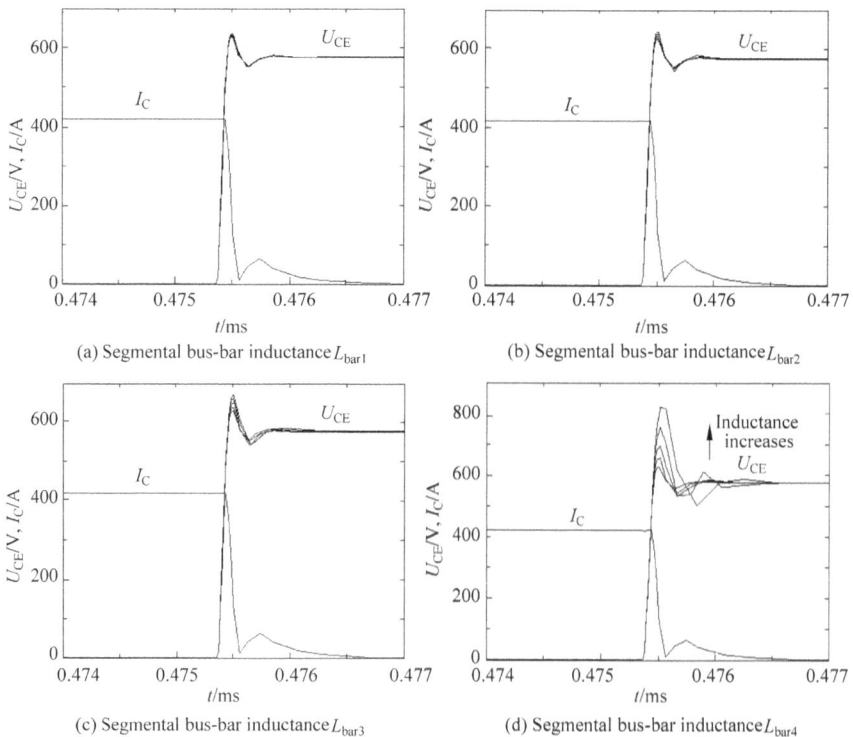

Fig. 4.37 Simulated influence of the stray inductance on the IGBT turn-off process

In addition, the simulation shows that the impact of capacitors close to the rectifier output on the bus-bar performance is negligible. Therefore, the optimization of the related part is less important. Copper strips could be adopted between the rectifier output and filtering capacitors instead of multilayer bus bars, which saves the space and material consumption of the overall converter.

We can define the evaluation index of stray parameters as

$$F_{Li} = \partial U_p / \partial L_{bari} (i = 1, 2, 3, 4) \tag{4.5}$$

Here U_P is the IGBT turn-off voltage peak, which could be replaced with other variables if needed. L_{bari} is the inductance of the ith-segment bus bar. It can be further expressed as the sensitivity of the IGBT turn-off voltage peak over the stray parameters. For the case in this section, we have

$$F_{L4} \approx 3.4(V/nH) >> F_{L3} > F_{L2} > F_{L1} \tag{4.6}$$

Based on such case analysis, different weights are given to bus-bar segments accordingly for the further optimization. Different from the IGCT converter, the optimization of the IGBT converter is not determined by the location, but the local details.

4.5.2 Bus-Bar Optimization for an IGBT Module Based Converter

Based on the stray-parameter evaluation index of the IGBT module based converter, let's revisit bus-bar parameters of two different structures again. The overall inductance of the three-layer bus bar is reduced by 24% compared to the two-layer structure, however, the fourth-segment inductance L_{bar4}, the largest indexed part, is 7% lower, which explains why the overall inductance drops without changing the turn-off voltage much. Therefore, following two basic rules need be complied with:

(1) Improve the insulation technology based on the basic insulation requirement;
(2) Focus on terminal connections of the IGBTs and bus bars and make such parts as compact as possible.

Overall bus bars could be optimized with the integrative design. The top layer is the positive bus bar and part of the zero bus bar, and the bottom layer is made of the negative bus bar and part of the zero bus bar. Instead of using the insulation pad between different bus bars like the two-layer and three-layer structures, the integrative bus bar adopts the dielectrics to secure the insulation between positive and negative bus bars as well as providing the fixture. The prototype sample of such bus bars is shown in Fig. 4.38, where printed circuit boards are adopted. Corresponding to the previous analysis, four segments are divided as well to provide a head-to-head comparison with conventional two-layer and three-layer bus bars, as shown in Table 4.5.

These three structures have the same locations of the DC-bus capacitors and IGBT connectors. Compared to the two-layer structure, the integrative bus bar has 59% lower overall inductance and 87% lower L_{bar4}, the part with the largest evaluation index. Compared to the three-layer structure, those numbers are 46 and 86% lower, respectively. Hence in addition to the reduction of the overall inductance, the weight of L_{bar4} in the overall inductance drops significantly as well.

Experiments are carried out for the three types of bus bars on the test bench, where power supply, mechanical terminals, temperature controller, control boards and current/voltage measurement are included. During the test, DC-bus capacitors, IGBTs, control pulses, DC-bus voltage and load current are consistent. The control signals are double pulses, the DC-bus voltage is 600 V and the load current is 450 A. The IGBT turn-off voltage with and without the snubber is shown in Fig. 4.39, respectively. Given the system assembly is so compact that no room is provided for current probes, the load current instead of the IGBT current is measured.

The IGBT turn-off voltage peaks are compared in Table 4.6. Without the snubber capacitor, the two-layer conventional design is used as the reference. The relative values of the overall stray parameters, the four-segment stray inductance and the voltage spike (peak minus the DC-bus voltage) are shown in Table 4.7. Even though both the three-layer and integrative designs are the improvement of the two-layer design, more advantages of the integrative bus bars are revealed once the evaluation index is applied.

For the no-snubber scenario shown in Table 4.7, under the same test condition, the difference of the turn-off voltage spike ($U_{peak} - U_{dc}$) results from the variation of the stray inductance. Compared to the two-layer structure, the overall stray inductance of the three-layer design drops to 76%, which only results in 88% of the original voltage spike. This is because the most important part L_{bar4} is only reduced to 93% of the original. For the integrative bus bar, the overall stray inductance drops to 41%,

Fig. 4.38 The optimized IGBT DC bus bars

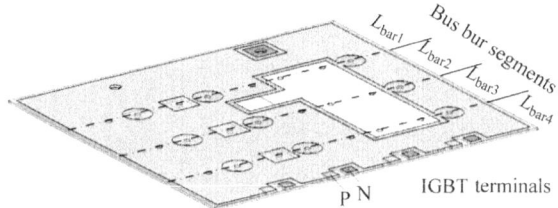

Table 4.5 Comparison of the optimized stray parameters

	L_{bar1} (nH)	L_{bar2} (nH)	L_{bar3} (nH)	L_{bar4} (nH)	Total (nH)
Two-layer	21.82	13.12	8.52	28.49	71.95
Three-layer	17.22	5.42	5.56	26.58	54.78
Optimized	11.85	11.07	2.93	3.68	29.53

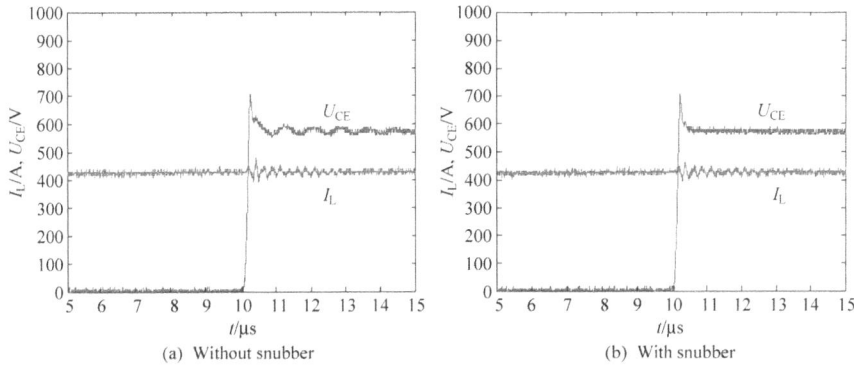

Fig. 4.39 Experimental results of the IGBT turn-off process with the optimized bus bars

Table 4.6 Experimental comparison among three design approaches

	Without snubber (V)	With snubber (V)
Two-layer	1000	725
Three-layer	950	725
Optimized	705	715

Table 4.7 Stray inductance versus the turn-off voltage spike

	Overall stray inductance (%)	L_{bar4} (%)	$U_{peak} - U_{dc}$ (%)
Two-layer	100	100	100
Three-layer	76	93	87.5
Optimized	41	13	26.3

yielding a 26% of the original voltage spike. This is mainly because L_{bar4} is further reduced. Such optimization essentially is based upon the evaluation index.

After the snubber circuit is equipped, the turn-off voltage peak value for both the two-layer and three-layer designs is 725 V, reduced significantly however still higher than the no-snubber integrative bus bar. For the integrative design, the IGBT voltage spike sees no change when equipped with the snubber circuit. On the contrary, the voltage oscillation is observed during the turn-off process. Therefore, the snubber circuit used for the conventional bus bars brings no benefits but negatives to the integrative bus bar.

The integrative bus bar effectively utilizes the space around the bus bar and component terminals to provide enough overlaps, which minimizes the local inductance thereby exhibiting the superior performance. The snubber circuit does not help reduce the voltage spike but introduces the oscillation, which means the stray inductance of the integrative design is already minimal and even less than the snubber parasitic inductance. It can meet the requirement without adding the snubber circuit. Without any voltage oscillation, the integrative design eliminates one potential EMI source and enhances the system EMC.

The comparison of three bus-bar design approaches reveals the superiority of the integrative bus bar, which reduces the stray inductance to the minimum due to the overlap of the critical zones, i.e., interface between the bus bar and components. For the conventional design using the multilayer bus bar, fixtures of the bus bar make it difficult to offer any overlaps. For those bus bars unable to use the multilayer structure, laminated connectors are recommended as shown in Fig. 4.40. Even though such laminated connectors can't compete with the integrative bus bar, the bus-bar performance can still be somewhat improved.

In the static electric field, laminated bus bars are equivalent to a planar capacitor, the capacitance of which is related to the dielectric between the bus bars. The integrative bus bar adopts the dielectric with the permittivity of 4.0, which guarantees the insulation and enlarges the stray capacitance. Such capacitance helps choke the turn-off voltage spike.

After meeting the insulation requirement, reducing the dielectric thickness will on one hand increase the mutual inductance of the positive and negative DC buses, and on the other hand enlarge the stray capacitance of the bus bar thereby enhancing the snubber performance. Selecting a high-permittivity dielectric also helps increase the stray capacitance. The problem is, the compact integrative bus bars are not good for the thermal dissipation. Tradeoffs must be made between the system performance and reliability to select the appropriate dielectric material and bus-bar distance.

As a summary, the design criterions of the bus bars in VSCs are:

(1) Due to the suppression of the upper-level bus-bar stray inductance by the filtering capacitors, that inductance between the rectifier output and the filtering capacitor affects little on the switch turn-off process;

(2) The stray inductance between the filtering capacitor and the switch terminals affects the most the switch turn-off voltage spike. Regardless of the bus-bar structure, the optimized design of this region is a must;

(3) If the component placement allows, shortening the distance between the filtering capacitor and the switch terminal needs be placed on the top priority, aiming at reducing the local inductance;

(4) Under the premise of the insulation, reducing the bus-bar distance and selecting the high-permittivity dielectrics will enlarge the mutual inductance and stray capacitance, which enhance the overall system performance;

(5) No turn-off snubber is needed for the integrative bus bars.

Fig. 4.40 The laminated connectors of the DC bus bar

4.5.3 Evaluation of the Stray Parameters in Clamping Pressed IGCT Based Converters

This section is focused on the evaluation of stray parameters in an IGCT based converter and its associated design rules. The goal is to simultaneously secure the switch reliability and improve switching transients through adjusting snubber parameters.

Figure 4.41 shows the amplified turn-off voltage of IGCTs T_{U1}, T_{U3}, T_{V1} and T_{V3}, respectively. Here the DC-bus voltage is 600 V and the load current is 65 A. To facilitate the test both the current and voltage are derated. Compared to the ideal IGCT turn-off where a voltage spike U_{DM} exists, an extra voltage spike U_{DSP} occurs at the early stage of the turn-off process. In addition, U_{DM} and the stable time are affected. Different stray inductances in different commutation loops result in different IGCT turn-off voltages, especially U_{DSP}.

The conventional analysis assumes the whole loop has only one stray inductance. In fact, the impact of inductance on each branch is different and even drastically different. Based on the TCT proposed in previous sections, commutation loops considering the stray inductance is shown in Fig. 4.42, where the stray inductance in five branches of the IGCT and diode is included. To clearly reveal the impact of each inductance on the switch performance, each inductance value is varied from 0.1 to 1 µH. Based on that we picture the IGCT turn-off voltage U_T and the clamping diode current I_{DCL} as Fig. 4.43.

As shown above, stray inductance of each branch highly influences the turn-off voltage waveform, e.g., peak and stable time for the commutation process. In details, it includes:

(1) At the first stage of the turn-off process, the IGCT current commutates to the FWD. A large current changing rate induces the first voltage spike U_{DSP} through the stray inductance L_{SIGCT}, L_{SDCL} and L_{SCCL}. Here the commutation loop is IGCT-D_{CL}-C_{CL}-FWD-IGCT. The larger the inductance the higher the voltage spike;

Fig. 4.41 Experimental turn-off voltage of one IGCT

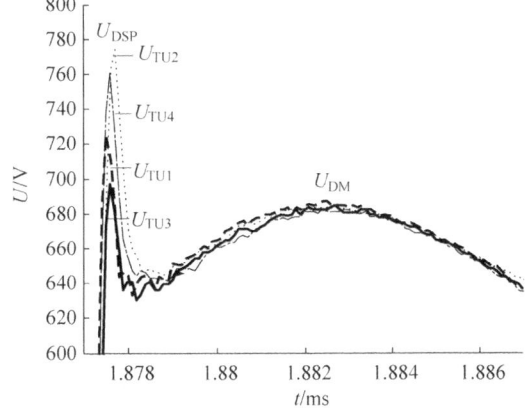

Fig. 4.42 The equivalent commutation loops with stray inductance

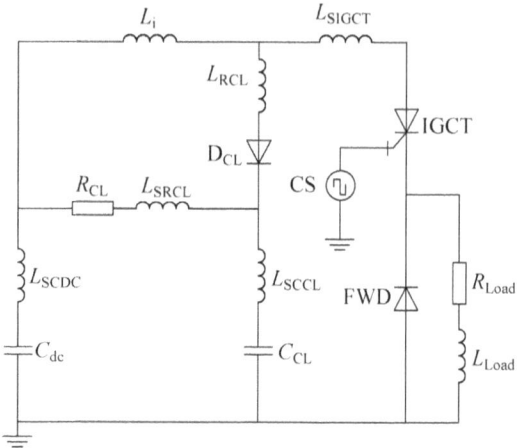

(2) At the second stage, the second voltage spike occurs due to the stray inductance L_{SRCL}, L_{SCCL} and L_{SCDC} in the loop of C_{DC}-R_{CL}-C_{CL}-C_{DC};

(3) Based on the current of the clamping diode, it can be seen that L_{SIGCT} does not affect the energy releasing time T_L of the current limiting inductor. However, all other inductances do, among which L_{SDCL} and L_{SRCL} affect the most, as shown in Fig. 4.43b and c.

To evaluate the impact of the stray inductance on the IGCT switching performance, the impact factor of the stray inductance is defined as below.

$$k = \Delta A / \Delta L_S \tag{4.7}$$

Here ΔL_S is the variation of the stray inductance, ΔA is the variation of IGCT turn-off parameters impacted by the stray inductance, e.g., U_{DSP}, U_{DM} and T_S. The related impact factors are k_{VDSP}, k_{VDM} and k_{TS} respectively.

Impact of the stray inductance on the IGCT turn-off voltage spike U_{DSP}, U_{DM} and T_S is simulated as Fig. 4.44. To facilitate the comparison, the stray inductance of each branch is varied from 0.01 to 0.2 L_i, i.e., 0.107 to 2.14 µH.

As shown in Fig. 4.44, the impact of the stray inductance in each branch on the IGCT turn-off performance is greatly diversified. Among all inductances, L_{SIGCT} and L_{SDCL} affect U_{DSP} the most, L_{SRCL} and L_{SDCD} affect U_{DM} the most, and L_{SRCL} and L_{SDCL} affect T_S the most.

Using the first-order linear equation $y = ax + by$ for the polynomials of above curves indicates the correlation coefficient is larger than 0.99. Therefore, when the stray inductance is 0.01–0.2 L_i, the IGCT turn-off parameters have linear relationships with the stray inductance. Based on Eq. (4.7), impact factors of the inductance in each branch are shown in Table 4.8. In this table, the impact factor of the stray inductance in the loop of IGCT-D_{CL}-C_{CL}-FWD-IGCT is $k_{VDMmax} = 140$ V/µH. L_{SDCL} and L_{SRCL} affect T_S the most with the maximum impact factor of $k_{TSmax} =$

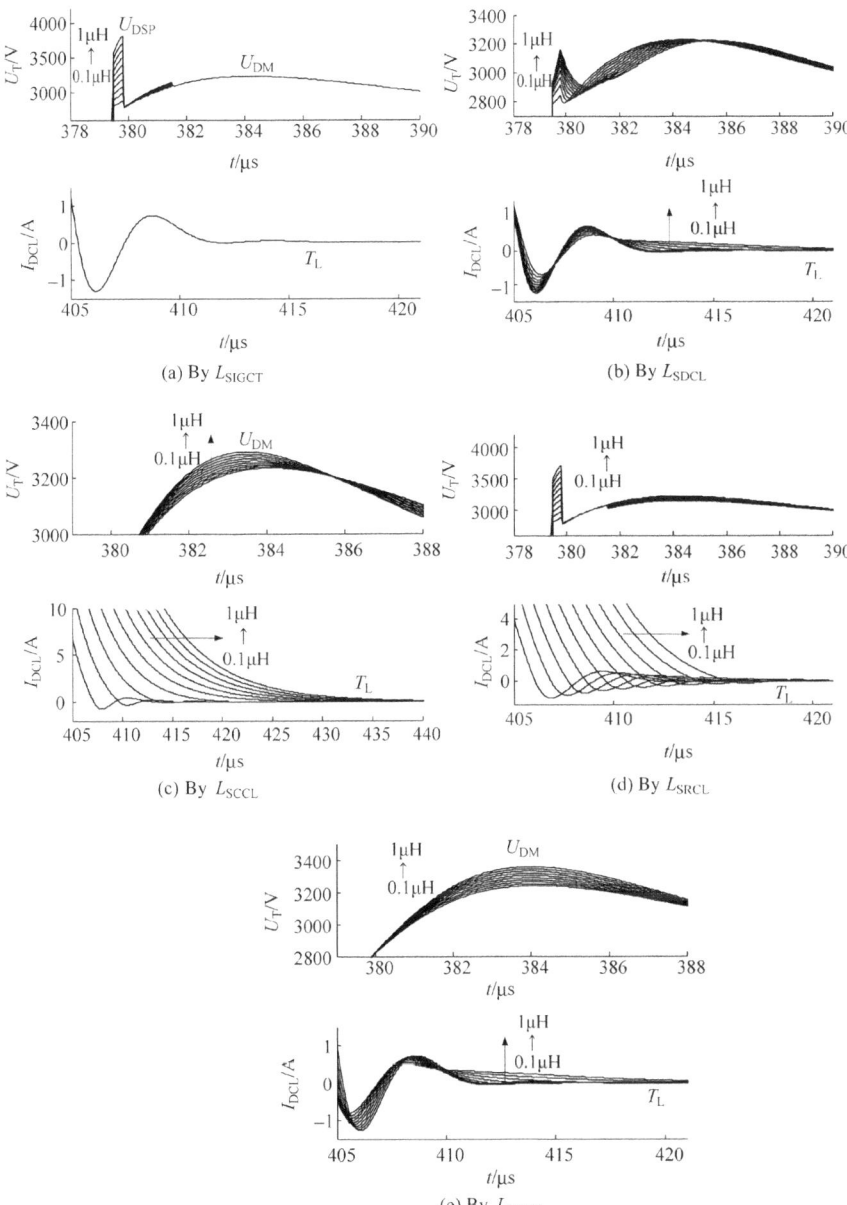

Fig. 4.43 U_T and I_{DCL} under the influence of stray inductance

8 μs/μH. Based on these impact factors, the correlation between the IGCT turn-off characteristics and stray inductance can be further evaluated.

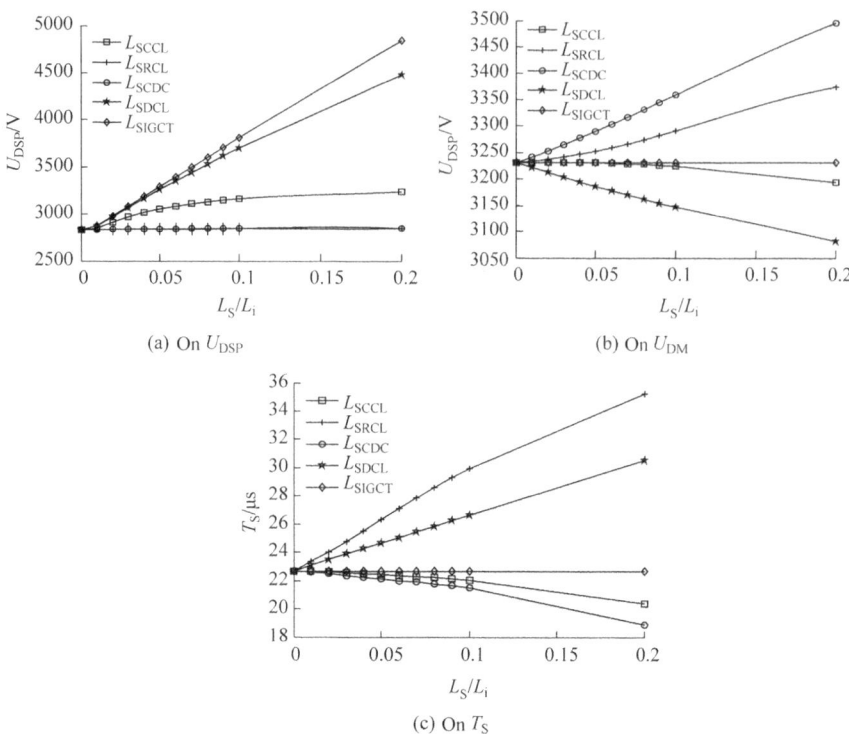

Fig. 4.44   Impact of the stray inductance on the IGCT turn-off performance

Table 4.8   Impact factors of stray inductance

	L_{SIGCT}	L_{SDCL}	L_{SCCL}	L_{SRCL}	L_{SCDC}
k_{VDSP} (V/μH)	1000	947	361	14	14
k_{VDM} (V/μH)	0	−91	−7	65	140
k_{TS} (μs/μH)	0	4.3	−0.7	8	−1.3

4.5.4   Bus Bar Optimization for the IGCT Based Three-Level Converter

For an IGCT based three-level high-power converter, bus bars undertake the high voltage and high current simultaneously, making the laminated bus bar an unqualified candidate for the requirements of voltage withstanding, insulation, mechanical and heat dissipation. In addition, the clamping structure of the IGCT gives high constraints on the assembly and thermal, further obstructing the application of laminated bus bars. Therefore, to connect the components in the IGCT based converter, the copper bus bars with the rectangular cross-section areas are the preference.

Abundant components plus the snubber circuit make such system more complex than the two-level converter. On the other hand, more bus bars applied further complicate the bus bar placement and design.

Take the Phase-U as an example. The leg commutation loop is shown in Fig. 4.45, when the first turn-off voltage spike of each switch appears. Current directions of each circuit branch are labelled. Here I_O is the bus-bar current and L_O is the bus-bar stray inductance. It is indicated that all commutations of the three-level converter happen between positive and zero bus bars, or between zero and negative bus bars. Outer-switch commutations and inner-switch commutations are two basic commutation modes.

(1) Outer-switch commutation, which is made of the IGCT and the clamping diode. Shown in Fig. 4.45a, c, e and g are the switching on/off commutation loops of outer switches T_{U1} and T_{U4}, respectively. Given such commutations involve outer switches, we call them outer-switch commutations.

(2) Inner-switch commutation, which involves the IGCT and the anti-parallel diode of the IGCT, as shown in Fig. 4.45b, d, f and h. They are related to switching actions of T_{U3} and T_{U2}, respectively, i.e., inner switches. Therefore, we name it as the inner-switch commutation.

Based on the analysis above, at the first voltage spike, each branch has the following features:

(1) All snubber bus-bar current increases in the certain direction, i.e., forward bias direction of the snubber diode;

(2) At the commutations of $P \leftrightarrow O$, when I_O flows into the neutral point O, the current amplitude reduces, otherwise, increases. Per commutations of $O \leftrightarrow N$, when I_O flows into the neutral point O, the current amplitude increases, otherwise, drops.

(3) At the commutations of $P \leftrightarrow O$, when the positive bus current flows into P, the current amplitude increases, otherwise, decreases. Per commutations of $O \leftrightarrow N$, when the negative-bus current flows into N, the current amplitude decreases, otherwise, increases.

Take the connecting bus bars B_P and B_O as examples, which connect P with T_{U1} and O with D_{U1}, respectively. Bus bars could be paralleled or overlapped. Here we define Fig. 4.46a as the bus-bar forward connection and Fig. 4.46b as the reverse connection.

The placement of the bus bar is aimed to minimize the impact of the stray inductance on the turn-off voltage spike. During the commutations of $P \leftrightarrow O$, four commutation loops shown in Fig. 4.45a–d all involve B_P and B_O. The stray inductance of these two bus bars when forward and reverse connected is shown in Table 4.9.

In Table 4.9, sign(M) represents the sign of the mutual inductance between two bus bars. When carrying the same-direction current, the sign is +, otherwise, −. Sign($dI_P/dt \cdot dI_O/dt$) represents the sign of the multiplication of two current changing rates. Only when both current changing rates are positive or negative, the sign is +, otherwise, −. "↓" for L_{eff} indicates the reduction of the stray inductance while "↑" means increment.

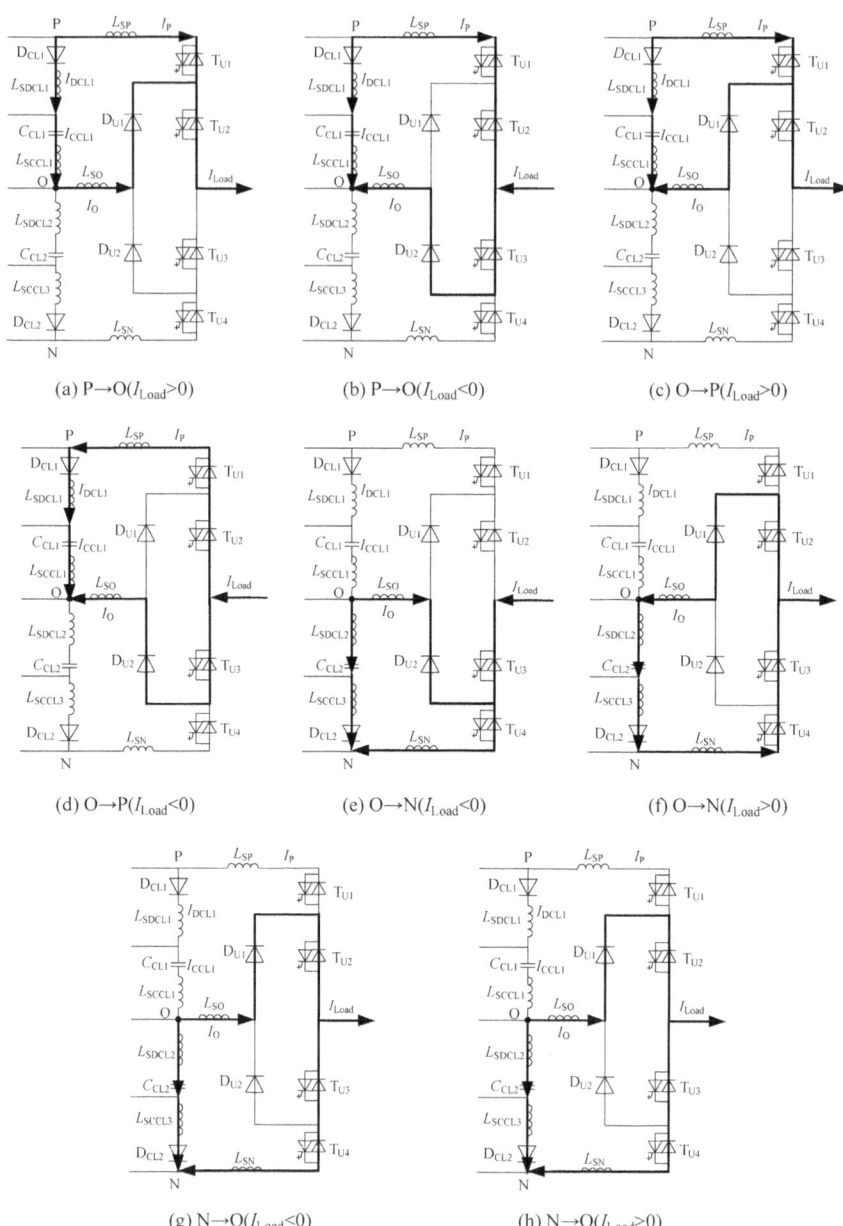

(a) P→O(I_{Load}>0)   (b) P→O(I_{Load}<0)   (c) O→P(I_{Load}>0)

(d) O→P(I_{Load}<0)   (e) O→N(I_{Load}<0)   (f) O→N(I_{Load}>0)

(g) N→O(I_{Load}<0)   (h) N→O(I_{Load}>0)

Fig. 4.45 Commutation loops of one leg in an IGCT based three-level NPC inverter

Therefore, the impact of the bus-bar placement on the stray inductance is determined by the current direction and changing rate, following the criterion as: when $\text{sign}(M)$ and $\text{sign}(dI_P/dt * dI_O/dt)$ have different polarities, the equivalent inductance is decreased, otherwise, increased.

As a summary, the proposed bus-bar design is shown as the flow chart of Fig. 4.47, where N represents the loop number involving two bus bars to be designed, K means commutation loops being analyzed, and L means the bus-bar layout method. $L = 0$ means no laminating design, 1 means the forward connection and 2 means the reverse connection.

The essence of the bus bar design is to first analyze the transient commutation processes, locate the first turn-off voltage spike, and understand the current direction and current changing rate. Secondly based upon the current of the bus bar we need decide the sign of the mutual inductance, in order to reduce the stray inductance. This will decide whether the laminated bus bars are required, forward or reverse connection.

It is worth pointing out that commutation processes of the three-level converter do not happen between positive and negative bus bars simultaneously, but between zero and positive or zero and negative bus bars. Therefore, the laminated design is only required between positive and zero bus bars, or negative and zero bus bars. No specific requirement is needed for positive and negative bus bars.

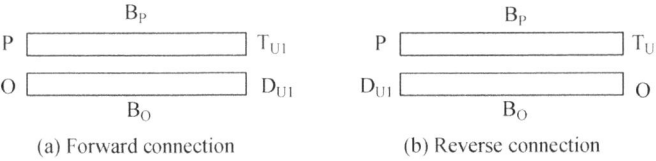

Fig. 4.46 Bus-bar placements

Table 4.9 Stray inductance with different bus-bar placements

Commutation loop	Connection	Current direction	di/dt	$\text{sign}(M)$	Sign $(dI_P/dt \cdot dI_O/dt)$	L_{eff}
(a)	Forward	$P \to T_{U1}$ $O \to D_{U1}$	$dI_P/dt > 0$ $dI_O/dt < 0$	$+$	$-$	$\downarrow$
	Reverse	$P \to T_{U1}$ $D_{U1} \leftarrow O$		$-$		$\uparrow$
(b) and (c)	Forward	$P \to T_{U1}$ $O \leftarrow D_{U1}$	$dI_P/dt < 0$ $dI_O/dt < 0$	$-$	$+$	$\downarrow$
	Reverse	$P \to T_{U1}$ $D_{U1} \to O$		$+$		$\uparrow$
(d)	Forward	$P \leftarrow T_{U1}$ $O \leftarrow D_{U1}$	$dI_P/dt > 0$ $dI_O/dt < 0$	$+$	$-$	$\downarrow$
	Reverse	$P \leftarrow T_{U1}$ $D_{U1} \to O$		$-$		$\uparrow$

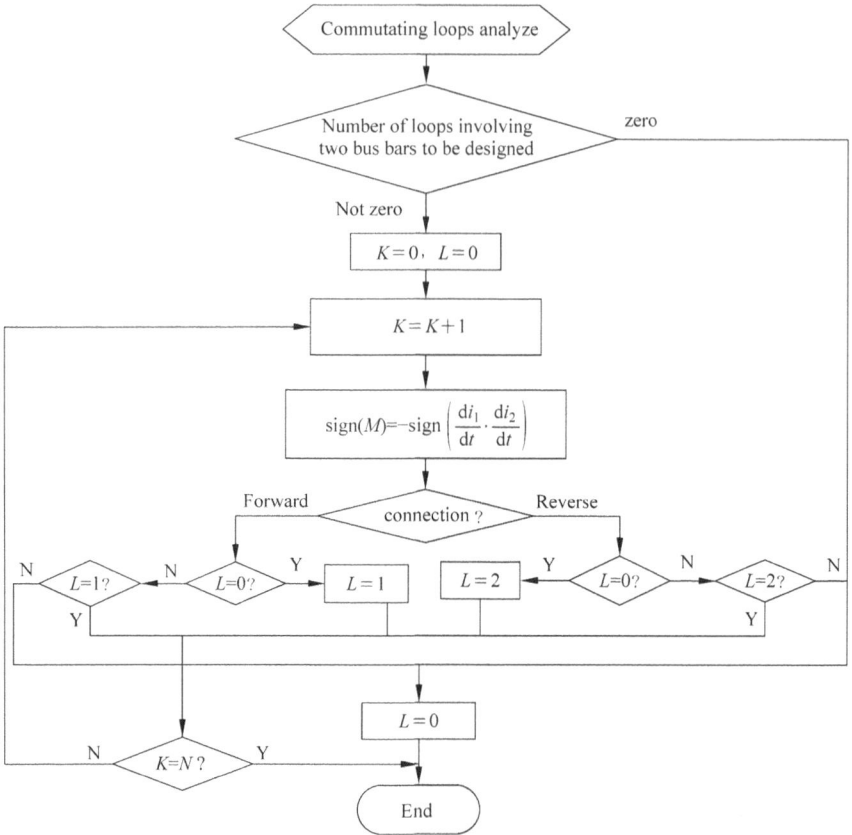

Fig. 4.47 Flow chart of the bus bar placement

Once the size and placement of bus bars are settled down, we need consider the connection between bus bars and components. Theoretically, any points on the bus bar could connectors. Usually we select connecting points based on the assembly convenience, distance and ease of the connector fabrication. Previous analysis indicates that stray inductance in different commutation loops has different impacts on transient characteristics. The connector of bus bars and components is exactly the electrical joint in the topology. It is possible for us to relocate the bus bars to different loops thereby changing their stray inductance, which is named as the relocation design of the bus bar. For example, on top of the bus bar S_2 located between the snubber capacitor C_{CL1} and snubber diode D_{CL1}, we need place the snubber resistor R_{CL1}. Two different design approaches are shown in Fig. 4.48.

The approach 1 assigns bus bar S_2 to the snubber diode branch, making L_{S2} become part of L_{SDCL}. The approach 2 assigns bus bar S_2 to the snubber capacitor branch, making L_{S2} become part of L_{SCCL}. According to the impact factors shown in Table 4.8, L_{SDCL} plays a far more important role than L_{SCCL}. Therefore, relocating the

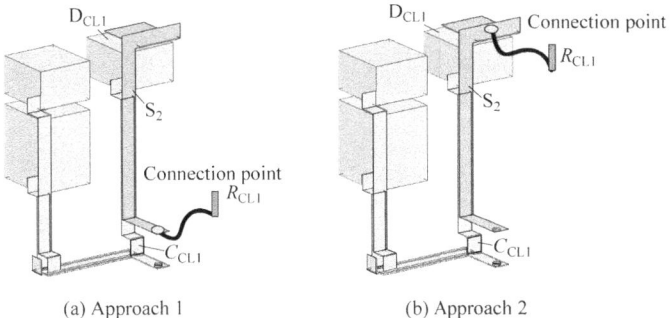

(a) Approach 1   (b) Approach 2

Fig. 4.48 Comparison of two design approaches

bus-bar inductance is highly critical. An optimized bus-bar relocation could assign appropriate stray inductance in the loop thereby minimizing its impact on the switch transient performance.

The relocation of bus bars based on the influence of the stray inductance roots in impact factors of stray inductance on the transient performance. Priorities of the stray inductance of all bus bars are studied first. During the optimization, the inductance is increased in low-priority bus bars and reduced in high-priority bus bars.

According to Table 4.8, priorities of L_{SIGCT}, L_{SDCL} and L_{SCCL} are shown as below:

(1) The top priority is L_{SDCL}, given U_{DSP} and T_S have strong positive correlations with L_{SDCL};
(2) The middle priority is L_{SIGCT}, which only impacts U_{DSP} without T_S;
(3) The lowest priority is L_{SCCL}, which affects all performance indexes the least.

Therefore, when relocating bus bars, if needed we could increase L_{SCCL} and minimize L_{SDCL}.

The conventional three-layer bus bar design and placement are shown in Fig. 4.49. Based on the above analysis, such design has following defects.

1. **Large stray inductance in the commutation loop.**

The commutation loop of the outer switch, Loop1, has the stray inductance mainly contributed by T_{U1} and L_{LU1}, the connecting bus bars between positive bus bars. Three layers of positive, zero and negative bus bars enlarge L_{LU1} with a relatively large area of Loop1. Such design meanwhile complicates the bus bar fabrication, assembly and maintenance.

The commutation loops of inner switches, Loop2 and Loop4, have inductance induced by L_{LU1} and phase bus bars between inner and outer switches, i.e., L_{PU4}, L_{PU6}, L_{PV4} and L_{PV6}. Due to the large distance between phase bus bars and L_{LU1}, loop areas are increased.

Fig. 4.49 The three-layer
bus bar design and placement

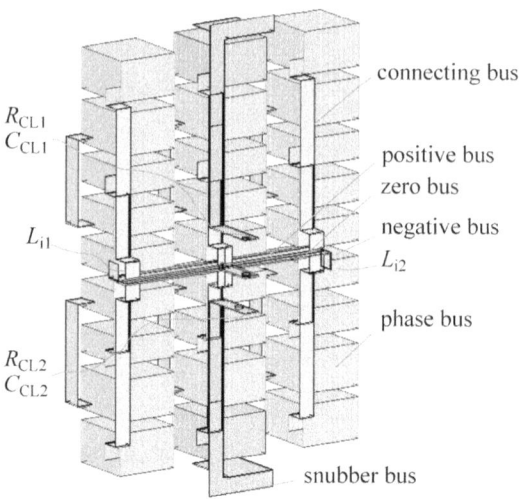

2. **Inappropriate relocation of the stray inductance.**

The lowest-priority inductance L_{SCCL} is the minimal in each loop. For the Loop1 and Loop3, the top-priority L_{SDCL} is the largest.

To verify the effectiveness of the proposed bus-bar relocation, we propose a novel two-layer laminated bus bar, with the placement and relocation shown in Fig. 4.50a. Different from the original three-layer bus bars, the newly proposed bus bars:

(1) Adopt two double-layer laminated bus bars, i.e., positive-zero bus bars and zero-negative bus bars. Such design minimizes L_{LU1} as well as the areas of Loop1 and Loop3. It also simplifies the bus-bar structure and facilitates the bus-bar fabrication and assembly;

(2) Overlap the phase bus bars on top of the connecting bus bars, further shrinking Loop2 and Loop4;

(3) Optimize locations of the snubber resistor, inductor and capacitor on the bus bar, based upon the bus-bar prioritization and relocation methodology. The equivalent circuit of Phase-U and Phase-V double-layer laminated bus bars is shown in Fig. 4.50.

All bus-bar inductance in the loops of three-layer and double-layer designs is calculated through the PEEC method and shown in Table 4.10. It can be seen that compared to the original three-layer bus bars, all high-impact-factor inductance is reduced, among which the inductance of Loop1 and Loop3 is reduced to 54%, respectively. For all three bus-bar loops using the double-layer approach with the inductance relocation, the top-priority bus-bar inductance L_{SDCL} is the minimal with the lowest-priority inductance L_{SCCL} increased. Therefore, the impact of the stray inductance on the switching transients is minimized.

To verify the effectiveness of the proposed double-layer laminated bus bars, switches T_{U1}, T_{U3}, T_{V1} and T_{V4} are selected for the multiple-pulse test. Simulation on the switching transients of such four switches with three-layer and double-layer bus bars is carried out as well.

The relationship between the first turn-off voltage spike ΔU_{DSP} induced by the stray inductance and the load current is pictured as Fig. 4.51. Shown in Fig. 4.51a are test results with three-layer bus bars. Figure 4.51b and c are simulation results of the voltage spike ΔU_{DSP} with the three-layer bus bars and two-layer bus bars, respectively.

Results of Fig. 4.51a and b are coincident with each other. Some minor difference is caused by the accuracy of the bus-bar parameter extraction and component modelling,

Fig. 4.50 The two-layer bus bar design and placement

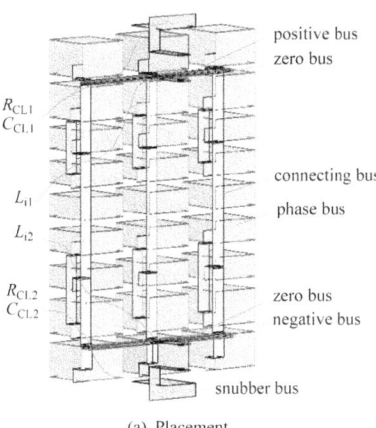

(a) Placement

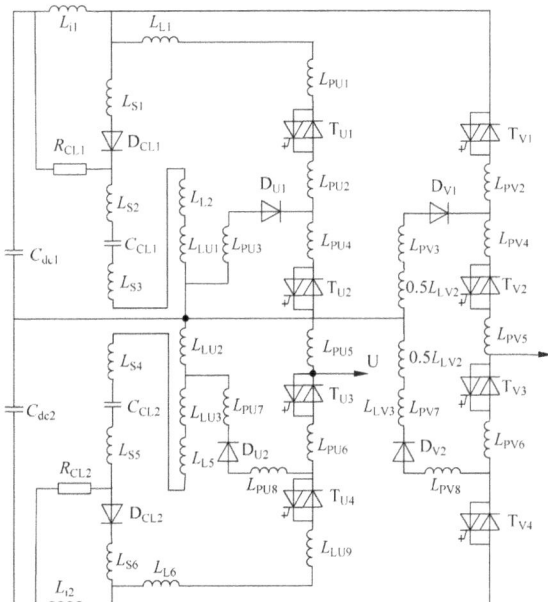

(b) Equivalent circuit of Phase-U and Phase-V

Table 4.10 Equivalent inductance of three-layer and two-layer laminated bus bars (nH)

Commutation loop	L_{SIGCT}		L_{SDCL}		L_{SCCL}		Overall	
	Three-layer	Two-layer	Three-layer	Two-layer	Three-layer	Two-layer	Three-layer	Two-layer
Loop1	134	115	185	8	31	69	350	192
Loop2	494	380	91	19	28	70	613	469
Loop3	41	67	191	5	35	70	267	142
Loop4	251	109	200	18	33	300	484	427

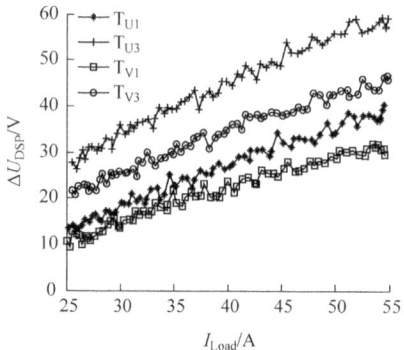

(a) Measurement results for three-layer bus bars

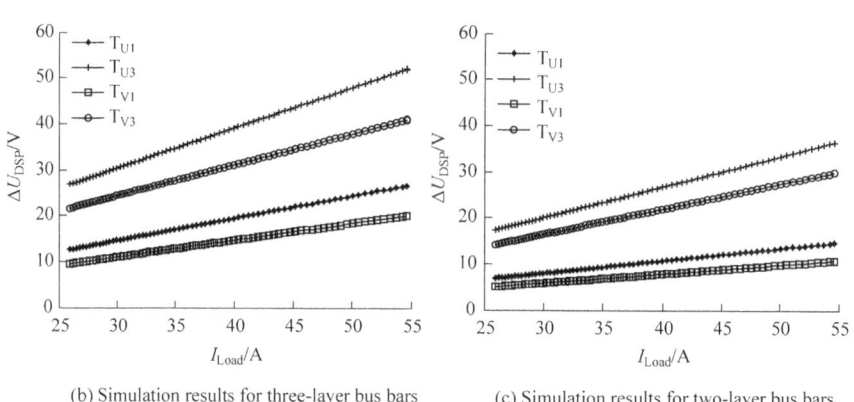

(b) Simulation results for three-layer bus bars   (c) Simulation results for two-layer bus bars

Fig. 4.51 The first turn-off voltage spikes under different bus bars

especially the modelling of the IGCT and diode. The comparison of Fig. 4.51b and c indicates that with the optimized two-layer laminated bus bars the IGCT turn-off voltage peak drops, where the turn-off voltage spike of outer switches is reduced by 50%.

Simulation and experimental results indicate that with the optimized bus bars the overall stray inductance reduces, the minimal of which drops to 54% of the original. Such effort further reduces the IGCT turn-off voltage spike, validating the effectiveness of the proposed bus-bar design in high-voltage high-power converters aiming at the high performance and high reliability.

As a summary, the TCT of the power electronics system is rooted in the analysis of short-timescale transients involving stray parameters. The critical steps, such as the extraction of stray parameters, need settle down the mechanical structure and commutation loops of the converter using the effective method. The key is to reasonably simplify the complex mechanical structure and effectively extract parameters. The complexity of the stray-parameter model is determined by the ultimate application. In such process, we need optimize the mostly-concerned objectives, quantify each impact factor, obtain the influential index and process the model simplification. The goal of minimizing the impact of the stray parameters, instead of minimizing all stray parameters, should be the criteria of the converter design and optimization.

Chapter 5
System Safe Operation Area Based on Switching Characteristics

The concept of the system safe operation area (SSOA) came from transients of power electronic converters. Such a concept is beneficial for the system reliability evaluation and optimization design. It is related to the device safe operation area (DSOA) with however obvious difference. The SSOA involves interactions between switches and other peripherals. From the aspect of the system analysis, the SSOA contains the power semiconductor devices, power conversion circuit and pulse control. It is based on the DSOA, emphasizes the interactions of DSOA, conversion circuit and pulse controls, and resolves the conflict between the reliability and maximum usage of power electronic converters from the aspect of the energy conversion and multi-timescale transients.

5.1 Definition of SSOA

In power electronics converters, semiconductor devices (switches) are the fundamental of the energy conversion. The switch reliability determines the system controllability. Traditional system designs refer to device datasheets, leaving enough allowance based on the empirical equations to avoid the switch damage. All experience roots from large amounts of faults, which is subject to change once a new device or new application emerges. Such empirical design prolongs the development period and is unable to accurately predict the device capability. Neither can it guarantee the system reliability if the allowance is misused. The industrial application of power electronic systems require the long-term operation, minimum amount of faults and high continuous power capability, which causes the conflict between the system power capability, requiring the long-run operation without frequent faults, and system reliability to avoid any device damage. To maximize the potential of semiconductors and secure the high reliability with continuous maximum power capability, an accurate analysis and design is a must.

© Tsinghua University Press and Springer Nature Singapore Pte Ltd. 2019   199
Z. Zhao et al., *Electromagnetic Transients of Power Electronics Systems*,
https://doi.org/10.1007/978-981-10-8812-4_5

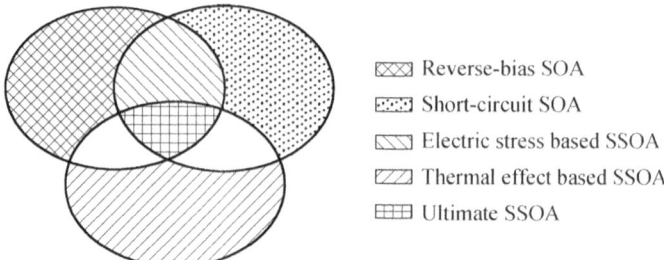

Reverse-bias SOA

Short-circuit SOA

Electric stress based SSOA

Thermal effect based SSOA

Ultimate SSOA

Fig. 5.1 Category of SSOAs and their relationships

5.1.1 Basics of SSOA

From the application point of view, SSOA is a tool for the quantitative analysis and design of power electronic converters. In theory, SSOA is a comprehensive analysis based on the all-timescale transients of power electronic converters.

The definition of SSOA is the region where power electronic systems can be safely operated under the influence of all major factors. Such region is usually represented by working points. Inside SSOA, we could assign the operation area (OA) for the converter, which usually is very straight forward and concise and can be directly used for the converter design, analysis and control.

Due to the significant behavioral difference between electric and thermal processes, two SSOAs can be independently considered, one of which is based on the electric stress and the other on the thermal effect. The ultimate SSOA is the overlap of these two.

(1) SSOA based on the electric stress

The DSOA contains the reverse-bias SOA (RBSOA) and short-circuit SOA (SCSOA). Similar approaches could be applied to the SSOA, e.g., RBSSOA and SCSSOA.

(2) SSOA based on the thermal effect

During the operation of the power converter, the temperature will rise gradually as well as the junction temperature. Over-temperature will deteriorate the switch performance and even cause the device failure. Therefore a temperature based SSOA is rewarding for the sake of the continuous operation and high reliability.

Different SSOAs are illustrated in Fig. 5.1.

SSOA describes the relationship between the semiconductor characteristics and other components in power electronic systems, reveals the impact of all system elements on the main-circuit performance, and evolves from the converter transient model considering actual system parameters and load conditions. The relationship between switch characteristics and peripheral elements needs be mathematically derived based upon the main-power circuit under various load conditions. Several major elements influencing the system safety should be included, e.g., power switches, gate-drive circuits, control units and main-loop parasitics.

5.1.2 DSOA Versus SSOA

For a DSOA, e.g. IGBT, the frequently used representative parameters are U_{CE}, I_C and T_j. The ultimate goal of an SSOA is to secure the device reliability, which is aligned with DSOA. Therefore the SSOA still uses voltage, current and temperature as variables. The difference is that, variables adopted by the SSOA are the DC-bus voltage, AC output current, heatsink temperature, etc., which are accessible during the operation of the converter and related to the DSOA as well. Therefore, all SSOA variables indirectly reveal the status of power switches thereby further helping implementation of protections.

For an IGBT, its RBSOA, SCSOA and the reverse recovery SOA (RRSOA) of the anti-parallel diode need be emphasized. For the RBSOA analysis, the IGBT turn-off behavior is the focus, including the follows.

(1) At the repetitive switching mode, the maximum turn-off current of the IGBT is twice of the rated current. When beyond such threshold, the IGBT failure is anticipated thereby deteriorating the reliable operation.
(2) At any conditions, the voltage drop across the switch, both in the steady state and in the transient process, should not exceed the ratings.
(3) In terms of the device package, leads and terminals, the stray inductance induces the voltage spike, leading to the IGBT actual turn-off voltage higher than its terminal voltage. When taking the switching-loss limitation into account, the DSOA based on the switch terminal voltage is not a regular rectangular.

Manufacturers provide different RBSOAs, even for the same rated 3.3 kV/1200 A IGBT, as shown in Fig. 5.2.

The SCSOA describes the short-circuit behavior of the IGBT. The SCSOA of one 3.3 kV/1200 A IGBT is shown in Fig. 5.3, where several main bullet points are to be considered.

(1) If the short-circuit fault happens when the IGBT turns on, the IGBT current will go from zero to the short-circuit current rapidly, where the IGBT turn-on behavior affects the whole process.
(2) If the short-circuit fault happens after the IGBT turns on, the IGBT current will go from some value to the short-circuit current rapidly. Such scenario has more impact on the IGBT gate than (1).
(3) Due to the rapidly increasing short-circuit current, the SCSOA is only true for very narrow gate pulses, e.g., 10 μs.

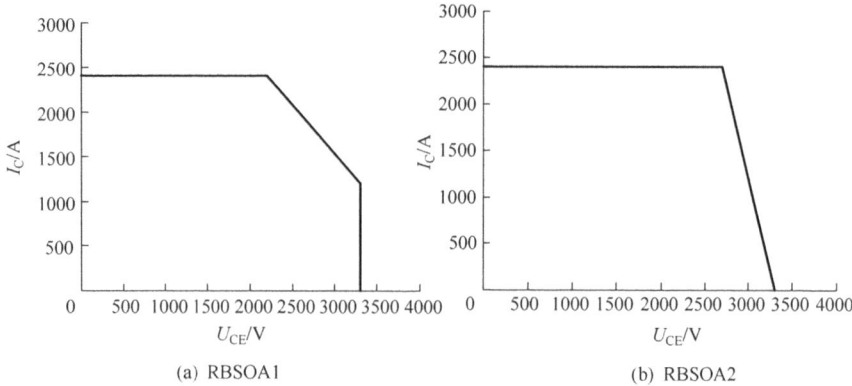

(a) RBSOA1   (b) RBSOA2

Fig. 5.2 IGBT RBSOAs

Fig. 5.3 The SCSOA of one
IGBT

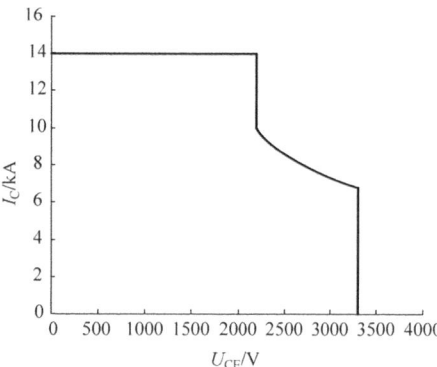

(4) The SCSOA has no repetitiveness. Only limited number of short-circuit faults
     are allowed.

The RRSOA is used to describe the switching behavior of the anti-paralleled
diode, especially the reverse recovery performance. The RRSOA of a 3.3 kV/1200 A
IGBT is shown in Fig. 5.4.

The comparison between the DSOA and SSOA indicates that the DSOA does not
change after the switch manufacturing and might only be subject to a minor variation
due to the switch diversity, while the SSOA varies with the mechanical structure,
heatsinks and control systems. Though their boundaries are both subject to change,
the variation of DSOA boundaries is random while the SSOA boundaries have certain
relations with the circuit topology, mechanical structure and main-power circuit.

In addition, the converter SSOA considers various operational conditions and
potential faults, which provides the reference for the protection and control during
the online operation, while the DSOA only provides the reference for the protection
at the early design stage. Compared to the DSOA, the SSOA is more valuable in
practice.

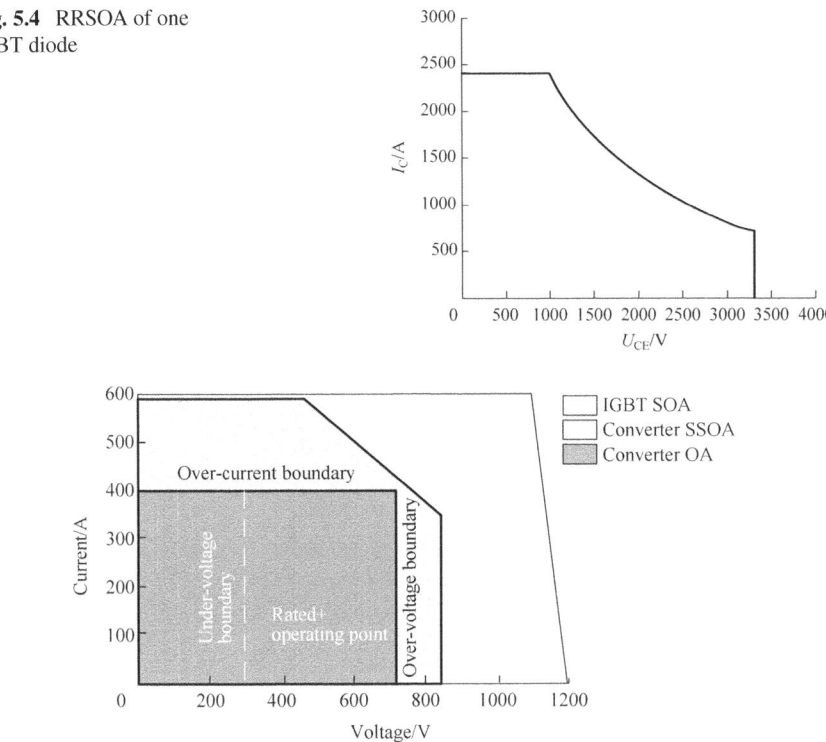

Fig. 5.4 RRSOA of one IGBT diode

Fig. 5.5 Illustration of the converter SSOA

Figure 5.5 illustrates the relationship between the converter SSOA and DSOA, based upon the RBSOA of one 55 kW/380 V inverter. The outermost boundary shows the device RBSOA. The light-color area is the converter SSOA and the dark area is the converter OA, which based on the definition is a subset of the SSOA. Hence the system OA is a typical application of SSOA.

The boundary of the switch DSOA is set based upon its maximum collector current and maximum reverse blocking voltage. The SSOA boundary is decided by the system DC-bus voltage and the output current. Therefore the converter SSOA is much smaller than the DSOA and becomes even narrower with the collector current increasing. The OA should not exceed the SSOA. Its boundary is determined by the over-voltage and over-current protections. If the DC-bus voltage needs provide the auxiliary power supply for the converter, the left-side boundary of the OA is determined by the under-voltage-protection threshold. The rated operation point could be set anywhere inside the OA.

Three boundary lines of the OA are included in Fig. 5.5, i.e., DC-bus over-voltage threshold, output over-current threshold and DC-bus under-voltage threshold, which altogether set the reference for the converter control and protections. Note the threshold refers to the actual sampled values, not the actual value at the protection moment, given these two values might not happen at the same moment thereby exhibiting quite a difference. In real applications, it is not wise to equal the threshold value to the actual protection value.

1. DC-bus over-voltage boundary

The DC-bus over-voltage is when the voltage sensor detects the DC-bus voltage exceeds the threshold. Such scenario will endanger the DC-bus capacitors and increase the failure rate of switches, given the switch turn-off voltage peak value (DC-bus voltage plus the voltage spike induced by the stray inductance) rises as well.

The DC-bus over-voltage usually happens when the large-inertia load changes abruptly, e.g., the motor regenerates the energy back to the DC-bus capacitor, or the input voltage has an abnormal variation. The over-voltage boundary is the reference for the control strategy and acts as the threshold value for the converter over-voltage protection.

2. Output over-current boundary

The output over-current occurs when the detected output current exceeds the OA current boundary. An over-current, exceeding the maximum repetitive current of the switch, might make the switch unable to turn off thereby causing the damage. On the other hand, such a large current will induce a voltage spike through the stray inductance thereby causing the switch breakdown.

Such scenario usually happens due to the abrupt load change, over-load operations and other abnormal operations, e.g., the leg shoot-through or output short-circuit. The output over-current boundary of the converter can be used as the over-current threshold as well as the limit of its power capability.

3. DC-bus under-voltage boundary

DC-bus under-voltage protection happens when the detected DC-bus voltage is lower than the OA under-voltage boundary value. A too low voltage will provide insufficient power to the load or other auxiliary circuits. For VSCs, the control system is powered by switched-mode power supplies, which for the safety purpose will receive part of the power from the DC-bus voltage. The under-voltage of the DC-bus will put its reliability in a vulnerable position and further cause the converter to lose the controllability.

When the converter output power has no particular requirements and the control system is solely powered by the external circuit, the under-voltage boundary could be set as zero. When the converter input voltage falls, the under-voltage happens and the DC-bus voltage will fall quickly if no further protections are implemented.

In theory, the OA could be anywhere inside SSOA. As long as the converter detected values are within SSOA, the component failure rate will be very low. In the real application, to avoid the complexity of the control strategy and protections, a regular boundary like Fig. 5.5 is assigned to the OA. For some special-purpose converter, to fully maximize its potentials we could use SSOA as the OA.

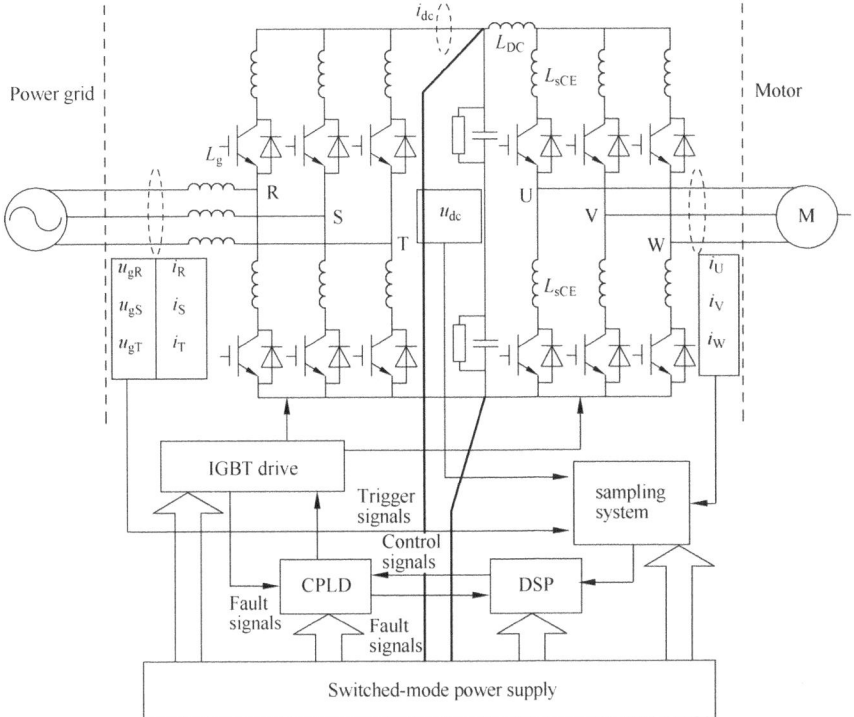

Fig. 5.6 A dual-PWM-inverter fed motor drive system

5.2 Mathematical Models of the SSOA

5.2.1 Key Components, Topology and Control Parameters

To derive the converter SSOA, a dual-PWM-inverter fed motor drive system is taken as an example, as shown in Fig. 5.6. The system contains the PWM rectifier, DC bus and the PWM inverter.

In Fig. 5.6, R, S and T are the grid-side input terminals, connected with the grid through three-phase inductors to reduce the current harmonics. U, V and W are the AC output terminals, directly connected with the three-phase motor. Voltage and current sensors are used to sample the DC-bus voltage, grid-side voltages and currents and the motor currents. All sampled values are fed to the control system where DSP generates PWM signals to trigger 12 IGBTs. At the same time fault signals are received for the protection implementation. The power of the control circuit is provided by the external switched-mode power supply, which directly obtains the power through the DC-bus in case the grid blackout. All parameters used for the SSOA model are defined as Table 5.1.

Table 5.1 Symbols for the SSOA model

Symbol	Definition
L_{sCE}	Lead inductance inside the IGBT module
t_f	Time interval for I_C to drop from 90 to 10% in the turn-off process
C_{res}	IGBT reverse capacitance
Δt	Time interval from when fault is detected to when the turn-off is implemented
I_{RBlim}	Current limit for the IGBT RBSOA
I_{SClim}	Current limit for the IGBT SCSOA
U_{lim}	Voltage limit for the IGBT
L_{DC}	Stray inductance between the DC-bus capacitor and the IGBT terminals
L_{SC}	Equivalent inductance of the output terminal during the short circuit
L_{ls}	Leakage inductance of the motor stator
f_{sw}	Switching frequency
E_{on}	One pulse IGBT turn-on energy at the rated operation
E_{off}	One pulse IGBT turn-off energy at the rated operation
L_g	Grid-side inductance
u_g	Grid-side voltage
U_{CE0}	IGBT on-state threshold voltage
r_{CE}	IGBT on-state resistance
$R_{th(j-c)Q}$	Thermal resistance of each IGBT from the junction to the case
$R_{th(c-h)Q}$	Thermal resistance of each IGBT from the case to the heatsink
P_{Qmax}	Maximum value of P_{Q_on} (on-state average loss) + P_{Q_sw} (transient average loss) per IGBT
T_{jmaxQ}	Maximum junction temperature of IGBT at the switching mode
T_h	Heatsink surface temperature
U_{D0}	Diode forward-biased threshold voltage
r_D	Diode forward resistance
E_{rec}	Diode one-time reverse recovery energy consumption at the rated operation
M	Modulation index, defined as the amplitude of the phase output over the DC-bus voltage
φ	Power factor angle

For the equivalent circuit shown in Fig. 5.6, the DC-bus current i_{dc} and the three-phase AC output current have the following relationship

$$\begin{cases} i_{dc} = S_R i_R + S_S i_S + S_T i_T \\ i_{dc} = S_U i_U + S_V i_V + S_W i_W \end{cases} \tag{5.1}$$

Here S_R–S_T and S_U–S_W are all switching functions, which are 1 when the top switch of the corresponding leg is on and 0 when the bottom switch is on. Therefore the candidates of i_{dc} are i_U, i_V, i_W, i_R, i_S, i_T and 0. To consider the extreme case, we have

$$i_{dc} = \max \left\{ |i_R| \; |i_S| \; |i_T| \; |i_U| \; |i_V| \; |i_W| \right\} \tag{5.2}$$

In later sections, i_{dc} is equivalent to the AC current of the faulty phase or the collector current i_C of the faulty IGBT.

5.2.2 Mathematical Model of SSOA

All the faults especially those extreme ones should be considered when constructing the SSOA boundaries. The electric stress based SSOA is similar to the DSOA, including the RBSSOA and SCSSOA. Such SSOAs target the grid-side or motor-side soft short circuit (when the short circuit happens at the large inductive load such as the AC inductors or motors) and the phase-phase short circuit. Given the motor-side module has a time constant longer than the grid-side module, we can use the ultra-low-frequency output at the motor side as an extreme case to derive the thermal based SSOA.

The critical step of modeling the SSOA is to locate its boundary. To do this, we partition the inverter as Fig. 5.6 finely into Fig. 5.7.

Once the electric stress undertaken at the protection moment is right on the boundary of the SSOA, the related voltage and current values can represent the boundaries. We in the follows use a one-time protection to analyze the relationship between the SSOA and other system elements.

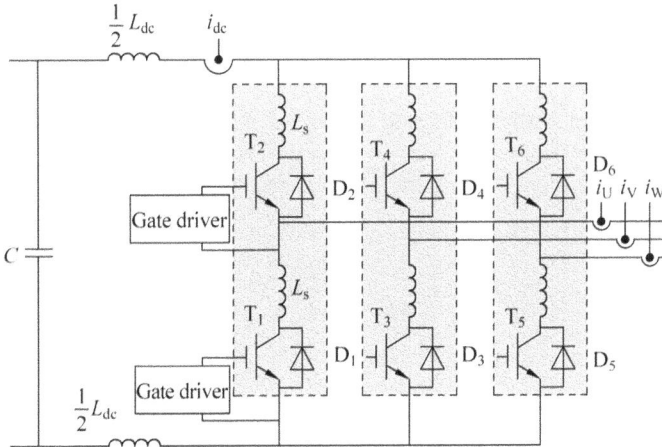

Fig. 5.7 The inverter partition

Assume the sensors sample the DC-bus voltage u_{dc} and current i_{dc} and detect the fault at the moment of t. After a time delay Δt, the switch turns off at the moment of $t + \Delta t$. The inverter operation points at these two moments are $[u_{dc}(t), i_{dc}(t)]$ and $[u_{dc}(t + \Delta t), i_{dc}(t + \Delta t)]$, respectively. Here

$$i_{dc}(t + \Delta t) = i_{dc}(t) + \frac{di_{dc}(t)}{dt} \cdot \Delta t + \frac{du_{dc}(t + \Delta t)}{dt} \cdot C_{res} \tag{5.3}$$

$$u_{dc}(t + \Delta t) = u_{dc}(t) + \frac{du_{dc}(t)}{dt} \cdot \Delta t \tag{5.4}$$

The current $i_{dc}(t + \Delta t)$ is made of $i_{dc}(t)$, current increment during Δt and the charging current through the junction capacitance when the switch turns off. The voltage $u_{dc}(t + \Delta t)$ consists of $u_{dc}(t)$ and the voltage increment during Δt. Given all electric stress should not exceed the limit, we have

$$i_c(t + \Delta t) \leq I_{lim}(T_j) \quad \& \quad u_{CE}(t + \Delta t) \leq U_{lim}(T_j) \tag{5.5}$$

After substituting Eqs. (5.3) and (5.4) to Eq. (5.5), we have

$$\begin{cases} i_{dc}(t) + \frac{di_{dc}(t)}{dt} \cdot \Delta t + \frac{du_{ce}(t+\Delta t)}{dt} \cdot C_{res} \leq I_{lim}(T_j) \\ u_{dc}(t) + \frac{du_{dc}(t)}{dt} \cdot \Delta t - \frac{di_c(t+\Delta t)}{dt} \cdot (L_{dc} + 2L_s) \leq U_{lim}(T_j) \end{cases} \tag{5.6}$$

Here $i_c(t + \Delta t) = i_{dc}(t + \Delta t)$ shown in Eq. (5.3). $u_{ce}(t + \Delta t)$ is $u_{dc}(t + \Delta t)$ plus the voltage induced by the loop stray inductance.

Now consider two most challenging scenarios for the inverter output, i.e., soft short circuit and hard short circuit. The former one happens when the inverter load is inductive while the latter one occurs when output terminals of the inverter are shorted through conductors.

1. Soft short circuit at the inverter output

When the converter output is shorted through an inductive load, the current rising rate is much higher than that of normal situation. For example, when T_2, T_3 and T_5 in Fig. 5.7 are on, load inductors will be charged. Based on the Kirchhoff current law (KCL),

$$u_{dc}(t) - (L_{dc} + \frac{3}{2}L_s + \frac{3}{2}L_{ls}) \cdot \frac{di_{dc}(t)}{dt} = 0 \tag{5.7}$$

which can be further derived as

$$\frac{di_{dc}(t)}{dt} = \frac{u_{dc}(t)}{L_{dc} + \frac{3}{2}L_s + \frac{3}{2}L_{ls}} \tag{5.8}$$

With the linearization of the voltage and current increment in the IGBT turn-off process, we have

$$\frac{du_{ce}(t + \Delta t)}{dt} = \frac{0.8u_{dc}(t + \Delta t)}{t_f} \tag{5.9}$$

$$\frac{di_c(t + \Delta t)}{dt} = -\frac{0.8i_{dc}(t + \Delta t)}{t_f} \tag{5.10}$$

Since the DC-bus voltage rises slowly while Δt is as tiny as $\sim\mu s$, the item $\Delta t \cdot du_{dc}(t)/dt$ is negligible compared to $u_{dc}(t)$, i.e.,

$$u_{dc}(t + \Delta t) \approx u_{dc}(t) \tag{5.11}$$

Substituting Eqs. (5.8)–(5.11) into Eq. (5.6) with some necessary linearization yields

$$\begin{cases} i_{dc}(t) + u_{dc}(t)[\frac{\Delta t}{L_{dc}+\frac{3}{2}L_s+\frac{3}{2}L_{ls}} + \frac{0.8C_{res}}{t_f}] \leq I_{lim\_RB}(T_j) \\ i_{dc}(t)\frac{0.8(L_{dc}+2L_s)}{t_f} + u_{dc}(t)[1 + \frac{0.8\Delta t(L_{dc}+2L_s)}{t_f(L_{dc}+\frac{3}{2}L_s+\frac{3}{2}L_{ls})} \\ \quad + \frac{0.64(L_{dc}+2L_s)C_{res}}{t_f^2}] \leq U_{lim}(T_j) \end{cases} \tag{5.12}$$

Here Eq. (5.12) is the mathematical model of the SSOA based upon actual system parameters. All the analysis above is for the continuous operation of the converter. Hence $I_{lim}(T_j)$ and $U_{lim}(T_j)$ are determined by the device RBSOA.

2. **Hard short circuit at the inverter output**

Due to the existence of the dead time and interlocked gate drive circuits, the occurrence of the leg shoot-through fault has very low possibility. Therefore such a fault is not under the consideration. However, due to the degradation of the insulation, mis-assembly and mis-operation, short-circuits of the inverter output terminals might occur, under which the rising rate of the DC-bus current is much higher. Assume T_2 and T_3 are on in Fig. 5.7 when the output short-circuit happens. With the short-circuit inductance of L_{sc}, we have

$$u_{dc}(t) - (L_{dc} + 2L_s + L_{sc}) \cdot \frac{di_{dc}(t)}{dt} = 0 \tag{5.13}$$

Furthermore the mathematical model of the converter SSOA is

$$\begin{cases} i_{dc}(t) + u_{dc}(t)[\frac{\Delta t}{L_{dc}+2L_s+L_{SC}} + \frac{0.8C_{res}}{t_f}] \leq I_{lim\_SC}(T_j) \\ i_{dc}(t)\frac{0.8(L_{dc}+2L_s)}{t_f} + u_{dc}(t)[1 + \frac{0.8\Delta t(L_{dc}+2L_s)}{t_f(L_{dc}+2L_s+L_{SC})} \\ \quad + \frac{0.64(L_{dc}+2L_s)C_{res}}{t_f^2}] \leq U_{lim}(T_j) \end{cases} \tag{5.14}$$

Under such circumstance, the safe operation of the converter still needs be secured by a timely protection. $I_{\lim}(T_j)$ and $U_{\lim}(T_j)$ in Eq. (5.14) are determined by the device SCSOA. The converter operation must comply with constraints of both Eqs. (5.12) and (5.14). Mathematically the SSOA now is the intersection of these two equations.

The SSOA reflected to the motor side is expressed as Eq. (5.15), corresponding to RBSSOA and SCSSOA.

$$
\begin{cases}
\mathbf{A}_{\text{RB}} \cdot \begin{pmatrix} i_c(t) \\ u_{dc}(t) \end{pmatrix} \leq \begin{pmatrix} I_{\text{RBlim}} \\ U_{\lim} \end{pmatrix} \\[3mm]
\mathbf{A}_{\text{SC}} \cdot \begin{pmatrix} i_c(t) \\ u_{dc}(t) \end{pmatrix} \leq \begin{pmatrix} I_{\text{SClim}} \\ U_{\lim} \end{pmatrix}
\end{cases}
\tag{5.15}
$$

The right side of the inequality is the boundary of the DSOA. The coefficient matrixes $\mathbf{A}_{\text{RB}}$ and $\mathbf{A}_{\text{SC}}$ are shown in Eqs. (5.15) and (5.16), respectively.

$$
\mathbf{A}_{\text{RB}} = \begin{pmatrix} 1 & \frac{\Delta t}{L_{dc}+\frac{3}{2}L_{sCE}+\frac{3}{2}L_{ls}} + \frac{0.8 C_{res}}{t_f} \\[3mm] \frac{0.8(L_{dc}+2L_{sCE})}{t_f} & 1 + \frac{0.8\Delta t(L_{dc}+2L_{sCE})}{t_f(L_{dc}+\frac{3}{2}L_{sCE}+\frac{3}{2}L_{ls})} + \frac{0.64(L_{dc}+2L_{sCE})C_{res}}{t_f^2} \end{pmatrix}
\tag{5.16}
$$

$$
\mathbf{A}_{\text{SC}} = \begin{pmatrix} 1 & \frac{\Delta t}{L_{dc}+2L_{sCE}+L_{sc}} + \frac{0.8 C_{res}}{t_f} \\[3mm] \frac{0.8(L_{dc}+2L_{sCE})}{t_f} & 1 + \frac{0.8\Delta t(L_{dc}+2L_{sCE})}{t_f(L_{dc}+2L_{sCE}+L_{sc})} + \frac{0.64(L_{dc}+2L_{sCE})C_{res}}{t_f^2} \end{pmatrix}
\tag{5.17}
$$

In dual PWM converters, building the SSOA for the grid-side converter is similar to the motor-side converter, with the SCSSOA nearly the same except that L_{SC} is the equivalent inductance when the grid is shorted. Due to the asymmetry of the grid-side and motor-side circuits, when deriving the RBSSOA, we need consider the grid impact, given the grid is a pure power supply and equipped with filtering inductors. Other than that, all derivations are similar. When the soft short-circuit happens, the equivalent circuit is shown in Fig. 5.8.

Assume the fault is detected at the moment t, when S_1, S_4 and S_5 are all on. At this moment, the DC-bus voltage u_{dc} and current i_{dc} follow

$$
2u_{dc}(t) + 3u_{gs}(t) = \frac{di_{dc}(t)}{dt}(2L_{dc} + 3L_{sCE} + 3L_g)
\tag{5.18}
$$

Through which the changing rate of i_{dc} is shown below

$$
\frac{di_{dc}(t)}{dt} = \frac{u_{dc}(t)}{L_{dc} + \frac{3}{2}L_{sCE} + \frac{3}{2}L_g} + \frac{u_{gs}(t)}{\frac{2}{3}L_{dc} + L_{sCE} + L_g}
\tag{5.19}
$$

With the linearization of U_{CE} and I_C during the IGBT turn-off, we have

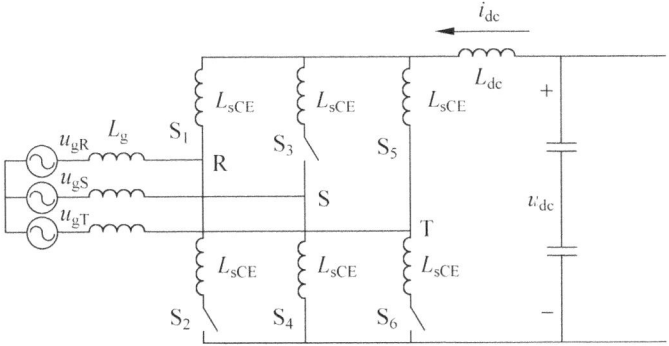

Fig. 5.8 Equivalent circuit for deriving the grid-side RBSSOA

$$\frac{du_{CE}(t + \Delta t)}{dt} = 0.8\frac{u_{dc}(t + \Delta t)}{t_f} \tag{5.20}$$

$$\frac{di_C(t + \Delta t)}{dt} = -0.8\frac{i_C(t + \Delta t)}{t_f} \tag{5.21}$$

Substituting Eqs. (5.19)–(5.21) to Eq. (5.6) yields

$$\begin{cases} i_{dc}(t) + u_{dc}(t)\left[\dfrac{\Delta t}{L_{dc} + \frac{3}{2}L_{sCE} + \frac{3}{2}L_g} + \dfrac{0.8C_{res}}{t_f}\right] \\ \qquad + u_{gs}(t)\dfrac{\Delta t}{\frac{2}{3}L_{dc} + L_{sCE} + L_g} \leq I_{RB\,lim}(T_j) \\[2mm] u_{dc}(t)\left[1 + \dfrac{0.8\Delta t(L_{dc} + 2L_{sCE})}{t_f(L_{dc} + \frac{3}{2}L_{sCE} + \frac{3}{2}L_g)} + \dfrac{0.64(L_{dc} + 2L_{sCE})C_{res}}{t_f^2}\right] \\[2mm] \qquad + i_{dc}(t)\dfrac{0.8(L_{dc} + 2L_{sCE})}{t_f} + u_{gs}(t)\dfrac{0.8\Delta t(L_{dc} + 2L_{sCE})}{t_f(\frac{2}{3}L_{dc} + L_{sCE} + L_g)} \leq U_{lim}(T_j) \end{cases} \tag{5.22}$$

The most severe fault happens when $u_{gs}(t + \Delta t)$ is the grid peak voltage U_{gm}. By replacing i_{dc} with i_C, the RBSOA is expressed as below.

$$\mathbf{A}_{RB\_g}\begin{pmatrix} i_C(t) \\ u_{dc}(t) \end{pmatrix} \leq \begin{pmatrix} I_{RBlim} - \dfrac{U_{gm}\Delta t}{\frac{2}{3}L_{dc} + L_{sCE} + L_g} \\[2mm] U_{lim} - \dfrac{0.8U_{gm}\Delta t(L_{dc} + 2L_{sCE})}{t_f(\frac{2}{3}L_{dc} + L_{sCE} + L_g)} \end{pmatrix} \tag{5.23}$$

where $\mathbf{A}_{RB\_g}$ could be obtained based on $\mathbf{A}_{RB}$, through neglecting the grid inductance and replacing L_{ls} with L_g, i.e.,

$$\mathbf{A}_{RB\_g} = \begin{pmatrix} 1 & \dfrac{\Delta t}{L_{dc} + \frac{3}{2}L_{sCE} + \frac{3}{2}L_g} + \dfrac{0.8C_{res}}{t_f} \\[2mm] \dfrac{0.8(L_{dc} + 2L_{sCE})}{t_f} & 1 + \dfrac{0.8\Delta t(L_{dc} + 2L_{sCE})}{t_f(L_{dc} + \frac{3}{2}L_{sCE} + \frac{3}{2}L_g)} + \dfrac{0.64(L_{dc} + 2L_{sCE})C_{res}}{t_f^2} \end{pmatrix} \tag{5.24}$$

For dual PWM converters, the electric stress based SSOA is the intersection of grid-side SSOA and motor-side SSOA. With different IGBT modules on each side, related SSOAs need be calibrated separately. Since what the datasheet presents is the DSOA at the maximum junction temperature, for simplicity of the analysis it is assumed that the SSOA boundary does not change with the temperature.

With the long-term operation of the converter, the IGBT module accumulates the power loss to increase the junction temperature. Over-temperature might disable and even fail the power module. Hence setting the over-temperature protection helps prolong the lifespan of the power module and enhance the system reliability. Generally it is difficult to directly measure the junction temperature, except the heatsink temperature. Therefore we could set the SSOA based upon the IGBT maximum junction temperature during actual operations, which is defined as the thermal related SSOA. It is highly related to the module power loss determined by the PWM strategy and switching frequency, and the heatsink design. In dual PWM converters, the output of the grid-side converter has the same frequency as the grid, e.g., 50 Hz. The output of the motor-side converter follows the fundamental frequency of the motor. The lower the motor speed, the lower the fundamental frequency. The inverter thermal limit is often hit at the low-speed-high-torque scenario. The time constant of the IGBT module is usually hundreds of milliseconds, much longer than the switching period and similar to the motor fundamental period. Generally, the IGBT junction temperature does not change within one fundamental period. Only the average power loss within one fundamental period needs be calculated when considering the thermal effect. In this process, we assume that:

(1)   The output current is a clean sinusoidal without ripples.
(2)   The IGBT junction temperature is kept at the maximum point.
(3)   The conduction and switching loss are both linear to the current.
(4)   The sampling frequency is infinite without the discrete error.

The IGBT loss contains three items, i.e., turn-on loss, turn-off loss and conduction loss, the first two of which are named as the switching loss. Given two IGBTs in one leg are interlocked, each IGBT only conducts half of the time during one fundamental period. Therefore we only need calculate the conduction loss within half of the period, as shown in Eq. (5.25) when using SPWM.

$$P_{Q\_on} = U_{CE0}I_{CQ}\left(\frac{1}{\pi} + \frac{M\cos\varphi}{4}\right) + r_{CE}I_{CQ}^2\left(\frac{1}{4} + \frac{2M\cos\varphi}{3\pi}\right) \qquad (5.25)$$

The IGBT switching loss is related to the DC-bus voltage, collector current, switching frequency, gate-drive voltage and the stray inductance. For a certain converter, its gate-drive voltage and stray inductance are known, yielding the switching loss only varied by the first three factors. For simplicity, the IGBT switching loss is assumed to be linear with U_{dc} and I_C, which after the normalization is

$$P_{Q\_sw} = \frac{2}{\pi}(E_{on} + E_{off}) \cdot f_{sw} \cdot \frac{U_{dc}}{U_{nom}} \cdot \frac{I_{CQ}}{I_{nom}} \qquad (5.26)$$

Fig. 5.9 Equivalent thermal loop of the IGBT

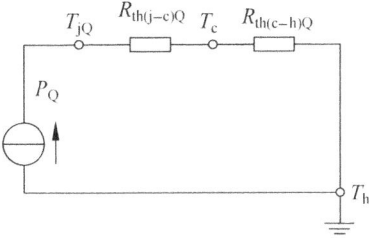

Here U_{nom} and I_{nom} are the nominal voltage and current.

To calculate the power-loss limit, the saturation voltage drops of the IGBT and its body diode are selected as maximum values under the harsh condition, i.e.,

$$\begin{cases} U_{\mathrm{CE0\_max}} = U_{\mathrm{CE0}} + U_{\mathrm{CE\_max}} - U_{\mathrm{CE\_typ}} \\ U_{\mathrm{D0\_max}} = U_{\mathrm{D0}} + U_{\mathrm{D\_max}} - U_{\mathrm{D\_typ}} \end{cases} \tag{5.27}$$

Here $U^*{}_{\mathrm{typ}}$ and $U^*{}_{\mathrm{max}}$ are the typical and maximum values of the on-state voltage drop, respectively, which can be obtained through the device datasheet.

The overall loss of one IGBT is

$$P_Q = P_{Q\_\mathrm{on}} + P_{Q\_\mathrm{sw}} \tag{5.28}$$

For the study of the reliability, we need calculate the maximum power loss within the tolerance of the junction temperature. Shown in Fig. 5.9 is the equivalent thermal loop of the IGBT, based on which the maximum power loss is defined as

$$P_Q \leq P_{Q\mathrm{max}} = \frac{T_{\mathrm{jmaxQ}} - T_{\mathrm{h}}}{R_{\mathrm{th(j-c)Q}} + R_{\mathrm{th(c-h)Q}}} \tag{5.29}$$

Substituting Eqs. (5.25)–(5.28) into Eq. (5.29) will generate the thermal based SSOA of the IGBT as

$$I_{\mathrm{CQ}} \leq I_{\mathrm{CQ\_max}} = \frac{\sqrt{B_Q^2 + 4 A_Q C_Q} - B_Q}{2 A_Q} \tag{5.30}$$

Here

$$\begin{cases} A_Q = r_{\mathrm{CE}}\left(\frac{1}{4} + \frac{2M \cos \varphi}{3\pi}\right) \\ B_Q = U_{\mathrm{CE0\_max}}\left(\frac{1}{\pi} + \frac{M \cos \varphi}{4}\right) + \frac{2 U_{\mathrm{dc}} f_{\mathrm{sw}}(E_{\mathrm{on}} + E_{\mathrm{off}})}{\pi U_{\mathrm{nom}} I_{\mathrm{nom}}} \\ C_Q = \frac{T_{\mathrm{jmaxQ}} - T_{\mathrm{h}}}{R_{\mathrm{th(j-c)Q}} + R_{\mathrm{th(c-h)Q}}} \end{cases} \tag{5.31}$$

For the anti-paralleled diode, its power loss includes the reverse recovery loss and the conduction loss. The average conduction loss within one fundamental period is

$$P_{D\_on} = U_{D0} I_{cd} \left(\frac{1}{\pi} - \frac{M \cos \varphi}{4} \right) + r_D I_{cd}^2 \left(\frac{1}{4} - \frac{2M \cos \varphi}{3\pi} \right) \tag{5.32}$$

The average reverse recovery loss within one fundamental period is

$$P_{D\_rr} = \frac{1}{\pi} E_{rec} \cdot f_{sw} \cdot \frac{U_{dc}}{U_{nom}} \cdot \frac{I_{cd}}{I_{nom}} \tag{5.33}$$

Therefore the thermal based SSOA of the anti-paralleled diode is

$$I_{cd} \le I_{cd\_max} = \frac{\sqrt{B_D^2 + 4A_D C_D} - B_D}{2A_D} \tag{5.34}$$

Here

$$\begin{cases} A_D = r_D \left(\frac{1}{4} - \frac{2M \cos \varphi}{3\pi} \right) \\ B_D = U_{D0\_max} \left(\frac{1}{\pi} - \frac{M \cos \varphi}{4} \right) + \frac{2U_{dc} f_{sw} E_{rec}}{\pi U_{nom} I_{nom}} \\ C_D = \frac{T_{jmaxD} - T_h}{R_{th(j-c)D} + R_{th(c-h)D}} \end{cases} \tag{5.35}$$

Overall the thermal based SSOA is the intersection of inequalities Eqs. (5.30) and (5.34). In Sect. 5.4.2, the SSOA incorporating the electric stress and thermal effect will be illustrated.

5.2.3 Design Examples Based on SSOA

The previous analysis indicates that the converter will be safe as long as working inside the SSOA, i.e., any $[u_{dc}(t), i_{dc}(t)]$ inside the SSOA could be set as the operation point.

For the motor drive system, with the rated power P and rated voltage U_N under other necessary restraints, the device ratings could be determined based upon the SSOA. Here the rated current of the motor is

$$I_N = \frac{P}{\sqrt{3} U_N \cos \varphi \cdot \eta} \tag{5.36}$$

Here $\cos \varphi$ is the motor rated power factor and η is the motor rated efficiency. Referring to the motor maximum current, we have

$$i_{dc}(t) = \sqrt{2}(k_1 k_2 \cdot I_N) = \frac{\sqrt{2}k_1 k_2 P}{\sqrt{3}U_N \eta \cos\varphi} \tag{5.37}$$

Here k_1 is the harmonics coefficient and k_2 is the starting-current coefficient, both of which are control settings. Considering the variation of the input power supply, we have

$$u_{dc}(t) = \sqrt{2}k_3 U_N \tag{5.38}$$

Substituting Eqs. (5.37) and (5.38) into the SSOA model generates the required IGBT device voltage and current ratings, as shown below.

$$\begin{cases} \begin{pmatrix} I_{\lim\_RB}(T_j) \\ U_{\lim}(T_j) \end{pmatrix} \geq \mathbf{A_{RB}} \begin{pmatrix} \frac{\sqrt{2}k_1 k_2 P}{\sqrt{3}U_N \eta \cos\varphi} \\ \sqrt{2}k_3 U_N \end{pmatrix} \\ \\ \begin{pmatrix} I_{\lim\_SC}(T_j) \\ U_{\lim}(T_j) \end{pmatrix} \geq \mathbf{A_{SC}} \begin{pmatrix} \frac{\sqrt{2}k_1 k_2 P}{\sqrt{3}U_N \eta \cos\varphi} \\ \sqrt{2}k_3 U_N \end{pmatrix} \end{cases} \tag{5.39}$$

The right side of the inequality Eq. (5.39) is fully determined by motor parameters, control parameters, device datasheets and the mechanical design, etc. Such a design methodology maximizes the device utilization thereby avoiding the over redundancy caused by the conventional empirical design.

Take one 160 kW/380 V general-purpose VSI as an example. The calculated device ratings using the SSOA model are shown in Table 5.2, including both motor and inverter parameters.

The power loss P_{ss} of the semiconductor switch can be calculated based on device parameters and the control strategy. With the temperature threshold given, we can reversely design the heatsink of the whole inverter. A low heatsink temperature tends to enhance the device current capability, however, increase the system cost. A low-cost heatsink will increase the system temperature thereby lowering the device current capability. A tradeoff is required between the performance and the cost, which is particularly determined by upper and lower limits of the heatsink temperature.

For the high-power converters, $I_{\lim}(T_j)$ and $U_{\lim}(T_j)$ calculated by Eq. (5.39) might not exactly match the actual switch, leading to the possible switch series/parallel connection required. When using the voltage or current balancing techniques to approach $I_{\lim}(T_j)$ and $U_{\lim}(T_j)$, the discount of the electrical capability caused by the voltage/current imbalance needs be taken into account. The diode and capacitor selection could be based on the conventional calculation, which will not be discussed in this chapter.

Table 5.2 Design parameters of the 160 kW/380 V inverter

Parameter	Value
P	160 kW
U_N	380 V
$\cos\varphi$	0.9
η	0.9
k_1	1.2
k_2	1.7
k_3	1.25
L_{DC}	10 nH
L_s(2 IGBTs in parallel)	10 nH
L_{ls}	0.19 mH
L_{SC}	2.0 μH
C_{res}	3.10 nF
Δt	4.1 μs
t_f	110 ns

Based on parameters in Table 5.2, two coefficient matrixes for the SSOA are

$$\mathbf{A_{RB}} = \begin{pmatrix} 1 & 0.0257 \\ 0.2182 & 1.0056 \end{pmatrix} \tag{5.40}$$

$$\mathbf{A_{SC}} = \begin{pmatrix} 1 & 2.0310 \\ 0.2182 & 1.4431 \end{pmatrix} \tag{5.41}$$

Therefore the IGBT selection should comply with

$$\begin{cases} \begin{pmatrix} I_{\lim\_RB}(T_j) \\ U_{\lim}(T_j) \end{pmatrix} \geq \begin{pmatrix} 883 \text{ A} \\ 864 \text{ V} \end{pmatrix} \\ \begin{pmatrix} I_{\lim\_SC}(T_j) \\ U_{\lim}(T_j) \end{pmatrix} \geq \begin{pmatrix} 2230 \text{ A} \\ 1158 \text{ V} \end{pmatrix} \end{cases} \tag{5.42}$$

Two IGBTs in parallel provide a feasible choice. Given the IGBT current limit of the RBSOA is more than double of the inverter rated value, two options are put on the table. One is to parallel two 1200 V/300 A IGBTs with high-performance heatsinks and strict current balancing techniques, the other is to parallel two 1200 V/450 A IGBTs with relatively low requirement on heatsinks and the current balancing. The ultimate device selection is up to the product requirements and working environment.

With the right device selection, the further performance evaluation and the protection settings are to be completed, aiming at the enhancement of the converter continuous running capability.

5.3 Impact Factors of the SSOA

The establishment of the SSOA in Sect. 5.2 involves the circuit, mechanical design, thermal dissipation and etc. Investigating impact factors of the SSOA allows the further system optimization.

From the equations above, influential factors of the SSOA boundaries include the control delay Δt, the DC-bus stray inductance L_{dc} and the AC filtering inductance L_g. The stray inductance inside the module L_{sCE} is determined by the package technology, which is 10–20 nH and can be treated as a constant. Eqs. (5.30) and (5.34) indicate impact factors of the thermal based SSOA are mainly the switching frequency f_{sw}, IGBT thermal resistance and switching loss. Generally IGBT modules with the same package could be assumed having the same thermal resistance. While the switching loss has complex correlations with other factors, here we focus on the impact of the switching frequency on the SSOA boundaries.

5.3.1 Impact of the DC-Bus Stray Inductance

For the load-side SSOA, we can reorganize Eq. (5.23) into the form of $i_c \leq f(u_{dc})$ and then obtain the partial derivative over L_{dc}, i.e.,

$$\begin{cases} \dfrac{\partial i_c(t)}{\partial L_{dc}} \leq \dfrac{\Delta t}{\left(L_{dc} + \frac{3}{2}L_{sCE} + \frac{3}{2}L_{ls}\right)^2} u_{dc}(t) \\[3mm] \dfrac{\partial i_c(t)}{\partial L_{dc}} \leq \left[\dfrac{t_f}{0.8(L_{dc} + 2L_{sCE})^2} + \dfrac{\Delta t}{\left(L_{dc} + \frac{3}{2}L_{sCE} + \frac{3}{2}L_{ls}\right)^2}\right] u_{dc}(t) \\[3mm] \qquad - \dfrac{t_f}{0.64(L_{dc} + 2L_{sCE})^2} U_{lim} \end{cases} \qquad (5.43)$$

$$\begin{cases} \dfrac{\partial i_c(t)}{\partial L_{dc}} \leq \dfrac{\Delta t}{(L_{dc} + 2L_{sCE} + L_{sC})^2} u_{dc}(t) \\[3mm] \dfrac{\partial i_c(t)}{\partial L_{dc}} \leq \left[\dfrac{t_f}{0.8(L_{dc} + 2L_{sCE})^2} + \dfrac{\Delta t}{(L_{dc} + 2L_{sCE} + L_{sC})^2}\right] u_{dc}(t) \\[3mm] \qquad - \dfrac{t_f}{0.64(L_{dc} + 2L_{sCE})^2} U_{lim} \end{cases} \qquad (5.44)$$

As shown in (5.43), the changing rate of the RBSSOA two boundaries has the positive correlation with L_{dc}. However, since the motor inductance is ~mH, the first-boundary changing rate is less obvious than the second boundary. Meanwhile the crossing point between the second boundary and the current axis drops with L_{dc} increasing, which further shrinks the area of the SSOA. Similarly, according to Eq. (5.44), two boundaries of the SCSSOA have the similar trend as the RBSSOA,

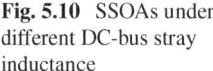

Fig. 5.10 SSOAs under different DC-bus stray inductance

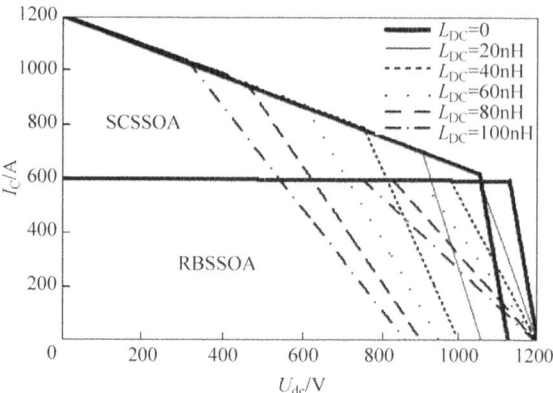

i.e., its area shrinks with L_{dc} increasing. So does the intersection of the above two electric-stress based SSOAs.

SSOAs under different DC-bus stray inductance are illustrated in Fig. 5.10, the taller one of which is the IGBT SCSSOA and the bold lines of which are the SSOA limit when $L_{dc} = 0$. Due to the existence of the module internal inductance, the final SSOA will somewhat shrink compared to the DSOA. At the maximum inductance of 100 nH, the SSOA area nearly shrinks 50%, which reveals the high impact of the stray inductance. Therefore during the design of the converter, a compact design of the bus bar such as the multilayer planar structure is beneficial, which minimizes the bus bar connected to the IGBT terminals and reduces the commutation-loop area thereby reducing the loop inductance and enlarging the SSOA area.

5.3.2 Impact of Control Parameters

1. **Control delay**

For the motor-side SSOA, partial derivatives over the time delay Δt are

$$\begin{cases} \dfrac{\partial i_C(t)}{\partial \Delta t} \leq -\dfrac{1}{L_{dc}+\frac{3}{2}L_{sCE}+\frac{3}{2}L_{ls}}u_{dc}(t) \\[3mm] \dfrac{\partial i_C(t)}{\partial \Delta t} \leq -\dfrac{1}{L_{dc}+\frac{3}{2}L_{sCE}+\frac{3}{2}L_{ls}}u_{dc}(t) \end{cases} \tag{5.45}$$

$$\begin{cases} \dfrac{\partial i_C(t)}{\partial \Delta t} \leq -\dfrac{1}{L_{dc}+2L_{sCE}+L_{sC}}u_{dc}(t) \\[3mm] \dfrac{\partial i_C(t)}{\partial \Delta t} \leq -\dfrac{1}{L_{dc}+2L_{sCE}+L_{sC}}u_{dc}(t) \end{cases} \tag{5.46}$$

As shown in Eq. (5.45), slopes of two RBSSOA boundary lines become more negative with Δt increasing, though crossing points with the current axis maintain the

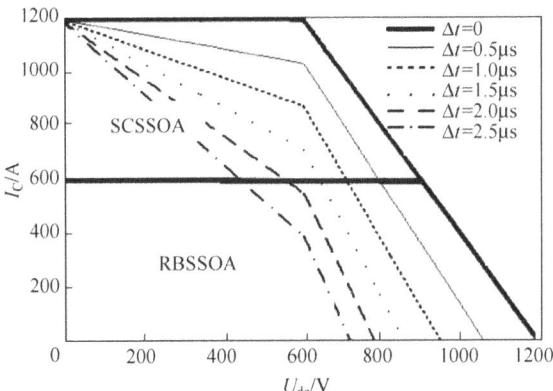

Fig. 5.11 SSOAs under different control delays

same. Therefore the area those boundary lines enclose has a negative correlation with Δt. A similar conclusion applies to the SCSSOA. Due to a larger motor inductance than the phase-phase short-circuit inductance, the area reduction of the RBSSOA is not significant.

SSOAs with different Δt are given in Fig. 5.11. The bold lines are for $\Delta t = 0$, where the right-side boundaries of the RBSSOA and SCSSOA are nearly overlapped. With Δt increasing, the RBSSOA boundary is barely altered while the SCSSOA varies significantly, which ultimately shrinks the area of the electric stress based SSOA. When $\Delta t = 2.5$ μs, an obvious area reduction is seen for the SSOA with a much lower DC-bus voltage limit. Any further increment of Δt will result in a lower DC-bus voltage. Since the DC-bus voltage in dual PWM inverters is higher than that in the diode rectifier, a stricter requirement is a must for Δt. Such a delay includes the fault-detection delay and the IGBT turn-off delay, which is related to the turn-off gate resistance and voltage value. It is possible to shorten the IGBT turn-off delay through imposing a more negative gate turn-off voltage or lowering the gate resistance, though it might bring the EMI challenges. Thus a tradeoff is required to further design the gate-drive circuit thereby widening the operational range and enhancing the system reliability.

2. **Impact of the switching frequency on the SSOA**

The major player of the thermal based SSOA is the switching frequency f_{sw}. The partial derivative of the current over f_{sw} is

$$\frac{\partial I_{CQ\_max}}{\partial f_{sw}} = \frac{B_Q - \sqrt{B_Q^2 + 4A_Q C_Q}}{2A_Q\sqrt{B_Q^2 + 4A_Q C_Q}} \cdot \frac{2U_{dc}(E_{on} + E_{off})}{\pi\, U_{nom} I_{nom}} \tag{5.47}$$

This equation reveals that with f_{sw} increasing the inverter maximum output current drops sharply. Such current reduction is even more obvious with the DC-bus voltage rising. Shown in Fig. 5.12 is a thermal based SSOA under a fixed DC-bus voltage and variable f_{sw}. The bold line is when f_{sw} is ultra-low. With the heatsink temperature (T_h)

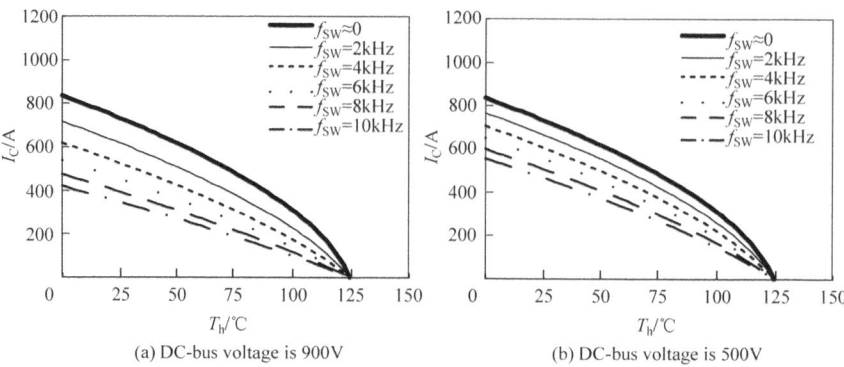

Fig. 5.12 SSOAs under different switching frequencies

rising, the inverter maximum output current reduces, which slides slowly before $T_h =$ 75 °C. After that the inverter power capability drops quickly. As shown in Fig. 5.12a, when the DC-bus voltage is a constant 900 V, the SSOA area reduces with f_{sw} increasing. At 10 kHz, the SSOA area is halved. Figure 5.12b shows the SSOA under a 500 V DC-bus voltage, indicating at a lower DC-bus voltage, the reduction of the power capability with f_{sw} increasing is less obvious than at a higher voltage. In practice, to maximize the system power capability, a lower switching frequency or DC-bus voltage is always favorable as long as the performance requirement is met. In addition, the heatsink temperature also limits the maximum output current significantly. Therefore in the process of designing the heatsink, a high thermal-conductivity material, an enhanced air flow and even a liquid cooling will lower the heatsink temperature during a continuous running, which enlarges the SSOA area.

5.3.3 Impact of External Parameters

The AC inductance alters the grid-side SSOA boundaries. The partial derivatives of the switch current over L_g can clearly reveal its impact on the RBSSOA.

$$
\begin{cases}
\dfrac{\partial i_C(t)}{\partial L_g} \leq -\dfrac{3\Delta t}{2\left(L_{dc}+\frac{3}{2}L_{sCE}+\frac{3}{2}L_g\right)^2}u_{dc}(t) + \dfrac{U_{gm}\Delta t}{\left(\frac{2}{3}L_{dc}+L_{sCE}+L_g\right)^2} \\[4mm]
\dfrac{\partial i_C(t)}{\partial L_g} \leq -\dfrac{3\Delta t}{2\left(L_{dc}+\frac{3}{2}L_{sCE}+\frac{3}{2}L_g\right)^2}u_{dc}(t) + \dfrac{U_{gm}\Delta t}{\left(\frac{2}{3}L_{dc}+L_{sCE}+L_g\right)^2}
\end{cases}
\tag{5.48}
$$

Similarly, we can derive the trend of the SCSSOA versus L_g.

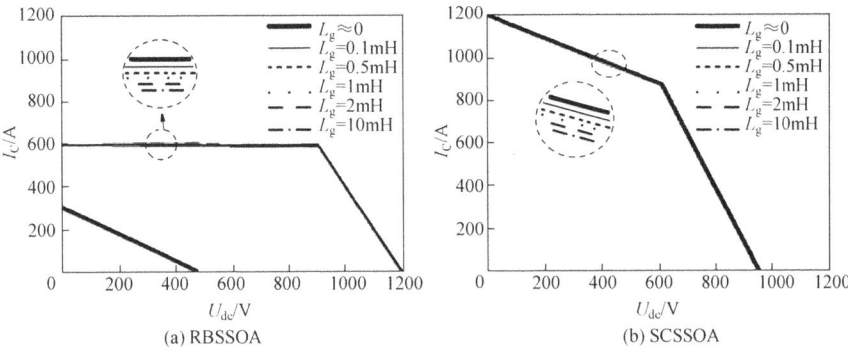

Fig. 5.13 SSOAs under different grid inductances

$$
\begin{cases}
\dfrac{\partial i_C(t)}{\partial L_g} \leq 0 \\[2mm]
\dfrac{\partial i_C(t)}{\partial L_g} \leq 0
\end{cases}
\tag{5.49}
$$

As shown in Eq. (5.48), with L_g increasing, the slew rates of the two RBSSOA boundary lines have similar trends, with their crossing points with the current axis increasing as well. Since the filtering inductance is ~mH, the slope increment is not obvious. However, the increment of the crossing points between the current-axis and RBSSOA boundaries enlarges the area of the RBSSOA, which is different from impact of the control delay and switching frequency. As shown in Eq. (5.49), the SCSSOA boundary is not altered with L_g, i.e., not sensitive to L_g. As a conclusion, the electric stress based SSOA will be widened with L_g increasing.

Shown in Fig. 5.13 are the SSOAs under different AC filtering inductances. When $L_g = 0.001$ mH, close to zero, RBSSOA shrinks significantly, shown as the bold line in Fig. 5.13a. When $L_g = 0.1$–10 mH, the RBSSOA is slightly enlarged. Regardless of L_g, the SCSSOA in Fig. 5.13b does not change, aligned with the previous analysis. Therefore L_g is not the most critical parameter of the SSOA. Even though a change is expected for the SSOA with L_g changing, such change is negligible with appropriately selected L_g. Hence in real applications, the impact of L_g not on the SSOA but on the grid-current harmonics is considered.

5.3.4 Impact of the Temperature

The switch current limit $I_{\lim}(T_j)$ and the voltage limit $U_{\lim}(T_j)$ are both functions of the junction temperature, especially $I_{\lim}(T_j)$ will drop sharply with T_j increasing. So is the SSOA. To facilitate the construction of SSOA, the heatsink temperature instead of the junction temperature is employed as one system characteristic parameter.

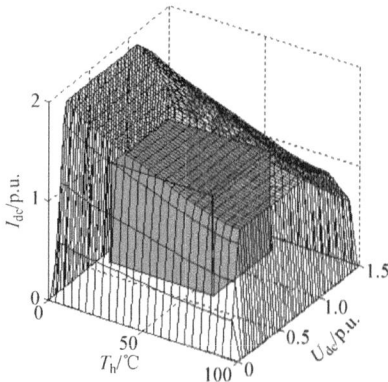

Fig. 5.14 The converter SSOC

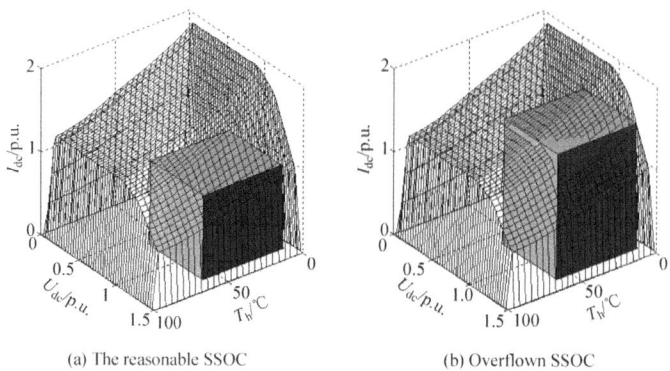

(a) The reasonable SSOC   (b) Overflown SSOC

Fig. 5.15 Temperature impact on the SSOC deployment

To differentiate from previous SSOA, the SSOA with the temperature information is named as systematic safe operating cube (SSOC), as shown in Fig. 5.14. It fully reveals the impact of the temperature on the converter SSOA. The areas enclosed by the grids are the SSOAs at different temperatures. The dark colored cube is the OA set upon the SSOA.

With the temperature rising, the switch-current limits shrink as well as the SSOA boundaries, which explains why the SSOA setting needs refer to the temperature. As shown in Fig. 5.15, the SSOA setting at the low temperature might not be applicable at the high temperature, given the current and voltage capability of the switch at the lower temperature is higher. The OA might overflow the SSOA at a higher temperature. In another word, without giving the full consideration to the temperature, the protection strategies based upon SSOA set at the low temperature might be invalid at the high-temperature operation, thereby not able to guarantee the system safe operation any more, as shown in Fig. 5.15b.

Fig. 5.16 Paralleled
switches in the converter

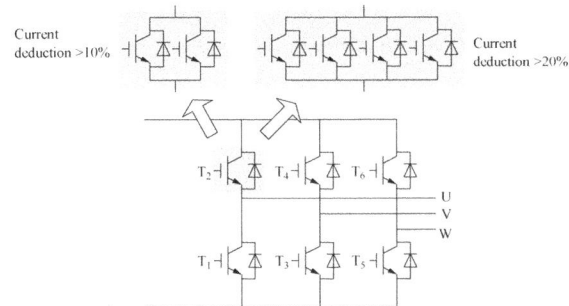

5.3.5 Impact of Paralleled Switches on SSOA

This section is mainly focused on the impact of paralleled switches on the SSOA deployment. In practice, we could parallel two and above small-current IGBTs to output high power, as shown in Fig. 5.16. Due to the potential current imbalance, current de-rating is usually a must to protect the paralleled switches, for instance, 10% of the current deduction for two in parallel and over 20% of the current deduction for 3 and more in parallel. The more switches in parallel, the more the current de-rating.

Such a current de-rating has a direct impact on the converter output capability, as shown in Fig. 5.17. Here 1200 V/500 A IGBTs are employed.

The two paralleled and four paralleled switches have been utilized in the actual 160 and 315 kW inverters, respectively. The de-rating of the switch current squeezes the SSOA thereby limiting the continuous power capability of the inverter. The well balanced switch current will widen the SSOA thereby lifting the system power capability. Therefore, the current balancing techniques are critical for the converter power capability and reliability. Since majority of the power electronic converters use the forced air cooling system, the imbalanced temperature distribution on the heatsink exists due to asymmetric cooling loops, which further causes the thermal imbalance among paralleled switches thereby putting the conventional current balancing techniques on a vulnerable position. Details have been discussed in Sect. 5.3.4, which will not be repeated here.

5.4 System Evaluation and Optimization Based on the SSOA

5.4.1 Procedure of the Evaluation and Optimization

For a typical AC–DC–AC power electronics converter, its structural blocks are shown in Fig. 5.18.

Descriptions of each block are shown as follows:

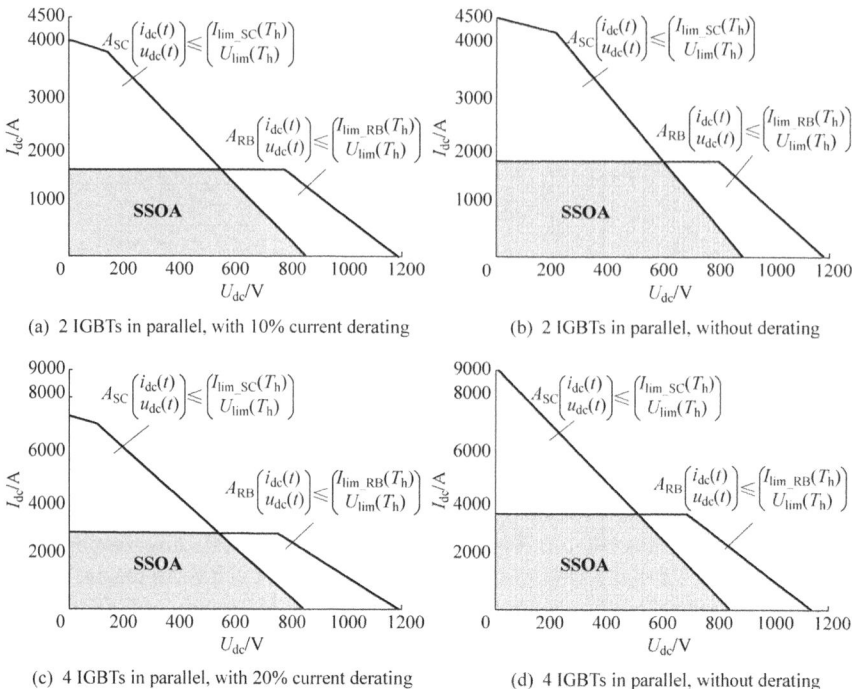

(a) 2 IGBTs in parallel, with 10% current derating

(b) 2 IGBTs in parallel, without derating

(c) 4 IGBTs in parallel, with 20% current derating

(d) 4 IGBTs in parallel, without derating

Fig. 5.17 Impact of current de-rating for the paralleled switches on the SSOA

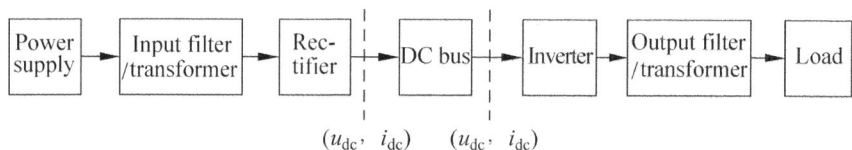

Fig. 5.18 A typical AC–DC–AC converter

(1) The power supply is either single-phase or three-phase. Its main characteristic parameters are phase numbers, rated voltage and frequency.

(2) The input filter/transformer is between the AC power supply and the rectifier, which is optional. It can be the direct connection, single filter, single transformer or filter + transformer.

(3) The rectifier could be single-phase or three-phase, two-level or multi-level, diode rectifier or PWM rectifier.

(4) The DC bus is usually a voltage type, made of DC capacitors. The voltage and current before and after the DC bus are main variables of the SSOA.

(5) The inverter part is a PWM inverter, either single-phase or three-phase, two-level or multi-level.

(6) The output filter/transformer is between the inverter and the load, which is
     optional. It could be direct connection, single filter, single transformer or filter
     plus transformer.
(7) The load is three-phase or single-phase, synchronous or asynchronous motors.
     It could be also the inductive or capacitive load.

The purpose of the converter evaluation and design based on SSOA is as follows.

(1) Establishing the SSOA and OA. The well-defined SSOA and OA provide accu-
     rate reference for the system protection.
(2) Reasonably selecting key components and evaluating the SSOA utilization. At
     this step we need comprehensively consider the circuit parameters, stray param-
     eters induced by the mechanical assembly, control parameters and interactions
     among components. A comparison between SSOA and OA reveals the actual
     utilization of components.
(3) Further optimizing system parameters. Based on the relationship between OA
     and SSOA, parameters of the circuit, mechanical assembly and control need be
     adjusted to optimize the overall system.

The evaluation and optimization flow chart is shown in Fig. 5.19, which could
be divided into two parts, i.e., internal evaluation and external optimization. For the
evaluation part, with the system specs and device selections, two separate routes
are laid out independently. One is related to power switches. Based on the switch
performance and related parasitics such as parasitic inductance, capacitance and
switching time, the DSOA could be determined. The other route considers the system
topology, control strategy, mechanical assembly and their derivatives, such as the
critical control parameters, circuit parameters and OA. By merging these two routes,
the SSOA could be quantitatively determined. The system evaluation could be carried
out by referring to relative positions of OA and SSOA. The evaluation part is the inner
loop of the optimization part. Design and evaluation results under various conditions
are substituted into a multi-objective optimization to obtain the optimal design.

Steps of applying the SSOA to the converter design are summarized as follows.

1) Based on the load and converter specs and the SSOA mathematical models,
    determine the ratings of the critical components and further select the appropriate
    components.
2) Based on component ratings and system specs, calculate the SSOA and test if the
    SSOA meets the load requirement and system performance. If so, move forward
    to design the prototype. If not, revisit component parameters based on the SSOA.
3) Based on the parameters fulfilling the SSOA, prototype the converter. Recalculate
    system parameters and furthermore the SSOA based on the final prototype. Settle
    down the ultimate OA, based on which the protection and control strategies are
    implemented.

The SSOA could secure the system reliability under all load conditions. Some
converters, due to mis-allocated parameters, limit the output power and remain distant
from the target SSOA. Thus, under the same component selection, enhancing the

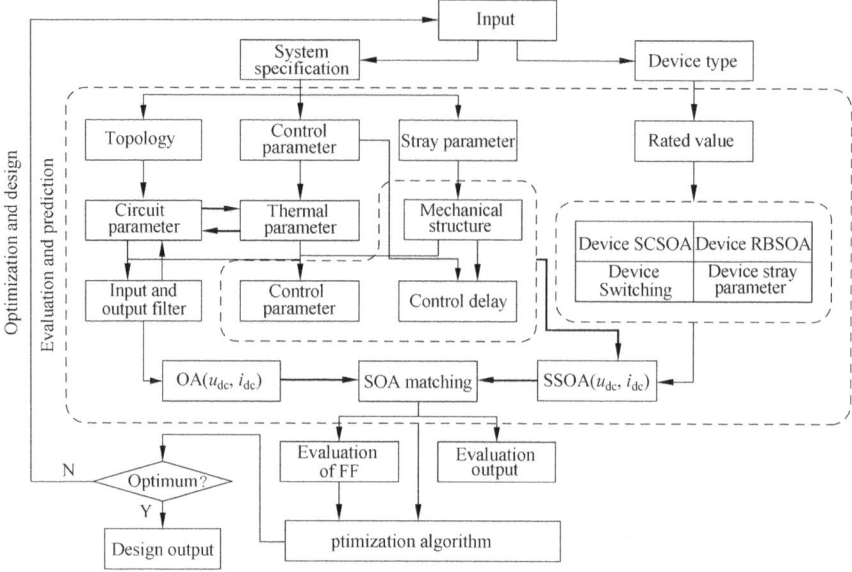

Fig. 5.19 The flow chart of the evaluation and optimization

system output capability is of importance. Based on the SSOA analysis, several factors impact the system output performance, as shown below.

(1) Switch application performance. The power switch is the building block of the power converter. Study of the switch application performance is the fundamental of the SSOA, which is influenced by the temperature, gate-drive circuits and etc., especially the techniques of series or paralleling power switches.

(2) Stray inductance of the DC bus bars, which along with the switch internal inductance limits the DC-bus voltage setting, according to the mathematical expression of the SSOA. Planar bus bars and snubber circuits will effectively mitigate the negativity of the stray inductance on the SSOA.

(3) Control delay, which is another major impact factor. The large current during the short-circuit fault is mainly caused by the control delay. Due to the delay of the control and pulse propagation, the switch cannot turn off the fault current in a timely manner, which causes the short-circuit current to rise rapidly during the delay time interval.

(4) System temperature. An appropriate design of the cooling system benefits the system output capability and reliability. As discussed in previous sections, the system temperature affects the converter performance. A high system temperature will reduce the current limit of the RBSSOA thereby shrinking the RBSSOA and weakening the system output power capability. Hence an appropriate design of the thermal system with the over-temperature protection is critical to the system reliability.

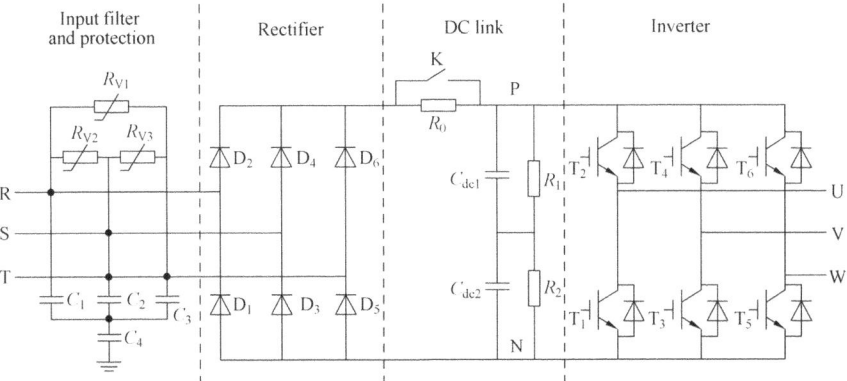

Fig. 5.20 The main-circuit topology of the two-level VSIs

5.4.2 SSOA Applications in Converter Seriations

To validate the effectiveness of the SSOA, here we apply the SSOA to three IGBT based two-level VSIs with different power ratings. The converters have the same topology, as shown in Fig. 5.20. Three major parts are included in the main circuit, i.e., rectifier, DC link and inverter. In addition, auxiliary circuits are needed such as the power supply, input filter and protection, sampling and protection circuits, cooling systems, etc.

Here the converter design using SSOA, including the component selection, protection strategy and system optimization, are to be demonstrated. With the target power ratings of 55, 160 and 315 kW, respectively, this section focuses on the component selection, SSOA and OA design and the experimental validation. Applying the SSOA to three different prototypes is not simply repeating the same procedure, given the SSOA of each prototype has its own emphasis. For the 55 kW system, the power switch in the main circuit is single IGBT module, different from the three-phase integrated power module (IPM). Therefore it requires the in-depth study of the single-module characteristics. For the 160 kW prototype, paralleled modules are employed. Under the forced air cooling, the thermal imbalance among modules is the focus. For the 315 kW system, more modules are to be paralleled, which requires even stricter on the SSOA design.

1. **The 55 kW/380 V inverter**

The component selection based on the SSOA is mathematically determined by Eq. (5.39). Based on such calculations of the IGBT ratings, the prototype parameters are settled as Table 5.3.

The calculated IGBT ratings are

Table 5.3 The design parameters of the 55 kW/380 V inverter prototype

Parameter	Value
P	55 kW
U_N	380 V
$\cos \varphi$	0.85
η	0.9
k_1	1.2
k_2	2.0
k_3	1.25
L	25 nH
L_σ	20 nH
L_{ls}	0.6 mH
L_{SC}	2.6 μH
C_{res}	1 nF
Δt	4 μs
t_f	90 ns

Table 5.4 The main components of the 55 kW/380 V inverter

Component type	Ratings	Quantity	Assembly manners
IGBT modules	1200 V/300 A	3	
Aluminum caps	5600μ/400 V	4	2 in series, 2 in parallel
Rectifier diodes	1200 V/100 A	3	
Pre-charge resistors	51 Ω/100 W	2	Parallel
DC-bus balancing resistors	5.1 kΩ/50 W	2	Series
Contactors	65 A × 3	1	
Cooling fans	22 W/220 V	3	Parallel

$$\begin{cases} \begin{pmatrix} I_{\lim\_RB}(T_j) \\ U_{\lim}(T_j) \end{pmatrix} \geq \begin{pmatrix} 379.7 \text{ A} \\ 891.1 \text{ V} \end{pmatrix} \\ \begin{pmatrix} I_{\lim\_SC}(T_j) \\ U_{\lim}(T_j) \end{pmatrix} \geq \begin{pmatrix} 1385 \text{ A} \\ 1472 \text{ V} \end{pmatrix} \end{cases} \tag{5.50}$$

Based on the calculation above, a 1200 V/300 A IGBT half-bridge module could meet the current requirement, though not the voltage requirement due to the DC-bus inductance L_{DC} and the control delay Δt. Further optimization of DC bus bars and gate-drive circuits is a must for the voltage rating. Actual component parameters of this inverter are shown in Table 5.4.

Given the selected IGBT modules cannot meet the system specs in Table 5.3, improvements of the DC-bus-bar structure and the gate-drive circuits are needed. Through such enhancement, the DC-bus equivalent stray inductance L_{dc} drops to

Fig. 5.21 The 55 kW/380 V
inverter and its SSOA

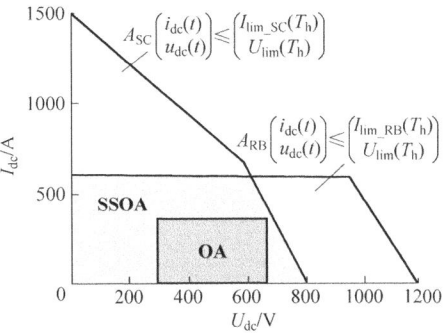

Fig. 5.22 The experimental
starting current of the 55 kW
inverter

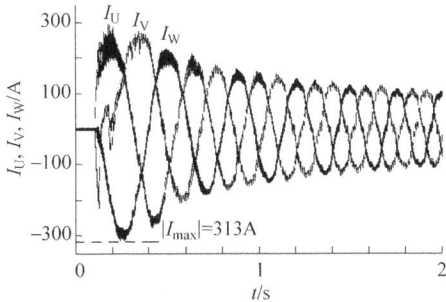

Fig. 5.23 The experimental
current of the 55 kW inverter
under 150% rated power

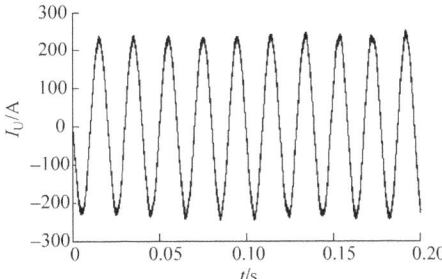

5 nH and the gate-drive delay is reduced to 3.7 μs, which makes the 1200 V/300 A
IGBT module an applicable candidate.

A 55 kW/380 V inverter was prototyped based upon the above specs, with the
actual SSOA shown in Fig. 5.21.

Inside such SSOA, the OA is the dark-colored zone with its three boundaries as the
over-voltage threshold of 700 V, the under-voltage threshold of 300 V and the over-
current protection threshold of 400 A. The experimental test is shown in Fig. 5.22
and 5.23. The three boundaries were used for the protection. As stated previously,
all these thresholds are the detected value for the protection, not the actual values
when the protection happens.

Parameter	Value
P	160 kW
U_N	380 V
$\cos \varphi$	0.9
η	0.9
k_1	1.2
k_2	2.0
k_3	1.25
L_{dc}	10 nH
L_s (2 IGBTs in parallel)	10 nH
L_{ls}	0.19 mH
L_{SC}	2.0 μH
C_{res} (2 IGBTs in parallel)	3.10 nF
Δt	3.8 μs
t_f	110 ns

Table 5.5 160 kW/380 V inverter parameters

For the experiments above, an inverter drove a 55 kW three-phase induction motor coupled with a 55 kW wind turbine, running at the open-loop mode. The starting current is shown in Fig. 5.22 and the steady current under 150% rated power is shown in Fig. 5.23. The conventional design sets the over-current protection threshold no higher than 300 A. With the enhancement of the DC-bus-bar structure and the control delay, such threshold could be increased to 400 A, widening the operation area. For the over-load operation shown in Fig. 5.23, the heatsink temperature is 75 °C, which further validated the effectiveness of applying the SSOA to the converter design, analysis and optimization.

2. **160 kW/380 V inverter**

Based on the component-selection formula, we could calculate the ratings of the IGBT used in the 160 kW/380 V inverter. The system design parameters are shown in Table 5.5.

The IGBT power ratings are calculated below.

$$\begin{cases} \begin{pmatrix} I_{\lim\_RB}(T_j) \\ U_{\lim}(T_j) \end{pmatrix} \geq \begin{pmatrix} 1042.7 \text{ A} \\ 899.3 \text{ V} \end{pmatrix} \\ \begin{pmatrix} I_{\lim\_SC}(T_j) \\ U_{\lim}(T_j) \end{pmatrix} \geq \begin{pmatrix} 2291.2 \text{ A} \\ 1171.7 \text{ V} \end{pmatrix} \end{cases} \tag{5.51}$$

Based on the calculation considering the potential current imbalance among the paralleled devices and the capability of the forced air cooling system, two paralleled 1200 V/450 A IGBT modules could meet the requirement. The component selections of the 160 kW/380 V inverter are shown in Table 5.6.

Table 5.6 The main components of the 160 kW/380 V inverter

Component type	Ratings	Quantity	Assembly manners
IGBT modules	1200 V/450 A	6	2 in parallel
Aluminum caps	5600µ/400 V	8	4 in parallel, 2 in series
Rectifier diodes	1200 V/380 A	3	
Pre-charge resistors	20 Ω/100 W	2	Parallel
DC-bus balancing resistors	10 kΩ/60 W	2	Series
Contactors	150 A × 3	1	
Cooling fans	25 W/220 V	3	Parallel

Fig. 5.24 The SSOA of the 160 kW inverter prototype

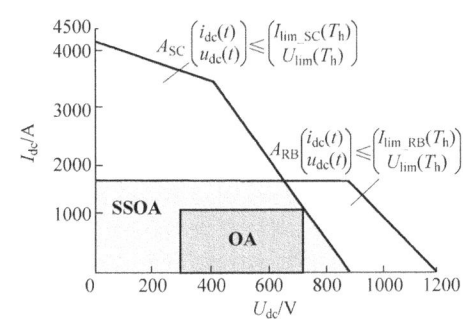

Fig. 5.25 The experimental starting current of the 160 kW inverter

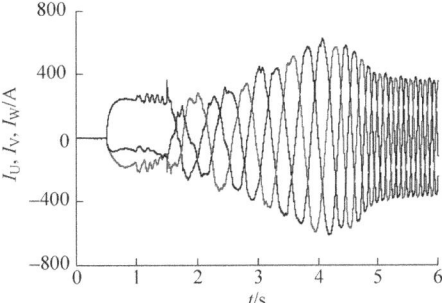

According to the specs and component selection above, a 160 kW/380 V inverter was prototyped, with the actual SSOA shown in Fig. 5.24.

Its three boundaries are the over-voltage threshold of 720 V, the under-voltage threshold of 300 V and the over-current threshold of 1150 A, respectively. The related load tests are shown in Fig. 5.25 and 5.26.

The inverter drove a 160 kW three-phase induction motor coupled with a DC generator as the load, which is further connected to a grid-tied four-quadrant DC/AC inverter. The speed/torque controller is employed for the DC machine to adjust the load. Figure 5.25 shows the experimental starting current of the 160 kW inverter. Figure 5.26 shows the experimental steady-state current of the induction motor under 150% rated load.

Fig. 5.26 The experimental current of the 160 kW inverter under 150% rated power

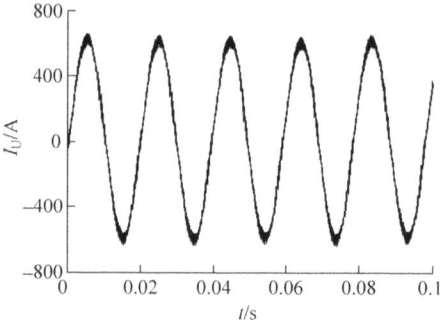

Compared to the 55 kW inverter design, in addition to lifting the power capability based upon the SSOA, the 160 kW design widens the SSOA region based on the study of paralleled switches to further increase the power capability. The conventional design requires the current de-rating of paralleled switches, e.g., 10% of the current de-rating for two paralleled 1200 V/450 A IGBTs, limiting the inverter output below 810 A. With the study of the current imbalance under the uneven thermal distribution, the inverter output-current limit is increased to 1150 A, which is a 40% increment compared to the conventional design.

3. **315 kW/380 V inverter**

The system parameters of the 315 kW/380 V inverter are shown in Table 5.7, based on which the IGBT ratings could be calculated with the SSOA model, as shown in Eq. (5.52).

Table 5.7 Parameters of the 315 kW/380 V inverter

Parameter	Value
P	315 kW
U_N	380 V
$\cos \varphi$	0.9
η	0.9
k_1	1.2
k_2	2.0
k_3	1.25
L_{dc}	10 nH
L_s(4 IGBTs in parallel)	5 nH
L_{ls}	0.10 mH
L_{SC}	1.5 μH
C_{res}(4 IGBTs in parallel)	6.20 nF
Δt	3.8 μs
t_f	110 ns

$$\begin{cases} \begin{pmatrix} I_{\text{lim\_RB}}(T_j) \\ U_{\text{lim}}(T_j) \end{pmatrix} \geq \begin{pmatrix} 2052.7 \text{ A} \\ 970.3 \text{ V} \end{pmatrix} \\ \begin{pmatrix} I_{\text{lim\_SC}}(T_j) \\ U_{\text{lim}}(T_j) \end{pmatrix} \geq \begin{pmatrix} 3715.1 \text{ A} \\ 1212.1 \text{ V} \end{pmatrix} \end{cases} \tag{5.52}$$

After taking the current imbalance among paralleled switches and the thermal capability of the forced air cooling system into account, we employed four 1200 V/450 A IGBT modules in parallel. The actual component selection of this 315 kW/380 V inverter is shown in Table 5.8.

The 315 kW/380 V inverter was prototyped based on the specs and design parameters above, with the actual SSOA shown in Fig. 5.27.

Its three boundaries as the over-voltage threshold of 700 V, under-voltage threshold of 300 V and the over-current threshold of 2100 A, respectively. Experimental results with the load are shown in Fig. 5.28 and 5.29, where Fig. 5.28 is the experimental waveform of the starting current and Fig. 5.29 is the steady-state current under 150% rated power.

Table 5.8 The main components of the 315 kW/380 V inverter

Component type	Ratings	Quantity	Assembly manners
IGBT modules	1200 V/450 A	12	4 in parallel
Aluminum caps	5600μ/400 V	18	9 in parallel, 2 in series
Rectifier diodes	1200 V/300 A	6	2 in parallel
Pre-charge resistors	24 Ω/100 W	3	3 in parallel
DC-bus balancing resistors	10 kΩ/60 W	4	2 in parallel, 2 in series
Contactors	260 A × 3	1	
Cooling fans	25 W/220 V	4	Parallel

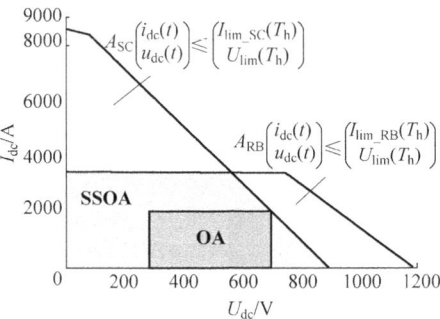

Fig. 5.27 The actual SSOA of the 315 kW inverter

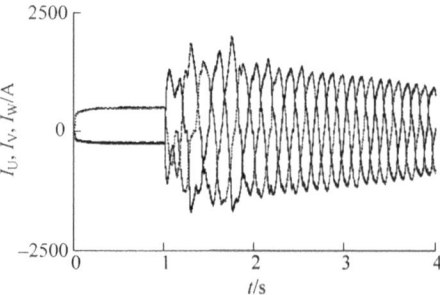

Fig. 5.28 The starting current of the 315 kW inverter with the load

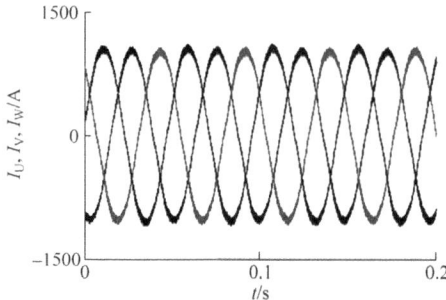

Fig. 5.29 The steady-state current of the 315 kW inverter under 150% rated power

Table 5.9 Performance comparison of three inverters

Rated power(kW)	Over-current capability(%)	Output-power enhancement(%)
55	258.9	33.3
160	271.0	42.0
315	251.3	45.8

    The major difference of such design is the temperature highly impacts the current distribution among paralleled switches, even though the cooling channel has been improved in terms of such thermal imbalance. When the inverter is running at the full load, differences remain among the case temperature of paralleled switches. When four modules are paralleled to form a 315 kW inverter, the uneven current distribution due to the thermal imbalance should not be overlooked. Based on the analysis in Chap. 3 for the paralleled switches, the compensation was made for the current imbalance thereby increasing the current capability of paralleled switches. Such effort benefits the inverter long-term output capability.

    A comparison of the three inverter prototypes is shown in Table 5.9. Here the over-current capability is a ratio of the short-term peak current over the rated current. The enhancement of the output power capability is the power increment using SSOA compared to the conventional design method.

5.4.3 Converter Evaluation and Protection Based on the SSOA

The SSOA can be employed for the system design and evaluation. It helps to maximize the device utilization, design the protection strategy for the off-the-shelf converters, and enforce the system reliability by locating the OA inside the SSOA.

1. **Evaluation of system parameters based on the SSOA**

The parameter evaluation is selecting appropriate components at the converter design stage through quantifying the system SSOA. For the dual PWM inverters with the rated power of 55 kW, the rated output current is 105 A and the rated voltage is 380 V. The IGBT options include 1200 V/300 A and 1700/300 A, considering the starting inrush current twice of the rated value. For the same type of IGBTs, different manufacturers and ID numbers are given. Here we use the volume of the SSOA as the evaluation matric, which is defined as the maximum power capability of a power electronic system under the given power devices. Such matric can be obtained through double integrals shown as follows.

$$C_{SSOA} = \iint i_C du_{dc} dT_h \qquad (5.53)$$

Once the main-circuit structure is settled down, the component selection complies with following procedures. Firstly, substitute device and main-circuit parameters into the SSOA mathematical model, calculate its boundaries, and conclude if such device could meet basic requirements of the SSOA, i.e., all possible running points are within the SSOA without overflowing. Secondly, for the components meeting the requirements, calculate the volume of the SSOA. For the same rated devices, select the one with the largest volume to maximize the SSOA region. For different rated devices, select the one with the minimum volume to maximize the device utilization.

Four types of 1200 and 1700 V dual-switch IGBT modules are compared in Table 5.10, all of which have similar packages and meet the SSOA boundaries. According to the principle of selecting the minimal SSOA volume for different rated switches, #II and IV are erased off. For the left two types, #I IGBT provides the largest SSOA volume, i.e., under the same system specs it provides the widest operating region thereby largest electrical allowance. Such IGBT needs be put on the top priority.

Therefore, in addition to evaluations of converters using different types of IGBTs, a further optimization on converter parameters can be carried out through employing the concept of SSOA volume.

2. **Design of the protection strategy based on the SSOA**

For the off-the-shelf converter, employing the SSOA will determine the safe operation regions thereby setting the applicable OA and protections. A reliable protection strategy secures the system reliability, though the conventional protection strategy

Table 5.10 SSOA comparison of different IGBTs

IGBT No.	I_{RBlim}/A	E_{on}/mJ	T_{jlim}/°C	$R_{th(j-c)Q}/(°C·W^{-1})$	$C_{SSOA}/(MVA·°C)$
	I_{SClim}/A U_{lim}/V	E_{off}/mJ E_{rec}/mJ		$R_{th(j-c)D}/(°C·W^{-1})$ $R_{th(c-h)}/(°C·W^{-1})$	
I	600	43	125	0.066	40.669
	1200	34		0.12	
	1200	15		0.015	
II	600	41	150	0.066	54.513
	1200	32		0.12	
	1200	22		0.015	
III	600	22	125	0.094	35.866
	1200	43		0.15	
	1200	26		0.009	
IV	600	100	150	0.083	89.296
	1400	110		0.13	
	1700	90.5		0.009	

Fig. 5.30 SSOA of 55 kW/380 V dual PWM converters

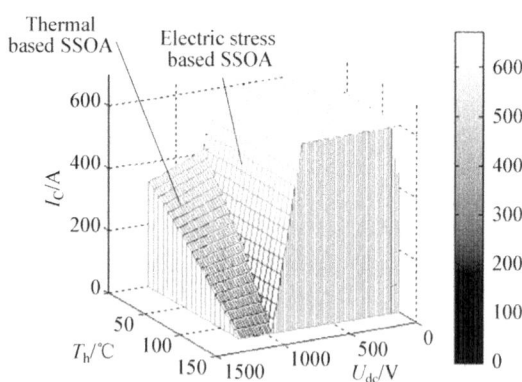

is solely based on the over-voltage, over-current and over-temperature without a comprehensive consideration of all related physical parameters.

Still take the 55 kW/380 V dual PWM inverters as an example. Some critical parameters in such a system are shown in Table 5.11. Here the DC-bus inductance is extracted using the PEEC method, the motor leakage inductance is the rated value, the short-circuit inductance is the experimental value through the phase-phase short-circuit test, and the IGBT thermal resistance and loss are obtained through the datasheet though not listed in the table. Based on the derivations above, its SSOA is shown in Fig. 5.30.

Based on the SSOA the system OA could be further derived, as shown in Fig. 5.31. In Fig. 5.31a, the OA overflows SSOA boundaries, unable to secure the system reliability thereby an irrational design. Figure 5.31b shows an applicable design.

Table 5.11 Key parameters of 55 kW dual PWM converters

Parameter	Value
U_{lim}/V	1200
I_{RBlim}/A	600
I_{SClim}/A	1200
C_{res}/nF	1
t_f/μs	0.13
L_{sCE}/nH	10
Δt/μs	1
L_{dc}/nH	57
U_g/V	380
L_g/mH	1
f_{sw}/kHz	6.4
L_{ls}/mH	0.6
L_{SC}/μH	1.8

When the operating point exceeds the OA boundaries, protections will be triggered to avoid any potential damage of IGBTs thereby securing the system reliability. Given its irregular shape does not facilitate the setting of the protections, in the actual applications we usually trim it into a regular shape. For instance, in this system, the DC-bus under-voltage threshold is 400 V, and the over-voltage threshold is 800 V with the rated value of 700 V. The OA for the current limitations is divided into two parts, one is 300 A when the heatsink temperature is 25–35 °C, double of the motor rated current of 105 A rms thereby securing the inverter full-load start, the other is 200 A for the steady-state operation when the heatsink temperature rises to 35–70 °C. Such settings prohibit the junction temperature from exceeding the maximum value and can be used as threshold values for the protection. The OA design is relatively flexible in different applications in order to maximize the utilization of the SSOA and power switches.

3. **Experimental validation of SSOAs**
(1) Electric stress based SSOA

IGBTs in the actual operations have similar switching behaviors, though the switching time has differences. To emulate actual operations, one IGBT in the 55 kW dual PWM converters is employed as the sample. As shown in Fig. 5.32, the test circuit is a Buck converter with the inductive load to emulate the motor leakage inductance. The over-current protection is to be enabled and tested. To vary the grid voltage, an adjustable transformer is located between the grid and the rectifier input.

Table 5.12 shows the actual turn-off current under the over-current protection when the DC-bus voltage varies. The over-current setting is 300 A. Once the current beyond the threshold is detected, the system immediately shuts down the switch. Due to the existence of the time delay, the actual turn-off current is always higher than the protection threshold, which further rises with the DC-bus voltage increasing.

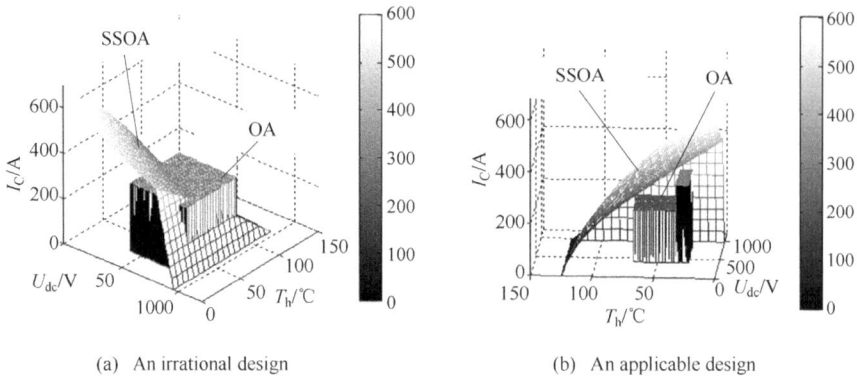

(a) An irrational design   (b) An applicable design

Fig. 5.31   SSOA and OA of 55 kW dual PWM converters

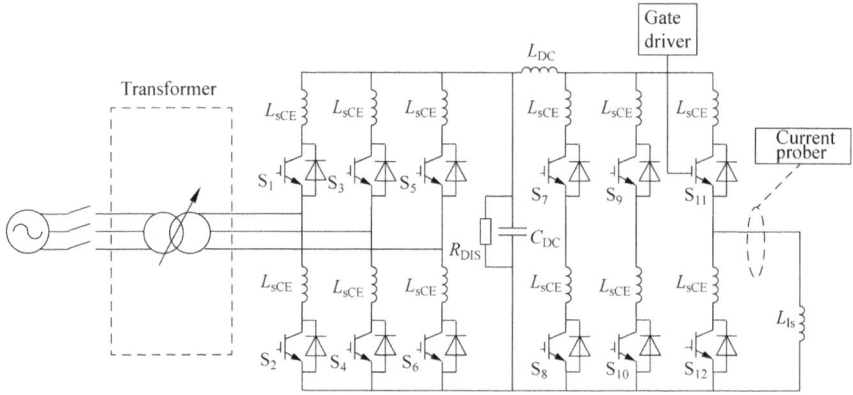

Fig. 5.32   Test circuit of the electric stress based SSOA

The experimental inductance is relatively large, yielding a slow current rising rate. Therefore the actual turn-off current has not much difference from the threshold value, 300 A. The measured turn-off current under different DC-bus voltages is shown in Fig. 5.33.

For the same IGBT above, we can further increase the over-current threshold to 700 A, replace the load inductor with a short-circuit inductor and implement the short-circuit test. With the DC-bus voltage varying from 100 to 600 V, the actual turn-off current with the short-circuit fault is shown in Fig. 5.34. It indicates that with smaller load inductance the actual turn-off current is far beyond the threshold value and becomes even worse with a higher DC-bus voltage. At the DC-bus voltage of 600 V, the actual turn-off current under the short-circuit fault approaches to 1200 A, the limit of the IGBT short-circuit current. Since the control delay Δt highly affects the actual turn-off current, a minimized time delay is rewarding to avoid the turn-off current higher than the SSOA boundary.

Table 5.12 Actual turn-off current under different DC-bus voltages

DC-bus voltage/V	Turn-off current/A
100	301
200	302
300	305
400	308
500	310
600	313
700	319

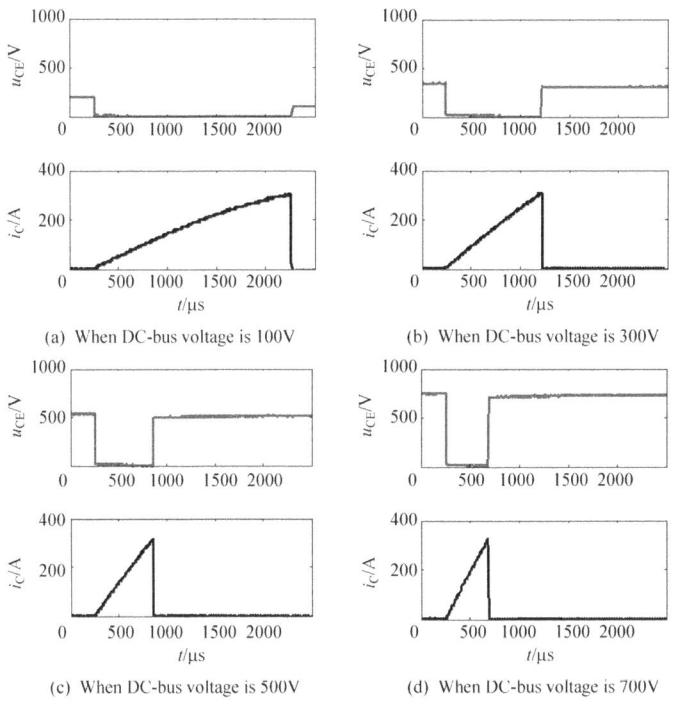

(a) When DC-bus voltage is 100V

(b) When DC-bus voltage is 300V

(c) When DC-bus voltage is 500V

(d) When DC-bus voltage is 700V

Fig. 5.33 The measured actual turn-off current under different DC-bus voltages

(2) Thermal based SSOA

Testing the thermal based SSOA design requires a long-term running. For the dual PWM inverters above, the cooling systems for the grid-side and motor-side modules are independent. To emulate the over temperature protection, a minor change is made on the cooling loop, i.e., plugging the inhaling port of thermal loop at the grid side while leaving the motor-side heatsink alone. A temperature sensor is employed to monitor the system temperature especially at the full load. After some running

Fig. 5.34 The actual
turn-off current under the
short-circuit fault with
different DC-bus voltages

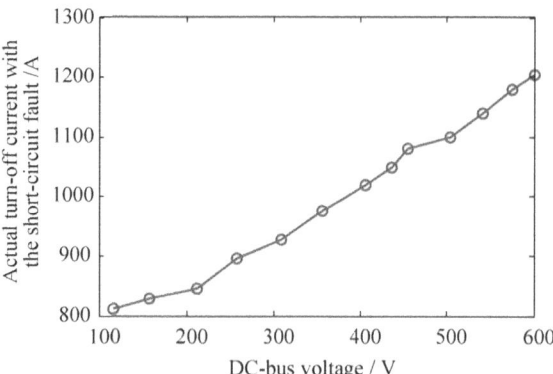

period, the grid-side inverter was heated up quickly to 93 °C when the damage of IGBT modules occurred. Even though the motor-side inverter undertook the reactive current thereby an overall current larger than the grid side, no module damage happened. Figure 5.35 shows the inverter running process tracked in OA. The DC-bus voltage was fixed at 700 V. Only the grid-side inverter current and the temperature are pictured. At $t = t_0$ the motor started with the full load and reached the steady state by $t = t_2$, during which the temperature gradually increased. A plugged cooling loop resulted in the system running to the point t_5, overflowing the SSOA thereby damaging the IGBT module. Shown in Fig. 5.36a is the IGBT module damaged due to the over temperature. Figure 5.36b shows the surrounding silicone gel around the module, which was squeezed out due to overheating. Even the full-load operation of the grid-side inverter at the DC-bus voltage of 700 V and current of 140 A did not exceed the electric-stress based SSOA with normal temperature, the heatsink over-temperature reduced the IGBT current capability sharply and further led to the running point overflowing the thermal based SSOA thereby causing the switch damage. Hence the temperature must be carefully considered when setting the protection strategy.

In addition, the IGBT module in Fig. 5.36a were damaged more in the middle than the right. It is because that the middle module was heated by the left and right module, yielding more heat cluster than its generated heat thereby a higher operational temperature and harsher ambient. The thermal distribution is given in Fig. 5.37, validating that the middle module bears the highest temperature. When setting the threshold value for the thermal protection, the middle-module temperature should be referred to.

As a destructive experiment, such test should be avoided in actual operations. An appropriate design of the cooling system and an unimpeded cooling loop facilitate the heat dissipation, lower the IGBT ambient temperature and prolong its lifespan.

As a summary, the SSOA is based on the DSOA, employs the DC-bus voltage, AC output current and heatsink temperature as the variables, takes the key factors such as parasitics, control delay and temperature into account, and bridges multi-

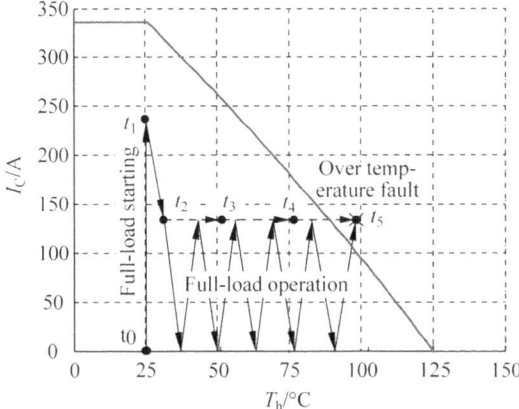

Fig. 5.35 State trajectory of the motor starting process

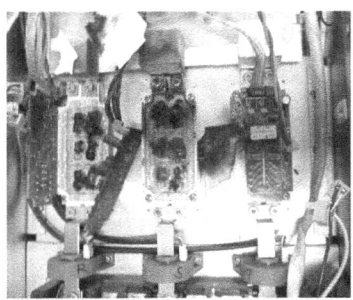

(a) Internal view of the damaged IGBT modules   (b) Silicone grease of the IGBT module

Fig. 5.36 The IGBT modules damaged by the over-temperature

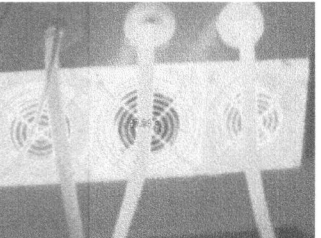

(a) Thermal distribution of modules   (b) Thermal distribution of air outlet

Fig. 5.37 The thermal distribution in a converter

timescale transients together. Based on time constants of different transients, the SSOA could be further divided. The stray inductance of DC bus bars, control time delays, switching frequency and etc. highly affects the SSOA boundary settings, which should be carefully deployed to widen the SSOA and maximize the device utilization. Compared to the DSOA, the SSOA comprehensively involves various influential factors thereby much closer to real applications. We could use the SSOA as the reference to quantify the component selection and system protection design, which secures the system reliability, maximizes the device utilization and enforces the long-term running capability simultaneously.

Chapter 6
Measurement and Observation of Electromagnetic Transients

Power electronic systems consist of interactions between the information and energy. To implement the effective energy propagation through the control system, the related energy-flow information needs be fed back to the control unit, i.e., measurement and analysis. From the control theory point of view, measurements and observations form the feedback loop of the closed-loop control, acting as the interface between the energy flow and information flow. Since majority of power electronic systems use the digital control algorithm with the discrete control system and information flow, while the main-power circuit carries the continuous energy flow, the observation system is the interface between the continuous and discrete parts of power electronics mixture systems.

The digitized power electronics control has its core strategy implemented in digital signal processors. The output of the observation system is the digital signal to manipulate the energy flow, the carrier of which is the control-unit firmware, i.e., weak-electricity parts in power electronic systems. At the same time, the input of the observation system is the energy flow, the carrier of which is the main-power circuit, i.e., strong-electricity part in power electronic systems. Therefore, in addition to providing the timely and accurate feedback information, the observation system needs meet requirements of the insulation, EMC and isolation between the strong and weak electricity circuits.

As shown above, the observation system integrates the energy with information, discrete with continuous signals, and software with hardware. Particularly in high-power electronic systems, the main circuit has high voltage and current, yielding more obvious transients. Meanwhile, the high-power semiconductor devices have limited switching frequency and control bandwidth. Variables in the observation system such as sampling frequency, control bandwidth, sampling points, delay, error, insulation, isolation and EMC will impact not only the system performance but also the reliability.

Based on the processing, transformation and transfer methods of physical variables in the power electronic converters, the observation system consists of the digital type and the analogue type. The former one aims at digital signals, feeding logic sig-

© Tsinghua University Press and Springer Nature Singapore Pte Ltd. 2019   243
Z. Zhao et al., *Electromagnetic Transients of Power Electronics Systems*,
https://doi.org/10.1007/978-981-10-8812-4_6

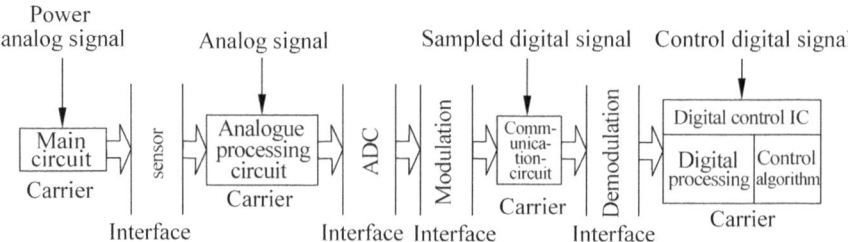

Fig. 6.1 A typical sampling system of power electronic converters

nals back to the control system, which however is not part of the core control strategy thereby not directly influencing the system control performance during normal operation. The latter one is also named as the sampling system, used to sample power signals, transform and process them into analogue signals acceptable to the control unit, and transfer and feedback them to the core control strategy. Such feedback signals are not only indispensable state variables for the closed-loop control but also important references for the system protection. Hence the performance of the sampling system highly impacts the power electronics control and reliability. With the progressing of power electronics technology, the demand for timely, accurate and reliable sampling systems is surging.

This chapter is focused on the structure and function of the sampling system, describes the difference between power signals and information signals, analyzes the impact of the sampling delay and error on the control performance, and improves the design of the sampling system.

6.1 Structure and Function of Sampling System

A typical sampling system of the power electronic converters is made of sensors, analogue processing circuit, analogue-to-digital conversion (ADC) ICs, digital control ICs, communication circuits, etc., as shown in Fig. 6.1.

Based on the characteristics of each element, the sampling system can be divided into the carrier and interface. The carrier is used to transfer and process physical variables without changing their properties. The interface, however, will transform the signal based on features and requirements of carriers, e.g., transforming power into information, analogue into digital, continuous into discrete and parallel into serial. Each part of the sampling system is detailed as follows.

1. **Sensors**

The International Electro-technical Committee (IEC) defines the sensor as "a front-end component to convert the input variable to measurable signals". Based on the application of sensors in power electronic systems, sensors in sampling systems can be defined as "devices or equipment to convert measured signals into related feasible electric signals". As the interface between the main circuit and the control system, sensors convert power analogue variables into information analogue signals with certain amplitude and accuracy, an indispensable element in the closed-loop control. The main specs of the sensors include measurable ranges, output range, accuracy, bandwidth, response time, insulating voltage, power loss, etc.

2. **Analogue processing circuits**

The output of the sensor will be converted into the digital signal through the ADC. However, it is usually impossible to perfectly match the sensor output with the ADC input. If the sensor output has a too wide range thereby exceeding the effective input range of the ADC, the saturation is expected. If the sensor output range is too narrow, it will result in a large error during the signal conversion. The ADC relative error is expressed as

$$
AD_{err} = \begin{cases} \left| \dfrac{u - AD_{in\_max}}{u} \right| & (u > AD_{in\_max}) \\[2mm] \left| \dfrac{u - AD_{in\_min}}{u} \right| & (u < AD_{in\_min}) \\[2mm] \left| \dfrac{AD_{in\_max} - AD_{in\_min}}{u(2^{AD_{bit}} - 1)} \right| & (AD_{in\_min} \leq u \leq AD_{in\_max}) \end{cases} \qquad (6.1)
$$

Here u is the input analogue signal of the ADC chip. AD_{in_min} and AD_{in_max} are the minimum and maximum input of the ADC, respectively. AD_{err} is the relative error of the ADC. One function of the ADC processing circuit is to adjust the amplitude and offset of the sensor output, covering the whole range of the ADC without overflow, i.e., minimizing the ADC relative error without sampling saturations. Besides, filters could be added to processing circuits to eliminate the noise and increase the precision.

3. **ADC ICs**

The digital control unit has been widely used in power electronic converters. The core control strategy and protection are all implemented in the digital control system, which requires the analogue states to be converted into digital signals and fed to the control unit. Majority of the actual state variables in power electronic systems are propagating in the form of power analogue signals. After sensors and analogue processing circuits, these signals are converted into information-level analogue signals, which later will be transformed by the ADC into discrete signals used in digital control processors.

The ADC process contains sampling, holding, quantifying and encoding. Sampling is to digitize continuous analogue signals and output pulse sequences, the amplitude of which is determined by the input analogue, as shown in Fig. 6.2. Usually, the sampling pulse width is very narrow. The holding circuit is to keep the present sample for the ADC before the next sampling pulse arrives, as shown in Fig. 6.3. The input analogue, after sampling and holding, is still a continuous trapezoidal analogue, which needs be quantified to a closest discrete number, given in

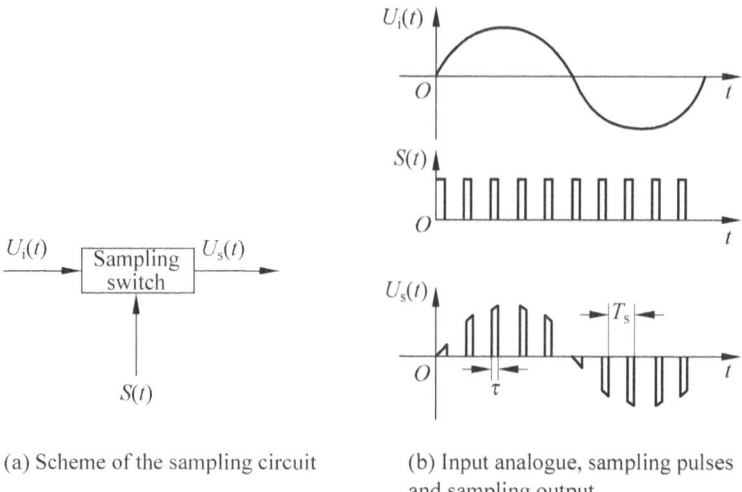

(a) Scheme of the sampling circuit

(b) Input analogue, sampling pulses and sampling output

Fig. 6.2 ADC processing

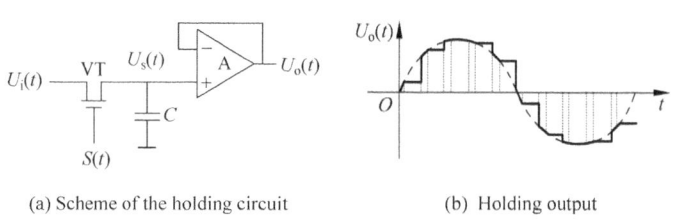

(a) Scheme of the holding circuit

(b) Holding output

Fig. 6.3 The sampling and holding of the ADC

the digital system an n-bit signal can only represent 2^n numbers. This step is called quantifying. Using digital signals to represent discrete signals is named as encoding.

Therefore in order to adapt digital control systems, the ADC digitizes the time and amplitude of analogue signals. Assume the ADC sampling rate is f_{sample}. Based on the sampling theory, converted digital signals all have the frequency lower than 1/2 f_{sample}. Therefore the signal bandwidth is narrowed, equivalent to adding a low-pass filter to the measured signal. The quantifying digitizes the signal amplitude. The signal amplitude before and after the quantifying is different due to the quantifying error, which lowers the conversion accuracy. Define the bit number of an ADC IC as AD_{bit} and the analogue input range as AD_{in_min}–Ad_{in_max}. The maximum quantifying error of the ADC is

$$\varepsilon_{max} = \frac{AD_{in\_max} - AD_{in\_min}}{2^{AD_{bit}} - 1}$$

(6.2)

which indicates that the sampling frequency and bit count determines the ADC bandwidth and accuracy. In addition, the input range and the channel count are also important indexes of the ADC.

4. **Digital Control Units**

For the power electronic system with the digital control unit, all the core control strategy is realized inside the digital control unit. Due to the data type and the unification, digital signals after the ADC cannot be directly fed to the digital control processor. All such data conversion, unification and amplitude transformation could be named as the digital signal processing. Meanwhile to further meet the requirement of the control unit on the feedback signals and improve the control performance, software filters and data diagnosis could be added.

5. **Communication**

The electromagnetic energy transfer and transformation is large in high-power electronic systems, which require large physical dimensions and cause strong EMI. To mitigate the impact of the EMI on the control system, we usually distant the control unit from the main-power circuit, which yields the long propagation distance of digital signals thereby bringing disturbance in the communication. To secure the reliable, accurate and timely sampling, high EMC and effective data diagnosis and retransmission are the must, with the certain communication rate. To lower the firmware cost, the serial communication is often recommended, which requires the highly efficient and reliable modem before and after the communication carrier. The data-bus resources need be allocated wisely based on data characteristics and usage to meet the communication requirement in different systems.

6.2   Difference of Power and Signals in Sampling System

A voltage-source three-level PWM rectifier is a perfect candidate to illustrate non-ideal characteristics of the feedback loop in power electronic systems. Compared to the diode rectifier, the PWM rectifier has features of relatively low input current harmonics, high power factor, controllable DC-bus voltage and bidirectional power flow, etc., which explains why it has been widely used in the motor control, DC-power transmission, reactive power compensation, active power filter, electric traction, grid-tied renewable power systems, and etc. As shown in previous chapters, compared to the two-level rectifier, the three-level PWM rectifier has lower voltage stress across the switch and lower switching frequency with the same amount of harmonics, which makes it well positioned for high-voltage applications. Shown in Fig. 6.4 is an NPC three-level PWM rectifier.

In Fig. 6.4, L_s and R_s are the equivalent inductance and resistance of the grid-tied inductor, respectively. C_{dc1} and C_{dc2} are two DC-bus capacitors. e_a, e_b and e_c are the three-phase voltage. u_a, u_b and u_c are three-phase inverter voltage. I_{dc} is the current

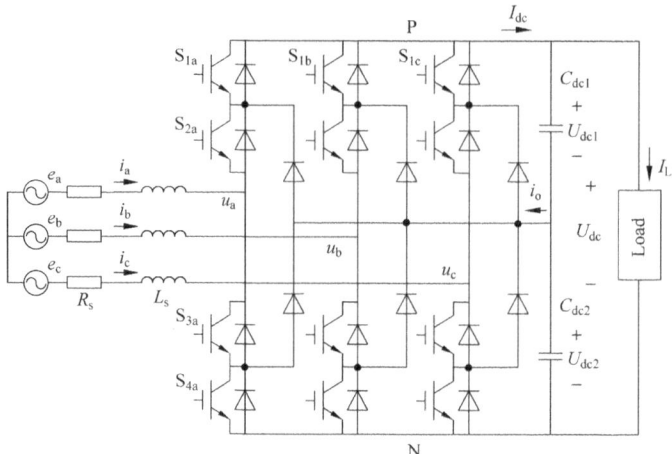

Fig. 6.4 NPC three-level PWM rectifier topology

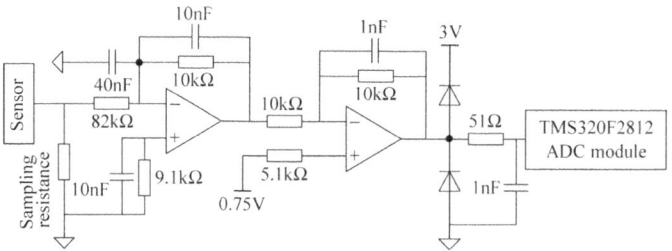

Fig. 6.5 The sampling system of the three-level PWM rectifier

flowing through the positive DC bus. i_o is the neutral-point current. U_{dc1} and U_{dc2} are the voltage across C_{dc1} and C_{dc2}, respectively. I_L is the load current.

Shown in Fig. 6.5 is the actual sampling circuit in the three-level PWM rectifier. Since the sampling circuit is very close to the control processor, no communication circuits or related modems are needed. The sensor and ADC module are the front and end carrier interfaces of the sampling system shown in Fig. 6.1, respectively. The analogue circuit in between is the carrier of the analogue signal processing.

During the hardware design, the sensor output current passes through the sampling resistor and gets converted to a voltage signal between -12 V and $+12$ V. Two OPAMs are used, with the first-stage as the inverting proportional circuit and the second-stage as the subtraction circuit, respectively. The voltage signal across the sampling resistor is converted to a voltage signal of 0–3 V, matching the ADC input range. The ADC is then executed by the ADC module of the DSP.

The software design needs set the appropriate ADC clock frequency and the sampling window. The ADC sampling is in the cascade sequencing mode and completes sampling of all channels, i.e., two DC-voltage channels, two grid-voltage channels

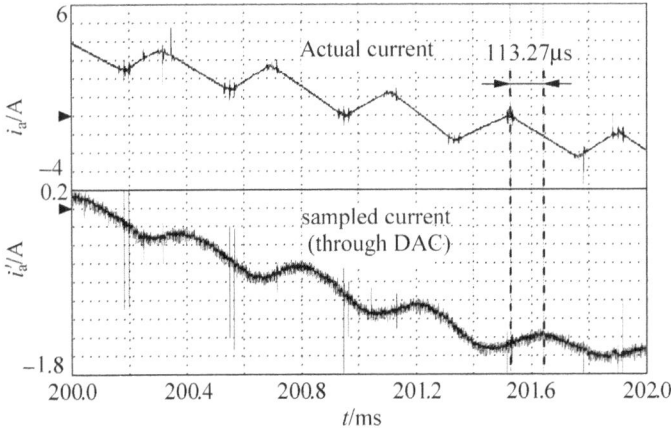

Fig. 6.6 The actual and sampled grid current

and two grid-current channels. The AD conversion is triggered upon the initiation of the sampling period, using the query mode. Once the related register is set to 0, the conversion is complete and ready for the data reading.

As shown in Fig. 6.1, the sampling system of power electronic converters usually adopts the serial structure. Delay and distortion of physical variables in their own carriers and interfaces are expected. Such non-linear delay and distortion are accumulated and overlapped in the serial system thereby forming the delay and error of the overall sampling system.

Experimental results show the obvious difference between the measured voltage/current and sampled feedback digital signals. As shown in Fig. 6.6, the actual phase-A current of the grid (measured value) and the sampled value (feedback signal) are compared. Here the sampled waveform is generated by the DAC module of the DSP after the ADC. The existence of the inductor between the grid and converter results in the variable current changing rate at different switching modes. With the rectifier sampling frequency and switching frequency much higher than the grid line frequency, it is assumed the voltage change between two switching actions is negligible thereby linearizing the current in between. It is clear that even after compensating the maximum transformation time of the DAC module the sampled current still shows a delay of about 100 μs from the actual value. When the measured value and the sampled current do not cross the zero at the same time, the sampling circuit has the DC offset. In addition, some high-frequency components of the actual current were filtered off, which generates the sampling error.

6.3 Impact of Sampling Delay and Error on Control Performance

As shown above, the measured analog signal through the sampling system is converted to the feedback digital signal, imposed with the delay and error. Namely, the information carried by the feedback signal has the distortion and hysteresis. This will directly impact the observation and control implementation of the control unit in power electronic systems and deteriorate the control performance. In this section, for the commonly used voltage vector oriented control (VOC) and the direct power control (DPC), the impact of the sampling delay and error on the control performance is quantified in the frequency and time domains, respectively.

This analysis is based on the mathematical model of the NPC three-level PWM rectifier. Define the transfer function of three-phase abc static coordinates to the two-phase $\alpha\beta$ static coordinates as

$$\begin{bmatrix} x_\alpha \\ x_\beta \end{bmatrix} = \sqrt{\frac{2}{3}} \begin{bmatrix} 1 & -\frac{1}{2} & -\frac{1}{2} \\ 0 & \frac{\sqrt{3}}{2} & -\frac{\sqrt{3}}{2} \end{bmatrix} \begin{bmatrix} x_a \\ x_b \\ x_c \end{bmatrix} \tag{6.3}$$

Here the α-axis in $\alpha\beta$ coordinates is aligned with the a-axis in the abc coordinates. The coefficient $\sqrt{2/3}$ guarantees the identical power before and after the transformation. The transfer function of $\alpha\beta$ coordinates to the rotatory dq coordinates is

$$\begin{bmatrix} x_d \\ x_q \end{bmatrix} = \begin{bmatrix} \cos\theta & \sin\theta \\ -\sin\theta & \cos\theta \end{bmatrix} \begin{bmatrix} x_\alpha \\ x_\beta \end{bmatrix} \tag{6.4}$$

Here θ is the angle between the d-axis in the dq coordinates and the α-axis in the $\alpha\beta$ coordinates.

Define the three-phase switching functions as follows ($i = a, b, c$):

(1) S_{1i} and S_{2i} on, S_{3i} and S_{4i} off, i.e., $s_{ip} = 1$, $s_{in} = 0$;
(2) S_{3i} and S_{4i} on, S_{1i} and S_{2i} off, i.e., $s_{ip} = 0$, $s_{in} = 1$;
(3) S_{2i} and S_{3i} on, S_{1i} and S_{4i} off, i.e., $s_{ip} = 0$, $s_{in} = 0$.

1. **In three-phase abc static coordinates**

$$\mathbf{Z\dot{X} = AX + Be} \tag{6.5}$$

Here,

$$
\begin{cases}
\mathbf{Z} = \mathrm{diag}\begin{bmatrix} L_s & L_s & L_s & C_{dc1} & C_{dc2} \end{bmatrix} \\[2mm]
\mathbf{X} = \begin{bmatrix} i_a & i_b & i_c & U_{dc1} & U_{dc2} \end{bmatrix}^{\mathrm{T}} \\[2mm]
\mathbf{A} = \begin{bmatrix}
-R_s & 0 & 0 & -(s_{ap} - \frac{s_{ap}+s_{bp}+s_{cp}}{3}) & (s_{an} - \frac{s_{an}+s_{bn}+s_{cn}}{3}) \\
0 & -R_s & 0 & -(s_{bp} - \frac{s_{ap}+s_{bp}+s_{cp}}{3}) & (s_{bn} - \frac{s_{an}+s_{bn}+s_{cn}}{3}) \\
0 & 0 & -R_s & -(s_{cp} - \frac{s_{ap}+s_{bp}+s_{cp}}{3}) & (s_{cn} - \frac{s_{an}+s_{bn}+s_{cn}}{3}) \\
s_{ap} & s_{bp} & s_{cp} & 0 & 0 \\
-s_{an} & -s_{bn} & -s_{cn} & 0 & 0
\end{bmatrix} \\[2mm]
\mathbf{B} = \mathrm{diag}\begin{bmatrix} 1 & 1 & 1 & -1 & -1 \end{bmatrix} \\[2mm]
\mathbf{e} = \begin{bmatrix} e_a & e_b & e_c & I_L & I_L \end{bmatrix}^{\mathrm{T}}
\end{cases}
\tag{6.6}
$$

2. In two-phase $\alpha\beta$ static coordinates

Applying Eq. (6.3) to the mathematical model under abc coordinates, we derive the mathematical model of the three-phase PWM rectifier under $\alpha\beta$ coordinates as follows.

$$
\begin{cases}
L_s \dfrac{di_\alpha}{dt} = -R_s i_\alpha - u_\alpha + e_\alpha \\[2mm]
L_s \dfrac{di_\beta}{dt} = -R_s i_\beta - u_\beta + e_\beta \\[2mm]
C_{dc1} \dfrac{dU_{dc1}}{dt} = s_{\alpha p} i_\alpha + s_{\beta p} i_\beta - I_L \\[2mm]
C_{dc2} \dfrac{dU_{dc2}}{dt} = -s_{\alpha n} i_\alpha - s_{\beta n} i_\beta - I_L
\end{cases}
\tag{6.7}
$$

Here

$$
\begin{cases}
\begin{bmatrix} s_{\alpha p} \\ s_{\beta p} \end{bmatrix} = \sqrt{\dfrac{2}{3}}
\begin{bmatrix} 1 & -\frac{1}{2} & -\frac{1}{2} \\ 0 & \frac{\sqrt{3}}{2} & -\frac{\sqrt{3}}{2} \end{bmatrix}
\begin{bmatrix} s_{ap} \\ s_{bp} \\ s_{cp} \end{bmatrix} \\[4mm]
\begin{bmatrix} s_{\alpha n} \\ s_{\beta n} \end{bmatrix} = \sqrt{\dfrac{2}{3}}
\begin{bmatrix} 1 & -\frac{1}{2} & -\frac{1}{2} \\ 0 & \frac{\sqrt{3}}{2} & -\frac{\sqrt{3}}{2} \end{bmatrix}
\begin{bmatrix} s_{an} \\ s_{bn} \\ s_{cn} \end{bmatrix} \\[4mm]
u_\alpha = s_{\alpha p} U_{dc1} - s_{\alpha n} U_{dc2} = \sqrt{\dfrac{2}{3}} \begin{bmatrix} 1 & -\frac{1}{2} & -\frac{1}{2} \end{bmatrix} \begin{bmatrix} u_a & u_b & u_c \end{bmatrix}^{\mathrm{T}} \\[4mm]
u_\beta = s_{\beta p} U_{dc1} - s_{\beta n} U_{dc2} = \sqrt{\dfrac{2}{3}} \begin{bmatrix} 0 & \frac{\sqrt{3}}{2} & -\frac{\sqrt{3}}{2} \end{bmatrix} \begin{bmatrix} u_a & u_b & u_c \end{bmatrix}^{\mathrm{T}}
\end{cases}
\tag{6.8}
$$

3. In *dq* rotatory coordinates

Assume the *d*-axis of the *dq* coordinates is aligned with the grid voltage vector. Its initial phase angle is 0 and the angular velocity is ω. Therefore

$$\theta = \omega t = 2\pi f_e t \tag{6.9}$$

Here f_e is the grid-voltage frequency. This forms synchronous *dq* coordinates. Applying Eq. (6.4) to the mathematical model under $\alpha\beta$ coordinates, we could derive the mathematical model under *dq* coordinates as follows.

$$\begin{cases} L_s\dfrac{di_d}{dt} = -R_s i_d + \omega L_s i_q - u_d + \sqrt{3}E \\[2mm] L_s\dfrac{di_q}{dt} = -R_s i_q - \omega L_s i_d - u_q \\[2mm] C_{dc1}\dfrac{dU_{dc1}}{dt} = s_{dp}i_d + s_{qp}i_q - I_L \\[2mm] C_{dc2}\dfrac{dU_{dc2}}{dt} = -s_{dn}i_d - s_{qn}i_q - I_L \end{cases} \tag{6.10}$$

Here E is the RMS value of the grid phase voltage. Relationships among other variables are

$$\begin{cases} \begin{bmatrix} s_{dp} \\ s_{qp} \end{bmatrix} = \begin{bmatrix} \cos\theta & \sin\theta \\ -\sin\theta & \cos\theta \end{bmatrix} \begin{bmatrix} s_{\alpha p} \\ s_{\beta p} \end{bmatrix} \\[4mm] \begin{bmatrix} s_{dn} \\ s_{qn} \end{bmatrix} = \begin{bmatrix} \cos\theta & \sin\theta \\ -\sin\theta & \cos\theta \end{bmatrix} \begin{bmatrix} s_{\alpha n} \\ s_{\beta n} \end{bmatrix} \\[4mm] u_d = s_{dp}U_{dc1} - s_{dn}U_{dc2} = \begin{bmatrix} \cos\theta & \sin\theta \end{bmatrix}\begin{bmatrix} u_\alpha & u_\beta \end{bmatrix}^T \\[3mm] u_q = s_{qp}U_{dc1} - s_{qn}U_{dc2} = \begin{bmatrix} -\sin\theta & \cos\theta \end{bmatrix}\begin{bmatrix} u_\alpha & u_\beta \end{bmatrix}^T \end{cases} \tag{6.11}$$

6.3.1 *Frequency-Domain Analysis*

VOC is one of the most exemplary and commonly used control strategies of the PWM rectifier. It has excellent stability, fixed switching frequency and ease of decoupling control, etc. Shown in Fig. 6.7 is the block diagram of the VOC, which is made of three control loops. The external loop controls the DC-bus voltage, while the internal two loops control the *d*-axis current and *q*-axis current, respectively. Three-level SVPWM is adopted as the fundamental modulation method. In addition, the PI

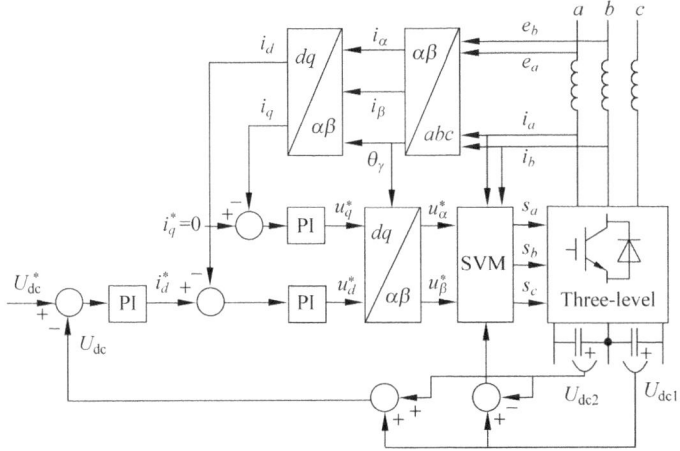

Fig. 6.7 A typical three-level VOC

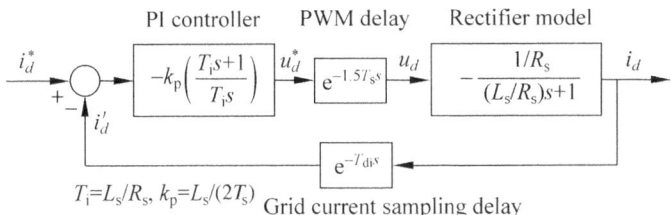

$T_i = L_s/R_s,\ k_p = L_s/(2T_s)$ Grid current sampling delay

Fig. 6.8 The frequency-domain model of the current loop in VOC considering the sampling delay

controllers have very certain transfer functions in the frequency domain, facilitating its frequency-domain analysis.

It can be seen that the three-level PWM rectifier needs sample the grid voltage, grid current and DC-bus voltage then forward them to the control algorithm. For the three-phase three-wire grid, at least two grid voltages, two grid currents and two DC-bus voltages need be sampled, which add up to 6 sampling channels. Here the grid voltages and currents are all in AC forms while the bus voltages are the DC ones.

1. Impact of the sampling delay on the control performance

Consider the sampling delay of the grid current, i.e., the sampled value falls behind the actual value. The frequency-domain model of the VOC used in the three-level PWM rectifier is shown as Fig. 6.8.

In this model, two delays are to be considered. One is the PWM delay, the other is the current sampling delay. Here T_s is the SVPWM control period and T_{di} is the sampling delay of the grid current. The reloading mechanism of the microprocessor causes the delay of one control period for the PWM generation. When using the center-aligned PWM generation, it is assumed that the present sampling current will be used at the center moment of the next control period. Therefore the overall control delay caused by the PWM generation is $1.5T_s$. When designing control parameters, we could refer to Fig. 6.8 for the current PI controller, i.e., adjusting the control loop as a typical I-type system without considering the sampling delay, which yields the optimum second-order dynamic response.

The open-loop transfer function of the current loop based on the frequency-domain model shown in Fig. 6.8 is

$$G_0(s) = \frac{1}{2T_s \cdot s} \cdot \exp(-1.5T_s \cdot s) \cdot \exp(-T_{di} \cdot s) \tag{6.12}$$

which can be further changed into the form of the logarithm, i.e.,

$$L(\omega) = -20\lg(2T_s \cdot \omega) \tag{6.13}$$

and its phase-frequency characteristics are

$$\Phi(\omega) = -90° - 180° \times \frac{(1.5T_s + T_{di})\omega}{\pi} \tag{6.14}$$

Furthermore the phase-angle margin is

$$\gamma = 90°(1 - \frac{1.5 + T_{di}/T_s}{\pi}) \tag{6.15}$$

A large sampling delay T_{di} will reduce the angle allowance, thereby weakening the system stability. Here the precondition for the system stability is

$$T_{di} < (\pi - 3/2)T_s \tag{6.16}$$

The bode plot of the current open loop when $T_s = 0.2$ ms is shown in Fig. 6.9. The impact of the sampling delay is illustrated.

As shown in Fig. 6.10, with the sampling delay increasing, the d-axis, q-axis and phase-a current of the three-level PWM rectifier are simulated. An increment of the sampling delay results in the decrease of the stability margin, e.g., increased oscillation of the d-axis and q-axis current, lower oscillation frequency and higher current harmonics. When $T_{di} > 0.33$ ms, Eq. (6.16) indicates the instability of the system.

The analysis above shows the impact of the sampling delay on the system steady-state performance based on the frequency-domain modelling of the three-level PWM rectifier. To facilitate the analysis of the dynamic performance, a linearization of the frequency-domain model is needed. Two non-linear delays are included in the model shown in Fig. 6.8, i.e., PWM block and the current sampling block. Since the time constant of the high-power PWM rectifier, L_s/R, is much larger than the delay of the sampling and the PWM link, we could use the first-order inertia to represent two

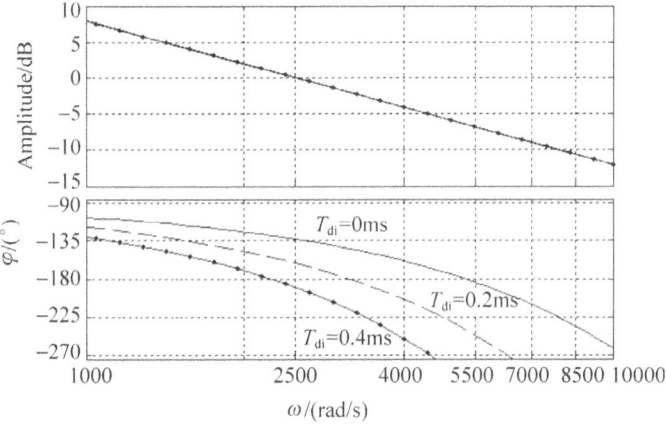

Fig. 6.9 The bode plot of the current open-loop control with variable sampling delays

time-delay blocks. The ultimate linear frequency model of the VOC current-loop is shown in Fig. 6.11.

Furthermore, its open-loop transfer function is

$$G_0(s) = \frac{1}{2T_s^2 T_{di} \cdot s(s + \frac{1}{T_s})(s + \frac{1}{T_{di}})} \tag{6.17}$$

with its closed-loop transfer function as

$$G(s) = \frac{T_{di}s + 1}{2T_s^2 T_{di}s^3 + 2T_s(T_s + T_{di})s^2 + 2T_s s + 1} \tag{6.18}$$

Based on Eq. (6.18), the root locus of the closed-loop transfer function considering the time delay is shown in Fig. 6.12. Three poles are exhibited, two of which are conjugate poles dominant over the system dynamic performance.

The variation of the system overshoot, modulation time and the unit step response versus the sampling delay is shown in Figs. 6.13, 6.14 and 6.15, respectively. It can be seen that the larger the sampling delay, the larger the overshoot and the longer modulation time thereby the more deteriorated dynamic performance. Simulation results of the current in Fig. 6.10 also verified the above analysis.

2. Impact of the sampling error on the control performance

In addition to the sampling delay, the sampling error has the impact on the control performance as well. Through comprehensively considering the sampling error of the voltage control loop and two current control loops in the VOC control, the frequency-domain model of the three-level PWM converter is shown in Fig. 6.16. Here i_{d_err} and i_{q_err} are the sampling errors of the d-axis current and q-axis current, respectively. U_{dc_err} is the sampling error of the DC-bus voltage. k_v and T_v are the proportional

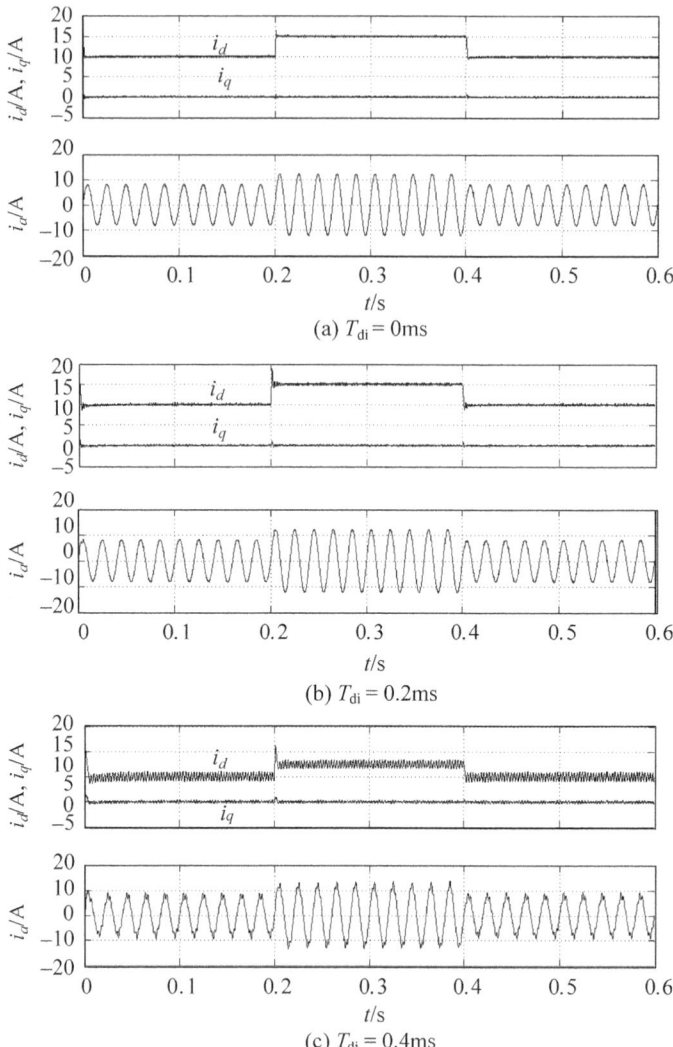

Fig. 6.10 d-axis current, q-axis current and phase-a current with different current sampling delays

coefficient and integral time of the voltage-loop PI controller, respectively. m and φ are the modulation index and initial phase angle of the AC fundamental voltage at the rectifier grid side, respectively. In the modeling process, the PWM harmonics are ignored with only the low-order harmonics considered. From such a model, the d-axis current in the frequency domain can be expressed as

$$i_{d}(s) = i_{di}(s) + i_{de}(s) \tag{6.19}$$

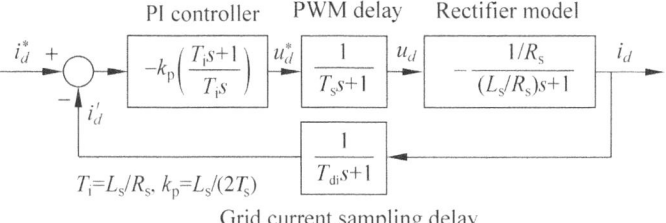

Fig. 6.11 The linear frequency model of the VOC current loop considering the current sampling delay

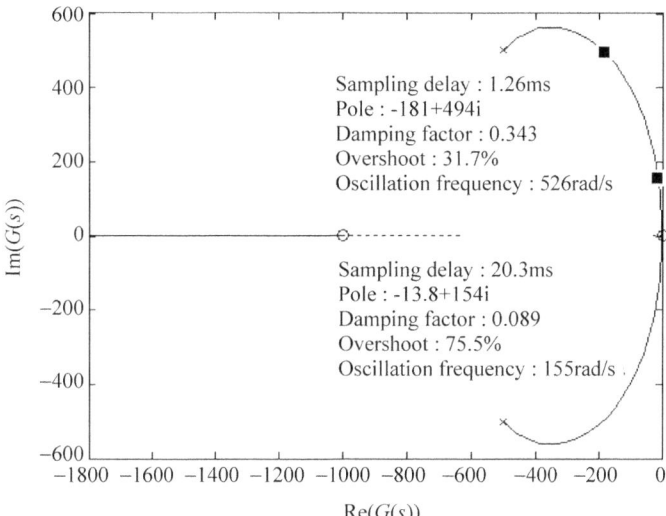

Fig. 6.12 The root locus of the VOC current loop considering the time delay

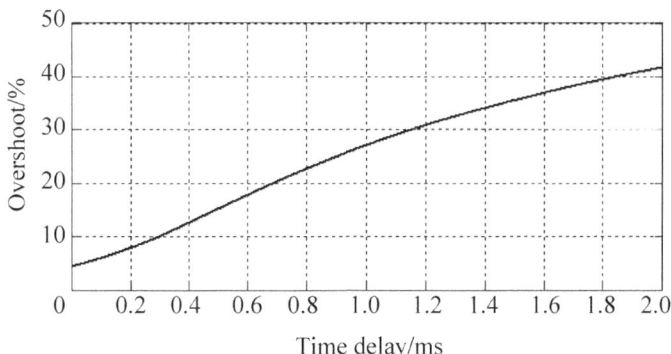

Fig. 6.13 Overshoot versus the sampling delay

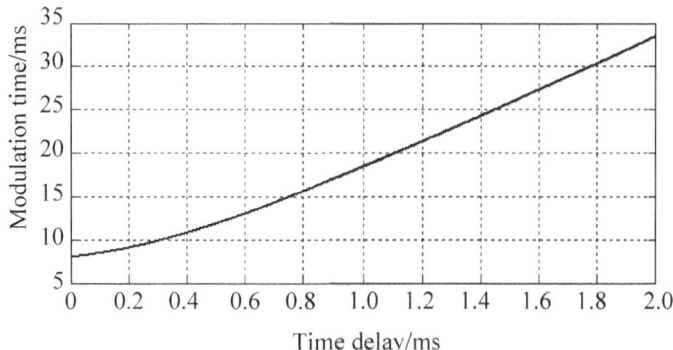

Fig. 6.14 Modulation time versus the sampling delay

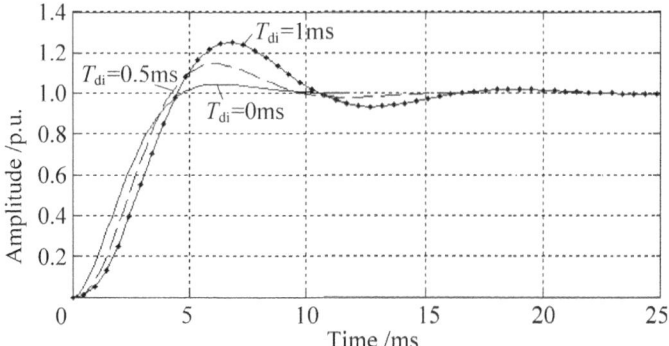

Fig. 6.15 The system step response versus the sampling delay

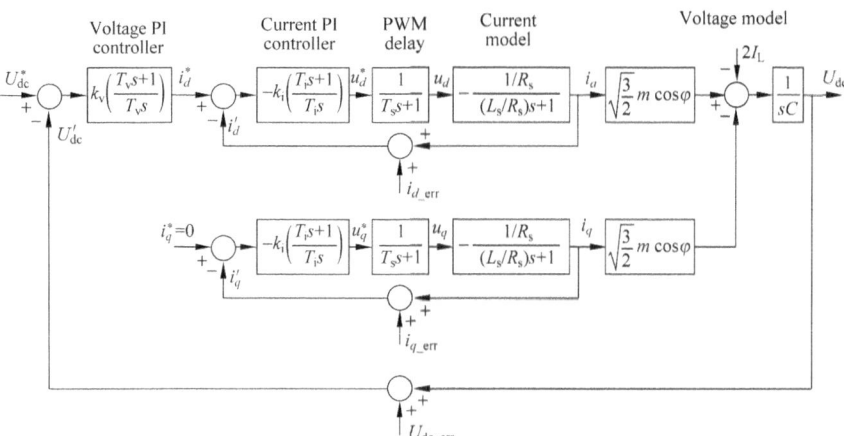

Fig. 6.16 The frequency-domain model of the three-level PWM converter with the sampling error

Here i_{di} is the actual d-axis current controlled by the reference value. i_{de} is the error current caused by the sampling error, i.e.,

$$i_{di}(s) = \frac{1}{2T_s^2 s^2 + 2T_s s + 1} \cdot i_d^*(s) \tag{6.20}$$

$$i_{de}(s) = -\frac{1}{2T_s^2 s^2 + 2T_s s + 1} \cdot i_{d\_err}(s) \tag{6.21}$$

In the steady state, based on the Laplace final value theorem, we have

$$\lim_{t \to \infty} i_{di}(t) = \lim_{s \to 0} \frac{s}{2T_s^2 s^2 + 2T_s s + 1} \cdot \frac{i_d^*}{s} = i_d^* \tag{6.22}$$

Therefore in the steady state,

$$i_d(t) = i_d^* + i_{de}(t) \tag{6.23}$$

Here $i_{de}(t)$ is the static error of the d-axis current.

With the DC offset involved in the sampled grid AC current, under the abc coordinates the sampled current in the time domain can be expressed as

$$\begin{cases} i_a'(t) = i_a(t) + i_{a\_err} \\ i_b'(t) = i_b(t) + i_{b\_err} \end{cases} \tag{6.24}$$

Here i_a and i_b are actual current values of Phase a and b, respectively. i_{a_err} and i_{b_err} are the DC offset of the sampled current. According to Eqs. (6.3) and (6.4), the sampling error of the d-axis current is

$$i_{d\_err}(t) = \sqrt{2} i_{a\_err} \sin\left(\omega t + \frac{\pi}{3}\right) + \sqrt{2} i_{b\_err} \sin \omega t \tag{6.25}$$

with its frequency-domain expression as

$$i_{d\_err}(s) = \frac{\sqrt{\frac{3}{2}} i_{a\_err} \cdot s + \frac{\omega}{\sqrt{2}}(i_{a\_err} + 2i_{b\_err})}{s^2 + \omega^2} \tag{6.26}$$

The substitution of Eq. (6.26) into Eq. (6.21) yields control errors of the d-axis current in the frequency domain and time domain, as shown below.

$$i_{de}(s) = -\frac{\sqrt{\frac{3}{2}} i_{a\_err} \cdot s + \frac{\omega}{\sqrt{2}}(i_{a\_err} + 2i_{b\_err})}{(2T_s^2 s^2 + 2T_s s + 1)(s^2 + \omega^2)} = \frac{a_1 s + b_1}{2T_s^2 s^2 + 2T_s s + 1} + \frac{c_1 s + d_1}{s^2 + \omega^2} \tag{6.27}$$

$$i_{de}(t) = \mathcal{L}^{-1}[i_d(s) - i_d^*(s)] = \mathcal{L}^{-1}\left(\frac{a_1 s + b_1}{2T_s^2 s^2 + 2T_s s + 1}\right) + \mathcal{L}^{-1}\left(\frac{c_1 s + d_1}{s^2 + \omega^2}\right) \tag{6.28}$$

Applying the Laplace FVT, we have

$$\lim_{t \to \infty} \mathcal{L}^{-1} \frac{a_1 s + b_1}{2T_s^2 s^2 + 2T_s s + 1} = \lim_{s \to 0} \frac{s(a_1 s + b_1)}{2T_s^2 s^2 + 2T_s s + 1} = 0 \tag{6.29}$$

Hence the static error of the d-axis current caused by the DC offset error is

$$i_d(t) - i_d^* = \mathcal{L}^{-1}(\frac{c_1 s + d_1}{s^2 + \omega^2}) = I_{md1} \sin(\omega t + \varphi_{d1}) \tag{6.30}$$

Here

$$\begin{cases} I_{md1} = \sqrt{\omega^2 + d_1^2}/\omega \\[2mm] \varphi_{d1} = \arctan(c_1 \omega / d_1) \\[2mm] c_1 = \dfrac{(\sqrt{6}T_s^2 \omega^2 + \sqrt{2}T_s \omega - \sqrt{\frac{3}{2}}) \cdot i_{a\_err} + 2\sqrt{2}T_s \omega \cdot i_{b\_err}}{4T_s^4 \omega^4 + 1} \\[4mm] d_1 = \dfrac{(\sqrt{2}T_s^2 \omega^3 - \sqrt{6}T_s \omega^2 - \frac{\omega}{\sqrt{2}}) \cdot i_{a\_err} + \sqrt{6}\omega(2T_s^2 \omega^2 - 1) \cdot i_{b\_err}}{4T_s^4 \omega^4 + 1} \end{cases} \tag{6.31}$$

Similarly the static error of the q-axis current caused by the DC-offset error is

$$i_q(t) - i_q^* = I_{mq1} \sin(\omega t + \varphi_{q1}) \tag{6.32}$$

Based on Eqs. (6.30) and (6.32), we can further derive the control error of the phase current caused by the DC offset error, as shown below.

$$\begin{cases} i_a(t) - i_a^*(t) = \dfrac{1}{\sqrt{6}}(I_{md1} \sin \varphi_{d1} - I_{mq1} \cos \varphi_{q1}) \\[3mm] \qquad\qquad + \dfrac{1}{\sqrt{6}}[I_{md1} \sin(2\omega t + \varphi_{d1}) + I_{mq1} \cos(2\omega t + \varphi_{q1})] \\[3mm] i_b(t) - i_b^*(t) = \dfrac{1}{\sqrt{6}}[I_{md1} \sin(\varphi_{d1} + \dfrac{2\pi}{3}) - I_{mq1} \cos(\varphi_{q1} + \dfrac{2\pi}{3})] \\[3mm] \qquad\qquad + \dfrac{1}{\sqrt{6}}[I_{md1} \sin(2\omega t + \varphi_{d1} - \dfrac{2\pi}{3}) + I_{mq1} \cos(2\omega t + \varphi_{q1} - \dfrac{2\pi}{3})] \\[3mm] i_c(t) - i_c^*(t) = \dfrac{1}{\sqrt{6}}[I_{md1} \sin(\varphi_{d1} - \dfrac{2\pi}{3}) - I_{mq1} \cos(\varphi_{q1} - \dfrac{2\pi}{3})] \\[3mm] \qquad\qquad + \dfrac{1}{\sqrt{6}}[I_{md1} \sin(2\omega t + \varphi_{d1} + \dfrac{2\pi}{3}) + I_{mq1} \cos(2\omega t + \varphi_{q1} + \dfrac{2\pi}{3})] \end{cases} \tag{6.33}$$

Here j_a^*, i_b^* and i_c^* are reference values of the phase currents, respectively, which in the steady state are symmetric sinusoidal values.

The analysis above reveals the existence of the DC component and second-order harmonics in the actual current, caused by the sampling error of the DC offset.

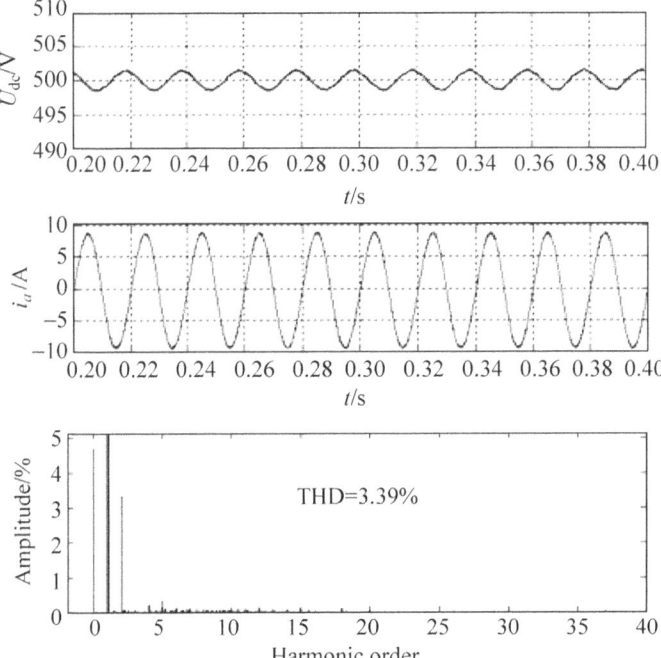

Fig. 6.17 Simulated voltage and current when the sampled grid current has the DC-offset error

Their amplitude could be quantified based on the equations above. Note the d-axis current loop is part of the DC-bus voltage loop, while Eq. (6.30) shows the DC-offset sampling error of the phase current could introduce another harmonics to the d-axis current, with the frequency of $\omega/(2\pi) = f_e$, which in return generates the same-frequency component to the DC-bus voltage. Simulation results are shown in Fig. 6.17, where 0.5 A DC-offset error is added to the phase current sampling.

The grid current spectrum indicates the existence of the DC offset and the second-order harmonics. Meanwhile an oscillation of 50 Hz is shown in the DC-bus voltage.

The main component of the grid current is still at the fundamental frequency. With the amplitude error, the sampled current can be formulated in the time domain as

$$\begin{cases} i_a'(t) = (1 + k_{a\_err})i_a(t) = i_a(t) + \sqrt{2}Ik_{a\_err}\cos\omega t \\ i_b'(t) = (1 + k_{b\_err})i_b(t) = i_b(t) + \sqrt{2}Ik_{b\_err}\cos(\omega t - 2\pi/3) \end{cases} \tag{6.34}$$

Here k_{a_err} and k_{b_err} are coefficients of the amplitude error for Phase a and b, respectively. I is the rms value of the grid fundamental current. The sampling errors of the d-axis current in the time domain and frequency domain are

$$i_{d\_err}(t) = Ik_{a\_err}[\sin(2\omega t + \frac{\pi}{3}) + \frac{\sqrt{3}}{2}] + Ik_{b\_err}[-\sin(2\omega t + \frac{\pi}{3}) + \frac{\sqrt{3}}{2}] \tag{6.35}$$

$$i_{d\_err}(s) = \frac{\sqrt{3}}{2} I k_{a\_err}\left(\frac{s + \frac{2}{\sqrt{3}}\omega}{s^2 + 4\omega^2} + \frac{1}{s}\right) + \frac{\sqrt{3}}{2} I k_{b\_err}\left(-\frac{s + \frac{2}{\sqrt{3}}\omega}{s^2 + 4\omega^2} + \frac{1}{s}\right) \qquad (6.36)$$

Substituting the above to Eq. (6.21) with the FVT, we could derive the static control error of the d-axis current caused by the amplitude sampling error as follows.

$$i_d(t) - i_d^* = \mathcal{L}^{-1}\left(\frac{c_2 s + d_2}{s^2 + 4\omega^2}\right) + I_{ed2} = I_{md2}\sin(2\omega t + \varphi_{d2}) + I_{ed2} \qquad (6.37)$$

Here

$$
\begin{cases}
I_{md2} = \sqrt{4\omega^2 + d_2^2}/2\omega \\[2mm]
\varphi_{d2} = \arctan(2c_2\omega/d_2) \\[2mm]
I_{ed2} = -\dfrac{\sqrt{3}}{2} I k_{a\_err} - \dfrac{\sqrt{3}}{2} I k_{b\_err} \\[3mm]
c_2 = \dfrac{(8T_s^2\omega^2 + \frac{4}{\sqrt{3}}T_s\omega - 1)\cdot(\frac{\sqrt{3}}{2}I k_{a\_err} - \frac{\sqrt{3}}{2}I k_{b\_err})}{64T_s^4\omega^4 + 1} \\[4mm]
d_2 = \dfrac{(\frac{32}{\sqrt{3}}T_s^2\omega^3 - 8T_s\omega^2 - \frac{4}{\sqrt{3}}\omega)\cdot(\frac{\sqrt{3}}{2}I k_{a\_err} - \frac{\sqrt{3}}{2}I k_{b\_err})}{64T_s^4\omega^4 + 1}
\end{cases}
\qquad (6.38)
$$

Similarly, the control error of the q-axis current caused by the amplitude difference is

$$i_q(t) - i_q^* = I_{mq2}\sin(2\omega t + \varphi_{q2}) + I_{eq2} \qquad (6.39)$$

Furthermore the error of the phase current can be translated into the *abc* coordinates, as shown below.

$$
\begin{aligned}
\iota_a(t) - i_a^*(t) = & \sqrt{\frac{2}{3}} I_{ed2}\cos\omega t - \sqrt{\frac{2}{3}} I_{eq2}\sin\omega t \\
& + \frac{1}{\sqrt{6}}[I_{md2}\sin(\omega t + \varphi_{d2}) - I_{mq2}\cos(\omega t + \varphi_{q2})] \\
& + \frac{1}{\sqrt{6}}[I_{md2}\sin(3\omega t + \varphi_{d2}) + I_{mq2}\cos(3\omega t + \varphi_{q2})] \qquad (6.40)
\end{aligned}
$$

Obviously the sampling error of the phase-current amplitude could introduce errors of the fundamental and third-order harmonics to the actual grid current, quantified as equations above. As shown in Eq. (6.37), the amplitude error introduces an error component with the frequency $2\omega/(2\pi) = 2f_e$ to the d-axis current, which further causes the same-frequency oscillation of the DC-bus voltage. Shown in Fig. 6.18 are the simulation waveform when the sampled grid current has 5% amplitude error.

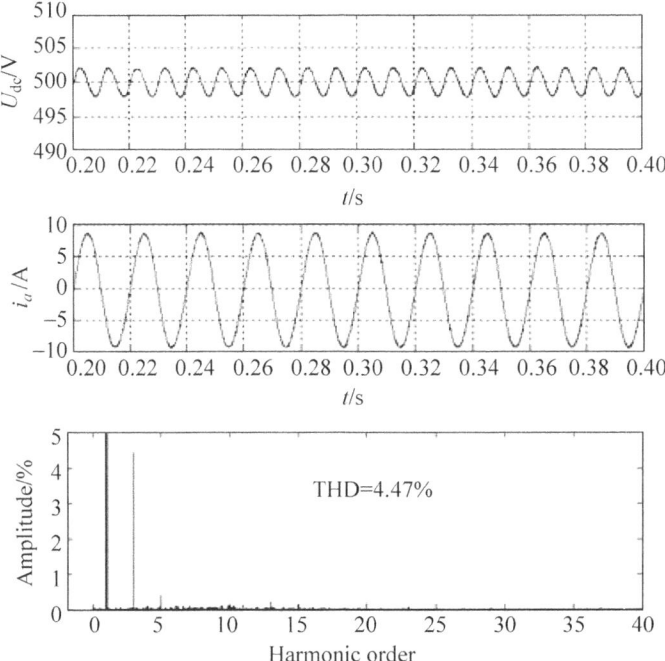

Fig. 6.18 Simulation waveforms when the grid sampled current has the amplitude error

The current spectrum indicates errors of the fundamental and 3rd-order harmonics. Meanwhile the DC-bus voltage has 100 Hz voltage ripple.

In the three-level PWM rectifier, the purpose of sampling the grid voltage is mainly to get the phase angle, which will be later used for the rotatory transformation. When the grid voltage sampling contains the DC-offset error, i.e.,

$$\begin{cases} e'_a(t) = e_a(t) + e_{a\_err} \\ e'_b(t) = e_b(t) + e_{b\_err} \end{cases} \tag{6.41}$$

after the $abc \rightarrow dq0$ transformation, an error of the d-axis current caused by the grid-angle error is shown as below, in the time and frequency domains, respectively.

$$i_{d\_err}(t) = \frac{\sqrt{2}I}{E} e_{a\_err} \sin(\omega t + \frac{\pi}{3}) + \frac{\sqrt{2}I}{E} e_{b\_err} \sin \omega t \tag{6.42}$$

$$i_{d\_err}(s) = \frac{\sqrt{\frac{3}{2}}\frac{I}{E} e_{a\_err} \cdot s + \frac{\omega}{\sqrt{2}}\frac{I}{E}(e_{a\_err} + 2e_{b\_err})}{s^2 + \omega^2} \tag{6.43}$$

When the grid-voltage amplitude has the sampling error, shown as follows,

$$\begin{cases} e'_a(t) = (1 + m_{a\_err})e_a(t) = e_a(t) + \sqrt{2}Em_{a\_err}\cos\omega t \\ e'_b(t) = (1 + m_{b\_err})e_b(t) = e_b(t) + \sqrt{2}Em_{b\_err}\cos(\omega t - 2\pi/3) \end{cases} \tag{6.44}$$

the resulted sampling error of the d-axis current in the time and frequency domains is

$$i_{d\_err}(t) = Im_{a\_err}[\sin(2\omega t + \frac{\pi}{3}) + \frac{\sqrt{3}}{2}] + Im_{b\_err}[-\sin(2\omega t + \frac{\pi}{3}) + \frac{\sqrt{3}}{2}] \tag{6.45}$$

$$i_{d\_err}(s) = \frac{\sqrt{3}}{2}Im_{a\_err}(\frac{s + \frac{2}{\sqrt{3}}\omega}{s^2 + 4\omega^2} + \frac{1}{s}) + \frac{\sqrt{3}}{2}Im_{b\_err}(-\frac{s + \frac{2}{\sqrt{3}}\omega}{s^2 + 4\omega^2} + \frac{1}{s}) \tag{6.46}$$

Equations (6.43) and (6.26) have similar forms. So do Eqs. (6.46) and (6.36). Using the same analyzing methodology as the grid-current sampling errors, we can conclude with the existence of the DC-offset error in the grid-voltage sampling, an extra DC offset and 2nd-order harmonics will appear in the actual grid current, with an oscillation at frequency $\omega/(2\pi) = f_e$ exhibiting in the DC-bus voltage. When the grid-voltage has the sampling error of the amplitude, the actual grid current will have error components of the fundamental and 3rd-order harmonics. Meanwhile an oscillation of the frequency $2\omega/(2\pi) = 2f_e$ can be observed in the DC-bus voltage.

Assume the DC-bus voltage sampling has the DC-offset error. Together with the DC-bus voltage loop, part of which is the current loop, we can formulate the frequency-domain DC-bus voltage as

$$U_{dc}(s) = \frac{\sqrt{\frac{3}{2}}mk_v\cos\varphi(T_vs + 1)[U^*_{dc}(s) - U_{dc\_err}(s)]}{T_vCs^2(2T_s^2s^2 + 2T_ss + 1) + \sqrt{\frac{3}{2}}mk_v\cos\varphi(T_vs + 1)} \tag{6.47}$$

Applying Laplace FVT, we could derive the static value of the DC-bus voltage with the influence of the DC-offset sampling error as follows.

$$U_{dc}(t) = U^*_{dc} - U_{dc\_err} \tag{6.48}$$

When the DC-bus voltage sampling contains the amplitude error, we have

$$U_{dc}(s) = \frac{\sqrt{\frac{3}{2}}mk_v\cos\varphi(T_vs + 1)\cdot U^*_{dc}(s)}{T_vCs^2(2T_s^2s^2 + 2T_ss + 1) + (1 + k_{dc\_err})\cdot\sqrt{\frac{3}{2}}mk_v\cos\varphi(T_vs + 1)} \tag{6.49}$$

After applying the FVT, in the steady state we have

$$U_{dc}(t) = U^*_{dc}/(1 + k_{dc\_err}) \tag{6.50}$$

Therefore for the grid-voltage sampling, both the DC-offset error and the amplitude sampling error will introduce the DC error on the actual bus voltage, as deter-

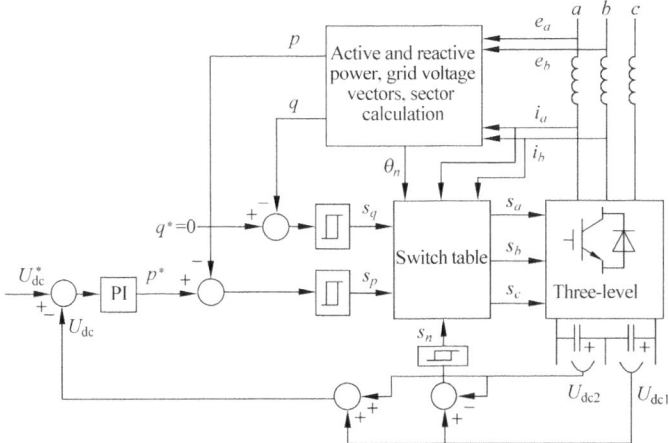

Fig. 6.19 The scheme of the three-level DPC

mined by Eqs. (6.48) and (6.50). It is worth pointing out that in the three-level NPC PWM converter, such sampling errors could potentially cause the imbalance of the natural point voltage.

6.3.2 Time-Domain Analysis

In addition to the VOC, DPC is another commonly used control method for PWM rectifiers. Based on the instantaneous power theory, the DPC makes rectifier exchange the instantaneous power with the grid, which is simpler and has better dynamic response compared to conventional PWM rectifier controls. The block diagram of such control is shown in Fig. 6.19, where three control loops are included, i.e., one external loop for the DC-bus voltage regulation using the PI controller to provide the reference for the active-power loop, and two inner loops for the active power and reactive power control, which adopt the Bang-bang control and the three-level vector tables. These two loops will compare the reference active and reactive power with actual values, select the appropriate voltage vectors based on their sectors, and regulate the active and reactive power.

Under the static $\alpha\beta$ coordinates and the synchronous dq coordinates, the instantaneous power flow from the three-phase grid to the rectifier is calculated as

$$\begin{cases} p = e_\alpha i_\alpha + e_\beta i_\beta \\ q = e_\beta i_\alpha - e_\alpha i_\beta \end{cases} \tag{6.51}$$

$$\begin{cases} p = e_d i_d \\ q = -e_d i_q \end{cases} \tag{6.52}$$

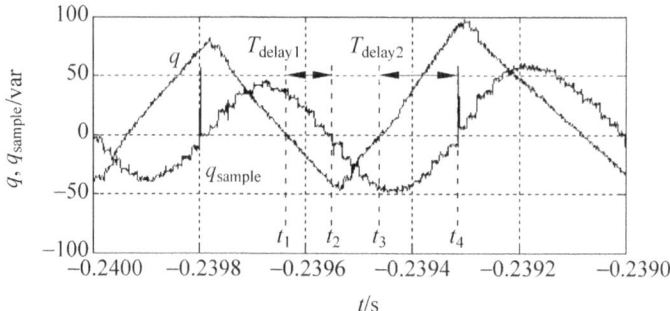

Fig. 6.20 Experimental waveforms of the reactive-power actual and reference values with the sampling delay

To avoid the trigonometric functions during the coordinate transformation thereby lowering the computation load, the active and reactive power of the DPC is finished under the static $\alpha\beta$ coordinates. Meanwhile given the Bang-bang control and the vector table looking-up are too nonlinear to be linearized, the previous frequency-domain analysis is unlikely feasible, unless the time-domain analysis is applied.

1. Impact of the sampling delay on the control performance

As shown in Fig. 6.6, the time delay exists between the actual grid voltage/current and the sampled values, which results in the delay of the calculated active and reactive power, i.e., p_{sample} and q_{sample}, respectively shown in Eq. (6.51). Such delayed power values when compared to the reference value will influence control pulses. Take the reactive power as an example. As shown in Fig. 6.20, at $t = t_1$, the actual reactive power is crossing zero towards negative. Theoretically control pulses need increase q. However, due to existence of the sampling delay, q_{sample} still remains positive, which with the hysteresis and the three-level vector table generates the pulse to keep reducing q. Such effort further enlarges the error between actual and reference values until q_{sample} crosses zero at $t = t_2$. In the steady state, the sampling delay introduces the oscillation for the actual power value and increases the current harmonics.

The time sequence in Fig. 6.20 indicates that the oscillation frequency of the active and reactive power caused by the sampling delay is

$$f_p - f_q \approx \frac{1}{(4T_{sample} + T_{delay1} + T_{delay2}) \times 2} \tag{6.53}$$

Here T_{sample} is the sampling period. Assume in the steady state the average value of the voltage vector is $[\bar{v}_d, \bar{v}_q]$. Based on Eqs. (6.10) and (6.52), the power oscillation amplitude is

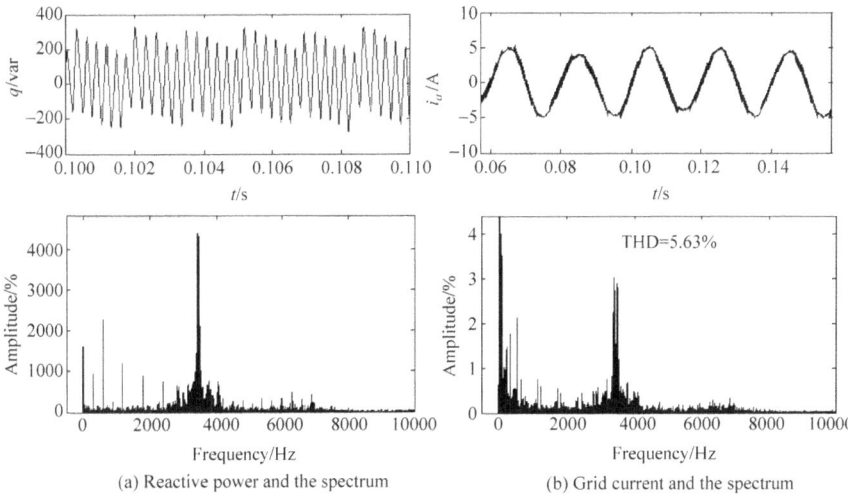

(a) Reactive power and the spectrum   (b) Grid current and the spectrum

Fig. 6.21 Simulation results of the three-level DPC when $T_{\text{sample}} = 0.01$ ms and $T_{\text{delay}} = 0.05$ ms

$$\begin{cases} \Delta p \approx \dfrac{\sqrt{3}E}{L_s}(\sqrt{3}E - \overline{u}_d)(4T_{\text{sample}} + T_{\text{delay1}} + T_{\text{delay2}}) \\[2mm] \Delta q \approx \dfrac{\sqrt{3}E\overline{u}_q}{L_s}(4T_{\text{sample}} + T_{\text{delay1}} + T_{\text{delay2}}) \end{cases} \qquad (6.54)$$

Through the coordinate transformation, the frequency of the current harmonics caused by the sampling delay is

$$f_i = f_q \pm f_e \qquad (6.55)$$

With $T_{\text{sample}} = 0.01$ ms, $T_{\text{delay1}} = T_{\text{delay2}} = 0.05$ or 0.1 ms, the simulated reactive power and the grid currents with the three-level DPC are shown in Figs. 6.21 and 6.22, respectively. The spectrum is aligned with the analysis above, i.e., the larger the sampling delay, the lower the oscillation frequency and the higher the oscillation amplitude of the power and the current harmonics.

Meanwhile, the sampling delay will cause the delay of the grid phase detection, as shown in Fig. 6.23.

When the actual grid voltage vector is aligned with the d-axis shown in Fig. 6.23, due to the sampling delay, the control system locates the sampled grid voltage vector at the d'-axis. The angle between d-axis and d'-axis is

$$\theta_{\text{de}} = 2\pi f_e T_{\text{delay}} \qquad (6.56)$$

In the steady state, the actual d-axis current is determined by the load of the rectifier, i.e.

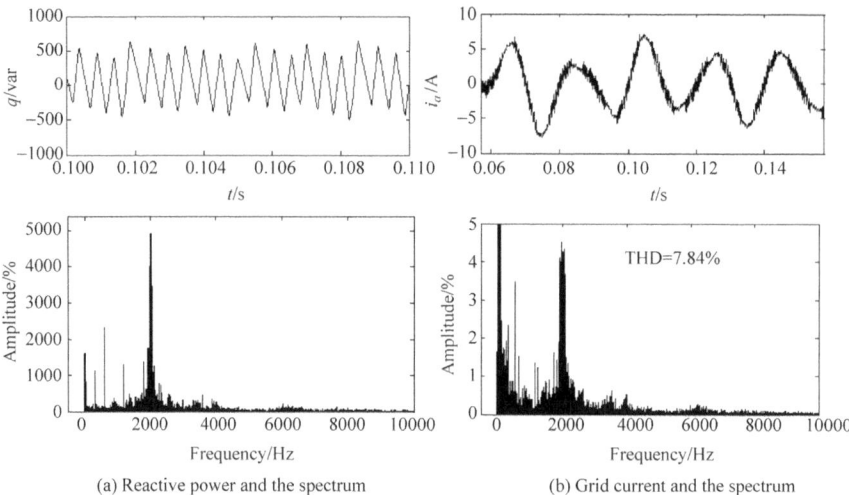

(a) Reactive power and the spectrum   (b) Grid current and the spectrum

Fig. 6.22 Simulation results of the three-level DPC when $T_{sample} = 0.01$ ms and $T_{delay} = 0.1$ ms

Fig. 6.23 Delay of the grid-voltage phase detection caused by the sampling delay

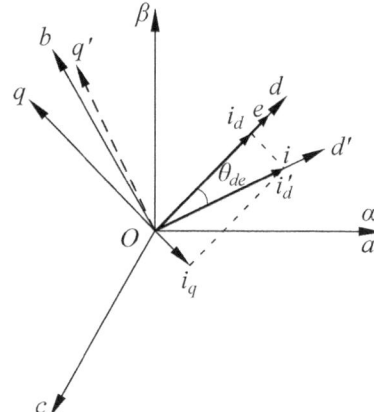

$$i_d = I_L U_{dc}/e_d \tag{6.57}$$

Thus the control strategy secures the phase alignment between the sampled grid voltage and current, i.e.

$$i'_q = 0 \tag{6.58}$$

Since the actual imposed current vector is at d'-axis, based on the geometrical relationships, we have

$$i'_d = i_d/\cos\theta_{de} \tag{6.59}$$

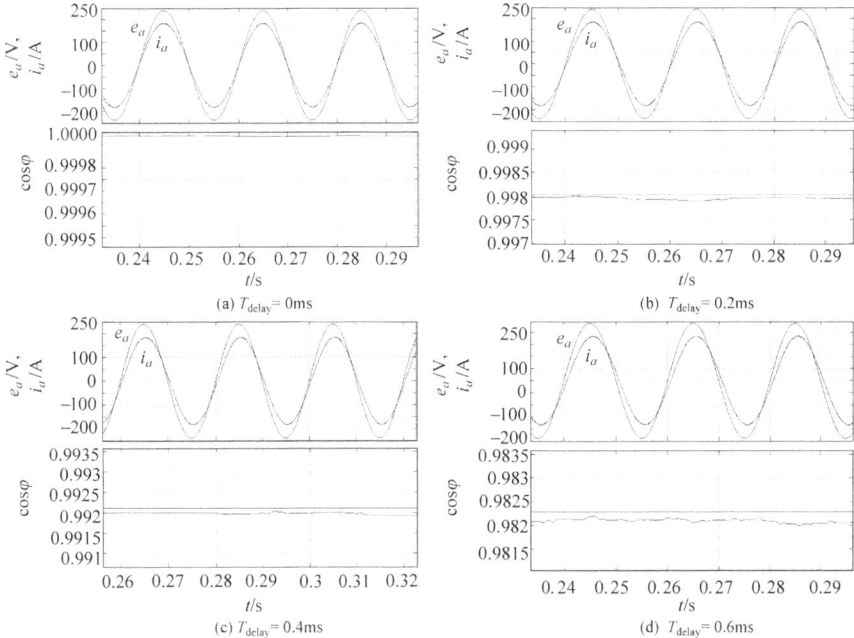

Fig. 6.24 Simulated grid voltage, current and power factor of the three-level DPC rectifier with the sampling delay increasing

$$i_q = -i'_d \sin \theta_{de} = -(I_L U_{dc}/e_d) \tan(2\pi f_e T_{delay}) \tag{6.60}$$

As shown in Eqs. (6.57) and (6.60), the sampling delay lowers the actual power factor of the rectifier, i.e.

$$\cos \varphi < \cos \theta_{de} = \cos(2\pi f_e T_{delay}) \tag{6.61}$$

Shown in Fig. 6.24 is the simulated grid voltage, current and power factor of the three-level DPC rectifier with the sampling delay increasing. The larger the sampling delay, the larger the angle error between the grid voltage and current thereby the lower the power factor error. Besides in the dynamic process the sampling delay causes control pulses to fall behind, which slows down the implementation of the reference power, i.e., deteriorating the dynamic response.

2. **Impact of the sampling error on the control performance**

Assume in the steady state the phase-a sampled current has the DC-offset error, i.e.,

$$i'_a(t) = i_a(t) + i_{a\_err} \tag{6.62}$$

Fig. 6.25 Simulated DPC
waveform when the current
sampling contains the
DC-offset error

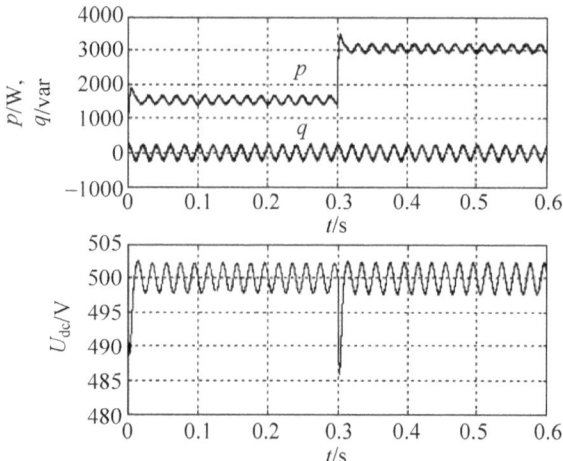

When transformed into the $\alpha\beta$ coordinates, we have

$$\begin{cases} i'_\alpha = i_\alpha + \sqrt{2/3} \cdot i_{a\_err} \\ i'_\beta = i_\beta \end{cases} \tag{6.63}$$

Assuming the grid voltage are symmetric sinusoidal waveforms, we have

$$\begin{cases} e'_\alpha = e_\alpha = \sqrt{3}E \, \cos \omega t \\ e'_\beta = e_\beta = \sqrt{3}E \, \sin \omega t \end{cases} \tag{6.64}$$

Substituting Eqs. (6.63) and (6.64) into Eq. (6.51), we have

$$\begin{cases} p_{sample} = p + \sqrt{2}E \, \cos \omega t \cdot i_{a\_err} \\ q_{sample} = q + \sqrt{2}E \, \sin \omega t \cdot i_{a\_err} \end{cases} \tag{6.65}$$

The sampled power overlaps one error on the actual value, which is oscillating
with the fundamental frequency. The control pulses generated to regulate the power
also vary periodically, which yields the actual power oscillating at the fundamental
frequency as well. So does the DC-bus voltage. When the sampled current of one
phase contains a 5% DC-offset error, the simulated active power, reactive power and
the DC-bus voltage are shown in Fig. 6.25. The load change of the rectifier happens
at $t = 0.3$ s.

It can be seen that the sampling error does not impact much on the system dynamic
performance. In addition, the oscillation of the power and the DC-bus voltage has
nothing to do with the load and input power while only depends on the DC-offset
sampling error and the grid voltage. Assume the instantaneous power flow from the
grid to the rectifier in the steady state is

Fig. 6.26 Simulated grid current and its spectrum when the current sampling has the DC offset

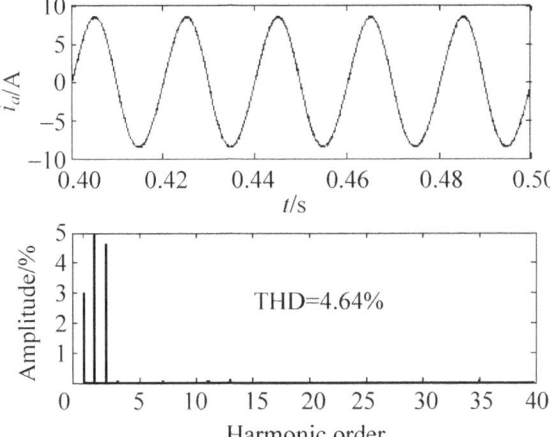

$$\begin{cases} p = p_0 + p_1 \cos(\omega t + \varphi_{p1}) \\ q = q_0 + q_1 \cos(\omega t + \varphi_{q1}) \end{cases} \tag{6.66}$$

Here p_0 and q_0 are DC components of the active and reactive power, respectively. p_1 and q_1 are oscillating components of the power. φ_{p1} and φ_{q1} are initial angles of the oscillating power.

From Eqs. (6.51) and (6.64),

$$\begin{cases} i_\alpha = \dfrac{e_\alpha \cdot p + e_\beta \cdot q}{e_\alpha^2 + e_\beta^2} = \dfrac{e_\alpha \cdot p + e_\beta \cdot q}{3E^2} \\ i_\beta = \dfrac{e_\beta \cdot p - e_\alpha \cdot q}{e_\alpha^2 + e_\beta^2} = \dfrac{e_\beta \cdot p - e_\alpha \cdot q}{3E^2} \end{cases} \tag{6.67}$$

Substituting Eqs. (6.51) and (6.66) into Eq. (6.67), we have

$$\begin{aligned} i_a &= \frac{\sqrt{2}}{3E}(p_0 \cos \omega t + q_0 \sin \omega t) + \frac{\sqrt{2}}{6E}\left(p_1 \cos \varphi_{p1} - q_1 \sin \varphi_{q1}\right) \\ &+ \frac{\sqrt{2}}{6E}[p_1 \cos(2\omega t + \varphi_{p1}) + q_1 \sin(2\omega t + \varphi_{q1})] \end{aligned} \tag{6.68}$$

It can be seen that the DC-offset sampling error introduces to the grid current not only a DC component but also a 2nd-order harmonics, both of which have the positive correlation with the sampling error. Simulation results shown in Fig. 6.26 verified the effectiveness of the theory above.

Assume in the steady state phase-*a* current sampling has the amplitude error,

$$i_a'(t) = (1 + k_{a\_err}) \cdot i_a(t) \tag{6.69}$$

Fig. 6.27 The three-level
DPC simulation waveform
when the sampled current
contains the amplitude error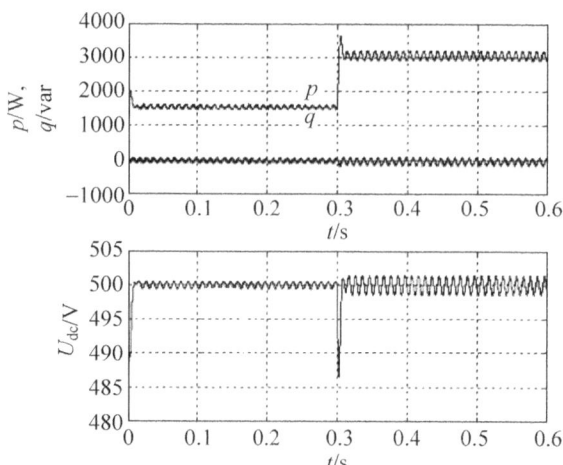

When transformed into the $\alpha\beta$ coordinates,

$$\begin{cases} i'_\alpha = i_\alpha + \sqrt{2/3}k_{a\_err} \cdot i_a \\ i'_\beta = i_\beta \end{cases} \tag{6.70}$$

After substituted into Eq. (6.51),

$$\begin{cases} p_{sample} = p + \sqrt{2}Ek_{a\_err}\cos\omega t \cdot i_a \\ q_{sample} = q + \sqrt{2}Ek_{a\_err}\sin\omega t \cdot i_a \end{cases} \tag{6.71}$$

Considering the fundamental current component aligned with the grid voltage, we have

$$\begin{cases} p_{sample} = p + EIk_{a\_err}(\cos 2\omega t + 1) \\ q_{sample} = q + EIk_{a\_err}\sin 2\omega t \end{cases} \tag{6.72}$$

With the existence of the sampled amplitude error, the calculated active and reactive power will overlap one 2nd-order harmonics on actual power values. So does the DC-bus voltage. When one-phase sampled current contains a 5% amplitude error, the simulated power and DC-bus voltage are shown in Fig. 6.27.

Assume in the steady state the power flow from the grid to the rectifier is

$$\begin{cases} p = p_0 + p_2\cos(2\omega t + \varphi_{p2}) \\ q = q_0 + q_2\cos(2\omega t + \varphi_{q2}) \end{cases} \tag{6.73}$$

Here p_2 and q_2 are the 2nd-order oscillating power. φ_{p2} and φ_{q2} are the initial angle of the oscillating power. With a derivation similar to the above, we have

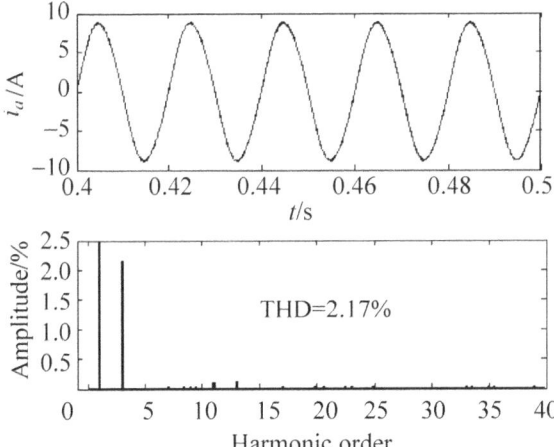

Fig. 6.28 Grid current and spectrum with the sampling amplitude error

$$i_a = \frac{\sqrt{2}}{3E}(p_0 \cos \omega t + q_0 \sin \omega t)$$
$$+ \frac{\sqrt{2}}{6E}[p_2 \cos(\omega t + \varphi_{p2}) + q_2 \sin(\omega t + \varphi_{q2})]$$
$$+ \frac{\sqrt{2}}{6E}[p_2 \cos(3\omega t + \varphi_{p2}) + q_2 \sin(3\omega t + \varphi_{q2})] \qquad (6.74)$$

As can be seen, the sampling amplitude error results in the error at the fundamental frequency and 3rd-order frequency in the grid current. The related simulation waveform is shown in Fig. 6.28.

A similar analysis could be extended to the scenario when the grid-current sampling error occurs on nth-order harmonics, where $n+1$th-order oscillation will appear on the active power, reactive power and DC-bus voltage. Ultimately nth-order and $(n+2)$th-order harmonics will appear in the grid current. With the superposition, it can be concluded that when the grid voltage/current sampling has both the DC-offset and amplitude error, the grid current will have the DC offset, error at the fundamental frequency, with the 2nd-order and 3rd-order oscillation.

For the DC-bus voltage, a relatively slowly changing variable, the sampling errors of the DC-offset and the amplitude will cause the DC-bus voltage error and the neutral-point imbalance in the steady state, with the minor impact on the DPC dynamic performance.

6.4 Reduction of Sampling Delay and Errors

As shown above, the sampling system of the power electronics converter has the cascade structure. The final sampling delay and error are caused and accumulated by nonlinear characteristics of each link. Therefore to eliminate the sampling delay and errors, improvements of each part are necessary, especially to suppress or compensate their ideal characteristics. Based on the analysis of the sampling delay and errors with their impacts in Sect. 6.3, characteristic analysis of each link in Sect. 6.1 and their non-ideal characteristics described in Sect. 6.2, the improvement of suppressing sampling delays and errors will be introduced in this section.

6.4.1 Hardware Development

The purpose of the hardware improvement is to reduce the delay and error of the signal transformation and transfer at the sensor interface and AD modulation circuit, before the signal goes into the ADC interface. This helps widen the hardware bandwidth and increase the accuracy.

1. **Sensor selection**

Based on the analogue signal to select the sensor with appropriate bandwidth and range is the key. Take the current sensor as an example. To reach the satisfactory control performance, not only the fundamental component but also high-frequency components are to be extracted. Therefore the selected sensor needs have the appropriate range, precision and sufficient bandwidth, which should be higher than the highest frequency of those useful signals.

The grid current of the PWM rectifier is controlled by voltage vectors of the strategy, regardless of the VOC or DPC. The highest control frequency is the sampling frequency. Therefore the bandwidth of the current sensor should not be lower than the sampling frequency f_{sample}. For example, when the sampling frequency of a VOC based PWM rectifier is 5 kHz while that of the DPC is 60 kHz, a current sensor with the bandwidth of 0–150 kHz is recommended.

For the grid voltage of the PWM rectifier, the control system needs detect the phase of its fundamental component. Therefore the requirement for the voltage sensor bandwidth is not high, i.c., no lower than the grid frequency f_e. In the real practice, we can select a voltage sensor with the bandwidth of 0–25 kHz.

The voltage sensor of the DC bus is determined by the requirement of the DC-bus dynamic response. In the real practice, a voltage sensor with the bandwidth of 0–25 kHz is recommended.

2. **Improvement of the analogue modulation circuits**

With the circuit analysis and tests, we can optimize parameters of the analogue conditioning circuits. Such circuits are mainly to transform the sensor output to the

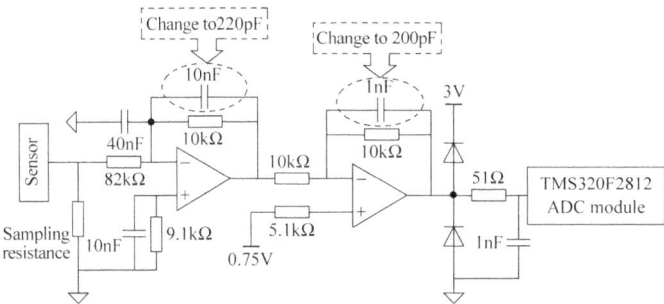

Fig. 6.29 Improvement of the analogue conditioning circuits in a three-level PWM rectifier

analogues matching the ADC input. Equipped with the low-pass filter, the circuit will eliminate noises and enhance the immunity to the disturbance. Theoretically the lower the cutoff frequency of the low-pass filter, the more noises eliminated. However a too low cutoff frequency attenuates and delays useful high-frequency signals. Similar to the bandwidth selection of the sensors, the cutoff frequency of each low-pass filter should be above the highest frequency of the useful signals. The parameters associated with filters in the analogue conditioning circuits need be determined through experiments. The final design a tradeoff between filtering the noise and keeping the useful signals.

Shown in Fig. 6.5 is one analogue conditioning circuit for a three-level PWM rectifier, which contains two first-order active low-pass filters and one first-order passive low-pass filter. Their cutoff frequency is 1.59 kHz, 15.9 kHz and 3.12 MHz, respectively. All three filters are cascaded in the circuit. To keep high-frequency useful signals, we need increase the cutoff frequency of those two narrow-band filters. Given the resistors determine both the cutoff frequency and the transfer coefficient, improvements of the analogue conditioning circuit is mainly focused on filtering capacitors. Based on experimental results, the final flow chart of the system improvement is shown in Fig. 6.29. With such modification, the cutoff frequency of two corresponding filters are lifted to 66.3 and 69.6 kHz, respectively, higher than the highest-frequency (60 kHz) component of the grid current using DPC.

The conditioning circuit has the higher bandwidth and shorter sampling delay. However, between the analogue signal and digital signal converted by the ADC module exists a time delay larger than one sampling period, which needs the further compensation through the software development.

6.4.2 Software Design

The software design includes the ADC setting, improvement of the ADC procedure, digital signal conditioning and selection of the sampling moment.

1. **Settings of the ADC module**

In theory, the lower the ADC clock frequency, the wider the sampling window, enhancing the ADC immunity to the disturbance as well as the precision. However, for the high-sampling-frequency DPC, each sampling period is relatively limited. A low clock frequency or a wide sampling window tends to prolong the ADC period. If the sum of the ADC time and control execution time is larger than the whole sampling period, it will affect the execution of the whole control strategy thereby deteriorating the control performance. Such scenario means the control strategy has not been fully executed upon the completion of the sampling period. When the next sampling interrupt is triggered, the software initiates the sampling without updating control parameters such as the duty cycle. Therefore we need appropriately assign the ADC settings, i.e., securing the overall time of the ADC and control execution shorter than the sampling period and widening the ADC sampling window.

For the cases above, the sampling frequency of the DPC is 60 kHz, i.e., the sampling period is 16.7 μs. Experimental results indicate the execution time of the DPC is ~15 μs. Considering the variation of the software execution, we need limit the ADC time within 1 μs. Therefore the ADC clock is set to 37.5 MHz, the sampling window is set to 5 ADC clocks, and the time to finish 6-channel ADC is 0.96 μs. Such effort maximizes the ADC sampling window while securing the real-time programming.

2. **Improvement of the ADC flow**

Experimental results show that some ADC modules misjudged the conversion completion moment. The ADC program might read conversion results prior to the conversion completion, while the data stored in the ADC register is still inherited from the previous sampling period, which creates the delay of one sampling period. To avoid such an ADC delay, we need improve the ADC module and its conversion flow, as shown in Fig. 6.30.

3. **Software correctness of the sampling error**

This part is finished within the digital-signal conditioning program, which is the last section of the ADC sampling. Such conditioning aims to correct and compensate the accumulated distortion, and in addition minimize the sampling error of feedback signals before sending them to the control unit.

Ideally the measured analogue x, after the sensor, conditioning circuit and ADC transformation, has the relationship with the digital signal y as follows.

$$y = ax + b \qquad (6.75)$$

Fig. 6.30 The improved
ADC conversion flow

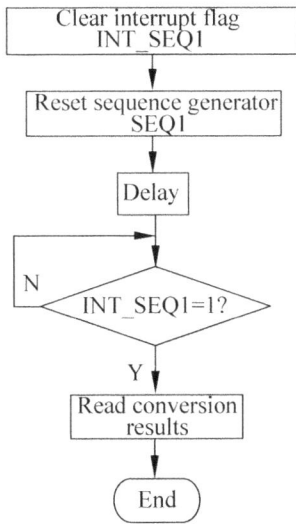

Here a is the multiplication of the sensor gain, conditioning-circuit gain and the ADC gain. For the sampling system in this chapter, $a = k_{sensor} \times k_{circuit} \times 4095/3$. For the DC sampling, $b = 0$. For the AC sampling, $b = 2048$. Due to non-ideal and non-linear characteristics of ADC sampling circuits, Eq. (6.75) is not always true. The actual relationship between x and y should be

$$y = a'x + b' + \text{err} \tag{6.76}$$

Here the difference between a', b' and a, b is mainly caused by the DC-offset error and the amplitude error, which are inherent sampling errors of the system. Hence a' and b' of each ADC channel can be assumed as constant. The variable, *err*, is the random error of the sampling system, which is unpredictable and inerasable. The best way to get rid of *err* is filtering. Since most of useful information in sampled parameters is at the low-frequency range, digital lower-pass filters could be adopted to suppress the high-frequency random error. Such filter design has been mature in the domain of the digital signal processing, which is not detailed here. The critical step of the filter design is to set the cutoff frequency, which complies with same rules of the analogue filter design, i.e., the cutoff frequency needs be higher than the frequency of all useful signal information.

To eliminate the DC-offset error and the amplitude error, the least square method (LSM) is utilized to identify a' and b' in Eq. (6.76). At the front end of the sampling, i.e., sensor inputs, some analogue signals for the sampling correction can be applied, which can be measured by the high-precision voltage meters, current meters or oscilloscopes. Such characteristic parameters can be recorded as x_i ($i = 1, 2, \ldots$ N). The corresponding digital values can be recorded at the ADC output as y_i ($i = 1, 2, \ldots$N). For AC channels, analogue signals for ADC corrections should be widely distributed within the actual operating range. For the DC channel, the analogue signals need focus on the rated operating point. Based on the LSM, a' and b' can be derived as

$$\begin{cases} a' = \dfrac{N \sum x_i y_i - \sum x_i \sum y_i}{N \sum x_i^2 - (\sum x_i)^2} \\ b' = \dfrac{\sum x_i^2 \sum y_i - \sum x_i \sum x_i y_i}{N \sum x_i^2 - (\sum x_i)^2} \end{cases} \tag{6.77}$$

Such coefficients calculated offline can further be used for the online correcting the DC-offset and amplitude errors, i.e.

$$z = \frac{y - b'}{a'} \tag{6.78}$$

4. **Selection of sampling moments**

While meeting the execution time of the control program, the sampling moment needs be distant away from switching moments to avoid potential noises. For the DPC the sampling period needs be as short as possible to enhance the control performance, while longer than the execution time of the control strategy to guarantee the completion of the control within one sampling period. Since the switching action happens at the completion of the control algorithm and the sampling period, to avoid switching actions, DPC based PWM rectifiers are recommended to sample and process the ADC in the middle of the sampling period.

6.4.3 Effectiveness of the Sampling-System Improvement

A 30 kW three-level PWM rectifier is used to verify the effectiveness of the presented improvement on the suppression of non-ideal characteristics of the feedback loop. Such rectifier has parameters shown in Table 6.1.

With the improved design shown in Sect. 6.4.2, the actual and sampled grid current are shown in Fig. 6.31, which indicates ~13 µs time delay between sampled and actual values. Such delay includes the DA transformation time. Thus the sampling delay is

Table 6.1 The main-circuit parameters of the 30 kW three-level rectifier

Parameters	Value	Parameters	Value
Grid inductance	10 mH	Inductor ESR	0.3 Ω
DC-bus upper/lower capacitor	680 µF	Load resistance	adjustable
Input phase-voltage rms	170 VAC	DC-bus rated voltage	500 V

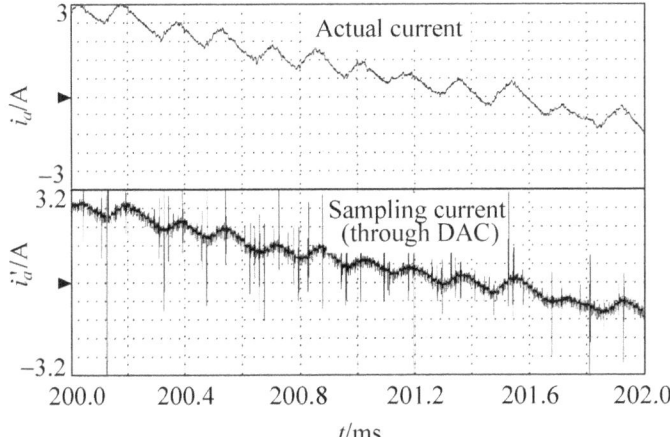

Fig. 6.31 The actual and sampled values of the grid current after the improvement

restrained within one sampling period (16.7 µs) firmly. The actual current and the sample current cross the zero nearly at the same time, namely the DC-offset sampling error is negligible. Calculated through the experimental waveforms, the improved DC-offset error and the amplitude error are below 1.5 and 1%, respectively. Compared to Fig. 6.6 before the improvement, the fidelity of the sampled waveform with the improvement is highly enhanced. The high-frequency useful information after the sampling does not fade, i.e., the control bandwidth meets the basic requirement of the control strategy. Meanwhile the delay and error between sampled values and actual values are reduced sharply, which enhance the dynamic response of the current control, given the high-frequency oscillation of the grid current diminishes while the oscillation frequency increases.

A head-to-head comparison of the grid current and its spectrum before/after the improvement is shown in Fig. 6.32. With the effective suppression of the sampling delay, DC-offset sampling error and amplitude sampling error, the grid current quality is significantly enhanced without changing any main-control parameters. The 2nd- and 3rd-order harmonics caused by sampling errors are reduced from 4.5 to 1.8% and from 2.9 to 0.4%, respectively. With other harmonics being reduced as well, the current THD after the improvement is 2.91%, less than half of 6.21% before the improvement. Meanwhile measurements through the digital power analyzer reveals the power factor is also increased from 0.993 to 0.998.

The same comparison is made for a VOC based three-level PWM rectifier, as shown in Fig. 6.33. Similar effects of reducing the grid-current THD and increasing the PF are achieved as well.

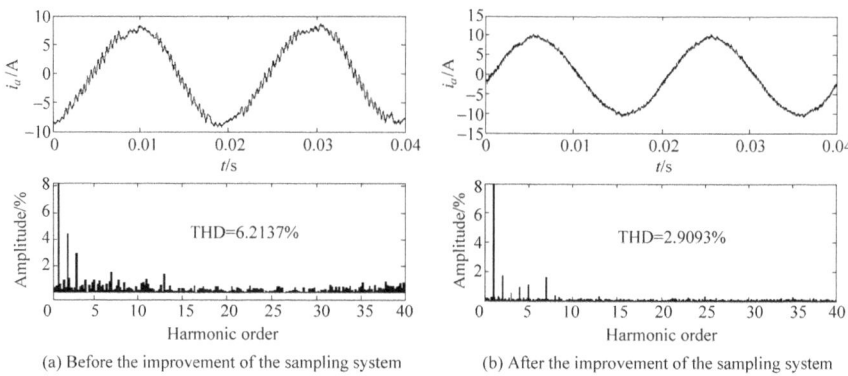

Fig. 6.32 The experimental comparison of the grid current before and after the improvement of the sampling system in a DPC based three-level PWM rectifier

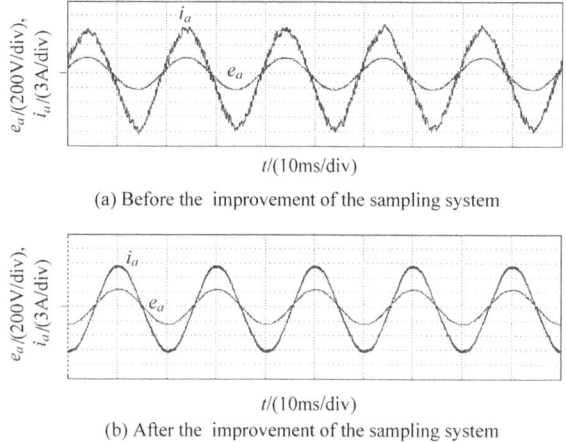

Fig. 6.33 The experimental comparison of the grid current before and after the improvement of the sampling system in a VOC controlled three-level PWM rectifier

More experimental results of the DPC and VOC controlled three-level PWM rectifiers are shown in Figs. 6.34 and 6.35, respectively, with the improved sampling system. Regardless of control strategies, each rectifier has highly sinusoidal current, which is tightly aligned with the grid voltage after the sampling system is improved. Meanwhile, neither the DC-bus voltage has any oscillation nor the neutral point imbalance occurs due to the sampling error.

As a summary, the measurement and observation in power electronic systems is of importance for the transient analysis. The distortion and delay in such system

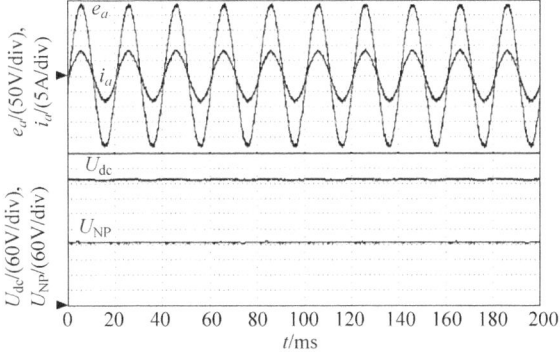

Fig. 6.34 Experimental results of the three-level DPC rectifier using the improved sampling system

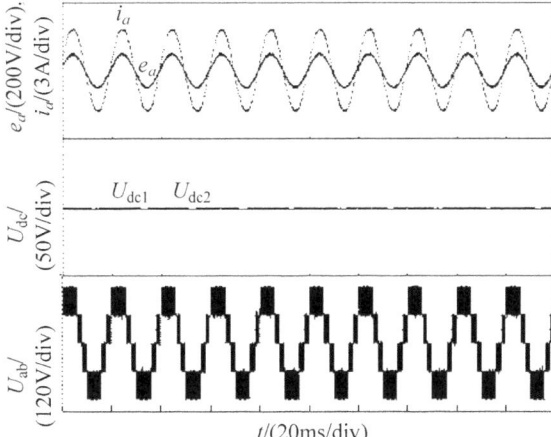

Fig. 6.35 Experimental results of the three-level VOC rectifier using the improved sampling system

will greatly impact the performance and reliability of the power electronic system, which however could be suppressed and compensated through improvement of the sampling system.

Chapter 7
Electromagnetic Pulses and Sequences in Main Circuit

Electromagnetic pulses and pulse sequences are the outcome of control and main circuits, which are also the carrier of energy conversion. From information to energy flow, the pulse and pulse sequence propagation includes signal generation, transfer, shaping, coordination, and transformation.

7.1 Mathematical Descriptions of Pulse and Pulse Sequences in Power Electronics Systems

Carriers of pulses in power electronic systems show great diversities. So do their shapes, time sequences, and energy characteristics. In order to accurately enhance control performance and to secure system reliability, we need firstly differentiate pulses in terms of their shape, time sequence, and behaviors, based upon which we can precisely describe such pulses.

7.1.1 Pulse Categories and Variations

Control signals generated by control strategies are all discrete at time and amplitude. The time discretization is determined by the control frequency while its amplitude discretization is up to the bits and accuracy of the microcontroller. Based on control signals, the microcontroller uses PWM methods to generate information pulses with the control circuit as the carrier. Such signals are time continuous but amplitude discrete, which means that they only have high-level and low-level values. Therefore, control pulses can be treated as ideal pulses, with the transition between high and low levels being finished instantaneously. No transients are considered.

© Tsinghua University Press and Springer Nature Singapore Pte Ltd. 2019
Z. Zhao et al., *Electromagnetic Transients of Power Electronics Systems*,
https://doi.org/10.1007/978-981-10-8812-4_7

After being shaped by the dead-band and minimum pulse width, which only adjust pulse edges and width, control pulses are then forwarded to the gate-drive circuit. Such pulses are still ideal pulses. However, compared to information pulses before shaping, the edges might be delayed with the width being varied. Note the signals in the control system can all be treated as ideal. Since majority of state-of-the-art power electronics systems use digital control, the control circuit is mainly to adjust and shape pulses at the macroscopic level, i.e., at the same level of the control period.

Pulses shaped after the gate-drive circuit and power switches form energy pulses in the main circuit, which are time and amplitude continuous. Electromagnetic transients are exhibited, e.g., neither the high-level nor low-level amplitude is flat, with the oscillations displayed. The pulse edges are non-ideal due to the rising process, falling process, overshoot and oscillations. Thus, energy pulses are significantly different from control pulses in terms of time sequence and shapes. All the non-ideal characteristics of energy pulses result from nonlinearity of gate-drive circuits and switches. Specifically, non-ideal electromagnetic transients of energy pulses are created by the gate-drive circuit, power switches, and main circuits, which adjust and shape energy pulses at the microscopic level with the time constant determined by gate-drive performance, power switch characteristics, and main circuit parasitics.

Formation of main circuit pulses is determined by information and energy, software and hardware, discrete systems and continuous systems, and macroscopic and microscopic timescales. Such pulses are also influenced by various non-linear factors in the system, and consequently contain strong nonlinearity and transients, which are even more dominant in high-power converters. Without proper control and design, their control performance and even reliability will deteriorate.

7.1.2 Mathematical Descriptions of Energy Pulses

The actual energy pulse in power electronic converters is shown in Fig. 7.1, where characteristic parameters are defined as below.

High-level (H):   The amplitude of the energy pulse when reaching the high steady state;

Low-level (L):   The amplitude of the energy pulse when reaching the low steady state;

Rising moment (t_{H0}):   The moment when the energy pulse begins to rise from the low level;

Falling moment (t_{L0}):   The moment when the energy pulse begins to fall from the high level;

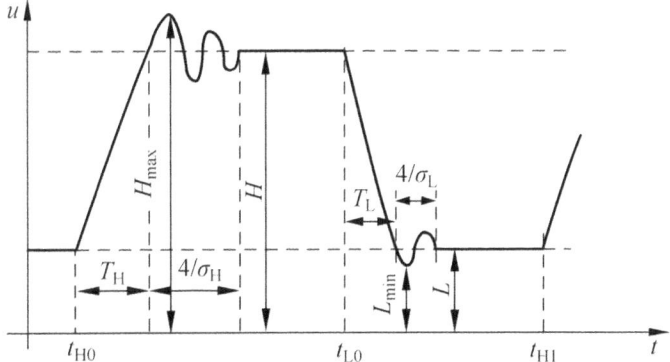

Fig. 7.1 The shape of one typical energy pulse

Rising time (T_H): The time interval for the pulse to rise from the low level to the high level at the first time;

Falling time (T_L): The time interval for the pulse to fall from the high level to the low level at the first time;

No-damped high-level overshoot (A_H): The maximum overshoot at the pulse rising edge before reaching the steady state, without damping;

No-damped low-level overshoot (A_L): The maximum overshoot at the pulse falling edge before reaching the steady state, without damping;

High-level oscillation damping coefficient (σ_H): The fading velocity of the oscillation at the rising edge;

Low-level oscillation damping coefficient (σ_L): The fading velocity of the oscillation at the falling edge;

High-level oscillation frequency (ω_H): The oscillation frequency at the rising edge;

Low-level oscillation frequency (ω_L): The oscillation frequency at the falling edge

With characteristic parameters defined above, energy pulses of power electronic converter can be mathematically formulated. For example, when from the low to high level,

$$u(t) = \begin{cases} L + \frac{H-L}{T_H} \cdot (t - t_{H0}) & (t_{H0} < t < t_H + t_{H0}) \\[2mm] A_H e^{-\sigma_H(t-t_{H0}-t_H)} \sin\left[\omega_H(t - t_{H0} - t_H)\right] + H & \left(t_H + t_{H0} \leq t < t_H + t_{H0} + \frac{4}{\sigma_H}\right) \\[2mm] H & \left(t_H + t_{H0} + \frac{4}{\sigma_H} \leq t \leq t_{L0}\right) \\[2mm] H - \frac{H-L}{T_L} \cdot (t - t_{L0}) & (t_{L0} < t < t_L - t_{L0}) \\[2mm] -A_L e^{\sigma_L(t-t_{L0}-t_L)} \sin\left[\omega_L(t - t_{L0} - t_L)\right] + L & \left(t_L + t_{L0} \leq t < t_L + t_{L0} + \frac{4}{\sigma_L}\right) \\[2mm] L & \left(t_L + t_{L0} + \frac{4}{\sigma_L} \leq t \leq t_{H1}\right) \end{cases}$$

$$(7.1)$$

It is worthwhile pointing out that A_H is not the overshoot from low to high levels, but the overshoot when no damping exists. To obtain the actual overshoot, we can differentiate $A_H e^{-\sigma_H(t-t_{H0}-t_H)} \sin[\omega_H(t - t_{H0} - t_H)] + H$ to get the peak-value moment, i.e.,

$$t_{pH} = \frac{\arctan(\omega_H/\sigma_H)}{\omega_H} + t_H + t_{H0} \qquad (7.2)$$

Furthermore we can get the actual overshoot at the rising edge, i.e.,

$$A_{H'} = A_H e^{-\sigma_H \left[\frac{\arctan(\omega_H/\sigma_H)}{\omega_H}\right]} \cdot \frac{\omega_H}{\sqrt{\omega_H^2 + \sigma_H^2}} \qquad (7.3)$$

and the peak value of the pulse,

$$H_{max} = A_H e^{-\sigma_H \left[\frac{\arctan(\omega_H/\sigma_H)}{\omega_H}\right]} \cdot \frac{\omega_H}{\sqrt{\omega_H^2 + \sigma_H^2}} + H \qquad (7.4)$$

Similarly we can get the minimum-value moment of the energy pulse as below.

$$t_{pL} = \frac{\arctan(\omega_L/\sigma_L)}{\omega_L} + t_L + t_{L0} \qquad (7.5)$$

The actual overshoot at the falling edge is

$$A_{L'} = A_L e^{-\sigma_L \left[\frac{\arctan(\omega_L/\sigma_L)}{\omega_L}\right]} \cdot \frac{\omega_L}{\sqrt{\omega_L^2 + \sigma_L^2}} \qquad (7.6)$$

and the minimum value of the pulse is

$$L_{\min} = -A_L e^{-\sigma_L \left[\frac{\arctan(\omega_L/\sigma_L)}{\omega_L}\right]} \cdot \frac{\omega_L}{\sqrt{\omega_L^2 + \sigma_L^2}} + L \qquad (7.7)$$

7.1.3 Mathematical Description of Information Pulses

The shape of information signals in power electronic converters is shown as Fig. 7.2.

The information pulses are ideal, with the mathematical description as one special case of energy pulses, i.e., $T_H = 0$, $t_L = 0$, $A_H = 0$, $A_L = 0$, $\sigma_H = \infty$ and $\sigma_H = \infty$.

$$u(t) = \begin{cases} H & (t_{H0} < t \le t_{L0}) \\ L & (t_{L0} < t \le t_{H1}) \end{cases} \qquad (7.8)$$

In the process of converting the information pulse to the energy pulse, all other pulses can be treated as special cases of the energy pulses as well, which can be described with characteristic parameters above. For example, the gate-drive signal of power switches can be simplified as

$$u(t) = \begin{cases} L + \frac{H-L}{t_H} \cdot (t - t_{H0}) & (t_{H0} < t < t_H + t_{H0}) \\ H & (t_H + t_{H0} \le t \le t_{L0}) \\ H - \frac{H-L}{t_L} \cdot (t - t_{L0}) & (t_{L0} < t < t_L + t_{L0}) \\ L & (t_L < t_{L0} \le t \le +t_{H1}) \end{cases} \qquad (7.9)$$

Fig. 7.2 Shape of the information pulses

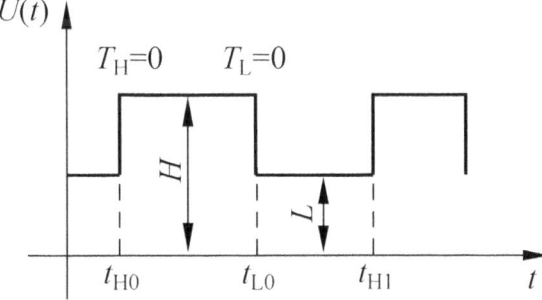

7.1.4   Mathematical Expression of the Energy Pulse Sequence

Based on the energy-pulse mathematical expression, we can define characteristic parameters of the energy pulse sequences as below.

Single-period interval: the time interval between two adjacent rising edges of the energy pulses;

Single-period duty cycle: ratio of the time interval between the adjacent rising and falling edges over the present period;

Length of the pulse sequence: amount of periods included in the pulse sequence.

Here the single-period interval and duty cycle are two one-dimension arrays, the amount of elements of which is equal to the length of the pulse sequence. Meanwhile the order of elements in these two arrays is aligned with the periods inside the energy pulse sequence.

With such characteristic parameters above, the energy pulse sequence can be uniquely represented. Such method applies to other pulses during the conversion of information pulses to energy pulses.

7.1.5   Mathematical Expression of Information Pulse Sequences

As mentioned above, the information pulse is a special case of the energy pulse. Since such pulse is the very initial output of control algorithm, to differentiate it from other pulses, we will define characteristic parameters of the information pulse sequence separately, as shown below.

Ideal single period: the time interval between two adjacent rising edges of the information pulses;

Ideal single-period duty cycle: the ratio of the time interval between the present rising edge and the following falling edge over the present time period;

Length of the sequence: number of the pulse periods within the whole sequence;

Lastly, we need pay attention that the characteristic parameters of pulses and pulse sequences above might not be constant, which could be the function of other variables such as current and switch temperature. To accurately describe pulse and pulse sequences, we need introduce related state variables with their time information to mathematically formulate the pulse and pulse sequence.

7.2 Impact of the Pulse Shape Variation with Related Solutions

After information pulses are generated by the microcontroller, each section of the control system tends to reshape pulses due to its nonlinearity. For example, in the SVPWM control, pulse shaping due to the nonlinearity exists in voltage vectors and transitions among vectors:

(1) Variations of pulse rising and falling edges result in the time variation of voltage vectors.
(2) Transitions between vectors are not finished instantaneously due to non-ideal pulse edges, but follow continuous trajectories between vectors.
(3) The overshoot and oscillation at pulse edges make the vector switching not unidirectional but moving back and forth.
(4) The steady-state high and low voltage levels are subject to change as well, thus varying locations of voltage vectors.

Such non-ideal characteristics of the space vector and space vector switching will further affect synthesized vectors, which directly relates to the control algorithm and control performance, and consequently causes error between actual energy pulses of the main circuit and initial information pulses.

7.2.1 Impact of Dead Band and Minimum Pulse Width Design

Dead band and minimum pulse width, among various non-linear parameters in power electronic systems, are the most commonly used factors. However, these two factors are not independent. Their impact on the pulse shaping is overlapped and coupled with each other. Therefore, when designing and analyzing power electronic systems, we need consider them together. In this section, we will use the snubber circuit shared by three phases together in a three-level VSI as an example to comprehensively analyze the impact of dead band and minimum pulse width. Related solutions will be discussed as well.

1. Dead band and current commutation

Actual turn-on and turn-off actions of the switch have the time delay t_{don} and t_{doff}, respectively, both of which need some time intervals to finish as well. If the complementary switches are triggered simultaneously, i.e., one turns on while the other turns off, in the switching process it is possible that two switches are both on for a short period. The dead band is set to prevent such shoot-through caused by the non-ideal characteristics of switches. The most common solution is to delay the turn-on edge of the gate signal for a dead-band interval, i.e., dead time, securing a time period for both switches to be off before one switch turns on.

Table 7.1 Phase-A switching states and their output

Switching logic	S_{a1}	S_{a2}	S_{a3}	S_{a4}	Output voltage U_{AO}	Switching state function S_a
P	1	1	0	0	E	$+1$
D_1	0	1	0	0	E or 0	$+1$ or 0
0	0	1	1	0	0	0
D_2	0	0	1	0	0 or $-E$	0 or -1
N	0	0	1	1	$-E$	-1

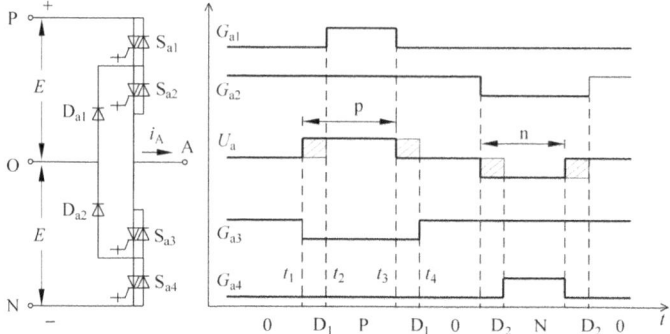

Fig. 7.3 Impact of the dead band on the phase output voltage

Table 7.2 Impact of the current polarity on the phase voltage within the dead band

Current polarity	U_{AO} during the dead band D_1	U_{AO} during the dead band D_2
Out of the leg	0	$-E$
Into the leg	E	0

The dead-band setting needs consider actual characteristics of the power switch, particularly the turn-off time delay and tailing time. Here we use T_{dt} to represent the dead time.

The transition of vectors is caused by switching actions, which must accompany the dead band. In return the dead band will affect space vectors. Take the three-level VSI as an example. The dead band introduces two extra switching states, as shown in Table 7.1. The switching states of one leg can only transition between two adjacent ones. The leg output voltage is still three-level, though the output within the dead band is determined by the leg output current.

The impact of the dead band on the output waveform is shown in Fig. 7.3, where G_{a1}–G_{a4} are gate signals of four switches from top to bottom, respectively. Based on the analysis of the current commutation process, the relationship between the leg output voltage and current within the dead band is shown in Table 7.2.

The leg output is the phase output. Therefore, Fig. 7.3 also indicates the impact of the dead band on the phase output voltage. Assume the current within the dead band does not change the polarity. For the vector p it generates E as the phase voltage. Even though one vector contains three phases, the vector transition only varies one phase while keeping the other two phases unchanged. Therefore, here we can use the variable phase to represent the whole vector. Based upon the current direction, the impact of the dead band on the actual vector is shown as below:

(1) t_1-t_3, when the vector does not change. Thus, the dead band has no impact on this vector. The output voltage is aligned with the original expectation;
(2) t_2-t_3, when the vector is deducted by one dead band, so is the phase output voltage;
(3) t_1-t_4, when the vector is increased by a dead band at the end, which expands the output voltage for one extra dead band as well;
(4) t_2-t_4, when the duration time of the vector does not change, however, the vector gets delayed for one dead band. Thus, the output voltage does not change the width but gets shifted for one dead band.

Here (1) & (4) occur when the current polarity varies during the time of vector p, which is quite rare. Various dead-band compensation algorithms also assume that the current polarity does not change within one vector. For (2) & (3), we could preprogram the width of vector p based on the current direction. Similarly, the impact of the dead band on the vector n can be analyzed, i.e., the phase output voltage is $-E$.

Once the current crosses zero within the dead band, it is possible that such current crosses zero for multiple times, due to the non-ideal sinusoidal waveform and existence of the harmonics. This yields oscillation of the output voltage, as shown in Fig. 7.4. Therefore, when the phase current is small and close to zero, the current polarity cannot be assumed unchanged for the dead-band compensation.

Within the vector transition, the current commutation could happen at either the front edge or the end edge of the dead band. Therefore, the dead band influences the moment of the current commutation. The relationship between vector transitions and

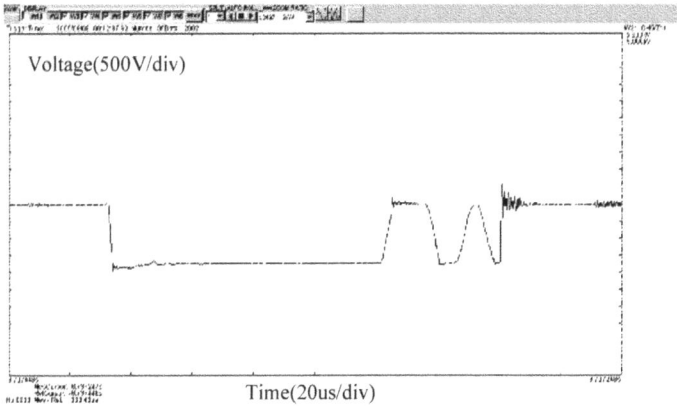

Fig. 7.4 The impact of the variable current polarity on the output voltage within the dead band

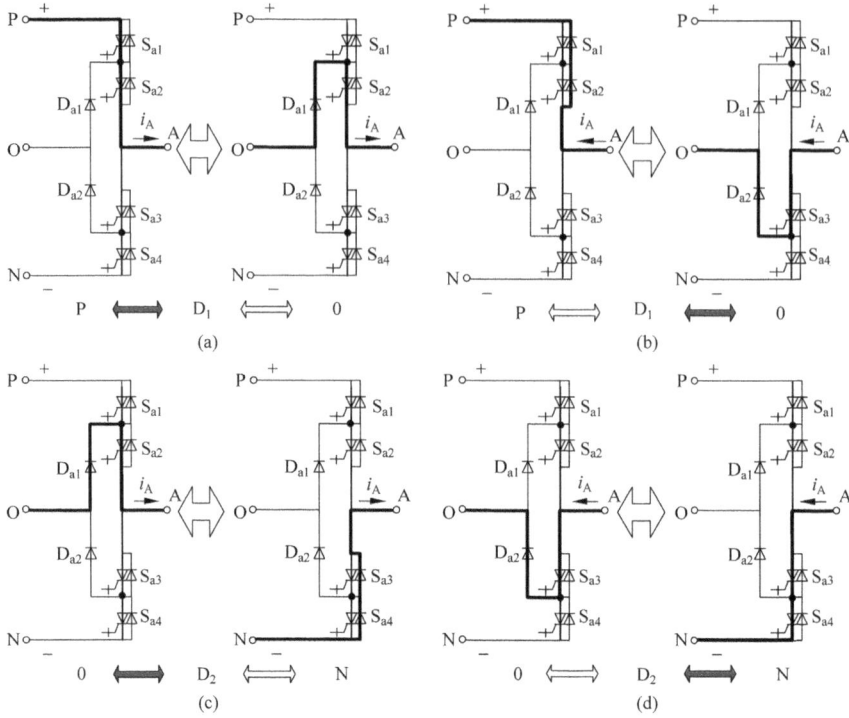

Fig. 7.5 Four commutation processes of Phase-A leg

current polarity can be categorized into four cases, shown as Fig. 7.5. The current commutation moment within the dead band is labelled as solid arrows.

Overall, the relationship between the dead band and the current commutation can be summarized as follows: when the phase output current is positive, the current commutation happens at the dead-band front edge, i.e., transitions between the high-voltage level and the dead band. Consequently, the output within the dead band is a low voltage. Otherwise, the commutation happens at the end edge of the dead band, i.e., between the dead band and the low voltage level.

2. **Single-switch minimum pulse width**

Due to the non-ideal characteristics of power switches, the pulse width of PWM output cannot be infinitely narrow. With different operating theories, different switches exhibit different switching performance as well.

(1) Voltage controlled device

Take one MOSFET, SKM 453A020 as an example. It has the turn-on delay less than 0.1 μs, turn-off delay less than 0.7 μs, the rising time less than 0.1 μs, and the falling time less than 0.25 μs.

For one IGBT, SKM 400GB123D, it has the turn-on delay less than 0.3 μs, turn-off delay less than 0.9 μs, the rising time less than 0.22 μs, and the falling time less than 0.1 μs.

(2) Current controlled device

Take one GTO, DG 858DW45 as an example. It has the turn-on delay of 50 μs, turn-off delay of 100 μs, the rising time of 2 μs, and the falling time of 2.5 μs. The turn-off time of the gate-drive signal is 28.5 μs and the charge restoration time is 26 μs.

For one IGCT, 5SHX 08F4502, it has the turn-on delay $t_{don} < 3$ μs, turn-off delay $t_{doff} < 6$ μs, the rising time t_r around 1 μs and the falling time t_f around 1 μs. Its minimum on time is $t_{on\_min} = 10$ μs, minimum off time is $t_{off\_min} = 10$ μs, and the minimum repetitive turn-off time as 60 μs.

Parameters above were given by switch datasheets based on standard test procedures, which indicate that voltage controlled switches do not specify the minimum on/off time. It is sufficient to only consider gate-voltage delay and rising/falling edges, both of which are very short. For instance, the minimum pulse width of MOSFET and IGBT gate drive signals is 1–2 μs, which make them seldom restrained by the requirement of minimum pulse width. The requirement of current-controlled devices is totally different, each type of which has the minimum on and off time, for instance, 10 μs on and off for 5SHX 08F4502.

The minimum pulse width is an important index for the PWM control strategy. Since actual gate signals imposed on the switch are subject to change due to the dead band, we also need take the impact of the dead band on the voltage vector into account. The minimum pulse width mentioned above originates from the non-ideal characteristics of the switches, particularly for the single switch to limit the time interval between adjacent turn-on and turn-off actions. Therefore, it is also defined as the single-switch minimum pulse width.

From the analysis above, the minimum pulse width M_{min} need be

$$M_{min} = \max \left\{ t_{on\_min}, t_{off\_min} \right\} \tag{7.10}$$

Based on principles of vector transitions, the gate-drive signal of any single switch is one dead time short of the control pulse width. Therefore, the minimum control pulse width should be

$$T \geq t_{dt} + M_{min} \tag{7.11}$$

A closer investigation into switching states shown in Fig. 7.6 reveals that the pulse with one dead time deduction must be the turn-on pulse, while the turn-off

Fig. 7.6 Scheme of the minimum pulse width

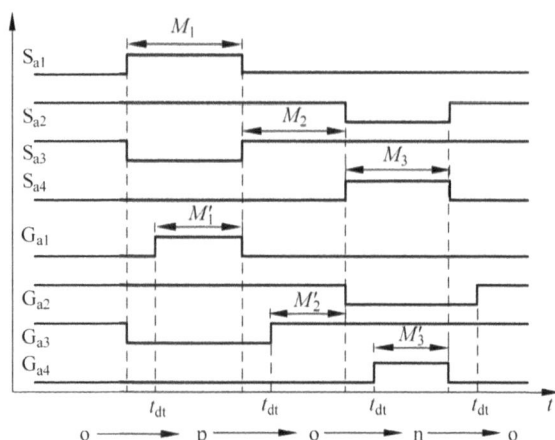

pulse is added with one dead time. Therefore, the correction of Eq. (7.11) is shown as Eq. (7.12)

$$T \geq t_{\text{on\_min}} + t_{\text{dt}}$$
$$T \geq t_{\text{off\_min}} - t_{\text{dt}} \tag{7.12}$$

3. **Vector minimum pulse width**

We need investigate two vectors adjacent to the present vector before concluding if the vector time represents the turn-on/off pulse width of a single switch. It is possible to link the vector sequencing with the potential minimum pulse width, which further allows us to adjust the vector sequence, thereby avoiding narrow pulses.

Only considering the minimum pulse width of a single switch is not enough. Given that the vector minimum pulse width is not only affecting the switch minimum pulse width, the vector sequence cannot essentially avoid the narrow pulse. For M_2 in Fig. 7.6, if the time interval is shorter than one dead band, the existence of the dead band will fully eliminate M_2'. Therefore, each vector time should be guaranteed longer than T_{dt}. Considering two aspects above, the ultimate vector time M_n need be

$$M_n > M_{\text{min}} + t_{\text{dt}} \tag{7.13}$$

Here t_{dt} is the dead time and M_{min} is the single-switch minimum pulse width determined by the switch characteristics. In the above scenario, narrow pulses could be avoided regardless of the current commutation before or after the vector.

Switches in high-power converters are relatively slow, namely M_{min} is usually large, around tens of μs. So is the dead time. Ultimately, the setting of the minimum pulse width could reach tens of or even hundreds of μs. The control error caused by the deduction or increment of the vector time is not negligible. Furthermore, while setting the vector minimum time, we should take into account the physical meaning behind and the dynamic process of two adjacent current commutations, especially the combination of the device characteristics, gate-drive performance, dead time, and snubber circuits. Application of a universal value for the minimum pulse width will inevitably deteriorate the control performance. However, if the value for the minimum pulse width is set too small or is not set at all, the vector pulse will overflow the SSOA, thereby damaging switches.

The vector minimum pulse width can be categorized into two scenarios: in one scenario, two current commutations happen within the same leg, i.e., commutation internal the phase; in the other scenario, two commutations happen between two different legs, i.e., phase-to-phase commutation.

(1) Current commutation internal the phase

The minimum pulse width of a single switch targets the commutation within the phase (the vectors before and after the minimum pulse width are exactly the same, making two switching actions happen at the same phase and yielding the phase internal commutation), we can name such corresponding vectors as phase internal type-I vectors. Now take a look at another scenario of $p \to o \to n$ and $n \to o \to p$, where the middle vector o is defined as phase internal type-II vectors. Back to M_2 in Fig. 7.6, if the related time is shorter than the dead time, such vector could totally be erased by the dead band. To detail its impact on the phase output voltage, we can refer to Fig. 7.7. Assume $M < 0.5\ T_{dt}$, vector o generated by the original PWM channel S_a is shorter than t_{dt}, with the related gate-drive signal and phase output voltage as G_a and U_A, respectively. When the current is flowing out of the leg, $D_1 = 0$ and $D_2 = -E$ as shown in Table 7.2. Therefore $D_0(0000) = -E$. When the current flows into the leg, $D_1 = E$ and $D_2 = 0$. Therefore $D_0(0000) = -E$. Regardless of the current direction, no voltage leaping between E and $-E$ is expected, but there is zero output with the pulsed width of M between E and $-E$, unless the current polarity changes during $D_0(0000)$. Such leaping could reverse bias anti-paralleled diodes of S_{a1} and S_{a2} and forward bias anti-paralleled diodes of S_{a3} and S_{a4}, causing the potential dynamic voltage imbalance and even damages to switches.

In consequence, a constraint should be placed upon the pulse width of the phase internal type-II vector, as principles of the multi-level converter are involved, i.e., if the current polarity changes, voltage leaping within $t_{dt} - M$ (as shown in Fig. 7.7) could occur even though the possibility is low. Such a rare occurrence could even cause dynamic voltage across switches to overflow the SOA, thereby damaging switches.

When $M > T_{dt}$, no D_0 will appear between D_1 and D_2. Instead, there will be (0110), securing no voltage leaping even when the phase current changes the polarity. Therefore, each phase internal type-II vector should be guaranteed to meet

$$T > T_{dt} \tag{7.14}$$

The corresponding simulation results are shown in Fig. 7.8. Here the vector changing in phase A is $p \to o \to n$. In subplot (a) the current is positive and in subplot

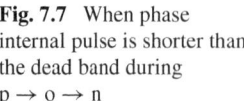

Fig. 7.7 When phase
internal pulse is shorter than
the dead band during
$p \rightarrow o \rightarrow n$

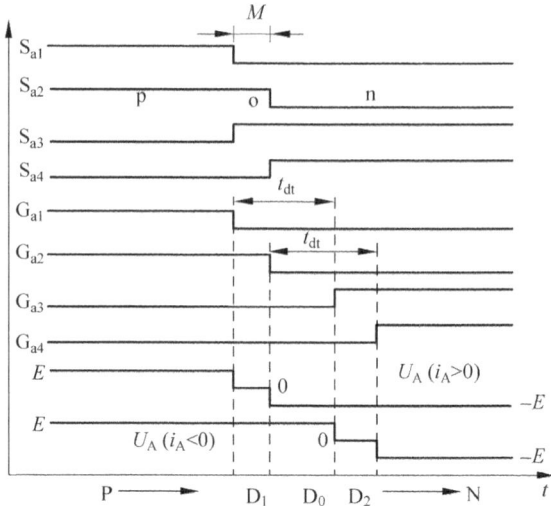

(b) the current is negative. As shown in Fig. 7.8(a), the phase output voltage $-E$ has
been overlapped with $-\Delta u_C$. In Fig. 7.8(b), the phase-A output voltage 0 has been
extended for another T_L, the on-time of the bottom switch S_{a4}. A similar waveform
applies to $n \rightarrow o \rightarrow p$.

(2) Phase-to-phase commutation

Such scenario contains two cases, one is $p \leftrightarrow o$, the other is $o \leftrightarrow n$. The corresponding
vector is named as phase-phase type-I vector. Though the VSI in this chapter has
three phases co-using the same snubber circuit, the transients of snubber circuits
could be ignored given the investigated switching actions are related to the upper
and lower snubbers, respectively. On the other hand, the dead-band impact on the
commutation still needs be considered.

Furthermore, we are introducing the concept of phase-phase commutation dead
band. Assume the investigated vector time is t_v. The first switching moment is t_1 and
the second switching is t_2. If $t_2 - t_1 = t_v$, no dead band is imposed. If $t_2 - t_1 = t_v$
$+ T_{dt}$, we define it as the positive dead band. If $t_2 - t_1 = t_v - T_{dt}$, we define it as a
negative dead band. From another aspect, if the first commutation is triggered by the
turn-off of one fully-controlled device and the second commutation is triggered by
the turn-on of the other fully-controlled device, the turn-on action will be delayed for
one dead band. Thus, the time interval between two commutations will be prolonged
for one extra dead band, i.e., a positive dead band. Otherwise, if the first commutation
is the turn-on of one device and the second commutation the turn-off of the other
fully-controlled device, the time between two commutations will be deducted by
one dead band, i.e., negative dead-band effect. If both switching actions are due to
the turn-on or turn-off, the commutation time does not change, i.e., zero dead-band

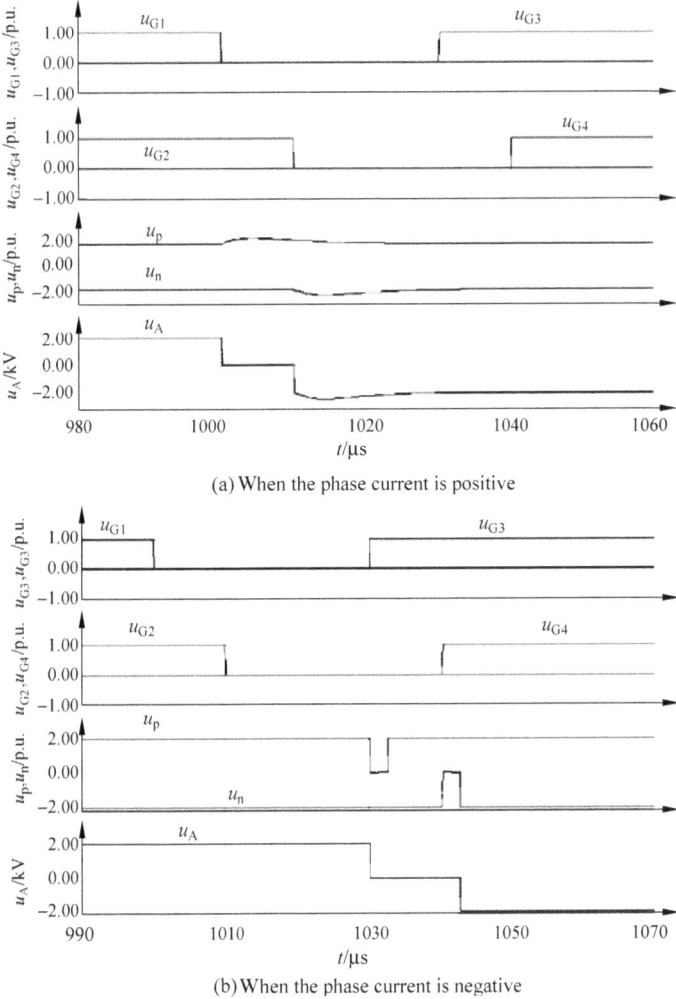

(a) When the phase current is positive

(b) When the phase current is negative

Fig. 7.8 Simulation waveform when the vector time is shorter than one dead band (p → o → n)

effect. In another word, the dead-band effect can be determined by the commutation of two legs, which further instructs the setting of the minimum pulse width between phases.

Table 7.3 Minimum pulse width of the phase-phase type-I vectors

Phase-phase commutation dead band	Minimum phase-phase pulse width	Phase-phase commutation dead band	Minimum phase-phase pulse width
On-state zero dead band	0	Positive dead band	0
Off-state zero dead band	0	Negative dead band	T_{dt}

We can conclude that the type-I phase-phase vector requires pulse width larger than a dead band when a negative dead band appears. Otherwise, abnormal pulses will appear on the line-line voltage. However, such pulses have no impact on the switch itself since two commutations employ different snubber circuits, respectively. Furthermore, the design rules of minimum pulse width of the phase-phase type-I vector are shown below in Table 7.3.

Another commutation between phases is when these two phases are both p ↔ o or o ↔ n. The related vector is named as the phase-phase type-II vector. p ↔ o involves the #1 and #3 switches, employing the top snubber circuit. o ↔ n involves the #2 and #4 switches, employing the bottom snubber circuit. The minimum pulse width of such vector needs take into account both the dead band and the dynamic process of the shared snubber circuits.

Assume Phase A and B are both transitioning p → o. The dead time is 30 μs. The dead-band effect could be simulated with different current polarities, shown as below.

p → o, $i_A > 0$, $i_B > 0$, with S_{a1} in Phase A off, S_{b1} in Phase B off and both phases experiencing the current commutation. As it is a zero dead band, we set the phase-phase pulse width as 1 μs. The simulation results are shown in Fig. 7.9, with the waveform from top to bottom as current of S_{a1} and S_{b1}, snubber diode current, snubber capacitor voltage, and voltage of S_{a1} and S_{b1}, respectively. The turn-off current of both phases is accumulating the charge into the snubber capacitor, with the inductor current $I = i_A + i_B$. Given the snubber circuit parameters are designed based on the single-phase maximum current, $i_A + i_B$ might exceed the maximum value, thereby over boosting the DC-bus voltage. Therefore we need settle the snubber circuit between the first and second current commutations. In this case the phase-phase minimum pulse width need be set as T_C, the time for the snubber circuit to reach the steady state after turning off the maximum current.

p → o, $i_A < 0$, $i_B > 0$, with S_{a3} in Phase A on, S_{b3} in Phase B on. It is still a zero dead band. We set the phase-phase pulse width as 1 μs with simulation results shown in Fig. 7.10. The waveform from top to bottom is current of anti-paralleled diodes of S_{a1} and S_{b1}, top snubber inductor current, positive DC-bus voltage, and voltage of S_{a1} and S_{b1}, respectively. Two turn-on currents are limited by the snubber inductor.

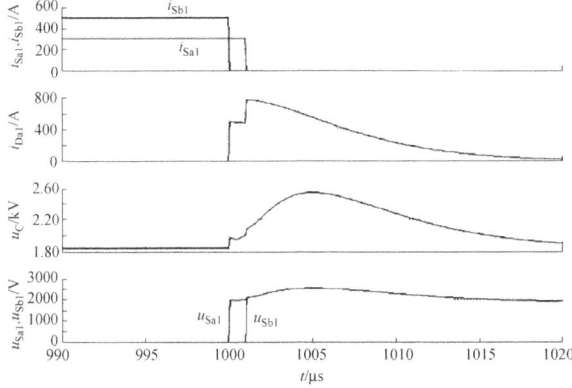

Fig. 7.9 Simulation waveform when two phases are commutating ($p \to o$, $i_A > 0$, $i_B > 0$)

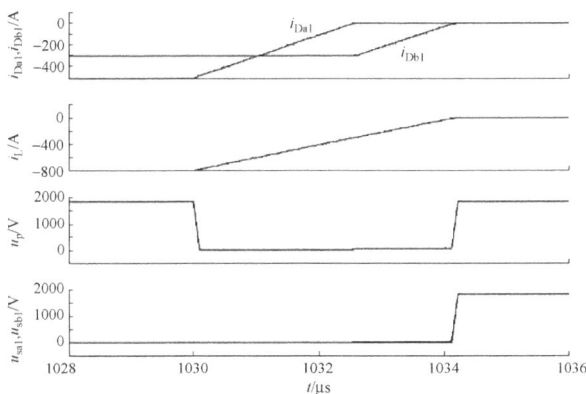

Fig. 7.10 Simulation waveform when two phases are commutating ($p \to o$, $i_A < 0$, $i_B > 0$)

When S_{a3} turns on, Phase A current goes through two paths: one as the positive DC-bus linearly reducing the current, and the other through the clamping diode between S_{a3} and S_{a4}. The positive DC-bus U_p is dragged down to 0, until all Phase-A current goes through the clamping diode. During this period, when S_{b3} is on, however, the Phase-B voltage is not enough to forward bias the clamping diode between S_{a3} and S_{a4}. Therefore a time gap is exhibited between Phase A and B commutations. When the Phase-A current is exactly equal to the load current, Phase-B current begins to increase with the inductor continuously limiting the switch turn-on current. In this case, the phase-phase minimum pulse width can be set as 0, the DC-bus voltage will be dragged down to 0, and top two switches in each phase begin to undertake the voltage only after the snubber inductor current drops to 0.

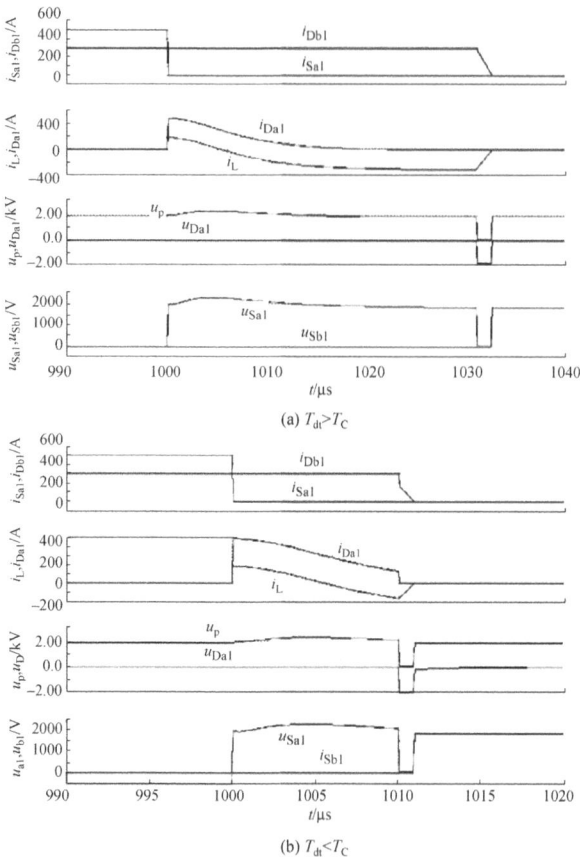

Fig. 7.11 Simulation waveform when two phases are commutating ($p \rightarrow o$, $i_A > 0$, $i_B < 0$)

$p \rightarrow o$, $i_A > 0$, $i_B < 0$, with S_{a1} in Phase A turning on and S_{b3} in Phase B turning on. It exhibits a positive dead band. The phase-phase pulse width is 1 μs. The simulation results are shown in Fig. 7.11a, where the Phase-B commutation is delayed for one dead band (30 μs) compared to Phase A. Therefore there is no need to set the minimum pulse width between phases, if $T_{dt} > T_C$. Otherwise similar to Fig. 7.11b, Phase B will employ the snubber circuit right after Phase A turns off, yielding the clamping diode reverse biased while conducting the current, i.e., reverse recovery process. This potentially could cause the diode damage.

$p \rightarrow o$, $i_A < 0$, $i_B > 0$, with S_{a3} in Phase A turning on and S_{b1} in Phase B turning off. It exhibits a negative dead-band effect. Vector width between two phases is deducted by one dead time. Therefore, we need the phase-phase vector time longer than one dead time. In addition, with S_{a3} turning on, its current linearly increases due to the snubber inductor L. Turning off S_{b1} at this time will force L to freewheel the current

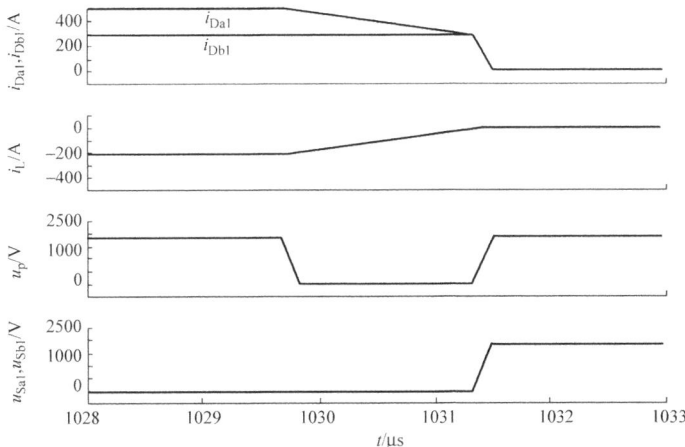

Fig. 7.12 Simulation waveform when two phases are commutating (p → o, $i_A < 0$, $i_B > 0$)

through the snubber diode, failing to limit the di/dt of S_{a3}, as shown in Fig. 7.12. This could cause damages to S_{a3}. Therefore, the phase-phase minimum pulse width need be $T_{dt} + T_L$.

All four cases above summarize the scenarios when two phases commutate and co-use the same snubber circuit. The phase-phase type-II minimum pulse width is determined by the dead-band effect when two phases are commutating, while the status of the snubber circuit and the dead-band effect are determined by time interval in-between. As a summary, related design rules are shown in Table 7.4. Compared to the type-I, the phase-phase type-II minimum pulse width involves and varies with snubber parameters.

Therefore, for a three-level NPC inverter with three phases sharing the same snubber circuit, the phase-phase type-II minimum pulse width needs meet the following requirement.

$$T > Max\{(T_{dt} + T_L), T_C\} \qquad (7.15)$$

Table 7.4 Minimum pulse width of the phase-phase type-II vectors

Phase-phase commutation dead band	Minimum phase-phase pulse width	Phase-phase commutation dead band	Minimum phase-phase pulse width
On-state zero dead band	0	Positive dead band	$T_C - T_{dt}$
Off-state zero dead band	T_C	Negative dead band	$T_L + T_{dt}$

Overall this section analyzes and derives several types of the minimum pulse width for such inverter, i.e.,

(1) The single-switch minimum pulse width (internal phase, type-I) $T \geq T_{\text{on\_min}} + T_{\text{dt}}$;
(2) The internal phase type-II minimum pulse width $T > T_{\text{dt}}$;
(3) The phase-phase type-I minimum pulse width $T > T_{\text{dt}}$;
(4) The phase-phase type-II minimum pulse width $T > \max\{(T_{\text{L}} + T_{\text{dt}}), T_{\text{C}}\}$.

Therefore, when designing the minimum pulse width, we need take into account not only the single switch performance and the dead time, but also the specific commutating process and the snubber-circuit dynamics. The phase-phase minimum pulse width corresponds to the phase-phase commutation dead band, which is further related to the snubber-circuit states. By fast and accurate sampling of the phase current, we could optimize the minimum pulse width per switching cycle based upon the current direction and synthesis of space vectors, with the dead-band effect being compensated as well.

7.2.2 Influence and Solution of Minimum Pulse Width

The minimum pulse width is determined by the non-ideal characteristic of the power switch, which will further affect the system output. Therefore, tracing back its generation and quantifying its impact are critical to system control performance and reliability. In this section, we will use three-level SVPWM control as an example.

1. Theory of the three-level SVPWM

There are only 12 non-zero voltage vectors that a three-level inverter can generate, which will be used to synthesize the reference voltage vector, control the motor flux linkage, and approach the circular flux when powered with the three-phase sinusoidal voltage. Through finely tuning the angle of the reference voltage vector, SVPWM can control the flux vector to rotate in a quasi-circle as well. Therefore, selecting basic vectors and assigning the vector time to synthesize the reference vector is the core of the SVPWM control.

Based on the amplitude and angle of the reference voltage vector, the control strategy will determine the triangle where the vector is located. Three space vectors located at vertices will be used with the time for each vector further assigned. Take the first sector as an example, as shown in Fig. 7.13. A similar approach could be applied to other sectors.

Assume the reference vector U_{ref} is located in zone 3 of Fig. 7.13. Three vertices correspond to the vectors U_{01}, U_{12}, and U_{02}, respectively. Further assume the time of each vector is t_{a0}, t_{b0} and t_{c0}. We have

$$U_{01} \cdot t_{\text{a0}} + U_{12} \cdot t_{\text{b0}} + U_{02} \cdot t_{\text{c0}} = U_{\text{ref}} \cdot T_{\text{s}} \qquad (7.16)$$

Fig. 7.13 Vector time
assignment in the first sector

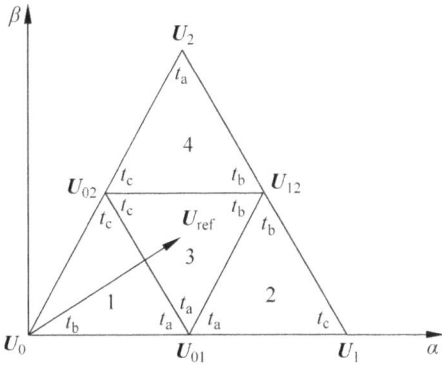

$$t_{a0} + t_{b0} + t_{c0} = T_s \tag{7.17}$$

Here T_s is the switching period of the SVPWM control.

Substituting each space vector into the above, we can calculate vector time as

$$t_{a0} = (1 - 2k \sin\theta) \cdot T_s$$
$$t_{b0} = \left[2k \sin\left(\theta + \frac{\pi}{3}\right) - 1\right] \cdot T_s$$
$$t_{c0} = \left[2k \sin\left(\theta - \frac{\pi}{3}\right) + 1\right] \cdot T_s \tag{7.18}$$

Here the amplitude modulation index $k = |U_{ref}|/|U_{12}|$.

Since U_{ref} does not have to be in zone 3, the calculated t_{a0}, t_{b0} and t_{c0} could be negative. By exhaustively attacking all possibilities, we can re-calculate the vector time as Table 7.5.

Once the reference location and the time of three vertex vectors are determined, we can further select the switching sequence and assign the time to each switch, during which we must obey the rule of vector transitions, i.e., only one-phase voltage can change per transition with a minimum voltage-level leap.

For the three-level SVPWM, there are three commonly used algorithms, i.e., asymmetric four-segment method, symmetric five-segment method, and seven-segment method. With the same PWM modulation frequency, the symmetric seven-segment method is more convenient and contains less harmonics than the four-segment method. Overall the seven-segment method inherits merits of both the four-segment and the five-segment methods. However, it also results in more switching actions per switching period.

Each sector can be divided into 6 subzones, as shown in Fig. 7.14. Selection of vectors to synthesize the reference vector needs comply following rules: (1) given the subsectors 0, 2, 4 share the same vector $U_{01,}$ and the subsectors 1, 3, 5 share the same vector U_{02}, the initial vector should be small vectors; (2) when ended, if

Table 7.5 Subsector judgment and time calculation in Sector 1

Subsector judgment	Vector position	Vector selection	Vector time
$\begin{cases} t_{a0} \geq 0 \\ t_{b0} \geq 0 \\ t_{c0} \geq 0 \end{cases}$	Zone 3	U_{01}, U_{12}, U_{02}	$\begin{aligned} t_a &= t_{a0} \\ t_b &= t_{b0} \\ t_c &= t_{c0} \end{aligned}$
$\begin{cases} t_{a0} < 0 \\ t_{b0} \geq 0 \\ t_{c0} \geq 0 \end{cases}$	Zone 4	U_2, U_{12}, U_{02}	$\begin{aligned} t_a &= -t_{a0} \\ t_b &= T_s - t_{c0} \\ t_c &= T_s - t_{b0} \end{aligned}$
$\begin{cases} t_{a0} \geq 0 \\ t_{b0} < 0 \\ t_{c0} \geq 0 \end{cases}$	Zone 1	U_0, U_{01}, U_{02}	$\begin{aligned} t_a &= T_s - t_{c0} \\ t_b &= -t_{b0} \\ t_c &= T_s - t_{a0} \end{aligned}$
$\begin{cases} t_{a0} \geq 0 \\ t_{b0} \geq 0 \\ t_{c0} < 0 \end{cases}$	Zone 2	U_{01}, U_{12}, U_1	$\begin{aligned} t_a &= T_s - t_{b0} \\ t_b &= T_s - t_{a0} \\ t_c &= -t_{c0} \end{aligned}$

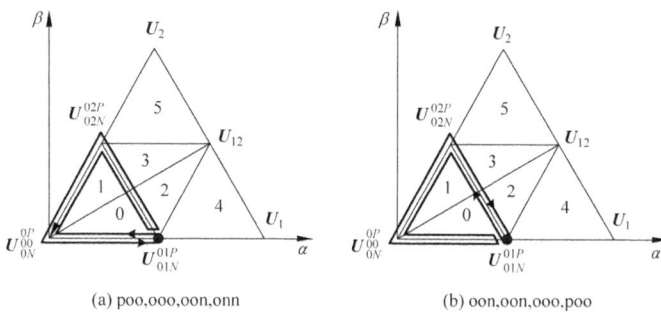

(a) poo,ooo,oon,onn   (b) oon,oon,ooo,poo

Fig. 7.14 Two vector selection methods in the subsector 0, with the seven-segment method

possible, small vectors or those zero/medium vectors adjacent to small vectors are preferred, which facilitates smooth transition among each sector.

Take the seven-segment method as an example. In Fig. 7.14 and 7.15, each segment starts with a small vector. #1, #7 and #4 segments located in the same position, #2 and #6 correspond to the same vector and #3 and #5 segments are the same vector.

To secure the symmetry, the time assignment is shown as Eq. (7.19). Here t_1 and t_7 together is the same as t_7, all of which are for the initial small vector. Since the reference vector is closer to the related triangle vertex, the time of such small vector t_a is the longest. The other small vector time t_b is relatively short.

$$t_1 = t_7 = \frac{1}{4}t_a, \quad t_4 = \frac{1}{2}t_a, \quad t_2 = t_6 = \frac{1}{2}t_b, \quad t_3 = t_5 = \frac{1}{2}t_c \qquad (7.19)$$

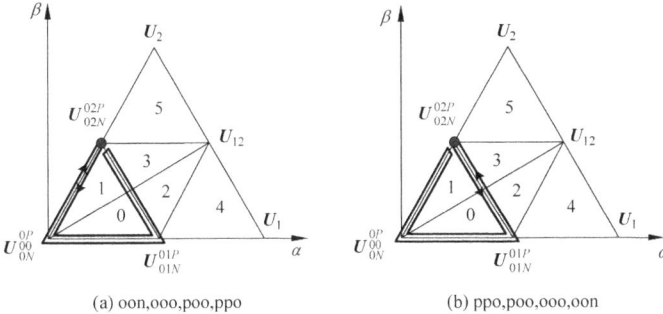

Fig. 7.15 Two vector selection methods in the subsector 1, with the seven-segment method

Shown in Fig. 7.14 and 7.15 are two vector selections in subsector 0 and 1, respectively. Although the vector selection could be various, when considering the sector transition, we still prefer following strategies.

Strategy (1). Combination of Fig. 7.14a and Fig. 7.15b. All initial vectors are positive small vectors, which prohibit any two phases from switching simultaneously. However, since the vector sequence in each sector is different, the waveform symmetry is not satisfactory;

Strategy (2). Combination of Fig. 7.14b and Fig. 7.15a. All initial vectors are negative small vectors. Similar to 1), no two phases will switch simultaneously, though the waveform symmetry is poor;

Strategy (3). Combination of Fig. 7.14a and Fig. 7.15a. Such vector selection yields a strict symmetry, given t_b is zero or medium vectors. Its demerit is when crossing sectors two phases might switch at the same time. For instance, poo-oon will result in abnormal pulses on the line-line voltage, where the voltage leap is the whole DC-bus voltage. Therefore, such strategy is not recommended. Neither is the combination of Fig. 7.14b and Fig. 7.15b.

In real practice, Strategies (1) and (2) are feasible. For Strategy (1), all vector selections in subsectors of Sector 1 with their time assignment shown in Table 7.6.

2. Restrained areas of minimum pulse width in three-level SVPWM

Based on the analysis in the previous section, vectors before and after the single-switch pulse are exactly the same. Therefore, the minimum pulse width only occurs in #1, #4, and #7 of all seven segments. Since vectors related to t_1 and t_7 are connected end to end, we only require t_1/t_7 larger than half of the minimum pulse width. t_4 does need be larger than the whole minimum pulse width. Note the phase-phase minimum pulse width occurs at #2, #3, #5, and #6 segments. It varies based upon the vector sequence. For instance, when starting with positive small vectors, six subsectors in the Sector 1 have restraints of minimum pulse width shown in Table 7.7. Furthermore, we can qualitatively picture impact zones of the minimum pulse width as Fig. 7.16a. When the reference vector is located in the shadow zone, the small vector U_{01} has very short time, which is subject to the restraint of the minimum pulse width. Therefore, the

Table 7.6 Vector sequence and time assignment in all subsectors of Sector 1

Subsector judgment	Subzone number	Vector sequence	Vector time
$\begin{cases} t_{a0} \geq 0 \\ t_{b0} < 0 , t_{a0} > t_{c0} \\ t_{c0} \geq 0 \end{cases}$	0	$U_{01P} \Leftrightarrow U_{00} \Leftrightarrow$ $U_{02N} \Leftrightarrow U_{01N}$	$\begin{cases} t_a = 1 - t_{c0} \\ t_b = -t_{b0} \\ t_c = 1 - t_{a0} \end{cases}$
$\begin{cases} t_{a0} \geq 0 \\ t_{b0} < 0 , t_{a0} < t_{c0} \\ t_{c0} \geq 0 \end{cases}$	1	$U_{02P} \Leftrightarrow U_{01P} \Leftrightarrow$ $U_{00} \Leftrightarrow U_{02N}$	$\begin{cases} t_a = 1 - t_{a0} \\ t_b = 1 - t_{c0} \\ t_c = -t_{b0} \end{cases}$
$\begin{cases} t_{a0} \geq 0 \\ t_{b0} \geq 0 , t_{a0} > t_{c0} \\ t_{c0} \geq 0 \end{cases}$	2	$U_{01P} \Leftrightarrow U_{12} \Leftrightarrow$ $U_{02N} \Leftrightarrow U_{01N}$	$\begin{cases} t_a = t_{a0} \\ t_b = t_{b0} \\ t_c = t_{c0} \end{cases}$
$\begin{cases} t_{a0} \geq 0 \\ t_{b0} \geq 0 , t_{a0} < t_{c0} \\ t_{c0} \geq 0 \end{cases}$	3	$U_{02P} \Leftrightarrow U_{01P} \Leftrightarrow$ $U_{12} \Leftrightarrow U_{02N}$	$\begin{cases} t_a = t_{c0} \\ t_b = t_{a0} \\ t_c = t_{b0} \end{cases}$
$\begin{cases} t_{a0} \geq 0 \\ t_{b0} \geq 0 \\ t_{c0} < 0 \end{cases}$	4	$U_{01P} \Leftrightarrow U_{12} \Leftrightarrow$ $U_1 \Leftrightarrow U_{01N}$	$\begin{cases} t_a = 1 - t_{b0} \\ t_b = 1 - t_{a0} \\ t_c = -t_{c0} \end{cases}$
$\begin{cases} t_{a0} < 0 \\ t_{b0} \geq 0 \\ t_{c0} \geq 0 \end{cases}$	5	$U_{02P} \Leftrightarrow U_2 \Leftrightarrow$ $U_{12} \Leftrightarrow U_{02N}$	$\begin{cases} t_a = 1 - t_{b0} \\ t_b = -t_{a0} \\ t_c = 1 - t_{c0} \end{cases}$

shadow area is impact zones of U_{01} due to the single-switch minimum pulse width. For any triangle, the edge opposite to the starting positive small vector is subject to the restraint of the single-switch minimum pulse width, as shown in Fig. 7.16a. Two edges connected with this vertex of the small positive vector and surrounding areas are subject to the restraint of the phase-phase minimum pulse width, as shown in Fig. 7.16b. We can see that all areas related to the minimum pulse width are close to triangle edges and are evenly distributed to the whole vector space.

A similar analysis applies to the minimum-pulse-width restraint on Sector 2, as shown in Tables 7.7 and 7.8. Differences of the phase-phase minimum pulse width exist from Sector 1, which are caused by selection of vector sequences. As shown in Fig. 7.17, with the positive small vector as the initial vector, when starting from U_{01}, U_{03}, and U_{05}, t_b is for the zero or medium vector. When starting from U_{02}, U_{04} and U_{06}, t_c is for the zero or medium vector. When starting from the negative small vector, the vector sequence is opposite.

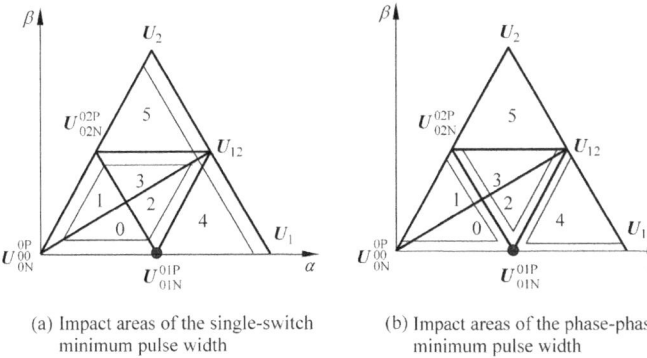

(a) Impact areas of the single-switch   (b) Impact areas of the phase-phase
    minimum pulse width   minimum pulse width

Fig. 7.16 Areas restrained by the vector minimum pulse width

After considering all vector types in different subsectors, we can show influential regions of the phase-phase minimum pulse width as in Fig. 7.18.

Table 7.7 Impact of the minimum pulse width on Sector 1

Subzone number	Vector sequence	t_1, t_7	t_2, t_6	t_3, t_5	t_4
0	poo-ooo-oon-onn-oon-ooo-poo	1/2S	PI	PII	S
1	ppo-poo-ooo-oon-ooo-poo-ppo	1/2S	PII	PI	S
2	poo-pon-oon-onn-oon-pon-poo	1/2S	PI	PI	S
3	ppo-poo-pon-oon-pon-poo-ppo	1/2S	PI	PI	S
4	poo-pon-pnn-onn-pnn-pon-poo	1/2S	PII	PI	S
5	ppo-ppn-pon-oon-pon-ppn-ppo	1/2S	PI	PII	S

Note S means the minimal pulse width of single device, PI is the first type of the minimal pulse width between phases, and PII is the second type of the minimal pulse width between phases

Table 7.8 Impact of the minimum pulse width on the Sector 2

Subzone number	Vector sequence	t_1, t_7	t_2, t_6	t_3, t_5	t_4
0	ppo-opo-ooo-oon-ooo-opo-ppo	1/2S	PII	PI	S
1	opo-ooo-oon-non-oon-ooo-opo	1/2S	PI	PII	S
2	ppo-opo-opn-oon-opn-opo-ppo	1/2S	PI	PI	S
3	opo-opn-oon-non-oon-opn-opo	1/2S	PI	PI	S
4	ppo-ppn-opn-oon-opn-ppn-ppo	1/2S	PI	PII	S
5	opo-opn-npn-non-npn-opn-opo	1/2S	PII	PI	S

Note S means the minimal pulse width of single device, PI is the first type of the minimal pulse width between phases, and PII is the second type of the minimal pulse width between phases

Fig. 7.17 Two types of
influential regions with
different vector sequences
Note: white region with the
odd-number small vector as
the initial; shadow region
with the even-number small
vector as the initial

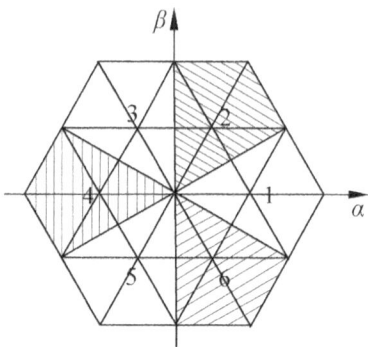

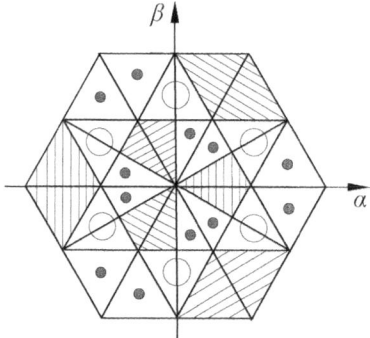

Fig. 7.18 Influential regions of the phase-phase minimum pulse width Note: shadow region-type
I and II; circle-type I; dot-type II and I

3. The impact of the minimum pulse width on the control strategy

The minimum pulse width sets the lower limit for the vector time, which directly
makes the synthesized vector deviate from its original. As shown in Fig. 7.19a,
once the reference vector falls into the shadow region, it will be influenced by the
minimum pulse width. In another word, with the restraint of the minimum pulse
width, the synthesized voltage vector cannot reach the shadow region.

Specifically the minimum pulse width influences the reference vector from two
aspects. As shown in Fig. 7.19b, the reference vector U has t_a too short thereby falling
into the influential region. With the minimum pulse width, t_c is unchanged while t_a is
enlarged to the minimum pulse width, which yields the ultimate synthesized vector
U' on the edge of the influential region with shifts of the amplitude and angle.

Shown in Fig. 7.19c are reference vectors U_2 and U_1 in subsector 0 and 1, respectively. A small modulation index yields small t_a and t_c for these two vectors. Through
adjustment of the minimum pulse width for t_a and t_c, all reference vectors in this
regions are changed to U', with the amplitude enlarged. When all synthesized vectors are at six particular locations, a hexagonal flux linkage and 6th-order torque

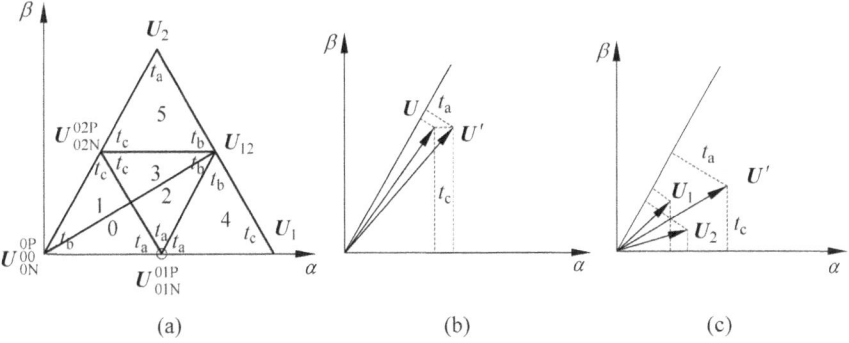

Fig. 7.19 The impact of the minimum pulse width on the synthesized vector

ripple will appear. This indicates that the minimum pulse width will mostly affect the low-frequency and low-modulation-index operations.

With the seven-segment method using positive small vectors and the control period as T_s, the minimum pulse width confines t_a and t_c both to M. We can further reversely calculate the modulation index k_0 through Eq. (7.20) for the reference vector U' in Fig. 7.19c, using Eq. (7.18) and Table 7.6. When the modulation index is lower than k_0, the reference vector falls into the influential region of the minimum pulse width, which distorts the flux linkage. Therefore, we can use k_0 as a reference to evaluate the impact of the minimum pulse width on the low-frequency output.

$$k_0 = \frac{2M}{T_s} \tag{7.20}$$

For a three-phase full-bridge inverter, the minimum pulse width will affect three aspects, i.e., output voltage THD, low-speed torque ripple, and starting in-rush current.

(1) Impact on the THD

As shown in Fig. 7.19, when reference vectors fall into the shadow regions, with adjustment of the minimum pulse width, final vectors will fall on the edge of these regions. With the deviation of the reference vector, the control strategy will be somewhat influenced. The vector amplitude will be changed at the boundaries and the angle will be stalled or subject to the abrupt change. This will make the ultimate flux linkage distorted from the original circular trajectory. Such distortion eventually will translated into the enlarged THD of the output voltage and current. The closer the reference vector stays to the subsector edge, the more influential the minimum pulse width and the more obvious the voltage and current distortion are.

(2) Torque ripple at the low-speed operation

With a low modulation index, synthesized voltage vectors can only be crowded in small shadow areas defined by the minimum pulse width, resulting in those vectors

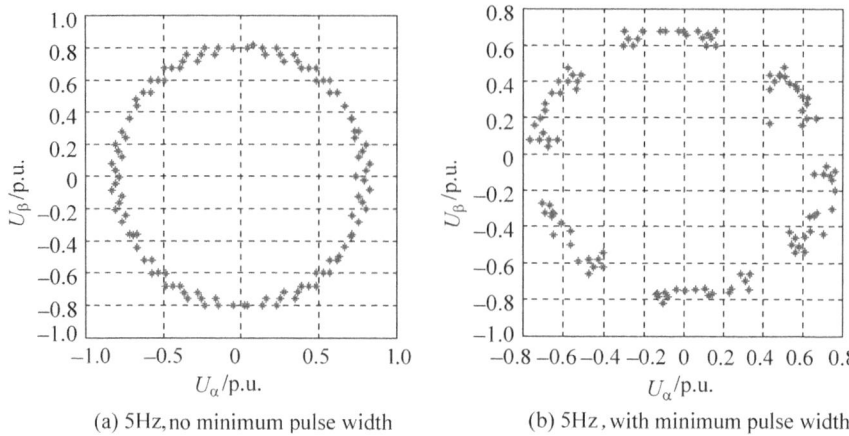

(a) 5Hz, no minimum pulse width

(b) 5Hz, with minimum pulse width

Fig. 7.20 Impact of the minimum pulse width on the synthesized voltage vector, at low speed

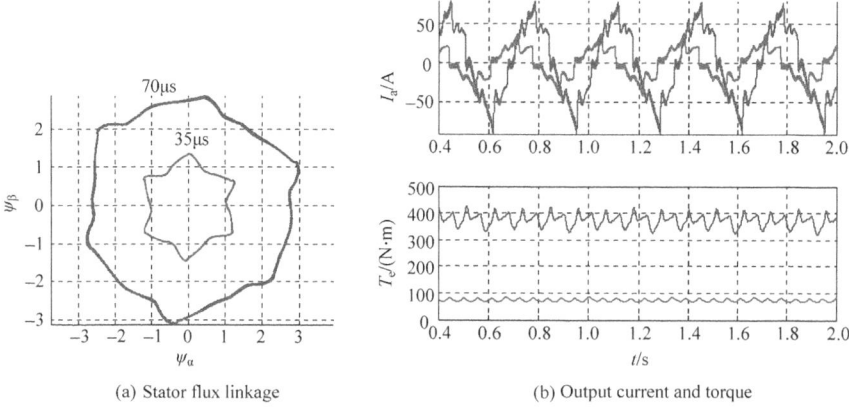

(a) Stator flux linkage

(b) Output current and torque

Fig. 7.21 Output waveform at $f = 5$ Hz and $k = 0.03$, with minimum-pulse-width settings of 70 and 35 μs, respectively

being clustered in the middle region of sectors every $60°$. As shown in Fig. 7.20, with the same 5 Hz operation, the minimum pulse width will unevenly distribute vectors, leading to the hexagonal flux linkage and furthermore, to the torque ripple. With $k = 0.03$ and $f = 3$ Hz, a comparison is made in Fig. 7.21a with the minimum pulse width of 35 μs and 70 μs, respectively. In both cases with the hexagonal flux linkage, the minimum pulse width of 70 μs yields a saturated flux due to enlarged voltage vectors. Shown in Fig. 7.21b are the output current and torque. Current is nearly doubled with the minimum pulse width of 70 μs. Its output torque and the torque ripple are also much higher than those at 35 μs. We can see that with the low modulation index and speed, the higher the minimum pulse width, the higher the torque ripple.

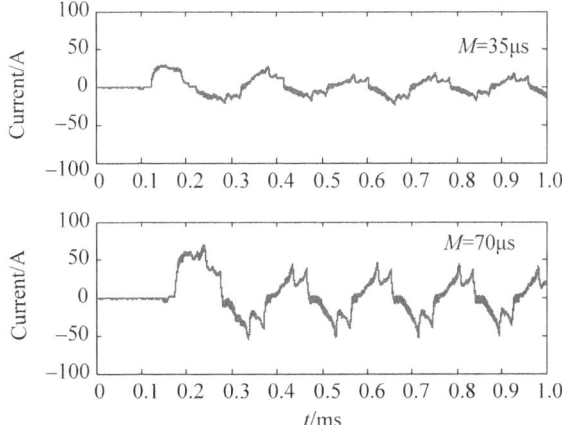

Fig. 7.22 Experimental starting current with minimum-pulse-width settings of 35 μs (top) and 70 μs (bottom)

(3) Large starting in-rush current

The lower the modulation index, the more severe the distortion caused by the minimum pulse width. As shown in Fig. 7.19c, the minimum pulse width amplifies the amplitude of the synthesized vector, prolongs the time of non-zero vectors, and increases the voltage-frequency ratio, which tends to create flux saturation and further enlarge the current. As shown in Fig. 7.22, the larger the minimum pulse width, the higher the starting current. This phenomenon is due to the relatively low modulation index in the starting process, making the system more vulnerable to the minimum pulse width, and thereby causing the in-rush current.

The analysis above shows that the minimum pulse width has more impact on the low modulation index. Specifically, it will cause the starting inrush and low-speed torque ripple. At the high modulation index, it will distort the flux trajectory and further increase the current THD. For the output voltage, the higher the operational frequency, the less impact of the minimum pulse width due to much larger vector amplitude and steps. No significant angle shift will occur even when the vector enters the influential region of the minimum pulse width.

4. **Comprehensive solutions for the minimum-pulse-width problem**

To alleviate the negative impact of the minimum pulse width, we can divide the whole space into multiple circular regions. Based upon the different characteristics of the minimum pulse width in those regions, we can adopt corresponding methods. Here we assume the single-switch minimum pulse width is M_1, the phase-phase minimum pulse width is M_2, and $M_2 < M_1$. $k_0 = 2 M_2/T_s$ and $k_1 = 2 M_1/T_s$.

(1) $k < k_0$

Based on the analysis above, the vector minimum pulse width is inevitable. Without the requirement of ultra-low speed operations, we can avoid such regions. Besides, the voltage drop across the motor leakage resistance and inductance is relatively high, which makes voltage compensation difficult as well.

(2) $k_0 \le k < k_1$

As shown in Fig. 7.23(a), inside subsector 0, a region C'-B'-D-O emerges, due to the restraint of the single-switch minimum pulse width on U_{02} and the phase-phase minimum pulse width on U_{01}. The influenced vectors will all move to the boundary D-B'-C'. Similarly, inside subsector 1, the impact region is A-B-D-O. Therefore, synthesized vectors are moving along B'-D-B, with the amplitude oscillating. This could further cause fluctuation of the flux amplitude, as seen in Fig. 7.21a.

The solution is to switch the vector type. Given no two-phase switching simultaneously at $k = k_0$ is allowed, when using positive small vectors in the white regions of Fig. 7.17, we need swap 000 and odd-number positive small vectors (U_{01P}, U_{03P} and U_{05P}) and rearrange subsequent vectors. When using negative small vectors in shadow regions of Fig. 7.17, we need swap 000 and even-number negative small vectors and rearrange left vectors.

(3) k slightly larger than 0.5

When $k \cong 0.5$, this region needs particular attention, given the single-switch minimum pulse width might occur in zero and medium vectors. This can only change the vector amplitude without much impact on the angle. When k is slightly larger than 0.5, it is possible that the reference vector will cross impact regions of the phase-phase minimum pulse width in subsectors 0, 2, 3 and 1, as shown in Fig. 7.23b. This can cause the vector shift to two opposite directions, thereby causing large fluctuation of the flux linkage. Illustrated in Fig. 7.24a is an experimental current waveform with the minimum pulse width of 70 μs and $k = 0.5, 0.51, 0.52$, respectively. Figure 7.24b shows the flux linkage at $k = 0.51$, indicating the largest distortion and fluctuation among all cases.

Since this region is heavily influenced by the minimum pulse width, a hysteresis of Δk can be added to avoid such region. As shown in Fig. 7.23b, the threshold of k to let U_{12} reach the boundary of the phase-phase minimum-pulse-width impact regions can be calculated as

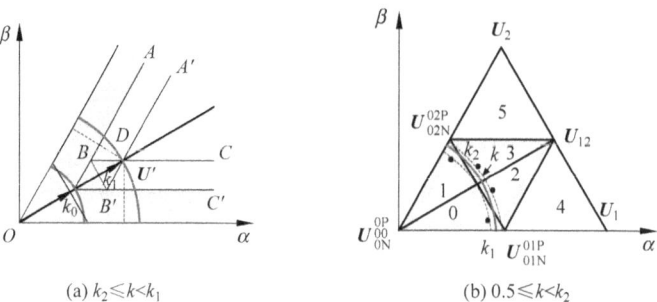

(a) $k_2 \le k < k_1$ (b) $0.5 \le k < k_2$

Fig. 7.23   The impact regions of the minimum pulse width at some particular cases

$$k_2 = \frac{M_2}{T_s} + 0.5 \qquad (7.21)$$

Therefore it is recommended to skip a zone of M_2/T_s when $k = 0.5$.

(4) $k > 1 - k_1$

Under this circumstance, the vector enters the impact zone of the minimum pulse width in subsectors 2 and 3. The system is running at the high speed/frequency since k is relatively large now. With a relatively low switching frequency in a high-power converter, the number of reference vectors within one circle is low as well. We can select the synchronous modulation to avoid impact zones of the minimum pulse width. Another merit of such method is to increase the modulation index to 1 and even higher.

As shown in Fig. 7.25, the maximum modulation index k_3 of the synchronous modulation is calculated below. With the symmetric feature of such control, when the angle of the reference vector is $\pi/12$ and $\pi/4$, the modulation index reaches the maximum. With $k' = 1 - M_1/T_s$, using geometrical knowledge we have

$$k_3 = \frac{1}{\cos(\pi/12)}\left(1 - \frac{M_1}{T_s}\right) = 1.035 \times \left(1 - \frac{M_1}{T_s}\right) \qquad (7.22)$$

Otherwise the impact of the minimum pulse width is minor. The conventional 7-segment method is sufficient. With the starting segment using the positive/negative small vector, the requirement of the single-switch minimum pulse width has been fulfilled. With solutions proposed above, we can globally minimize the negative impact of the minimum pulse width in whole-modulation-index zones.

Essentially the comprehensive solutions above are to optimize the minimum pulse widths based on the regions located. The proposed methods inherit advantages of multiple methods and exhibit ease and feasibility. As a summary, a voltage-frequency

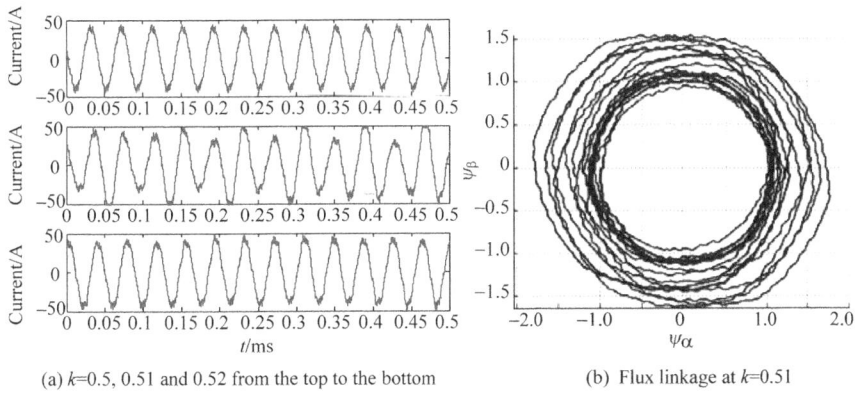

(a) k=0.5, 0.51 and 0.52 from the top to the bottom   (b) Flux linkage at k=0.51

Fig. 7.24 Experimental validation of the minimum-pulse-width impact when k is slightly larger than 0.5

Fig. 7.25 The maximum k at the synchronous modulation

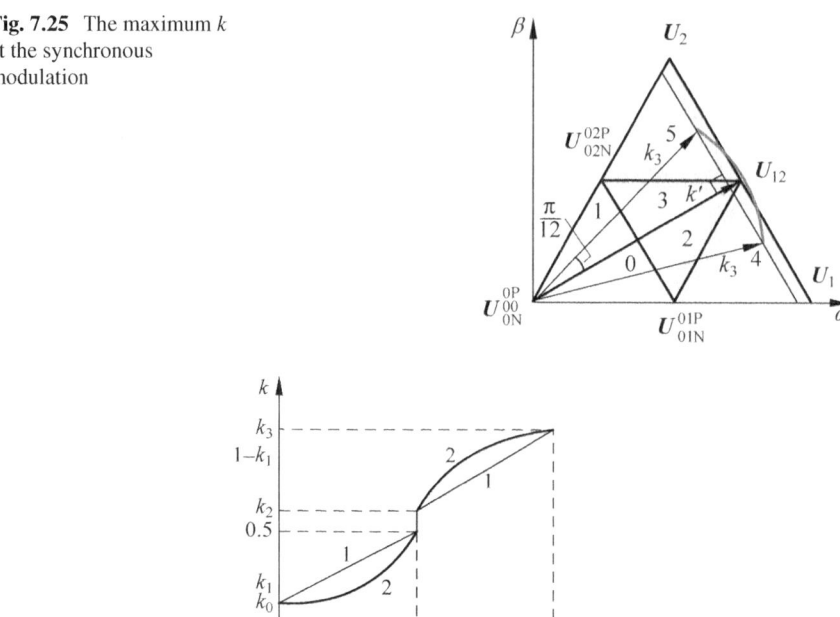

Fig. 7.26 The overall proposed V-f curve

curve can be set as Fig. 7.26. Curve 1 is linear while Curve 2 considers the low-frequency voltage compensation and high-speed operation. All solutions for different zones are summarized in Table 7.9.

5. Coordination of the minimum pulse width and the dead band

Such two factors are tightly coupled. The dead band sets a higher bar for the minimum-pulse-width setting, affects the output voltage, and varies the vector time, which causes neutral point imbalance in the three-level inverter due to variation of the small vector time. While a dead-band compensation will ease the neutral-point imbalance, the output current polarity needs be detected for a reasonable compensation. Furthermore, as shown in Sect. 7.2.1, with detection of the phase current, we could determine the phase-phase commutating dead band, thereby setting the optimal value of the phase-phase minimum pulse width. This helps alleviate the side impact of the minimum pulse width. Therefore, it is possible to coordinate the dead band and the minimum pulse width together based on the phase current direction, i.e., compensating the dead band, balancing the neutral point, and optimizing the minimum pulse width together.

Based on a real-time current sampling, the 7-segment SVPWM method is used meeting Eq. (7.19), i.e., $t_7 = t_1$, $t_6 = t_2$, $t_5 = t_3$. Meanwhile $t_a = t_1 + t_4 + t_7$, $t_b = t_2 + t_6$, and $t_c = t_3 + t_5$. Shown in Table 7.10 are variations of the vector time and

Table 7.9 Comprehensive solution of the minimum pulse width

Region location	Solution	Key parameters
$k < k_0$	Avoided	$k_0 = 2M_2/T_s$ $k_1 = 2M_1/T_s$ $k_2 = \dfrac{M_2}{T_s} + 0.5$ $k_3 = 1.035 \times \left(1 - \dfrac{M_1}{T_s}\right)$
$k_0 \leq k < k_1$	Switched the vector type	
$k_1 \leq k < 0.5$	Unchanged	
$0.5 \leq k < k_2$	Avoided	
$k_2 \leq k < 1 - k_1$	Unchanged	
$1 - k_1 \leq k \leq k_3$	Synchronous modulation	
$k > k_3$	Avoided	

types of phase-phase commutating dead band in subsector 0, Sector 1. Here segments 1, 4, and 7 are related to the single-switch minimum pulse width. Time difference from original values is given. Because other four vectors are related to phase-phase vectors, the phase-phase commutating dead band is present. With the impact of the dead band, all vector time is subject to change. To coordinate the dead band with the minimum pulse width, we need

(1) Consider the dead-band compensation and recover each vector time;
(2) Balance the neutral point by setting $t_4 = t_1 + t_7 = 2t_1$;
(3) Impose restraints of the minimum pulse width. Here t_1 needs be larger than ½ of the single-switch minimum pulse width, t_4 needs be larger than the single-switch minimum pulse width, and t_2 and t_3 should be set based on the phase-phase vector type, phase-phase commutating dead band, and current direction.

Detailed vector time optimization is processed based on six current combinations, as shown in Table 7.10.

(1) With the current direction $++-$, the actual phase time are $t_b + 2\,T_{dt}$ and $t_c - 2\,T_{dt}$.

With the dead-band compensation, $t_2' = t_2 - 2\,T_{dt}$ and $t_3' = t_3 + 2\,T_{dt}$;
With the neutral-point balancing, $t_4' = t_4 - T_{dt}$ and $t_1' = t_1 + 0.5T_{dt}$;
With the minimum pulse width, $t_4' > M$, $t_2' > 0$, $t_3' > T_L + T_{dt}$ ($t_3 > T_L$) and $t_1' > 0.5\,M_1$.

(2) With the current direction $+-+$, the actual phase time are $t_a - 2\,T_{dt}$ and $t_c + 2\,T_{dt}$.

With the dead-band compensation, $t_1' = t_1 + 0.5\,T_{dt}$, $t_4' = t_4 + T_{dt}$ and $t_3' = t_3 - T_{dt}$;

Table 7.10 Vector equivalent time in subsector 0, sector 1

Time	Vector	Current ++−	Current +−+	Current +−−	Current −++	Current −+−	Current −−+
t_1	poo	Unchanged	Unchanged	Unchanged	$+T_{dt}$	$+T_{dt}$	$+T_{dt}$
t_2	ooo	Positive dead-band I	Zero dead-band off I	Positive dead-band I	Negative dead-band I	Zero dead-band on I	Negative dead-band I
t_3	oon	Negative dead-band II	Positive dead-band II	Zero dead-band off II	Zero dead-band II	Negative dead-band II	Positive dead-band II
t_4	onn	$+T_{dt}$	$-T_{dt}$	$-T_{dt}$	$+T_{dt}$	$+T_{dt}$	$-T_{dt}$
t_5	oon	Negative dead-band II	Positive dead-band II	Zero dead-band on II	Zero dead-band on II	Negative dead-band II	Positive dead-band II
t_6	ooo	Positive dead-band I	Zero dead-band on I	Positive dead-band I	Negative dead-band I	Zero dead-band off I	Negative dead-band I
t_7	poo	$-T_{dt}$	$-T_{dt}$	$-T_{dt}$	Unchanged	Unchanged	Unchanged

No need for the neutral-point balancing;

With the minimum pulse width, $t_4' > M_1$, $t_2' > 0$, $t_3' > T_C - T_{dt}$ $(t_3 > T_C)$ and $t_1' > 0.5M_1$.

(3) With the current direction $+--$, the actual phase time are $t_a - 2\,T_{dt}$ and $t_b + 2\,T_{dt}$.

With the dead-band compensation, $t_1' = t_1 + 0.5\,T_{dt}$, $t_4' = t_4 + T_{dt}$ and $t_2' = t_2 - T_{dt}$;

No need for the neutral-point balancing;

With the minimum pulse width, $t_4' > M_1$, $t_2' > 0$, $t_3' > T_C$ and $t_1' > 0.5\,M_1$.

(4) With the current direction $-++$, the actual phase time are $t_a + 2\,T_{dt}$ and $t_b - 2T_{dt}$.

With the dead-band compensation, $t_1' = t_1 - 0.5\,T_{dt}$, $t_4' = t_4 - T_{dt}$ and $t_2' = t_2 + T_{dt}$;

No need for the neutral-point balancing;

With the minimum pulse width, $t_4' > M_1$, $t_2' > T_{dt}$ $(t_2 > 0)$, $t_3' > T_C$ and $t_1' > 0.5M_1$.

(5) With the current direction $-+-$, the actual phase time are $t_a + 2T_{dt}$ and $t_c - 2T_{dt}$.

With the dead-band compensation, $t_1' = t_1 - 0.5\,T_{dt}$, $t_4' = t_4 - T_{dt}$ and $t_3' = t_3 + T_{dt}$;

No need for the neutral-point balancing;

With the minimum pulse width, $t_4' > M_1$, $t_2 > 0$, $t_3' > T_C$ and $t_1 > 0.5\,M_1$.

(6) With the current direction $--+$, the actual phase time are $t_b - 2T_{dt}$ and $t_c + 2T_{dt}$.

With the dead-band compensation, $t_2' = t_2 + T_{dt}$ and $t_3' = t_3 - T_{dt}$;

With the neutral-point balancing, $t_4' = t_4 + T_{dt}$ and $t_1 = t_1 - 0.5\,T_{dt}$;

With the minimum pulse width, $t_4' > M$, $t_2 > T_{dt}$ $(t_2 > 0)$, $t_3' > T_C - T_{dt}$ $(t_3 > T_C)$ and $t_1' > 0.5M_1$.

As shown above,

(1) The turn-on and turn-off zero dead bands are in pairs. To incorporate the phase-phase type-II zero dead band, for the sake of symmetry, we can set all minimum pulse width as T_C;

(2) The dead band results in one of the three vectors added with double dead time and another one losing double dead time. The third vector remains unchanged, but will have discounted neutral-point balancing if such vector is the initial small vector;

(3) Based on the current direction we can set the minimum pulse width accordingly to minimize its side impact. The dead-band compensation can mostly alleviate the impact of the dead band on the minimum pulse width, allowing a smaller-value phase-phase minimum pulse width;

(4) If t_1' is less than the single-switch minimum pulse width, we can add one ΔT. Meanwhile for the sake of the neutral-point balancing, an extra $2\ \Delta T$ needs be added to t_4'. Similarly, if t_4' is added with ΔT, t_1' needs be increased by $0.5\ \Delta T$.

Shown in Table 7.11 is the vector time assignment incorporating the dead-band compensation, neutral point balancing, and the optimal minimum pulse width. Subsectors 0 and 1 are given as an example. 1 and 0 represent the positive and negative current, respectively, with overall six possibilities. Given the sum of the three-phase current is zero, there is no 111 or 000, though the current value could be small. Such a small-current case might result in a scenario where the current changes polarity within one switching period. In real practice such case is labelled as 0 without any compensation.

7.2.3   Discrete Error with Its Compensation

Within one control period, the alteration of pulse edges caused by the dead band and minimum pulse width greatly impacts pulse logic combination and control performance. Particularly in a high-power system, switching actions take longer time, and the dead time and minimum pulse width are larger, which causes significant difference between the energy pulse and information pulse, with control performance highly influenced. In addition, original information signals generated by the control loop also contain errors, one of which is caused by the discrete control. In this section, we will use the three-level PWM rectifier as an example to analyze the discrete control error and its compensation.

1. Cause and impact of discrete errors

The mathematical description of the energy pulse in Sect. 7.1.2 indicates that the main circuit of a power electronic converter is a continuous system, with all variables being time variant. The commonly used digital control system, however, has the control method implemented discretely in the microcontroller. In every control period, the controller implements the control strategy once, calculating control variables and generating control pulses or pulse sequences. Essentially a discrete controller controls a continuous system. For a continuous system, regardless of control periods, all state variables and control parameters are continuous. For a discrete control, those variables and parameters, however, are assumed constant within one control period. This generates a discrete error. Without suppressing or eliminating such error after the control pulse generation, the ultimate energy pulse sequence of the converter will contain the discrete error as well, which lowers control precision. In a high-power power electronic converter, due to the limitation of switching frequency, the control period is longer, and the discrete characteristics are more distinct, and the control and state variable change more within one control period. Therefore, the discrete control error has more impact on control performance.

Table 7.11 The vector time assignment incorporating with the dead-band compensation, neutral point balancing and the optimal minimum pulse width

Subsector	110(6)	101(5)	100(4)	011(3)	010(2)	001(1)	000(0)
0-0	$t_1' = t_1 + 0.5\,T_{dt}$, $t_2' = t_2 - T_{dt}$, $t_3' = t_3 + T_{dt}$, $t_4' = t_4 - T_{dt}$, $t_2' > 0, t_3' > T_L + T_{dt}$	$t_1' = t_1 + 0.5\,T_{dt}$, $t_4' = t_4 + T_{dt}$, $t_3' = t_3 - T_{dt}, t_2$ $> 0, t_3' > T_C - T_{dt}$	$t_1' = t_1 + 0.5\,T_{dt}$, $t_4' = t_4 + T_{dt}$, $t_2' = t_2 - T_{dt}$, $t_2' > 0, t_3 > T_C$	$t_1' = t_1 - 0.5\,T_{dt}$, $t_4' = t_4 - T_{dt}$, $t_2' = t_2 + T_{dt}$, $t_2' > T_{dt}, t_3 > T_C$	$t_1' = t_1 - 0.5\,T_{dt}$, $t_4' = t_4 - T_{dt}$, $t_3' = t_3 + T_{dt}, t_2$ $> 0, t_3' > T_C$	$t_1' = t_1 - 0.5\,T_{dt}$, $t_2' = t_2 + T_{dt}$, $t_3' = t_3 - T_{dt}$, $t_4' = t_4 + T_{dt}$, $t_2' > T_{dt}, t_3' > T_C - T_{dt}$	–
0-1	$t_1' = t_1 + 0.5\,T_{dt}$, $t_4' = t_4 + T_{dt}$, $t_3' = t_3 - T_{dt}, t_2$ $> T_C, t_3' > 0$	$t_1' = t_1 - 0.5\,T_{dt}$, $t_4' = t_4 - T_{dt}$, $t_2' = t_2 + T_{dt}$, $t_2' > T_L + T_{dt}, t_3$ > 0	$t_1' = t_1 - 0.5\,T_{dt}$, $t_2' = t_2 + T_{dt}$, $t_3' = t_3 - T_{dt}$, $t_4' = t_4 + T_{dt}$, $t_2' > T_L + T_{dt}, t_3$ $t_3' > 0$	$t_1' = t_1 + 0.5\,T_{dt}$, $t_2' = t_2 - T_{dt}$, $t_3' = t_3 + T_{dt}$, $t_4' = t_4 - T_{dt}$, $t_2' > T_C - T_{dt}, t_3$ $t_3 > T_{dt}$	$t_1' = t_1 + 0.5\,T_{dt}$, $t_4' = t_4 + T_{dt}$, $t_2' = t_2 - T_{dt}$, $t_2' > T_C - T_{dt}, t_3$ > 0	$t_1' = t_1 - 0.5\,T_{dt}$, $t_4' = t_4 - T_{dt}$, $t_3' = t_3 + T_{dt}, t_2$ $> T_C, t_3' > T_{dt}$	–

Note 1-positive current, 0-negative current, 000-one-phase current is zero or close to zero

For a closed-loop control system containing the integral part, such as the VOC or SVM-DPC in the PWM rectifier, the discrete control error can be limited by the related integral component. In other systems with a higher demand of control accuracy and pulse-width modulation, the discrete control error might continuously despair control performance.

For instance, the predictive direct power control (PDPC) introduces the predictive theory into the DPC of the PWM rectifier, with the aim of enhancing both dynamic response and steady performance. Based on low-level information pulse generations, the PDPC can be categorized into two, one being the variable switching frequency (VSF) PDPC based on the vector optimal selection, and the other the constant switching frequency (CSF) PDPC based on the PWM method. Comparatively, CSF-PDPC has merits of fixed switching frequency, low sampling frequency, low computation load, and excellent power/current control performance. The high-level block diagram of the three-level CSF-PDPC is shown in Fig. 7.27.

Similar to the three-level DPC, the three-level CSF-PDPC contains three control loops as well, i.e., one outer loop for the DC-bus voltage control using the PI controller, and two inner loops for the active and reactive power control. The difference is that the CSF-PDPC applies power prediction in the power loop to directly calculate control variables. Such controller is based on Eq. (7.23), derived through the mathematical model of the PWM rectifier to connect the voltage vector with the power rate. The power command and sampled power value can be adopted to accurately calculate the amplitude and angle of needed control voltage vector, which is used to generate control signals through SVPWM algorithm.

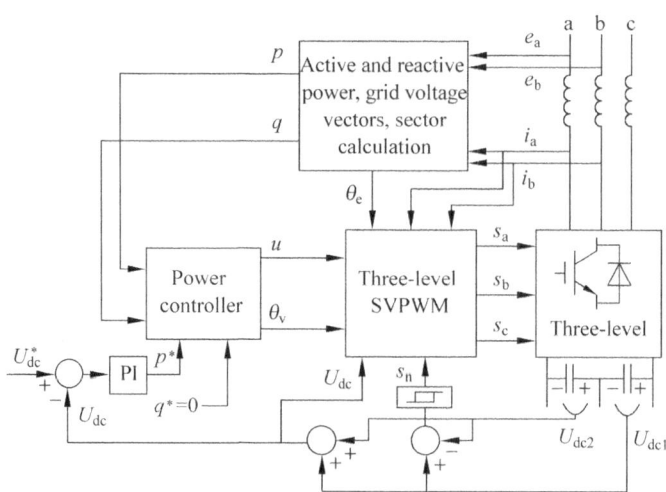

Fig. 7.27 The block diagram of the three-level CSF-PDPC

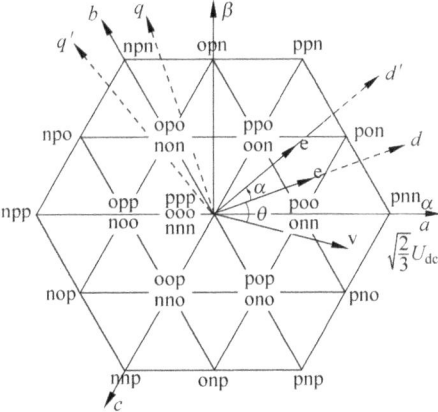

Fig. 7.28 The grid-voltage vector variation during one control period

$$
\begin{cases}
L_s \dfrac{dp}{dt} = e_d(e_d - u\cos\theta_v) - R_s p - \omega L_s q \\
L_s \dfrac{dq}{dt} = e_d u \sin\theta_v - R_s q - \omega L_s p
\end{cases}
\tag{7.23}
$$

Here u is the amplitude of the control vector based on the equal-power conversion. θ_v is the angle of the control vector ahead of the d-axis in synchronous rotatory coordinates. The power equation of the three-level CSF-PDPC can be derived as below based on Eq. (7.24).

$$
\begin{cases}
u = \dfrac{\sqrt{\left[L_s(p^* - p) - T_s\left(e_d^2 - \omega L_s q - R_s p\right)\right]^2 + [L_s(q^* - q) - T_s(\omega L_s p - R_s p)]^2}}{e_d T_s} \\
\theta_v = \arctan\left[\dfrac{L_s(q^* - q) - T_s(\omega L_s p - R_s q)}{-L_s(p^* - p) + T_s\left(e_d^2 - \omega L_s q - R_s p\right)}\right]
\end{cases}
\tag{7.24}
$$

Therefore, the power loop of the CSF-PDPC does not contain an integral part, which will not effectively eliminate the discrete error in the PWM rectifier control, thereby deteriorating control performance.

Such control strategy samples grid voltage at the beginning of each control period for the voltage orientation, which locates the d-axis, as shown in Fig. 7.28. Given grid voltage changes continuously, the angle between the d-axis and a-axis in the abc coordinates changes continuously as well. In the abc coordinates, the rotatory angle of the d-axis per switching period is

$$
\alpha = 2\pi f_e T_s
\tag{7.25}
$$

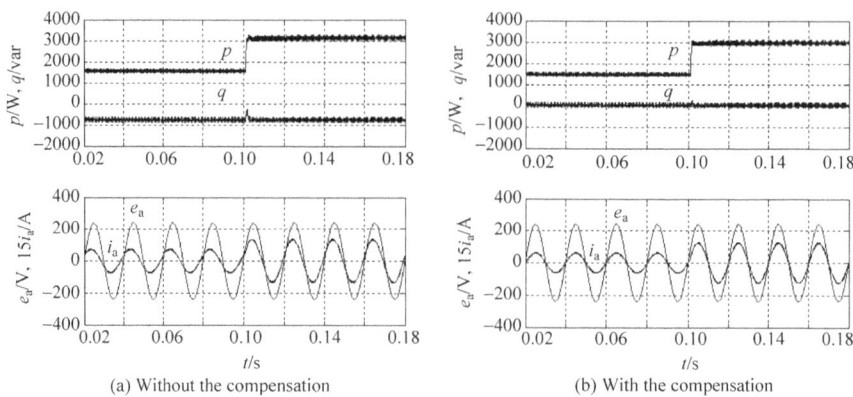

Fig. 7.29 Simulation comparison with/without discrete error compensated

Namely the d-axis will be d'-axis by the end of the control period, as shown in Fig. 7.29.

In the discrete control system, the control algorithm based on the initial d-axis location calculates present control vectors and impose such vectors on the rectifier output. Under the d-q coordinates, the discrete control regards the vector angle θ_v as a constant within one control period. However, due to continuous change of the grid voltage, the angle between d-axis and the control vector is variable, i.e., the control vector angle under the d-q coordinates is time variant, as shown below.

$$\theta_v(t) = \theta_v - 2\pi f_e t \tag{7.26}$$

Substituting Eqs. (7.24) and (7.26) into Eq. (7.23) yields

$$\begin{cases} p^* - p = -\dfrac{e_d u T_s}{L_s}\cos\theta_v + \dfrac{1}{L_s}\int_0^{T_s} e_d u \,\cos(\theta_v - \omega t)\cdot \mathrm{d}t \\ \qquad = \dfrac{2e_d u T_s}{L_s}\sin(\dfrac{2\pi f_e T_s}{4})\sin(\theta_v - \dfrac{2\pi f_e T_s}{4}) \\ q^* - q = -\dfrac{e_d u T_s}{L_s}\sin\theta_v - \dfrac{1}{L_s}\int_0^{T_s} e_d u \,\sin(\theta_v - \omega t)\cdot \mathrm{d}t \\ \qquad = \dfrac{2e_d u T_s}{L_s}\sin(\dfrac{2\pi f_e T_s}{4})\cos(\theta_v - \dfrac{2\pi f_e T_s}{4}) \end{cases} \tag{7.27}$$

Shown in Eq. (7.27) is the power error caused by the discrete control error. In the steady state,

$$\begin{cases} u \approx \sqrt{3E^2 + (2\pi f_e L_s I)^2} \\ \theta_v \approx \arctan(\dfrac{2\pi f_e L_s I}{\sqrt{3}E}) \end{cases} \tag{7.28}$$

Substituting Eq. (7.28) into Eq. (7.27) results in the steady-state power error due to the discrete control error as below.

$$
\begin{cases}
p^* - p \approx \frac{2\sqrt{3}ET_s\sqrt{3E^2+(2\pi f_eL_sI)^2}}{L_s}\sin(\frac{2\pi f_eL_s}{4})\sin\left[\arctan(\frac{2\pi f_eL_sI}{\sqrt{3}E}) - \frac{2\pi f_eT_s}{4}\right] \\
q^* - q \approx \frac{2\sqrt{3}ET_s\sqrt{3E^2+(2\pi f_eL_sI)^2}}{L_s}\sin(\frac{2\pi f_eL_s}{4})\cos\left[\arctan(\frac{2\pi f_eL_sI}{\sqrt{3}E}) - \frac{2\pi f_eT_s}{4}\right]
\end{cases}
\tag{7.29}
$$

We can see that the discrete error caused by the voltage variation within one control period will generate the DC offset between the actual and commanded power. Such offset increases with the control period, which ultimately misaligns the grid voltage and current, thereby lowering the power factor.

2. Compensation of the discrete error

To eliminate the side impact of the discrete error, when generating control pulses and pulse sequences, we will not treat continuous variables and parameters as constant, but as time-variant function. For instance, we can still use Eq. (7.23) to derive the power for the CSF-PDPC, but the angle of the vector within each control period is not constant any more. To take the angle variation as shown in Eq. (7.26) into account, we have

$$
\begin{cases}
L_s(p^* - p) = \int_0^{T_s} e_d[e_d - u\cos(\theta_v - 2\pi f_e t)] \cdot dt - R_s pT_s - \omega L_s qT_s \\
L_s(q^* - q) = \int_0^{T_s} e_d u\sin(\theta_v - 2\pi f_e t) \cdot dt - R_s qT_s + \omega L_s pT_s
\end{cases}
\tag{7.30}
$$

Furthermore, power equations of the CSF-PDPC with the discrete control error being compensated are

$$
\begin{cases}
u = \frac{\omega\sqrt{[L_s(p^*-p)-T_s(e_d^2-\omega L_sq-R_sp)]^2+[L_s(q^*-q)-T_s(\omega L_sp-R_sp)]^2}}{2e_d\sin(\omega T_s/2)} \\
\theta_v = \arctan[\frac{L_s(q^*-q)-T_s(\omega L_sp-R_sq)}{-L_s(p^*-p)+T_s(e_d^2-\omega L_sq-R_sp)}] + \frac{1}{2}\omega T_s
\end{cases}
\tag{7.31}
$$

The simulation comparison between with and without discrete error compensated is shown in Fig. 7.29, which reveals the power error when the discrete control error is not compensated.

Such discrete error has more impact on the reactive power control. When the actual reactive power is significantly different from the command, the large angle between the grid voltage and current emerges, which lowers the power factor. Besides, such discrete error does not fully decouple the reactive power and the active power. In the dynamic process of the active power control, the reactive power has been introduced with the fluctuation as well. After the error is compensated, power control accuracy is enhanced, the grid voltage and current are more aligned with the power factor increased, and the control of the reactive and active power is decoupled.

Fig. 7.30 The frequency-domain model of the power loop in the PDPC considering the pulse delay

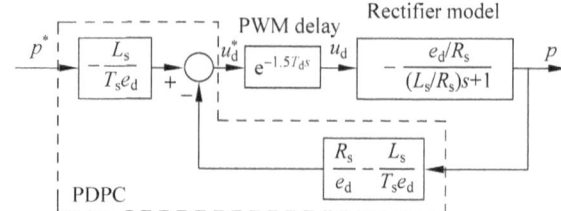

7.3   Variation of the Pulse Time Sequence and Its Solutions

As shown in Fig. 2.15, in addition to reshaping information signals, energy pulses exhibit time delay as well due to nonlinearity of the power electronic system. Similar to the control error, the pulse delay is also caused by accumulation and overlapping of various nonlinearities. In particular,

(1)   The PWM-register reloading mechanism of the DSP causes a delay of half or one whole control period between the information pulse and the control pulse;
(2)   The input-output performance of the pulse management chip, such as CPLDs, results in the delay as well;
(3)   Gate-drive signal delay is caused by the gate-drive circuit;
(4)   The threshold gate-drive voltage of the power switch yields the energy pulse delay.

Over others, the control pulse delay caused by the control chip is dominant. This is particularly true in the high-power system where the control period is large due to limited switching frequency. The half or whole switching period is the major delay of the energy pulse. Such delay directly postpones implementation of the control algorithm, slows down the dynamic response, and in addition, amplifies fluctuation of control variables, thereby deteriorating the steady-state performance.

7.3.1   Impact of the Control Pulse Delay on the Control Performance

In this section we will focus on the CSF-PDPC used in a three-level PWM rectifier, as show in Fig. 7.27. The impact of the pulse delay on the system steady-state performance and dynamic response will be investigated.

Firstly, we can establish the frequency-domain model of the power loop in the PDPC after considering the pulse delay, as shown in Fig. 7.30. Here the delay module is employed to represent the pulse delay, T_s is the control period, and T_d is the delay between the information pulse and energy pulse.

Based upon the model above, the open-loop transfer function can be derived as below.

$$G_0(s) = \frac{L_s - T_s R_s}{T_s(L_s \cdot s + R_s)} \cdot \exp(-1.5T_d \cdot s) \approx \frac{1}{T_s \cdot s} \cdot \exp(-1.5T_d \cdot s) \quad (7.32)$$

which has the amplitude and phase-angle characteristics as

$$L(\omega) = 20\lg(\frac{L_s - T_s R_s}{T_s\sqrt{L_s^2\omega^2 + R_s^2}}) \approx -20\lg(T_s \cdot \omega) \quad (7.33)$$

$$\Phi(\omega) = -\arctan(\frac{L_s\omega}{R_s}) - 180° \times \frac{1.5T_d\omega}{\pi} \approx -90° - 180° \times \frac{1.5T_d\omega}{\pi} \quad (7.34)$$

Further, the phase-angle margin is

$$\gamma = 90°(1 - \frac{3T_d}{\pi T_s}) \quad (7.35)$$

Therefore, with the increment of T_d, the delay between the information pulse and the energy pulse, the angle margin shrinks and system reliability worsens. The precondition of system reliability is

$$T_d < \frac{\pi T_s}{3} \approx 1.047T_s \quad (7.36)$$

In most cases, to gain sufficient time for the software exaction, the control system samples analogue signals at the beginning of the control period and reloads PWM comparators at the end of the control period. Thus the control pulse of the IC is delayed for one control period compared to the information pulse. With all other delays in the control loop, it is very difficult to meet the requirement of Eq. (7.36) for system reliability.

Shown in Fig. 7.31 in the open-loop Bode-plot of the PDPC power loop with $T_s = 0.2$ ms and various T_d. It indicates a reduced phase margin when the control delay increases. From Eq. (7.36), the system becomes unstable once $T_d > 0.2094$ ms. When the PWM generation introduces a time delay of a control period to information and control pulses, i.e., $T_d = T_s = 0.2$ ms, the system reaches the boundary of instability with very little phase margin.

When the pulse delay increases, the simulated active and reactive power of the PDPC based PWM rectifier is shown in Fig. 7.32. The increment of such delay causes oscillation of the power with a reduced oscillation frequency. As shown in Fig. 7.32a, even though the system is still stable, a static error has emerged due to the control pulse delay.

Assume the majority of the time delay in the energy pulse is one control period caused by the control IC, i.e., $T_d = T_s$. Such delay imposes the actual calculated voltage vector on the PWM rectifier by the next switching period, instead of the present period. Through the mathematical model of the measurement and observation introduced in Chap. 6, we can conclude that generation of the current command (related to the power command) based on the present grid current is delayed for one control period, as shown in Fig. 7.33. Assume two adjacent sampling points

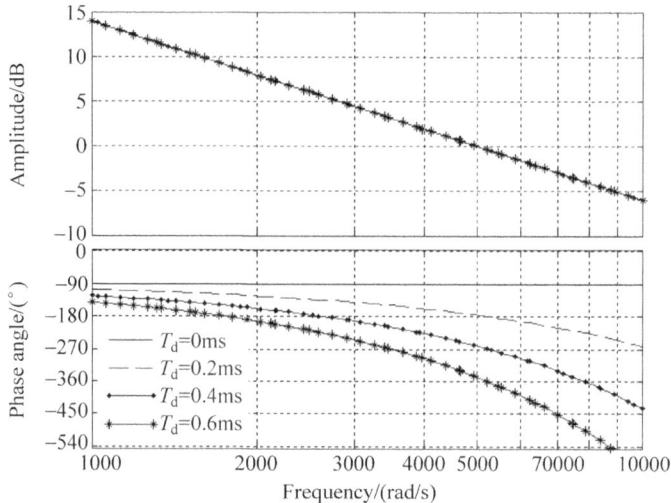

Fig. 7.31 The open-loop Bode-plot of the PDPC power loop with $T_s = 0.2$ ms and various T_d

have current instantaneous values of i_0 and i_1, respectively. With the time-domain derivation, the grid current comes back to i_0 after 6 control periods to start the next line period. Therefore, the oscillating frequency of the steady-state current caused by the time delay of one control period is

$$f_i \approx \frac{1}{6T_s} = \frac{f_s}{6} \tag{7.37}$$

Here $f_s = 1/T_s$, which is the control frequency. With the instantaneous power calculated through Eq. (6.46), the power oscillating frequency at the steady state is

$$f_p = f_q \approx \frac{f_a}{6} \pm f_e \tag{7.38}$$

With $f_s = 5$ kHz and $T_d = T_s$, the simulated grid current and reactive power along with the frequency spectrum is shown in Figs. 7.34 and 7.35, respectively. As can be seen, the maximum harmonics of the grid current occurs at $f_s/6 = 833$ Hz while the two power peaks happen at $833 - 50 = 783$ Hz and $833 + 50 - 883$ Hz. This aligns with the theoretical analysis.

To study the impact of the pulse delay on the system dynamic response, a first-order inertia is used to represent the time delay. Then the frequency-domain model of the power loop in Fig. 7.30 can be further linearized as Fig. 7.36.

The linear model of the open-loop and closed-loop transfer functions is

$$G_0(s) = \frac{L_s - T_s R_s}{T_s(L_s s + R_s)(T_d s + 1)} \tag{7.39}$$

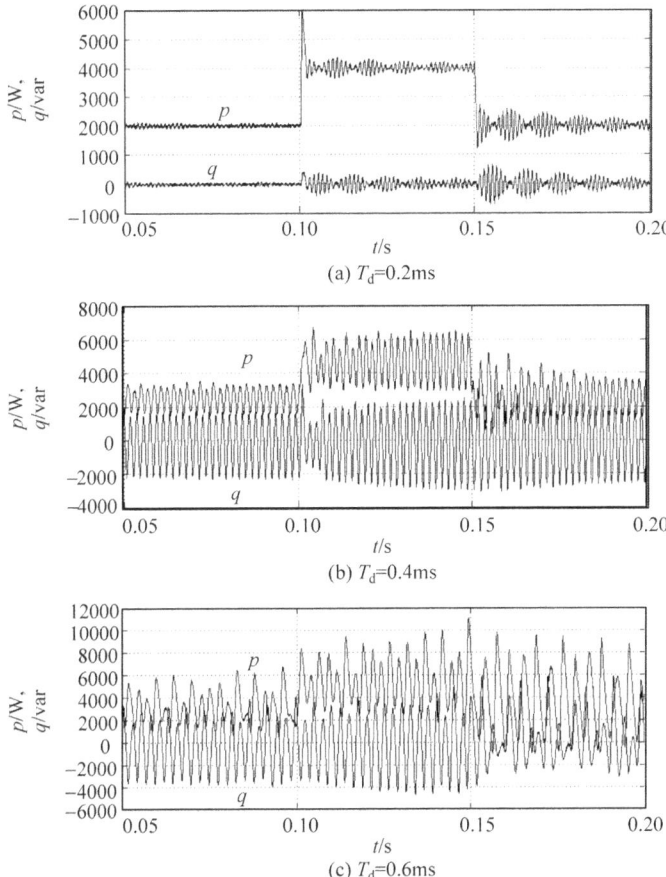

Fig. 7.32 The active and reactive power of a PDPC based three-level PWM rectifier with different T_d (simulation)

Fig. 7.33 Steady-state current oscillation caused by the pulse delay

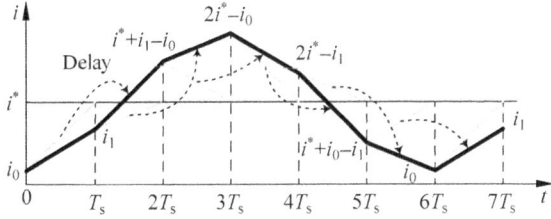

$$G(s) = \frac{L_s}{T_s T_d L_s s^2 + T_s(L_s + T_d R_s)s + L_s} \approx \frac{1}{T_s T_d s^2 + T_s s + 1} \quad (7.40)$$

Fig. 7.34 Simulated grid current with the spectrum of the PDPC with $T_d = T_s$

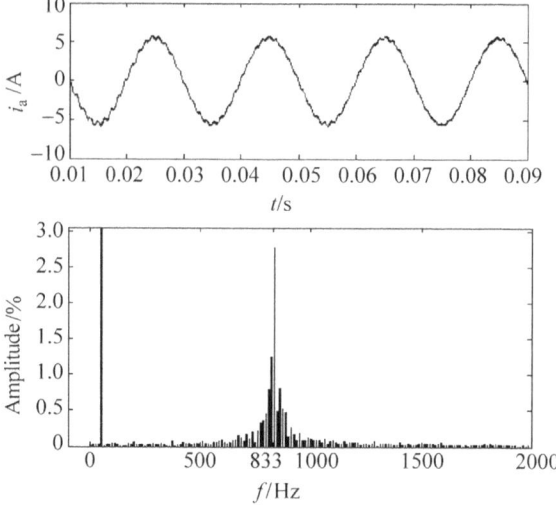

Fig. 7.35 Simulated reactive power with the spectrum of the PDPC with $T_d = T_s$

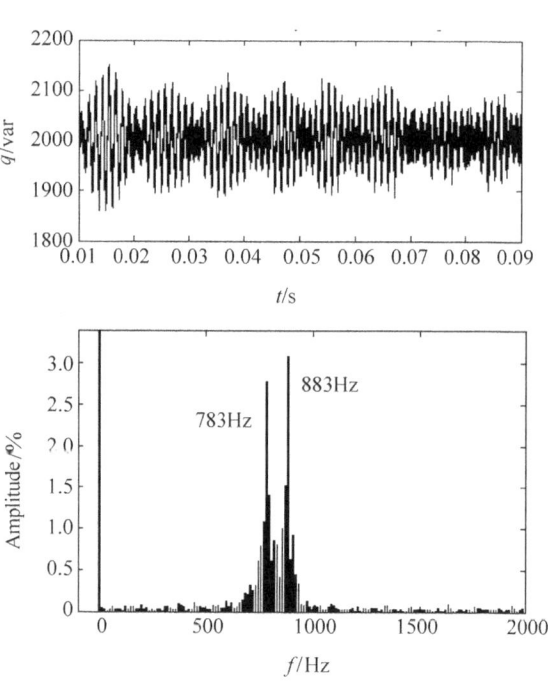

Based on the closed-loop transfer function, the system root locus with various pulse delays is shown in Fig. 7.37. Here the system time constant and damping coefficient are

$$T = \sqrt{T_s T_d} \tag{7.41}$$

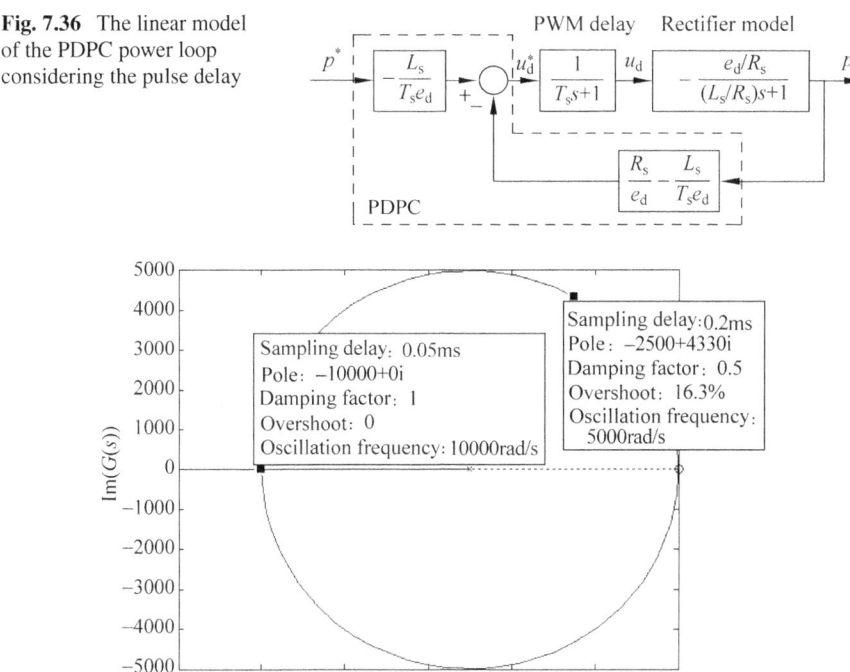

Fig. 7.36 The linear model of the PDPC power loop considering the pulse delay

Fig. 7.37 The root locus of the PDPC closed-loop power control with various pulse delays

$$\zeta = \sqrt{T_s \ / \ 4T_d} \tag{7.42}$$

The larger the pulse delay, the higher the system time constant and the smaller the damping coefficient. When $T_d < T_s/4$, the system is over damped without overshoot. The modulation time is

$$t_s = \frac{8T_d}{1 - \sqrt{1 - 4T_d/T_s}} \tag{7.43}$$

When $T_d > T_s/4$, the system is under damped, with the overshoot and the modulation time as

$$\delta = \exp(-\frac{\pi}{\sqrt{4T_d/T_s - 1}}) \tag{7.44}$$

$$t_s = 8T_d \tag{7.45}$$

Therefore, with the increment of the pulse delay, the system dynamics begin to exhibit overshoot and oscillation. Both the overshoot and modulation time increases,

Fig. 7.38 The system step response with various time delays

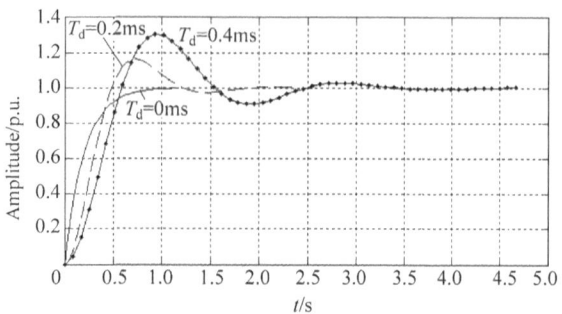

Table 7.12 The time-delay impact of one control period on the control dynamic performance

Performance index	Ideal SVM-DPC	SVM-DPC	Ideal PDPC	PDPC
Closed-loop transfer function	$\frac{1}{2T_s s+1}$	$\frac{1}{2T_s^2 s^2+2T_s s+1}$	$\frac{1}{T_s s+1}$	$\frac{1}{T_s^2 s^2+(\frac{T_s^2 R_s}{L_s}+T_s)s+1}$
Time constant	$2T_s$	$\sqrt{2}T_s$	T_s	T_s
Damping coefficient	/	$1/\sqrt{2}$	/	$\frac{1}{2}(\frac{T_s R_s}{L_s}+1) \approx \frac{1}{2}$
Modulation time	$8T_s$	$8T_s$	$4T_s$	$8T_s$
Overshoot	0	$e^{-\pi} = 4.32\%$	0	$e^{-\pi/\sqrt{3}} = 16.30\%$

while the system dynamic performance deteriorates. The system step response with various time delays is shown in Fig. 7.38.

For the commonly used $T_d = T_s$, the impact of the time delay on the dynamic performance of CSF SVM-DPC and PDPC can be summarized as Table 7.12, along with the analysis above. Here the ideal CSF SVM-DPC and ideal PDPC present no time delay in the control loop. With one-control-period time delay in the control loop, the overshoot of the SVM-DPC increases by 4.32%, but the modulation time remains the same. For PDPC, the time delay results in 16.3% increment of the overshoot and double of the modulation time. Therefore, for the control strategy calculating the accurate vector and pulse width without any integral, such as PDPC, the impact of the time delay on the dynamic performance is more severe.

7.3.2 Compensation of the Time Delay

To study the discrete control error generated by the digital control system and investigate the accurate compensation of the time delay in the power control process, this section selects the PDPC of the three-level PWM rectifier as the objective.

As shown in Fig. 7.33, due to the existence of the time delay, the control vector calculated based on present samplings is only implemented to the system in the next control period, when the system states and variables have already changed. Hence the vector generated cannot make control variables follow the reference accurately, unless at the present period the control system could predict the sampled feedback information in the next period, i.e., effectively compensating the one-period time delay. Furthermore, based on the mathematical model of the PWM rectifier, calculations of the next-period control variables rely on the present state and control variables, which can be sampled through the ADC and known through the last-period vector selection, respectively, namely known parameters.

The time-delay compensation is summarized as follows: at the present control period, based on present samplings and calculated vectors in the last control period, we can predict the next-control-period system state variables. Together with system commands, control variables of the system in the next control period can be further calculated. Specifically in the three-level PDPC, the grid voltage and current need be sampled in the beginning of the control period, based on which and with the last-period voltage vector we can calculate the next-period grid voltage and current. With the coordinate transformation and power calculation, the instantaneous active and reactive power can be calculated at the beginning of the next control period. At the end, the voltage vector for the next control period can be generated through power control equations. The related operation is illustrated in Fig. 7.39.

To reduce the computation load of the current calculation based time-delay compensation, based on Eq. (7.23), with the instantaneous power (p_0, q_0) in the present period and the last-period calculated voltage vector (u_0, θ_0), we can directly calculate the next-period instantaneous power (p_1, q_1) as below.

$$
\begin{cases}
\begin{aligned}
p_1 &= p_0 + \frac{T_s}{L_s}(e_d^2 - \omega L_s q_0 - R_s p_0) - \frac{1}{L_s}\int_0^{T_s} e_d u_0 \cos(\theta_0 - \omega t)\cdot dt \\
&= p_0 + \frac{T_s}{L_s}(e_d^2 - \omega L_s q_0 - R_s p_0) - \frac{2e_d u_0}{\omega L_s}\sin(\frac{\omega T_s}{2})\cos(\theta_0 - \frac{\omega T_s}{2}) \\
q_1 &= q_0 + \frac{T_s}{L_s}(\omega L_s p_0 - R_s q_0) + \frac{1}{L_s}\int_0^{T_s} e_d u_1 \sin(\theta_1 - \omega t)\cdot dt \\
&= q_0 + \frac{T_s}{L_s}(\omega L_s p_0 - R_s q_0) + \frac{2e_d u_0}{\omega L_s}\sin(\frac{\omega T_s}{2})\sin(\theta_0 - \frac{\omega T_s}{2})
\end{aligned}
\end{cases} \tag{7.46}
$$

Furthermore, the next-period voltage vector (u_1, θ_1) is

Fig. 7.39 The time sequence of the pulse delay compensation

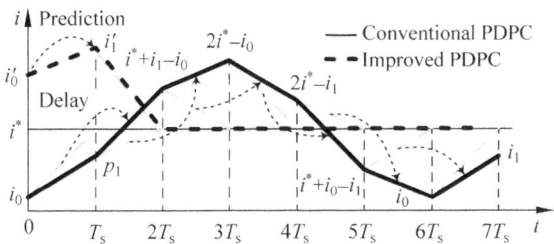

$$\begin{cases} u_1 = \dfrac{\omega\sqrt{[L_s(p^*-p_1)-T_s(e_d^2-\omega L_s q_1-R_s p_1)]^2+[L_s(q^*-q_1)-T_s(\omega L_s p_1-R_s q_1)]^2}}{2e_d\sin(\omega T_s/2)} \\ \theta_1 = \arctan[\dfrac{L_s(q^*-q_1)-T_s(\omega L_s p_1-R_s q_1)}{-L_s(p^*-p_1)+T_s(e_d^2-\omega L_s q_1-R_s p_1)}]+\dfrac{3}{2}\omega T_s \end{cases} \tag{7.47}$$

Note Eqs. (7.46) and (7.47) both compensate the discrete error caused by the grid-voltage sampling.

Shown in Fig. 7.40 is the block diagram of the delay compensation for the three-level PDPC. Compared to the conventional CSF-PDPC shown in Fig. 7.37, the proposed strategy adds one more power prediction, which is related to Eq. (7.46). The time delay is compensated by predicting the instantaneous power at the beginning of the next control period.

To derive the performance index of the time-delay compensated three-level PDPC, the linear frequency-domain model is established as Fig. 7.41.

The closed-loop transfer function of the time-delay compensated PDPC is

$$G(s) = \frac{1}{T_s^2 s^2 + 2T_s s + 1} \tag{7.48}$$

The overshoot is calculated as zero. The modulation time is

$$t_s = 4T_s \tag{7.49}$$

Referring to the dynamic index shown in Fig. 7.12, we can conclude that with the accurate time-delay compensation, the PDPC can reach the same dynamic response as the ideal control.

The simulation results with one-control-period time delay are shown in Fig. 7.42, where SVM-DPC, conventional DPC, and delay-compensated PDPC are compared.

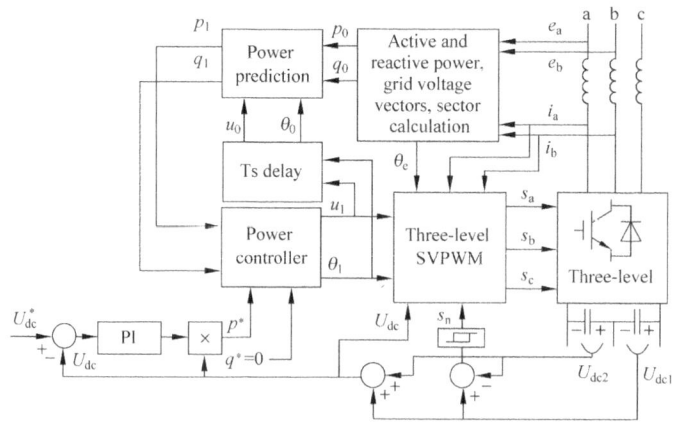

Fig. 7.40 Block diagram of the delay compensation for the three-level PDPC

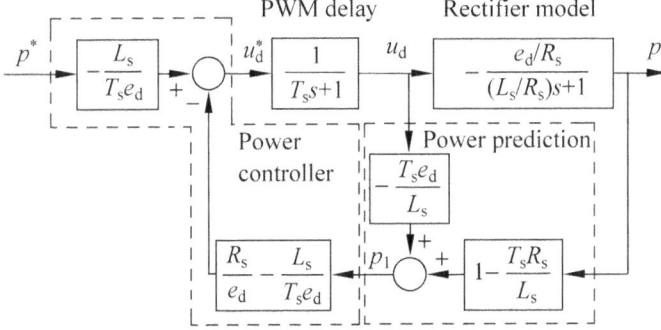

Fig. 7.41 The linear frequency-domain model of the time-delay compensated three-level PDPC

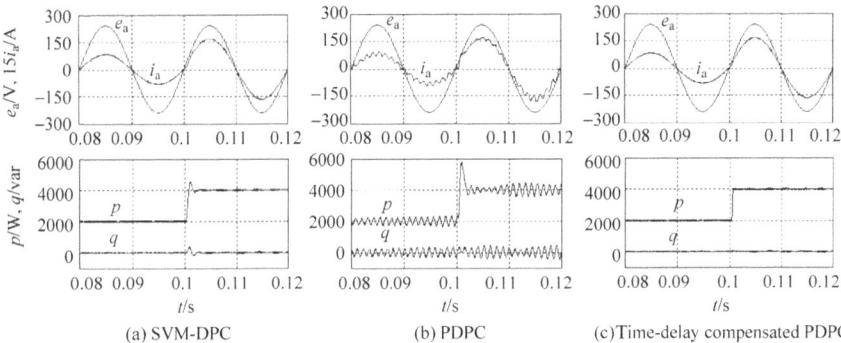

Fig. 7.42 Simulation comparison of various control strategies

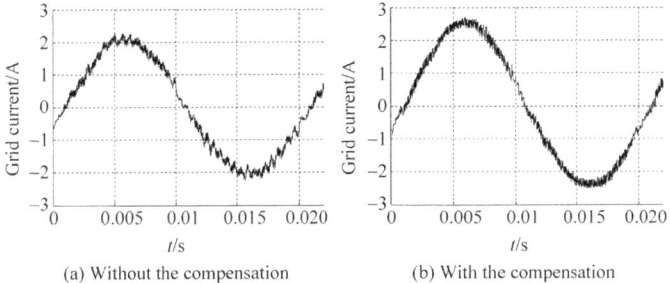

Fig. 7.43 Experimental grid current of PDPC with and without delay compensation

Experimental results of the grid current with PDPC are shown in Fig. 7.43, where the control performances with and without delay compensation are compared. Figure 7.44 shows the steady-state performance and load step response of the time-delay compensated PDPC.

We can see that with the accurate delay compensation, the grid current and power oscillation due to the pulse time delay is eliminated for the three-level PDPC. Mean-

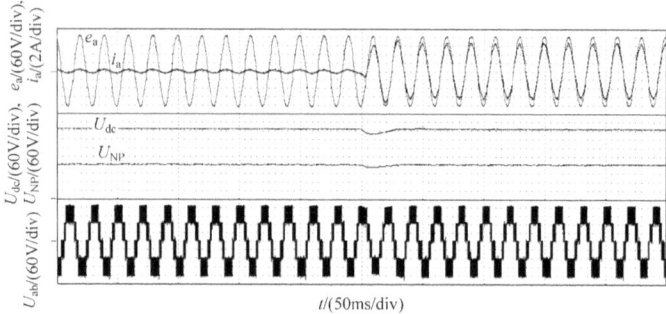

Fig. 7.44 Experimental results of the time-delay compensated PDPC

while, the dynamic process has nearly no overshoot with the dynamic response much faster than that in the SVM-DPC and conventional PDPC.

Chapter 8
High-Performance Closed-Loop Control and Its Constraints

Pulse generation through turning on and off power switches is the fundamental of power electronics conversion. At the macroscopic level, two steady states exist for power switches, i.e., on and off, determining that electromagnetic variables of the power electronics conversion are in the form of pulses, pulse overlapping and combinations, and pulse sequences, e.g., output voltage of the VSI and output current of the CSI. To differentiate from information pulses in the control system and gate-drive circuits, energy-level voltage and current pulses are named as energy pulses. The controllable energy pulse and pulse sequence are basic behaviors of the power electronics conversion.

8.1 Closed-Loop Control System and Its Constraints

To reach the high-performance control is the ultimate goal of the power electronic system, which requires the closed-loop control. As a mixture system of continuous and discrete signals, the closed-loop control system of power electronic converters is different from the conventional control system.

8.1.1 The Structure of the Closed-Loop Control System

An exemplary power electronic system is shown in Fig. 8.1, which consists of the control system and the main-power circuit. The main-power circuit is made of power switches, energy storage components and energy consuming parts. The input power is processed by power switches, part of which is output as energy pulses or stored in the energy storage component, other of which is generated as the heat when passing through the lossy component during the power transfer and transformation. In addition to the core control strategy, the control system of the power electronic converter

© Tsinghua University Press and Springer Nature Singapore Pte Ltd. 2019
Z. Zhao et al., *Electromagnetic Transients of Power Electronics Systems*,
https://doi.org/10.1007/978-981-10-8812-4_8

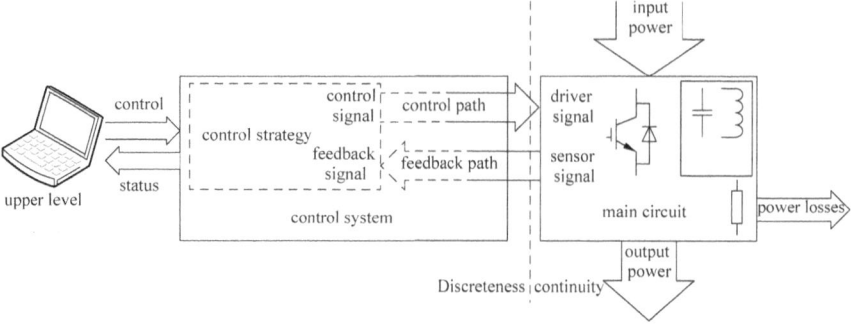

Fig. 8.1 A typical power electronic system

consists of the control loop and feedback loop. Information pulses generated by the control algorithm yield energy pulses ultimately. Therefore the power electronic conversion is a process of using the weak-electricity signal of the control system to control the strong-electricity signal of the main-power circuit, i.e., information controls energy.

Compared to the open-loop control, the closed-loop control will reach better performance of the power electronic converter, which however further requires the feedback loop to sample energy-level variables, convert them into information-level signals and forward them to the control algorithm. From this perspective, the power electronics conversion is the interaction of the information and energy. In this chapter, energy-level signals in the main-power circuit are named as energy analogues, to be differentiated from those in the control system.

With the development of microcontrollers, more and more power electronic systems adopt digital microprocessors. Note the digital control system is usually discrete while the main-power circuit is continuous, which complies with fundamental rules of the energy conversion, i.e., energy conservation and no sudden energy change. Therefore, the digital control system based power electronics conversion is a typical process of the discrete system controlling the continuous system.

As shown in Fig. 8.1, the whole power electronic system contains the feedback loop and control loop, as discussed in Chaps. 6 and 7, respectively. As described in the preface, the closed-loop control is an exemplary case embracing the software and hardware, energy and information, linearity and nonlinearity, discrete and continuous, and coordination of multi-timescale subsystems.

8.1.2 The Limitation of the Conventional Control Theory

The conventional power electronics research usually makes some ideal assumptions, in order to facilitate the analysis and derivation, such as

(1) The feedback signal of the control system has the same time sequence and shape as the energy-level variable in the main-power circuit;

(2) The information pulse generated by the control system has the same time sequence and shape as the energy pulse in the main-power circuit;

(3) With a high enough control frequency, system state variables within one discrete control period remain unchanged;

(4) The converter can handle any switching states, with the switching transition finished instantaneously.

The characteristics of the time sequence are the timing relationship among points of physical variables, including the pulse-edge position, pulse width, etc. The shape characteristics are focusing on the pulse shape, including the pulse-edge shape, pulse steady-state amplitude, overshoot, decaying process, etc. With the assumption above, the ideal control method should accurately calculate then generate control pulses through the power reference based on the system mathematical model. However, in the actual power electronic converter non-ideal characteristics exist in both the control system and main-power circuit. Specifically,

(1) The non-ideal factors in the feedback loop of the control system will result in the sampling delay and error, causing the delay and shape difference between the feedback signal and the actual variable. Meanwhile the reliability of the communication in the feedback loop is deteriorated due to the strong EMI, potentially interrupting the sampling.

(2) Influenced by the pulse generating mechanism, dead band, minimum pulse width, control-loop input-output characteristics, power switch characteristics and stray parameters, the energy pulse exhibits the delay in time and distortion in shape compared to the ideal information pulse.

(3) The system state variables will change significantly within one control period when the control frequency is relatively low. A large control error is expected.

(4) Non-ideal switching performance and topology parameters yield continuous electromagnetic variables. Transient processes exist between transitions. For the multi-level or other special topology based converter, random switching among states might cause abnormal pulses harmful and even destructive to the system.

Impacted by the non-ideal characteristics above, adopting the ideal control method, i.e., using the mathematical model to calculate and generate information pulses based on the power reference in the actual system will lower the control accuracy, slow down the dynamic response, increase the overshoot and generate the abnormal pulse. To suppress or compensate such non-ideal characteristics, the conventional control theory introduces the proportional-integral (PI), bang-bang, sliding-mode, fuzzy logic, neural network controllers, etc. However, such inaccurate controller can only somewhat compensate or suppress the non-ideal characteristics of the power electronic system. Compared to the low-power system, the high-power converter has sensors with lower speed, narrower bandwidth and lower sampling frequency, with more significant electromagnetic transients, longer energy-pulse dynamics, larger propagation delay of the gate-drive circuit and lower switching frequency. Thus the non-ideal characteristics mentioned above are more obvious, which potentially become the bottleneck of the control performance and reliability.

The conventional power electronics control adopts inaccurate controllers, resulting in unsolvable conflicts, for instance, the conflict between the dynamic response and overshoot of the PI controller. Increasing I-coefficient will increase the dynamic response as well as the overshoot. Decreasing I-coefficient will lower the overshoot but slow down the dynamic response. For the bang-bang controller, conflicts exist between the control accuracy and the power loss. With the hysteresis becoming narrower, the control accuracy is enhanced. However, the equivalent switching frequency and the power loss are increased. A wide hysteresis lowers the equivalent switching frequency and the power loss, however, sacrifices the control accuracy. Such conflicts in conventional controllers obstruct the power electronic system from further improving the control performance.

At the same time, control objectives of conventional controllers are usually the voltage or current, with one control variable for one control objective. It is hard for the conventional controller to coordinate multiple different types of objectives.

8.2   Invalid Pulses Caused by Control Strategy and Related Solutions

The invalid pulses of power electronic systems are those unable to meet the control requirement. Except the difference of the pulse width, their shapes are similar to normal pulses. A single invalid pulse will not cause the device failure or system fault. Multiple or periodic abnormal pulses, however, can deteriorate the system control performance and reliability.

8.2.1   Invalid Pulses Caused by the Control Coupling

1. **Cause and impact**

Based on the mathematical model and instantaneous power equations of the three-level PWM rectifier, the overall power control equations are shown below. Here the grid-inductor resistance is assumed very small, i.e., $R_s \approx 0$.

$$
\begin{cases}
L_s \dfrac{dp}{dt} = e_d(e_d - u_d) - \omega L_s q \\[2mm]
L_s \dfrac{dq}{dt} = e_d u_q + \omega L_s p
\end{cases}
\tag{8.1}
$$

Imposing the appropriate control variable u_d and u_q can result in the actual active and reactive power follows the command closely. As shown in Eq. (8.1), the active-power control equation contains one item directly proportional to the reactive power. At the same time, the reactive-power control equation contains one similar item in direct proportion to the active power as well. Namely the active and reactive power control is coupled with each other. With control algorithms based on synthesized vectors, such as VOC, SVM-DPC and CSF-PDPC, SVPWM is employed as the bottom-level strategy to generate information pulses, which can add the decoupling into the control variable thereby making the active and reactive power control independent, aiming at the decoupling control. However, for the bang-bang control and switching-table based DPC, the vector within each control period is fixed in the table, which is unable to add decoupling to the control variable.

For the conventional three-level DPC shown in Fig. 6.19, with the coupling ignored, power control equations are

$$\begin{cases} L_s \dfrac{dp}{dt} = e_d(e_d - u_d) \\ L_s \dfrac{dq}{dt} = e_d u_q \end{cases} \tag{8.2}$$

The DPC switching table then can be generated as Table 8.1.

It can be seen that vectors U_{13}–U_{24} emerge at six different locations in pairs. For the pair of vectors in the same location, they have the same u_p and u_q thereby the same effect on the instantaneous power control, however, different commutation loops with the same grid current. Their impact on the neutral-point voltage is totally opposite. The three-level DPC then can select any vector of this pair to control the active and reactive power and compensate the neutral-point voltage, based on the grid current direction and voltage across upper and lower capacitors.

To reduce switching actions caused by vector transitions, the conventional DPC can select vectors within a quadrilateral where the grid voltage vector is located, for instance, the shadow area of Fig. 8.2. Four vectors (U_1, U_2, U_{13}/U_{14} and U_{15}/U_{16}) form a quadrilateral when the grid voltage vector $\mathbf{e}$ is located in the sector θ_2.

In the steady state of the three-level PWM rectifier using conventional DPC, the reactive power reference is set as zero, resulting in a minimal actual reactive power.

Table 8.1 The switching table of the conventional DPC

S_p	S_q	θ_1	θ_{2k}	θ_{2k+1}	θ_{12}
1	0	U_{23}/U_{24}	U_{2k+11}/U_{2k+12}	U_{2k+11}/U_{2k+12}	U_{23}/U_{24}
1	1	U_{13}/U_{14}	U_{2k+13}/U_{2k+14}	U_{2k+13}/U_{2k+14}	U_{13}/U_{14}
0	0	U_{12}	U_{2k-1}	U_{2k}	U_{11}
0	1	U_1	U_{2k}	U_{2k+1}	U_{12}

Here S_p and S_q are the switching states. $S_p = 1$ means increasing the active power. $k = 1$–5 Vector locations are shown in Fig. 8.2

Fig. 8.2 Vector selections
for the three-level DPC

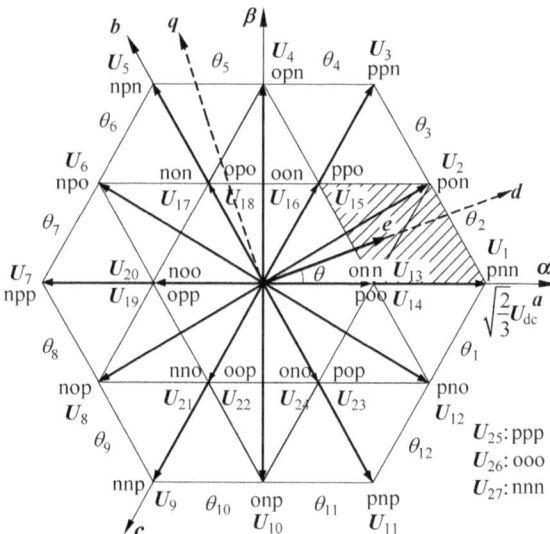

Based on Eq. (8.1), the influence of the reactive power on the active power is minor. However, given that the input power is equal to the output power, i.e.,

$$p \approx I_L U_{dc} \tag{8.3}$$

With a large input/output active power, the coupling between the active and reactive power can highly influence the reactive power control.

When the three-phase grid forwards the active power to the rectifier, i.e., $p > 0$ while the control system samples the positive reactive power ($q > 0$), the bang-bang controller will output $S_q = 0$. The switching table then selects vectors to make $u_q < 0$, aiming at reducing the reactive power to follow $q^* = 0$. If vectors in the switching table do not meet

$$u_q \leq -\frac{\omega L_s p}{e_d} \tag{8.4}$$

Then based on Eq. (8.1), we have

$$\frac{dq}{dt} = \frac{e_d u_q + \omega L_s p}{L_s} > 0 \tag{8.5}$$

The coupling item in the direct proportion to the active power will enlarge the reactive power, making it deviate from the command. Now output pulses of the control system are invalid to meet the control requirement, given the reactive power is uncontrollable.

Geometrical locations of vectors yield restraints of generating invalid pulses in odd-number sectors when $p > 0$ as

$$p > \frac{e_d}{\omega L_s} \cdot \frac{U_{dc}}{\sqrt{2}} \sin \theta \tag{8.6}$$

with invalid pulses located at

$$\theta < \arcsin \left(\frac{\sqrt{2} \omega L_s p}{e_d U_{dc}} \right) \tag{8.7}$$

Here θ is the angle between the grid voltage vector $\mathbf{e}$ and the sector edge, when rotating clockwise.

The restraint of generating invalid pulses in the even-number sectors when $p > 0$ is

$$p > \frac{e_d}{\omega L_s} \cdot \frac{U_{dc}}{\sqrt{6}} \sin \theta \tag{8.8}$$

with invalid pulses located at

$$\theta < \arcsin \left(\frac{\sqrt{6} \omega L_s p}{e_d U_{dc}} \right) \tag{8.9}$$

Within the invalid-pulse region shown in Eqs. (8.7) and (8.9), the conventional three-level DPC is unable to reduce the reactive power. Therefore q will increase continually until the grid voltage vector is out of those regions.

Similarly, the region where invalid pulses emerge in the odd-number sectors when $p < 0$ is

$$\theta > \frac{\pi}{6} + \arcsin \left(\frac{\sqrt{6} \omega L_s p}{e_d U_{dc}} \right) \tag{8.10}$$

When $p < 0$, the region where invalid pulses are generated in the even-number sectors is

$$\theta > \frac{\pi}{6} + \arcsin \left(\frac{\sqrt{2} \omega L_s p}{e_d U_{dc}} \right) \tag{8.11}$$

In summary, the invalid-pulse region will be widened when the active power increases, due to the coupling between the active and reactive power.

The simulated active and reactive power when the grid forwards the power to the rectifier ($p > 0$) is shown in Fig. 8.3. As shown in Fig. 8.3a, the impact of the coupling on the active power is minor while causing the fluctuation of the reactive

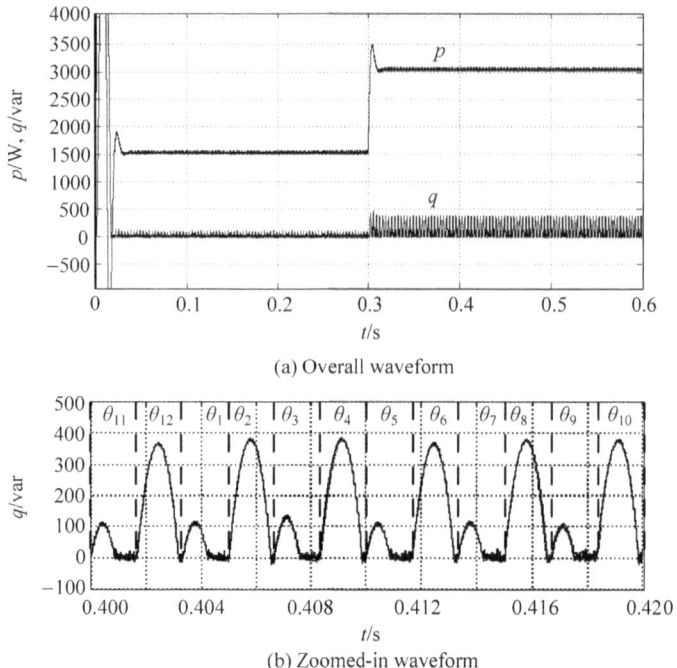

Fig. 8.3 Simulation waveform of the conventional three-level DPC

power, which is aligned with the analysis before. With the active power increasing, the invalid-pulse zone is widened, which further amplifies the fluctuation of the reactive power. Figure 8.3b is the zoomed-in reactive power of Fig. 8.3a, indicating each sector has some invalid-pulse region where the monotonically decreasing reactive power is caused by the control coupling. Such invalid pulses eventually will amplify the grid current harmonics and lower the power factor.

2. **Suppression of invalid pulses caused by the control coupling**

As shown above, the control coupling can be traced back to Eq. (8.1), showing that the voltage vector is unable to compensate the coupling item. As shown above, the control vector is the fixed one generated by the underlying switch table, making it impossible to suppress the coupling item through only adding the compensation. To enforce the control of the three-level DPC on the reactive power, we can enrich the switch table to diversify the vector selection in the switch table, e.g., selecting the vector with larger q-axis component under d–q coordinates.

The enhanced three-level DPC switch table is shown in Table 8.2, where $p > 0$.

It can be seen that selectable vectors in Table 8.2 do not necessarily form the quadrilateral any more. The number of transitions among non-adjacent vectors is increased significantly. To secure the reliability of the converter during vector transitions, the intermediate vectors are inserted, which as a result increases the number of switching actions. To reduce such switching actions thereby the switching frequency and avoid the fluctuation of the steady-state power due to the over modulation, Table 8.2 is only adopted for invalid-pulse regions, as shown in Eqs. (8.7) and (8.9) when $p > 0$. In other regions the conventional switch table (Table 8.1) is still preferred. Based on such dynamic switching principles, the polygon formed by selected vectors in sector θ_2 is shown as the shadow in Fig. 8.4, when $p > 0$.

A comparison between Figs. 8.4 and 8.2 shows that with the dynamic switch table for $p > 0$, the q-axis voltage, u_q, can be more negative thereby effectively compensating the impact of the active power on the reactive power caused by the coupling item. Similarly, for $p < 0$, the enhanced switch table is shown as Table 8.3.

Similar to $p > 0$, for $p < 0$ we will only utilize the enhanced Table 8.3 within the invalid-pulse region shown as Eqs. (8.10) and (8.11). Otherwise the conventional three-level DPC table is preferred. Selectable vectors for $p < 0$ within sector θ_2 are shown as the shadow polygon in Fig. 8.5.

Table 8.2 The enhanced three-level DPC switch table to suppress the invalid pulses caused by the control coupling ($p > 0$)

S_p	S_q	θ_1	θ_2	θ_{2k-1}	θ_{2k}	θ_{11}	θ_{12}
1	0	U_{23}/U_{24}	U_{23}/U_{24}	U_{2k+9}/U_{2k+10}	U_{2k+9}/U_{2k+10}	U_{21}/U_{22}	U_{21}/U_{22}
1	1	U_{13}/U_{14}	U_{15}/U_{16}	U_{2k+11}/U_{2k+12}	U_{2k+13}/U_{2k+14}	U_{23}/U_{24}	U_{13}/U_{14}
0	0	U_{11}	U_{12}	U_{2k-3}	U_{2k-2}	U_9	U_{10}
0	1	U_1	U_2	U_{2k-1}	U_{2k}	U_{11}	U_{12}

Note k = 2, 3, 4, 5

Fig. 8.4 Selectable vectors when $p > 0$ in sector θ_2 to suppress invalid pulses generated by the control coupling

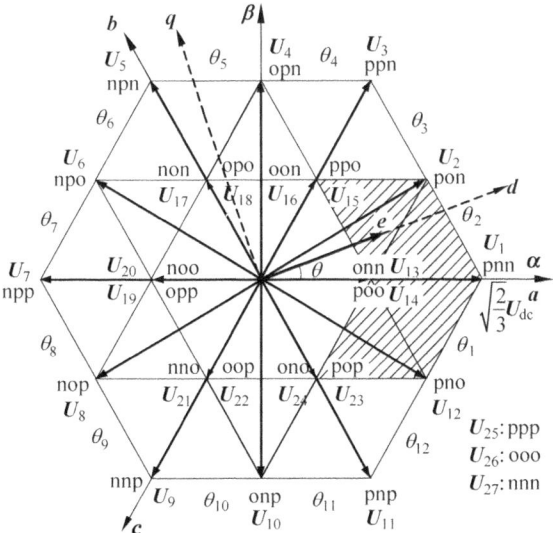

Table 8.3 The enhanced three-level DPC switch table to suppress the invalid pulse caused by the control coupling ($p < 0$)

S_p	S_q	θ_1	θ_2	θ_{2k-1}	θ_{2k}	θ_{11}	θ_{12}
1	0	U_{23}/U_{24}	U_{13}/U_{14}	U_{2k+9}/U_{2k+10}	U_{2k+11}/U_{2k+12}	U_{21}/U_{22}	U_{23}/U_{24}
1	1	U_{15}/U_{16}	U_{15}/U_{16}	U_{2k+13}/U_{2k+14}	U_{2k+13}/U_{2k+14}	U_{13}/U_{14}	U_{13}/U_{14}
0	0	U_{12}	U_1	U_{2k-2}	U_{2k-1}	U_{10}	U_{11}
0	1	U_2	U_3	U_{2k}	U_{2k+1}	U_{12}	U_1

Note k $= 2, 3, 4, 5$

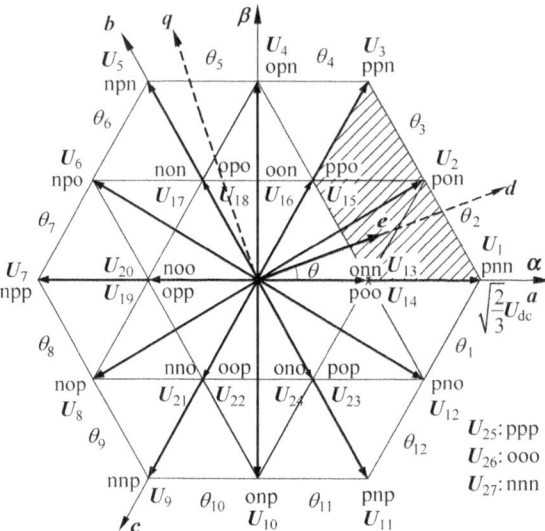

Fig. 8.5 Selectable vectors when $p < 0$ in sector θ_2 to suppress invalid pulses generated by the control coupling

A comparison between Figs. 8.5 and 8.2 shows that with the dynamic switch table for $p < 0$, the q-axis voltage, u_q can be more positive thereby effectively compensating the impact of the negative active power on the reactive power caused by the coupling item in Eq. (8.1).

Shown in Fig. 8.6 is the simulation comparison between w/n the dynamic switch table for the three-level PWM rectifier. Figure 8.7 is the comparison of the experimental grid current and its spectrum before and after suppressing control-coupling-caused invalid pulses.

With the proposed method above, invalid pulses generated by the power coupling item are effectively suppressed. So is the periodic fluctuation of the steady-state

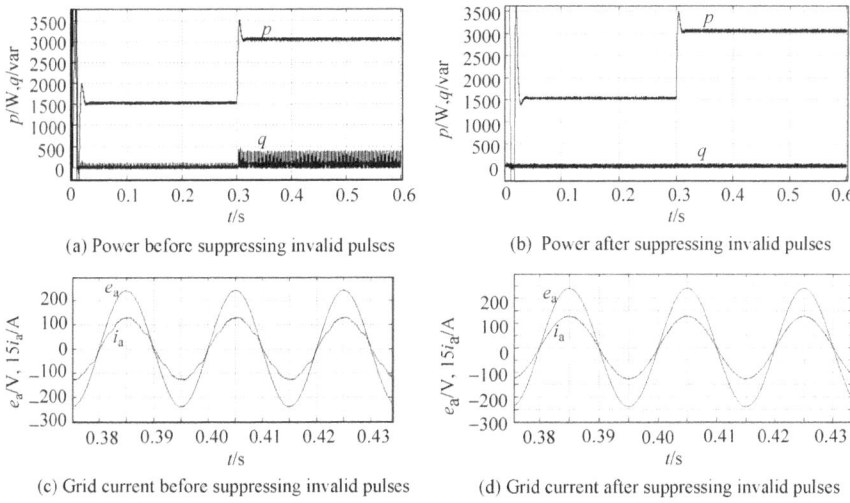

Fig. 8.6 Simulation comparison between w/n the dynamic switch table for the three-level PWM rectifier

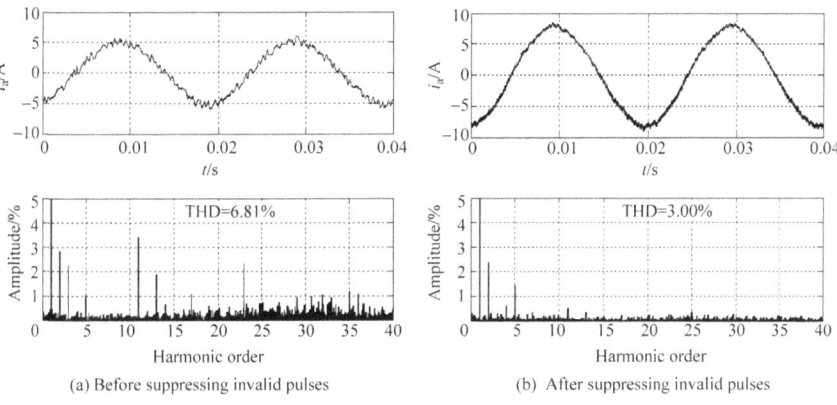

Fig. 8.7 Experimental grid current before and after suppressing the control-coupling-caused invalid pulses

reactive power. With the decoupling of the active and reactive power control, the grid current harmonics is smaller and the power factor is increased.

In addition, with the revised dynamic switch table enhancing the power modulation, the dynamic response of the instantaneous power is faster. The simulation result is shown in Fig. 8.8.

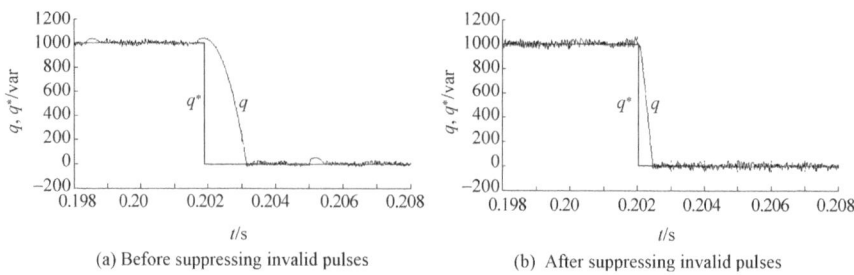

(a) Before suppressing invalid pulses   (b) After suppressing invalid pulses

Fig. 8.8 Simulation results of the reactive-power dynamic response before and after suppressing control-coupling-caused invalid pulses

8.2.2 Invalid Pulses Caused by the Saturation of the Controller

1. Cause and impact

Various controllers are adopted in the control of power electronic converters, which based on the time sequence of the input and output can be categorized as memory-less system and memory system. The former one is called the instantaneous system, with the output only related to the present input. The latter one is called the dynamic system, with the output related to the previous input. For the power electronic converter using the discrete control, the "present input" means the feedback or command within the present control period, while "previous input" means the feedback or command within some of previous control periods. Based on such definition, the hysteresis controller of the three-level DPC is a memory-less system, while the PI controller in the three-level VOC and SVM-DPC is a memory system.

For the memory system in power electronic converters, if the controller output at the present control period is related to the previous abnormal feedback status, i.e., the controller memorizes the past abnormal information, it is possible for the controller to unfollow the command due to those abnormal states thereby generating abnormal pulses and influencing operational conditions later.

The relationship between the input and output of the continuous PI controller is

$$y(t) = k_p[x^*(t) - x(t)] + k_i \int_0^t [x^*(t) - x(t)] \cdot dt \tag{8.12}$$

where k_p is the proportional coefficient and k_i is the integral coefficient. With constraints of the physical system, limitations are imposed on the PI output, i.e.,

$$\begin{cases} y(t) = \min[y(t), y_{max}] \\ y(t) = \max[y(t), y_{min}] \end{cases} \tag{8.13}$$

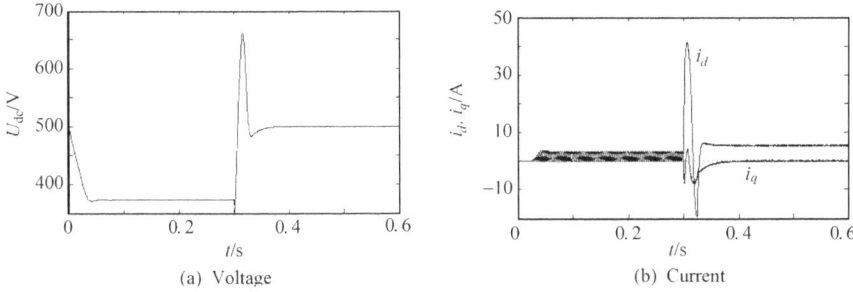

Fig. 8.9 Simulated starting process with invalid pulses caused by the PI saturation

For the discrete PI controller, its input and output have the relationship as

$$\begin{cases} u(k) = u(k-1) + k_i[x^*(k) - x(k)] \\ y(k) = k_p[x^*(k) - x(k)] + u(k) \end{cases} \tag{8.14}$$

$$\begin{cases} y(k) = \min[y(k), y_{max}] \\ y(k) = \max[y(k), y_{min}] \end{cases} \tag{8.15}$$

Here $u(k)$ is the integral part of the discrete PI controller, indicating that all present and past errors between command and feedback values are accumulated. Such memory effect will continually affect the controller output.

Take the VOC based three-level PWM rectifier as an example, as shown in Fig. 6.7. Before starting the motor, all switches remain off, with the grid pre-charging the DC-bus capacitor through anti-paralleled diodes of switches. To detect the grid voltage, the control system needs run before generating the control pulse. The DC-bus voltage under such uncontrollable rectifying mode is lower than the target. Since switches are all locked off, the d-axis and q-axis current can't track the reference value, which makes the d-axis-current controller, q-axis-current controller and DC-bus-voltage controller saturate quickly, based on Eqs. (8.14) and (8.15). The controllers will reach their upper or lower limits.

Once the rectifier starts, due to the memory effect the output of the PI controller stays at the upper or lower limits before quitting the saturation. Information pulses generated by the controller will not effectively follow the DC-bus, d-axis current and q-axis current commands, yielding invalid pulses to further cause the starting inrush current and DC-bus over voltage, as shown in Fig. 8.9.

2. Elimination of invalid pulses

To eliminate invalid pulses above, we need clear the abnormal feedback memorized by the controller before the system starts, i.e., zero the integral part of the PI controller. This ensures the PI controller not saturated before starting. Therefore control pulses

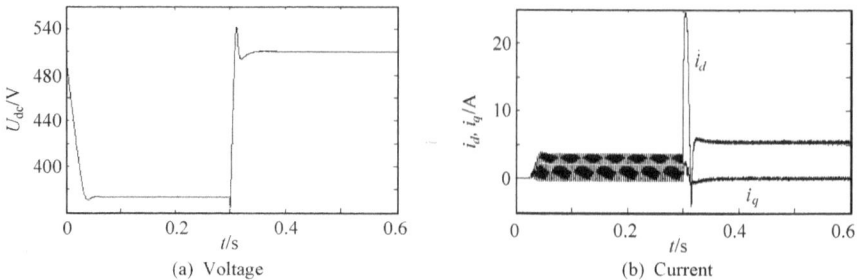

Fig. 8.10 Simulated starting process after eliminating invalid pulses caused by the PI saturation

used to start the system will not be impacted by system states before starting, as simulated in Fig. 8.10. It can be seen that the DC-bus voltage and the grid current in the starting process have much lower overshoot, which reduces the electrical stress on power switches and other components.

8.2.3 Invalid Pulses Generated at Some Special Operational States of the Converter

The special operational states include over/under input voltage, over/under output voltage, etc. Under those special states, control pulses might miss the control target due to output limits of the controller. Namely the system generates invalid pulses and deteriorates the control performance.

1. **Cause and impact of invalid pulses with the input under-voltage**

Take the DPC based three-level PWM rectifier shown in Fig. 6.19 as an example. With the conventional DPC switch table, as shown in Table 8.1, selectable vectors form a quadrilateral shadow in Fig. 8.11, when the grid voltage vector **e** is located in the section θ_1. Meanwhile the traditional DPC has the control Eq. (8.2), which requires the grid voltage vector to follow (8.16) for the sake of the bidirectional power flow.

$$\begin{cases} U_{13/14d} < e_d \\ U_{13/14q} > 0 \\ U_{23/24d} < e_d \\ U_{23/24q} < 0 \\ U_{1d} > e_d \\ U_{1q} > 0 \\ U_{12d} > e_d \\ U_{12q} < 0 \end{cases} \qquad (8.16)$$

Fig. 8.11 Selectable vectors with the effective input-voltage range under the conventional three-level DPC (sector θ_1)

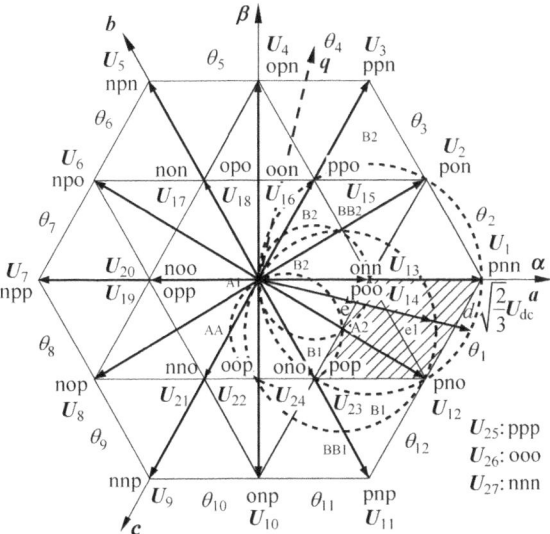

Here U_{id} and U_{iq} are d-axis and q-axis components of the fixed vector U_i, respectively.

Assume M_1 is the common region outside two circles with U_{13}/U_{14} and U_{23}/U_{24} as the diameter, respectively, M_2 is the common region inside two circles with U_1 and U_{12} as the diameter, respectively, and M_3 is the sector θ_1. Based on Eq. (8.16), grid voltage vectors allowing the bidirectional power flow are located in the intersection of M_1, M_2 and M_3, shown as the bold-line region in Fig. 8.11.

Furthermore, with the relationship between the grid voltage vector and the grid voltage rms value shown as below

$$e_d = \sqrt{3}E \tag{8.17}$$

and the geometrical relationship among vectors and the symmetry of sectors, the effective input voltage range for the conventional three-level DPC is

$$\frac{1}{3\sqrt{2}}U_{dc} < E < \frac{1}{2\sqrt{2}}U_{dc} \tag{8.18}$$

A too low input voltage value outside the range shown in Eq. (8.18), for instance, e_2 in Fig. 8.11 can only reduce the active power when $S_p = 1$, regardless of using U_{13}/U_{14} ($S_q = 1$) or using U_{23}/U_{24} ($S_q = 0$). Such invalid pulses due to the input under-voltage will further reduce the DC-bus voltage.

2. **Elimination of invalid pulses caused by the input under-voltage**

To eliminate invalid pulses caused by the input under-voltage of the three-level PWM rectifier, we can enforce the active-power modulation by enriching selectable vectors.

Table 8.4 The three-level DPC switch table with wide input-voltage range

S_p	S_q	dp	θ_1	θ_2	θ_{2k-1}	θ_{2k}	θ_{11}	θ_{12}
1	0	+	U_{23}/U_{24}	U_{13}/U_{14}	U_{2k+9}/U_{2k+10}	U_{2k+11}/U_{2k+12}	U_{21}/U_{22}	U_{23}/U_{24}
1	0	−	U_{23}/U_{24}	U_{23}/U_{24}	U_{2k+9}/U_{2k+10}	U_{2k+9}/U_{2k+10}	U_{21}/U_{22}	U_{21}/U_{22}
1	1	+	U_{13}/U_{14}	U_{15}/U_{16}	U_{2k+11}/U_{2k+12}	U_{2k+13}/U_{2k+14}	U_{23}/U_{24}	U_{13}/U_{14}
1	1	−	U_{15}/U_{16}	U_{15}/U_{16}	U_{2k+13}/U_{2k+14}	U_{2k+13}/U_{2k+14}	U_{13}/U_{14}	U_{13}/U_{14}
0	0	X	U_{12}	U_1	U_{2k-2}	U_{2k-1}	U_{10}	U_{11}
0	1	X	U_1	U_2	U_{2k-1}	U_{2k}	U_{11}	U_{12}

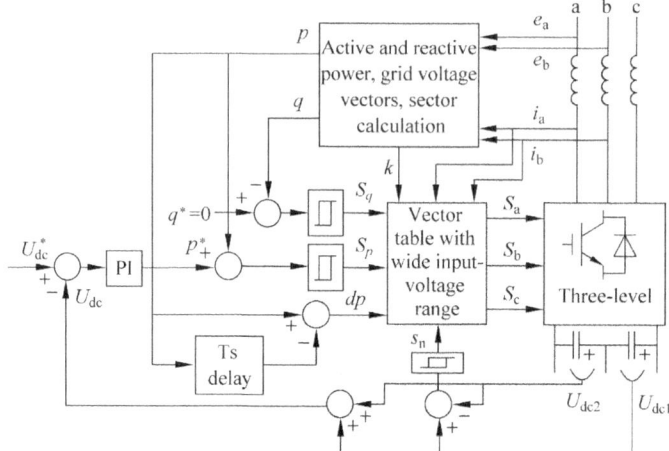

Fig. 8.12 The three-level DPC with the wide input-voltage range

The improved switch table with the wide input-voltage range is shown as Table 8.4. Here the "wide-input-voltage-range" control strategy means with the large variation of the grid voltage, the DC-bus and the grid current can be effectively controlled.

In Table 8.4, dp represents the active power trend by comparing the previous control period to the present. "+" means the active power increases and "−" means decreasing. "±" means either "+" or "−".

The three-level DPC with a wide input-voltage range is illustrated as Fig. 8.12.

Compared to the conventional DPC shown in Fig. 6.19, the improved DPC with widened input-voltage range calculates the active and reactive power based on samplings. In addition, it predicts the power trend in the previous control period, which together with the hysteresis controller merges into the switch table. With $S_p = 1$ and a positive change of the active power, the grid voltage is within the effective input range. No invalid pulses are generated. Thus the enhanced switch table adopts same voltage vectors as the conventional one. With $S_p = 1$ and a negative change of the active power, the grid voltage goes beyond the effective input range thereby generating invalid pulses. To further control the active power bi-directionally, the enhanced switch table adopts voltage vectors with the smaller d-axis component,

Fig. 8.13 Selectable vectors for a wide-input-range three-level DPC rectifier

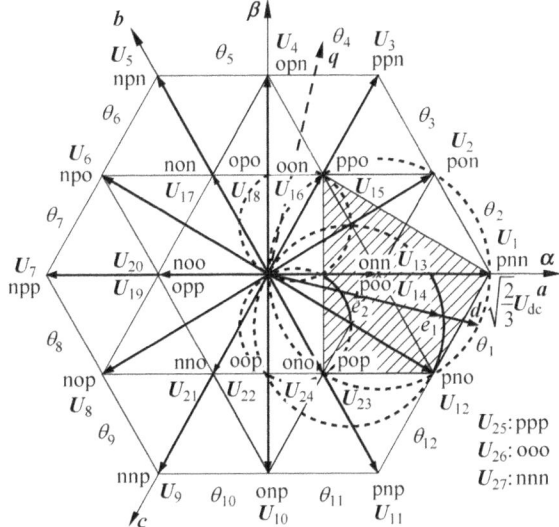

i.e., enforcing the modulation of the active power. Still for the grid voltage vector **e** located in the sector θ_1, the enhanced DPC switch table replaces U_{13}/U_{14} in the conventional switch table with U_{15}/U_{16}, when $S_p = 1$, $S_q = 1$ and dp is negative, as shown in Table 8.4. Therefore selectable voltage vectors for sector θ_1 form a quadrilateral shadow in Fig. 8.13.

Similar to the derivation of the input voltage range when building the conventional DPC switch table, the input voltage range of the enhanced switch table for sector θ_1 can be derived as well to control the bidirectional power flow, as shown in the bold-line area in Fig. 8.13. It can be seen that the voltage vector e_2 when the input voltage is low is within the effective zone, indicating that the enhanced switch table can still meet the control target by generating valid pulses. Furthermore the effective voltage range of the wide-input-range three-level DPC is

$$\frac{1}{2\sqrt{6}}U_{dc} < E < \frac{1}{2\sqrt{2}}U_{dc} \tag{8.19}$$

which compared to Eq. (8.18) for the conventional three-level DPC widens the input range by 26.8%.

The conventional DPC and the wide-input-range DPC are simulated, as shown in Figs. 8.14 and 8.15, respectively. The grid voltage drops from the rated value of 170 VAC (phase RMS) to zero within 1 s. With the conventional DPC, invalid pulses caused by the low input voltage emerge when the grid voltage drops to 63% of the rated value, revoking the control of the active power and the DC-bus voltage and yielding them both dropping with the grid voltage. With the wide-input-voltage DPC control, no invalid pulses appear even when the grid voltage drops to 40% of

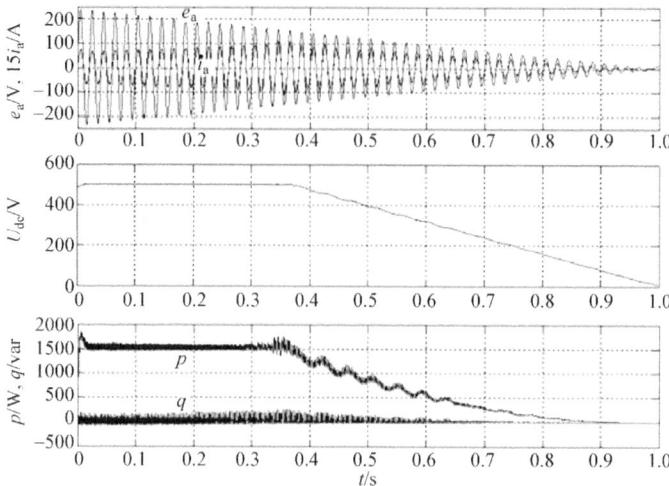

Fig. 8.14 Simulation waveform of the conventional three-level DPC when the grid voltage drops

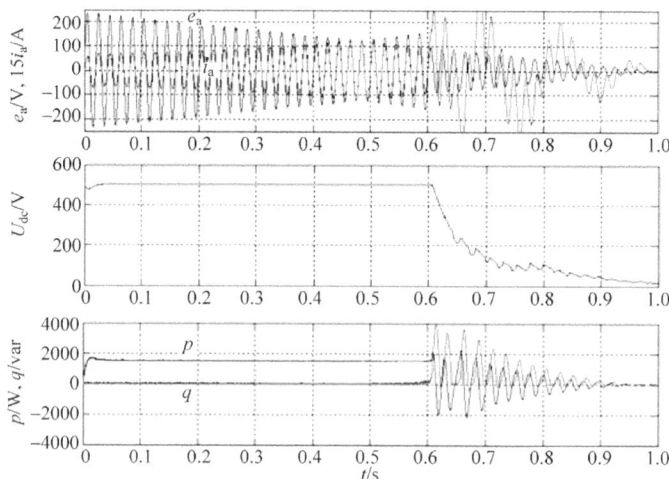

Fig. 8.15 Simulation waveform of the enhanced three-level DPC when the grid voltage drops

the rated value (note such number is related to operational states), with the DC-bus voltage and active power controllable.

Figures 8.16 and 8.17 are the simulated steady-state operation of the conventional DPC and the enhanced wide-input-range DPC, respectively. Here the phase voltage RMS is 100 VAC, which has exceeded the range of Eq. (8.18) thereby yielding the conventional DPC unable to stabilize the DC-bus voltage around the command

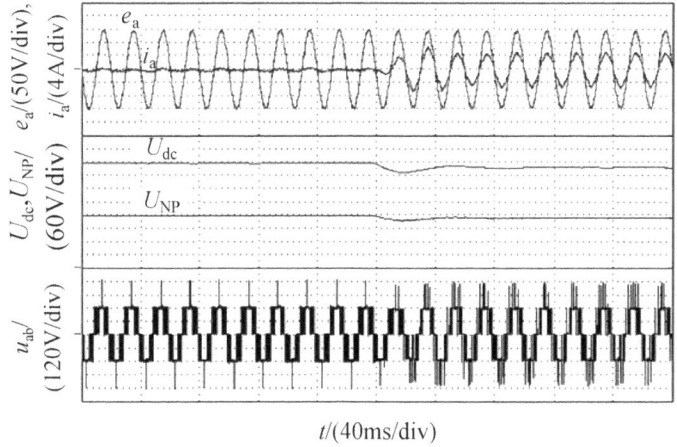

Fig. 8.16 The experimental waveform of the conventional DPC with a low input voltage

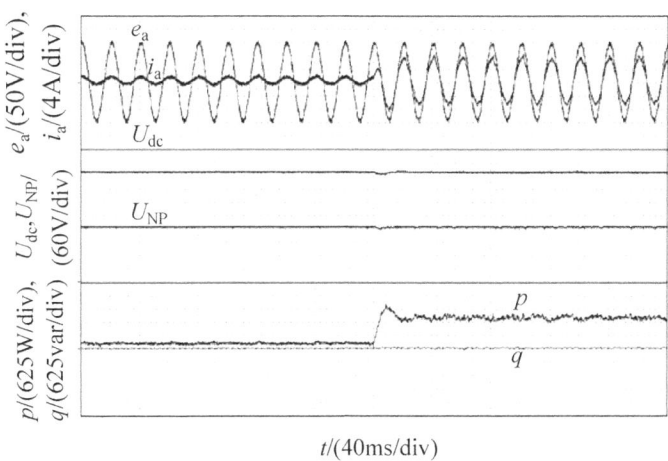

Fig. 8.17 The experimental waveform of the wide-input-range DPC with a low input voltage

(500 V). For the enhanced wide-input-range DPC, invalid pulses are effectively eliminated with the DC-bus voltage effectively controlled.

Furthermore, the experimental grid current and its THD are compared, as shown in Fig. 8.18. With the enhanced wide-input-range three-level DPC, the current can be well controlled even when the grid voltage collapses. Its current THD is also lower than the conventional three-level DPC.

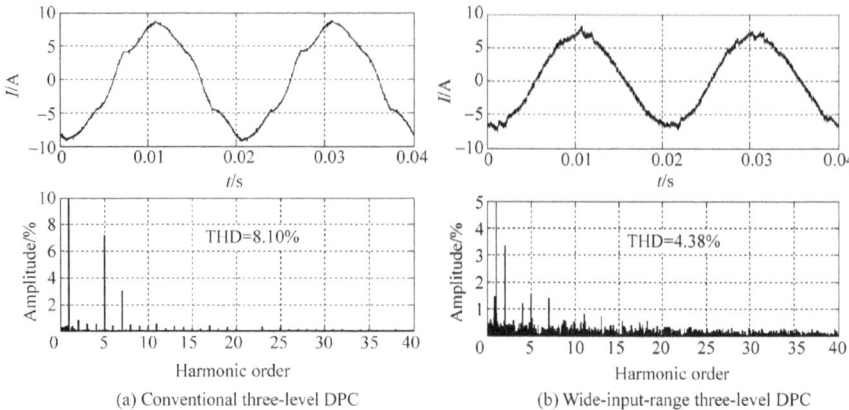

Fig. 8.18   The grid-current experimental waveform with the low input voltage

Finally, it should be noted that "low input voltage" is related to the DC-bus voltage, as shown in Eqs. (8.18) and (8.19). With the rated input voltage, the enhanced three-level DPC yields a wider range of the DC-bus voltage.

8.3   Active Control of Short-Timescale Pulses

The previous analysis did not cover the interaction between power switches and control strategies, while in reality power-switch short-timescale transients centered high-precision control is critical, given the power switch is the utmost important device in power electronic systems.

8.3.1   Classification of the Active Control for Main-Power-Circuit Electromagnetic Pulses

At the present voltage level of power semiconductors, three manners are effective to increase the voltage rating of power electronic systems, i.e., transformers, multi-level topology and switches in series connection. The adoption of the transformer will significantly increase the system volume and space while the multi-level topology will complicate the converter structure and control. Comparably, the most direct, feasible and low-cost method is series connecting switches, which is equivalent to forming a higher-voltage-rating switch.

IGBT series connection requires the voltage balance during the steady on/off states and switching transients, which are named as the steady voltage balancing

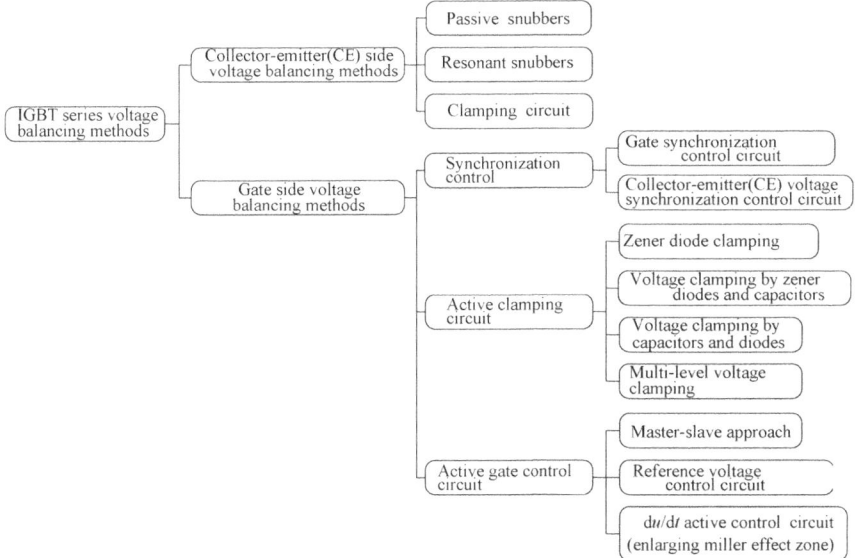

Fig. 8.19 The conventional dynamic voltage balancing for IGBTs in series connection

and dynamic voltage balancing, respectively. The steady-state voltage imbalance is mainly caused by the diversity of the off-state leakage current, which can be compensated by paralleling appropriate resistors across C and E terminals of IGBTs. The dynamic voltage imbalance is rather complex, which could be caused by the parameter diversity of the IGBT, commutation loop, gate-drive circuit and gate-drive signal, etc. Therefore, compared to the steady voltage balancing, the dynamic voltage balancing is more challenging.

The state-of-the-art voltage balancing control of IGBTs in series connection can be categorized as Fig. 8.19, all of which have their pros and cons and are difficult to meet all requirements of the voltage balancing, switch loss, balancing-circuit loss, switching speed and reliability. For high-voltage IGBTs (> 3300 V) in the series connection, the voltage and current are both high, exhibiting more critical electromagnetic transients. Therefore the requirement of the voltage balancing, power loss and reliability is strict. Besides, the component selection for the high-voltage IGBT balancing circuit is more difficult and the EMI is more severe. Therefore it is also required for the balancing method to be more feasible and anti-disturbance.

Regardless of the detailed voltage-imbalance mechanism of series-connected IGBTs, physically it always attributes to the imbalance of the energy stored inside IGBTs in the dynamic process. From the energy perspective, the conventional load-side dynamic voltage balancing control transfers the imbalanced energy stored inside the IGBT to the external snubber circuit and dissipates such energy into the heat through the lossy component. This increases no IGBT loss or switching time, but the loss of the voltage balancing circuit. The gate-side voltage balancing circuit mostly increases the gate voltage in the switching process, prolongs the time for the IGBT working in the active region, and dissipates the imbalanced energy inside the IGBT. Thus such method yields low loss of the voltage balancing circuit, however, increases the IGBT switching loss and switching time.

As a summary, most of conventional voltage balancing methods for the IGBTs in series connection passively transfer or dissipate the imbalanced energy, once such imbalanced energy emerges among IGBTs. It increases the loss of either the balancing circuit or the IGBT, which are called passive balancing methods. Despite its effectiveness of balancing the voltage across IGBTs, it will create the loss imbalance across series connected IGBTs or the balancing circuit, resulting in the local loss cluster of some components and deteriorating the system reliability.

8.3.2 The Active Control of the Main-Power-Circuit Electromagnetic Pulses

From the energy perspective, an active control of the imbalanced energy among series-connected IGBTs can avoid the energy absorption or dissipation, which in addition to realizing the voltage balancing avoids both the power-loss increment of the balancing circuit or switches and prolonged switching time due to staying in the active region for extra time. Based upon this assumption along with the consideration of the design feasibility, EMC and reliability, a main-power-circuit pulse feedback based dynamic balancing control for series connected HV IGBTs was proposed, as shown in Fig. 8.20. The dashed arrow represents the optical-fiber communication. Such method comprises four parts, i.e., the steady-state balancing circuit, dynamic balancing circuit, gate-control balancing circuit and active balancing control algorithm. Here steady-state and dynamic balancing circuits are at the load end, i.e., paralleled to the C and E of IGBTs, the gate-control balancing circuit is at the gate side, i.e., paralleled to the C and G terminals of IGBTs, and the active balancing control is realized by the software embedded in the microcontroller. Such a design embodies interactions between switches and control.

1. Steady-state balancing circuit. It consists of the steady-state balancing resistor R_{static}, far less than the IGBT off-state equivalent resistance $R_{\text{off}} = U_{\text{CES}}/I_{\text{CES}}$ (U_{CES} is the IGBT voltage rating and I_{CES} is the IGBT leakage current at the rated voltage). This resistor helps balance the IGBT voltage in the off state.

2. Dynamic balancing circuit, which is made of the dynamic balancing resistor and capacitor, i.e., R_{dynamic} and C_{dynamic}. The theory is similar to the RC snubber circuit, which suppresses the voltage spike and lowers the voltage imbalance in the switching process. As mentioned before, the RC snubber circuit is a load-side passive balancing circuit. The loss of the balancing circuit will be intolerable if series-connected IGBTs only rely on such circuit to balance the voltage. Thus in the real practice, the dynamic capacitance is usually set to a small value to lower the loss of the load-side balancing circuit thereby the loss of the whole

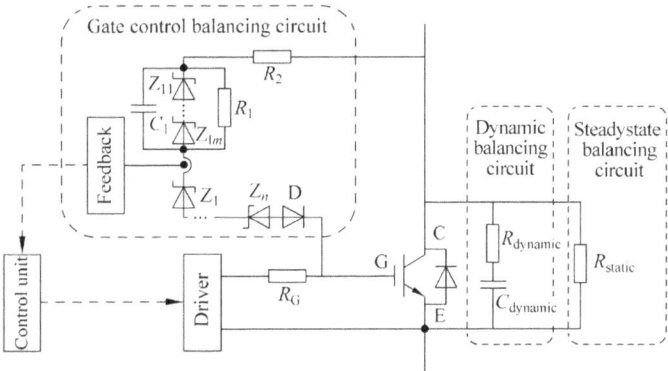

Fig. 8.20 The main-power-circuit pulse feedback based active balancing topology for IGBTs in series connection

balancing circuit. This, however, makes the balancing effect very minor while only benefiting the current-tailing stage. Thus the dynamic voltage balancing of series connected HV IGBTs mainly relies on the gate-control active balancing circuit and the active control software.

3. Gate-control balancing circuit, which is made of the active clamping circuit and the feedback circuit, as shown below:

(1) Active clamping circuit

As shown in Fig. 8.20, the active clamping circuit is made of transient voltage suppressors (TVSs, similar to zener diodes but with much faster response) Z_1–Z_n and Z_{11}–Z_{1m}, current-limiting resistor R_2, anti-reversing diode D, capacitor C_1 paralleled to Z_{11}–Z_{1m} and the discharging resistor R_1. Define the overall breakdown voltage of Z_1–Z_n and Z_{11}–Z_{1m} as U_{Z1} and U_{Z2}, respectively. Since the HV IGBT has the gate-voltage range of -15 to 15 V while U_{Z1} is much higher than $+15$ V, we have $U_{CG} \gg U_{GE}$ when IGBT turns off and U_{CG} close to U_{Z1}. Furthermore $U_{CE} = U_{CG} + U_{GE} \approx U_{CG}$. Based on the analysis above, the turn-off process of the HV IGBT with the active clamping circuit can be divided into three stages, as shown in Fig. 8.21.

Stage 1: $U_{CE} < U_{Z1}$, when all TVS diodes in the clamping circuits are off. At this stage the active clamping circuit will not affect the IGBT turn-off behavior, with U_{CE} rapidly increasing;
Stage 2: $U_{Z1} < U_{CE} < U_{Z1} + U_{Z2}$, when Z_1–Z_n are broken down and Z_{11}–Z_{1m} remain off. The active clamping circuit is energized, with the IGBT collector injecting the current through R_2, C_1, Z_1–Z_n and D to the gate. This lifts the gate voltage and meanwhile charges C_1, slowing down the increment of V_{CE} thereby limiting the voltage imbalance;
Stage 3: $U_{CE} > U_{Z1} + U_{Z2}$, when all TVSs Z_1–Z_n and Z_{11}–Z_{1m} are broken down. The active clamping circuit continues conducting the current. The IGBT

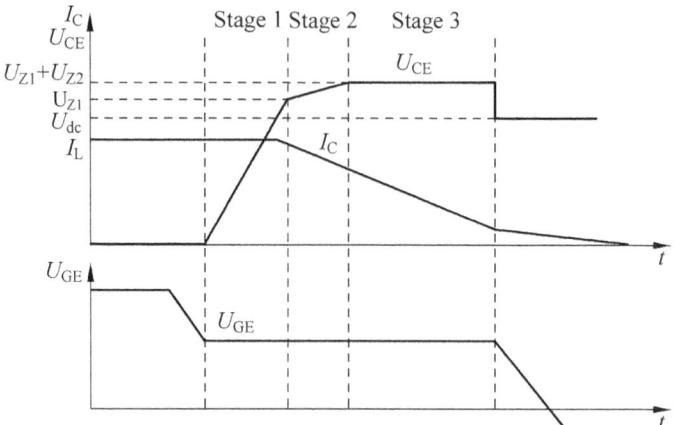

Fig. 8.21 The turn-off process of the HV IGBT with the active clamping circuit

collector injects the current through R_2, Z_{11}–Z_{1m}, Z_1–Z_n and D to the gate, lifting the gate voltage. With the voltage drop across R_2, U_{CE} is clamped to a value slightly higher than $U_{Z1} + U_{Z2}$ and will drop to the DC-bus voltage when entering the current-tailing stage. This effectively limits the voltage spike and the voltage imbalance of balancing circuits.

As a summary, the active balancing method adopts two-stage active clamping circuits. When any of series connected IGBTs undertakes the voltage higher than U_{Z1}, the active balancing circuit will first limit its voltage increasing rate. When the IGBT voltage reaches $U_{Z1} + U_{Z2}$, the active balancing circuit clamps its voltage around $U_{Z1} + U_{Z2}$. Therefore, the active balancing circuit can effectively limit the voltage spike and realize the dynamic balancing. However, it is still a gate-side passive balancing method. At Stages 2–3, the active clamping circuit turns on, injects the current to the IGBT gate and increases the gate voltage. This prolongs the stay in the active region of the IGBT, which rapidly increases the switching loss and slows down the switching speed. Therefore, it is recommended for the clamping circuit not to conduct too long or continually conduct. On the other hand, to better reduce the voltage spike and voltage imbalance, a clamping circuit with a lower breakdown voltage is preferred, which however will creates longer conducting time and higher switching loss. Therefore, only relying on active balancing circuits to realize the dynamic voltage balancing of series connected IGBTs will create a conflict between the voltage balancing effect and the IGBT switching loss and speed.

(2) Feedback circuit

In switching transients of series connected IGBTs, the clamping circuit with a higher IGBT voltage will conduct for longer time. Therefore the status of the clamping circuit could reveal the status of its corresponding IGBT to some extent. If the control system could obtain the status of IGBT clamping circuits and adjust the gate signal accordingly, the closed-loop control of main-power-circuit pulses can be realized, i.e., active balancing control, which can achieve the voltage balancing after several pulses. In this process, the clamping circuit is only activated when the modulation is needed, which avoids the long-time or continual conduction. No extra switching loss or time of the IGBT is added once the active balancing is realized.

To realize the active balancing above, a feedback circuit is needed to sample and forward the status of the clamping circuit to the control system, i.e., the feedback loop of the active balancing closed-loop control. Most of the active clamping circuit has very short conducting time, e.g., nano ~ sub-micro seconds. To secure the real-time signal, the feedback circuit needs to fast respond and transfer signals. Besides, the fidelity of the feedback signal highly impacts the performance of the closed-loop active balancing control and the reliability of series-connected circuits. Therefore a high precision is required for the feedback circuit. Lastly, the feedback circuit is close to the gate of the HV IGBT, the interface between strong and weak electricity circuits. A high EMC is required for the feedback circuit.

As a summary, in the active balancing process, the gate-control balancing circuit will (1) secure the IGBT reliability before the series connected circuit reaches the steady state through suppressing voltage spikes over the clamping threshold, and (2) provide feedback signals revealing the main-power-circuit status to the active control software.

4. Active balancing control algorithm

The feedback circuit forwards signals to the control IC. A low-level signal means the clamping circuit is off while a high-level signal means the clamping circuit conducts. Thus the control IC during switching transients could sample and count feedback signals from the IGBT gate-control balancing circuit, i.e., counting the overall conduction time of each IGBT clamping circuit.

During switching transients, the earlier turned-on or faster switched-on IGBT will experience the voltage drop earlier, making other IGBTs undertake the higher voltage with the overall voltage across IGBTs as a constant. On the other hand, the later-turned-on IGBT's clamping circuit will remain conduction for a longer time. If the control IC could advance its next switching-on signal for Δt_{on}, the voltage dynamic balancing could be realized in the next switching-on process.

Similarly, during switching-off transients, the earlier turned-off or faster switched-off IGBTs will undertake a higher voltage. Accordingly, such IGBTs' clamping circuits will remain conduction for a longer time in switching-off transients. If the control IC could delay its next switching-off signal for Δt_{off}, the voltage dynamic imbalance in the next switching-off process could be alleviated.

Take two IGBTs (T_1 and T_2) in series connection as an example. Based on active balancing methods proposed above, we can design the forward loop of the closed-loop active balancing strategy. During switching-on transients of the kth pulse, the control IC samples the status of gate-control circuits of T_1 and T_2 and calculates the conduction time as $t_{AC1}(k)$ and $t_{AC2}(k)$, respectively. Upon the completion of the switching-on process, a PI controller is used to calculate the advancement of switching moments of T_1 versus T_2, i.e., $\Delta t_{\text{on}}(k + 1)$.

$$\begin{cases} P_{\text{on}}(k) = k_{\text{P}}[t_{\text{AC1}}(k) - t_{\text{AC2}}(k)] \\ I_{\text{on}}(k) = I_{\text{on}}(k - 1) + k_{\text{I}}[t_{\text{AC1}}(k) - t_{\text{AC2}}(k)] \\ \Delta t_{\text{on}}(k + 1) = I_{\text{on}}(k) + P_{\text{on}}(k) \end{cases} \qquad (8.20)$$

Similarly, at switching-off transients of the kth pulse, the control IC samples the status of gate-control circuits of T_1 and T_2 and calculates the conduction time as $t_{\text{AC1}}(k)$ and $t_{\text{AC2}}(k)$, respectively. Upon the completion of the switching-off process, a PI controller is used to calculate the delay of the switching moment of T_1 versus T_2, i.e., $\Delta t_{\text{off}}(k + 1)$.

$$\begin{cases} P_{\text{off}}(k) = k_{\text{P}}[t_{\text{AC1}}(k) - t_{\text{AC2}}(k)] \\ I_{\text{off}}(k) = I_{\text{off}}(k - 1) + k_{\text{I}}[t_{\text{AC1}}(k) - t_{\text{AC2}}(k)] \\ \Delta t_{\text{off}}(k + 1) = I_{\text{off}}(k) + P_{\text{off}}(k) \end{cases} \qquad (8.21)$$

Based upon Eqs. (8.20) and (8.21), $\Delta t_{\text{on}}(k + 1)$ and $\Delta t_{\text{off}}(k + 1)$ are calculated for the next-period switching moment of each IGBT. Such a closed-loop active balancing will effectively improve the voltage imbalance during switching-on and switching-off transients.

With the active voltage balancing control, series connected HV IGBTs could achieve the voltage balance after several pulses. After that, with the breakdown threshold of the clamping circuit higher than the IGBT voltage spike, the balanced IGBT will not conduct its clamping circuit any more. No switching transients will be affected. All IGBTs will work under the natural switching loss and speed, avoiding the increment of the switching loss or switching time. Meanwhile the feedback signal is not the collector voltage but the voltage-level signal representing the on/off of the clamping circuit. It will remain low once the voltage balancing is achieved with no clamping circuit conducting any more, i.e., no further adjustment of the gate signal is needed. The balancing circuit can keep the present status of the voltage balancing for the continuous operation.

In the modulation process of the active balancing control, some IGBTs might undertake higher voltage before series connected circuits reach the steady state. The existence of the active clamping circuit will reduce potential voltage spike during this modulation, securing its operation within the SOA and reliability of series voltage balancing circuits. Meanwhile, the conduction of the clamping circuit will alter the level of the feedback signal, with its pulse width representing part of the main-power-circuit characteristics thereby providing the reference for actively controlling IGBT gate signals.

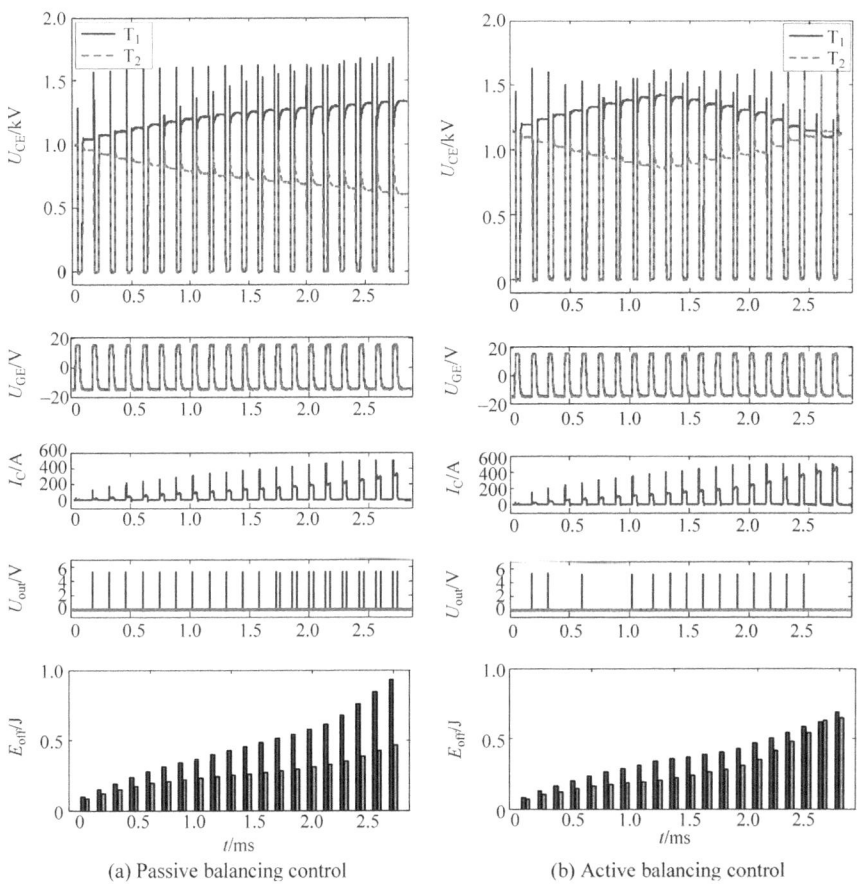

Fig. 8.22 Experimental comparison of the passive and active balancing methods ($U_{dc} = 2$ kV)

8.3.3 Effectiveness of the Active Control Method

The active balancing control method proposed above can be experimentally verified through a test bench for two HV IGBTs in series connection. Furthermore it can be compared with the passive balancing method.

The gate-side passive balancing with only clamping circuits employed is compared to the active balancing control at the DC-bus voltage of 2 and 5 kV, as shown in Figs. 8.22 and 8.23, respectively. The subplots from the top to the bottom are U_{CE}, U_{GE}, I_C, gate-control feedback signal U_{out}, turn-on loss E_{on} and turn-off loss E_{off} of the IGBTs, respectively. For the passive balancing control, U_{out} is only used to monitor the conduction time of the clamping circuit, not for the feedback control.

As shown in Figs. 8.22a and 8.23a, the passive balancing method can reduce the voltage spike and dynamic imbalance with limited effectiveness. The voltage

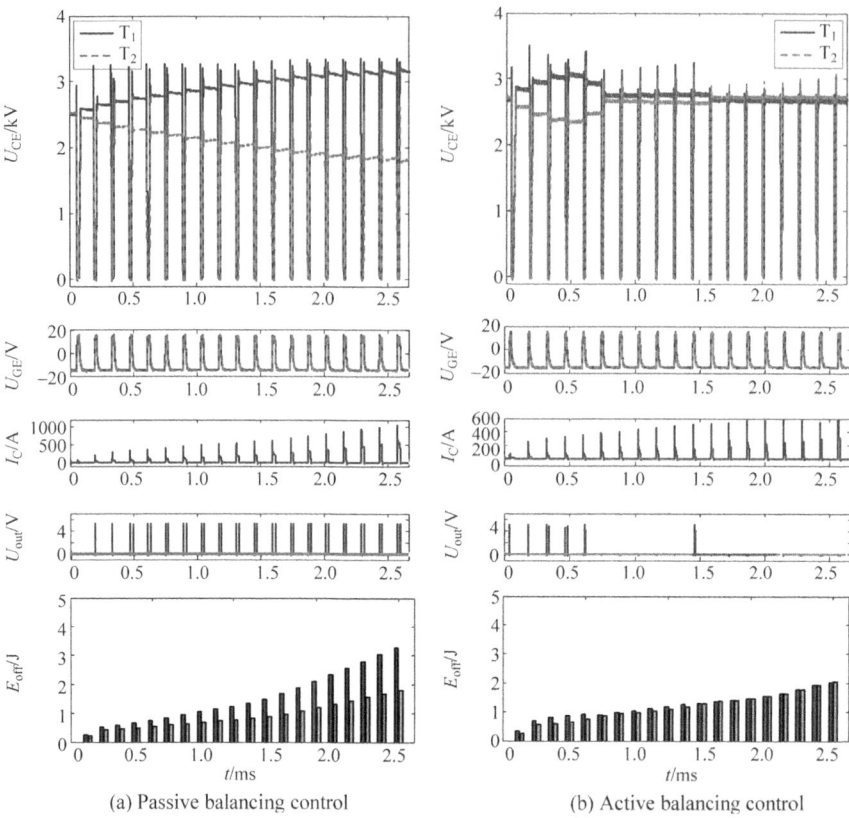

Fig. 8.23 Experimental comparison of passive and active balancing methods ($U_{dc} = 5$ kV)

imbalance worsens as the load current increases. In addition, with the passive balancing method, significant switching-loss difference exists among IGBTs. This is because under the imbalanced voltage one IGBT undertakes higher voltage, causing its clamping circuit to conduct longer time. This IGBT then stays in the active region for a longer time. Therefore the passive balancing method causes the imbalance of the switching loss and creates the loss cluster to some IGBTs thereby prone to damaging switches.

As shown in Figs. 8.22b and 8.23b, with the active balancing control, the voltage balance will be achieved after several pulses with much better balancing performance. After reaching the balanced voltage, the feedback signal U_{out} shows the clamping circuit of each IGBT gate-control unit stops conducting, with no extra switching loss or time added. Therefore the IGBT switching loss is balanced as well, avoiding the power-loss cluster on some specific IGBT thereby enhancing the system reliability. Before fully realizing the voltage balancing, the clamping circuit of the gate-control

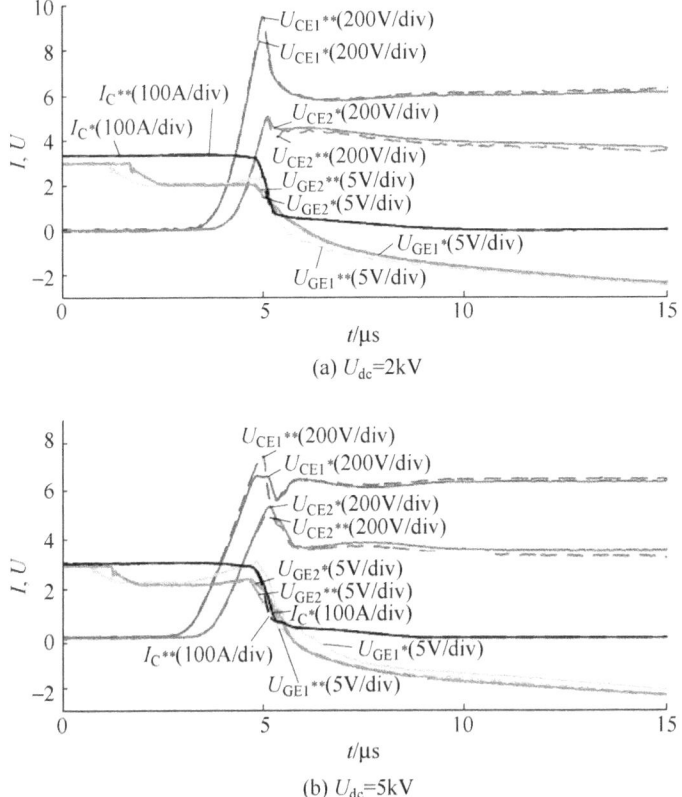

Fig. 8.24 The voltage spike when using the active balancing method

balancing circuit is effective to suppress the voltage spike and secure the system safety, as shown in Fig. 8.24, where "*" means with the gate balancing circuit and "**" means without the gate-control balancing circuit.

As shown in Fig. 8.24, with a DC-bus voltage of 2000 V, the clamping circuit in the gate-control balancing circuit reduced the voltage spike of the higher-voltage IGBT from 1800 to 1680 V. With a DC-bus voltage of 5000 V, the clamping circuit in the gate-control balancing method reduced the voltage spike of the IGBT from 3720 to 3300 V.

The experimental voltage imbalance and switching loss are compared between gate-side passive balancing and active balancing methods in Table 8.5, where two HV IGBTs were in series connection. It can be seen that compared to the conventional gate-side passive balancing method, a better balancing is achieved by the main-circuit pulse feedback balancing control, which in addition could balance the IGBT switching loss, avoid the loss clustering, effectively reduce the overall loss of series balancing circuits and increase the system efficiency.

Besides, to reach the same balancing performance as the active balancing control, the load-side balancing for IGBTs in series connection yields the loss of 9.93 kW under a 5000 V DC-bus voltage, in contrast to < 200 W loss of the active balanc-

Table 8.5 Experimental comparison of the passive and active balancing control methods

	ΔU_{CE} (kV)	$E_{on\text{-}T1}$ (J)	$E_{on\text{-}T2}$ (J)	$E_{on\text{-}T1}$ + $E_{on\text{-}T2}$ (J)	$E_{off\text{-}T1}$ (J)	$E_{off\text{-}T2}$ (J)	$E_{off\text{-}T1}$ + $E_{off\text{-}T2}$ (J)
$U_{dc} = 2$ kV passive	0.57	0.953	0.364	1.317	0.928	0.460	1.388
$U_{dc} = 2$ kV active	0.03	0.712	0.560	1.272	0.689	0.648	1.346
$U_{dc} = 5$ kV passive	0.67	4.639	2.791	7.430	3.264	1.800	5.064
$U_{dc} = 5$ kV active	0.09	3.692	3.439	7.131	2.112	2.132	4.244

ing. Therefore compared to the conventional load-side passive balancing, the active balancing will greatly reduce the loss of the balancing circuit, shrink the size of the balancing circuit and increase the system efficiency.

8.3.4   Integration of the Active Balancing with the Main-Power Circuit

To integrate the active balancing control based on the main-power pulse feedback with more complex topology where HV IGBTs are in series connection, it is possible to use the centralized active balancing control, as shown in Fig. 8.25 where all feedback signals of gate-control active balancing circuits and IGBT gate-drive signals need be connected with the central control unit through the gate drive circuit. Such design has cumbersome connectors, complex structures and poor scalability. In addition, the centralized control unit has the control circuit distant from the main circuit, deteriorating the real-time performance of feedback pulses and IGBT gate-drive signals. Such long wire connections between the main-power circuit and control circuit increase the system complexity as well.

For the software design, with such centralized active balancing, all feedback and balancing commands need be realized within one microprocessor, with high computation load and high demand on the computation speed and resources. It is very difficult to meet the real-time and high-precision control within short-timescale transients. In addition, multi-channel feedback signals and IGBT gate-drive signals are coupled inside one control unit. So are the short-timescale active balancing control, the large-timescale ADC samplings, ADC conversion, protections and communications. Such couplings are prone to the competition and risk at the time sequence and logic, which lowers the system reliability.

Therefore, the centralized active balancing has drawbacks at both the software and hardware design. It is difficult to be integrated with the main-power circuit, increases the system complexity and the demand on the resources, limits the real-time performance and scalability of the balancing control, and more importantly, cannot secure the accuracy and reliability of the time sequence and logic when applied to the complex topology or multi-switch-series-connection scenarios. Scalability of such a centralized active balancing control is restrained.

To resolve conflicts above, a chain-structure active balancing circuit was proposed for IGBTs in series connection, as shown in Fig. 8.26 where dashed arrows represent the optical fiber communication. For the n-switch-series-connection case, the gate of each IGBT T_i ($i = 1 \sim n$) is equipped with one active balancing module M_i, which receives the original gate signal from M_{i-1} and transmits the active clamping status and protections (F_B & F_O) to M_{i-1}. At the same time, M_i sends the original gate signal to M_{i+1} and receives the active clamping status and protections (F_B & F_O) of T_{i+1} from M_{i+1}.

Now we can start to design the balancing module. Similar to the gate-control balancing shown in Fig. 8.20, each balancing module contains the active clamping circuit and status feedback, responsible for suppressing the voltage spike above the clamping threshold during the modulation and providing the feedback signal of the clamping circuit, respectively.

Different from the centralized active balancing circuit, each module is equipped with one microcontroller, i.e., realizing the active balancing within multiple microcontrollers. The module M_i samples the status of the clamping circuit of T_i as F_B, extracts the clamping-circuit status of T_{i+1} through F_B & F_O, calculates and adjusts the T_i gate signal based on samplings above along with the original gate signal of T_i transmitted from M_{i-1}, and forwards it to the gate-drive circuit. Meanwhile the active balancing control module M_i combines the clamping-circuit status signal F_B

Fig. 8.25 Control blocks of the centralized active balancing control

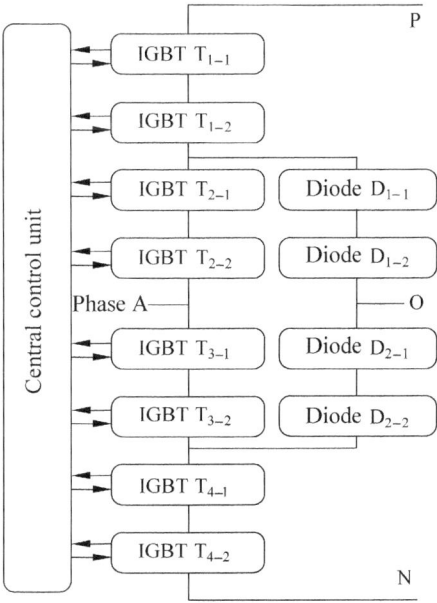

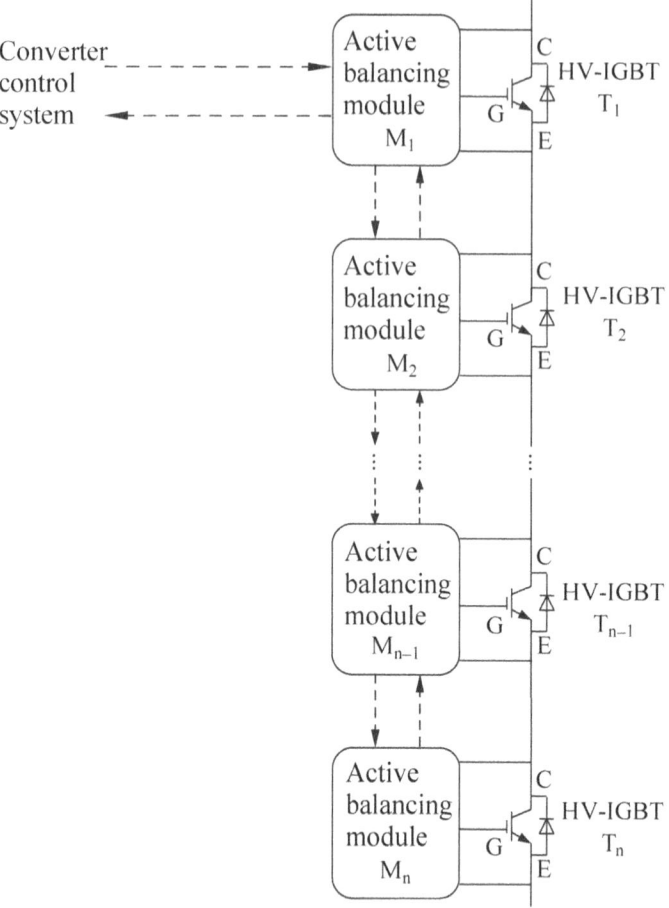

Fig. 8.26 The chain-structure topology of the active balancing for HV IGBTs in series connection

with the protection signal from the gate-drive circuit F_O as one signal (F_B & F_O) and sends it to the module M_{i-1}.

As shown above, the chain-structure topology has limited input and output for each module. The active balancing control only acquires the present and adjacent IGBT status and only balances its related IGBT. Therefore the decoupling of the multi-channel feedback and control signals is a must for such a distributive balancing system.

In the switching transient, the module M_i samples and counts feedback signals F_{Bi} and F_{Bi+1}. Here $F_B = 1$ means the clamping circuit is on, otherwise off. The conduction time of clamping circuits for T_i and T_{i+1} is counted as A_{Ci} and A_{Ci+1}, respectively. Sending the subtraction of A_{Ci+1} from A_{Ci} to a PI controller yields the gate-signal delay P_{on} or P_{off} for T_i, within the limitation of $-T_{max}$ and $+T_{max}$ in case the over modulation or oscillation. At last, the calculated gate-signal delay is imposed on following switching-on/off moments. Such an active adjustment of the IGBT gate signal will advance the turn-on moment for those IGBTs undertaking higher voltage in the turn-on process, and delay the turn-off moment for those IGBTs undertaking higher voltage in the turn-off process, namely the dynamic balancing. The control of the T_i gate signal is only related to the clamping-circuit status of T_i and T_{i+1} and the original gate signal of T_i. Other parts of the circuit are irrelevant. Thus it realizes the decoupling of control and feedback signals, distributive control and modular design, which allows it to be scaled to multi-switch series connection.

Such a chain-structure topology with the software & hardware design of the balancing module integrates the distributive balancing control with the IGBT centered main-power circuit, facilitates the timely and effective control of IGBT short-timescale switching transients, and secures the real-time control of the active balancing. From the converter control perspective, all IGBTs in series connection and their balancing modules can be seen as one circuitry unit, similar to one IGBT. The converter only forwards one gate signal to the whole IGBT string and receives one feedback signal from the IGBT string. Thus the converter inherits the original control unit without adding IO terminals. The top-level converter control does not overlap with the bottom-level active balancing control, which realizes the hierarchical design of the converter and balancing control and the decoupling of multi-level control processes. Such design guarantees the sufficient software and hardware resources for every level of control and enhances the system reliability. In addition, such a distributive and modular design enforces the real time and scalability of the active balancing control and resolves the conflict of the centralized active balancing control in the complex topology or multi-switch series connection.

8.3.5 Validation of the Distributive Active Balancing

A three-level test bench using HV IGBTs in series connection is built to verify the conventional gate-side passive balancing (only using the active clamping circuit) and the distributive active balancing, as shown in Figs. 8.27 and 8.28, respectively. Under the DC-bus voltage of 6 and 10 kV, experimental results of U_{CE} and I_L are displayed from the top to the bottom. Here each subplot represents the U_{CE} of T_{1-1} & T_{1-2}, T_{2-1} & T_{2-2}, T_{3-1} & T_{3-2} and T_{4-1} & T_{4-2}, respectively. Using the conventional passive balancing control yields a limited balancing capability. With the load current increasing, the voltage imbalance deteriorates. With the active balancing control integrated in the main-power circuit, all four sets of series connected IGBTs reach the balanced voltage after several pulses, resulting in much better balancing performance. Given the threshold voltage of the clamping circuit is higher than the IGBT turn-off peak voltage, the clamping circuit will not conduct once the balanced voltage is achieved. No extra switching loss or time is exhibited.

Table 8.6 Experimental imbalanced voltage of the passive and active balancing control on a three-level test bench

	$T_1-\Delta U_{CE}$ (kV)	$T_2-\Delta U_{CE}$(kV)	$T_3-\Delta U_{CE}$(kV)	$T_4-\Delta U_{CE}$(kV)
$U_{dc}=6$ kV passive	1.08	0.98	1.12	1.01
$U_{dc}=6$ kV active	0.12	0.13	0.01	0.01
$U_{dc}=10$ kV passive	2.28	2.16	2.33	2.18
$U_{dc}=10$ kV active	0.16	0.18	0.03	0.04

Table 8.6 shows the imbalanced voltage of the passive and active balancing control on a three-level test bench, which indicates that the active balancing integrated with the main-power circuit can achieve better balancing effect.

In the modulation process of the active balancing before four sets of IGBTs in series connection reached the balancing, the clamping circuit of each balancing module effectively reduced the voltage spike thereby securing the device safety, as shown in Table 8.7.

Four IGBTs in series connection were tested using the conventional gate-side passive balancing (only using the active clamping circuit) and the active balancing integrated with the main-power circuit, as shown in Figs. 8.29 and 8.30 where the

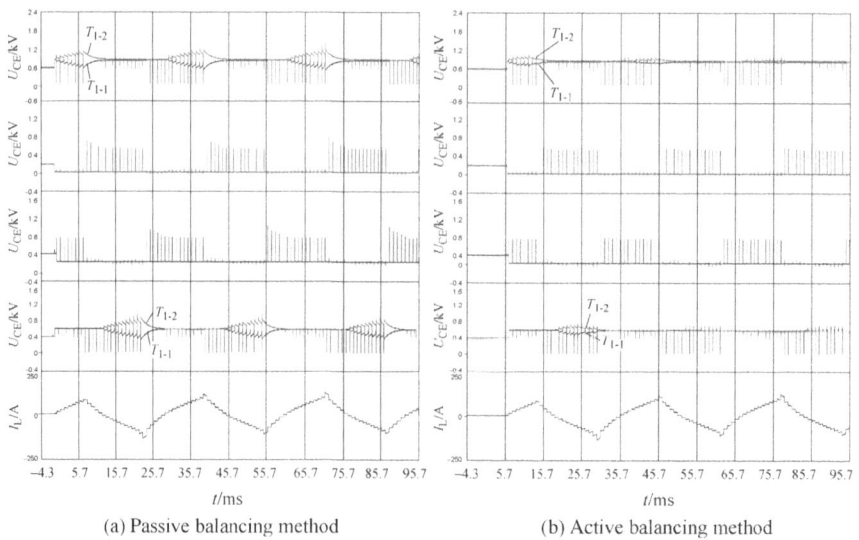

(a) Passive balancing method   (b) Active balancing method

Fig. 8.27 Experimental comparison of the passive and active balancing control on a three-level test bench ($U_{dc}=6$ kV)

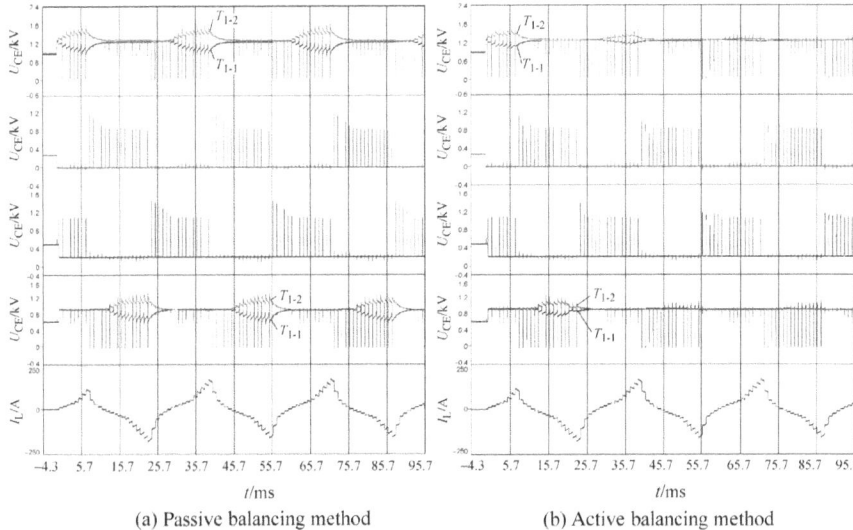

(a) Passive balancing method   (b) Active balancing method

Fig. 8.28 Experimental comparison of the passive and active balancing control on a three-level test bench ($U_{dc} = 10\,kV$)

Table 8.7 The voltage-spike reduction of the IGBT during transients using the active balancing

	T_1(kV)	T_2(kV)	T_3(kV)	T_4(kV)
$U_{dc} = 6\,kV$ passive	2.64	2.56	2.61	2.64
$U_{dc} = 6\,kV$ active	2.08	2.04	2.15	2.08

DC-bus voltage is 4500 and 6000 V, respectively. Experimental waveforms are U_{CE}, IGBT current I_C, the feedback signal of the active clamping circuit (U_{out}) and the IGBT turn-off loss (E_{off}). For the passive balancing, U_{out} is only used to monitor the conduction time of the active clamping circuit, without being sampled by the control IC. It can be seen clearly that the active balancing integrated with the main-power circuit achieved much better balancing effect than the conventional passive balancing. After several pulses all IGBTs in series connection undertook the balanced voltage. U_{out} shows that once the balanced voltage was realized the active clamping circuit did not conduct any more, with no impact on the IGBT switching loss and the switching time. Thus the IGBT switching loss was balanced as well, avoiding the potential loss cluster on one particular IGBT. Before entering the steady-state, the active clamping circuit effectively reduced the IGBT voltage spike and secured the switch safety, as shown in Fig. 8.31.

Table 8.8 listed the experimental data of voltage imbalance and switching loss when using the gate-side passive balancing and active balancing control on a

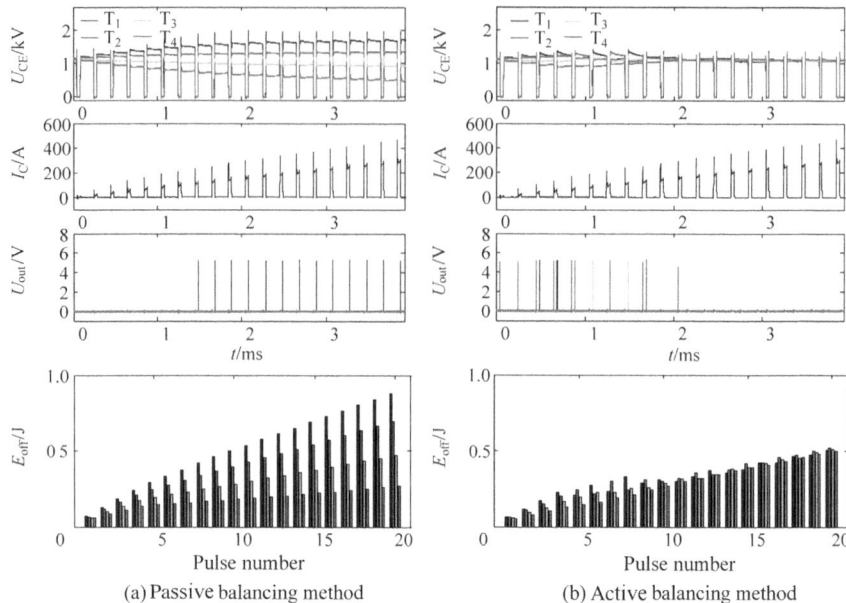

Fig. 8.29 Experimental comparison of the passive and active balancing control for four IGBTs in series connection ($U_{dc} = 4.5$ kV)

Table 8.8 Experimental data of voltage imbalance and switching loss when using the gate-side passive balancing method and active balancing method on a four-IGBT-series test bench

	ΔU_{CE} (kV)	$E_{off\text{-}T1}$ (J)	$E_{off\text{-}T2}$ (J)	$E_{off\text{-}T3}$ (J)	$E_{off\text{-}T4}$ (J)	$E_{off\text{-}T1}$ + $E_{off\text{-}T2}$ + $E_{off\text{-}T3}$ + $E_{off\text{-}T4}$ (J)
$U_{dc} = 4.5$ kV, passive	1.58	0.878	0.693	0.469	0.267	2.307
$U_{dc} = 4.5$ kV, active	0.08	0.503	0.516	0.508	0.498	2.025
$U_{dc} = 6.6$ kV, passive	1.74	1.888	1.600	1.216	0.760	5.464
$U_{dc} = 6.6$ kV, active	0.16	1.267	1.347	1.314	1.224	5.152

four-IGBT-series-connection test bench. It indicates that, compared to the gate-side passive balancing, the active balancing integrated with the main-power circuit realized a much better balancing effect, which further balanced the switching loss thereby avoiding the loss clustering, reduced the overall switching loss and increased the system efficiency.

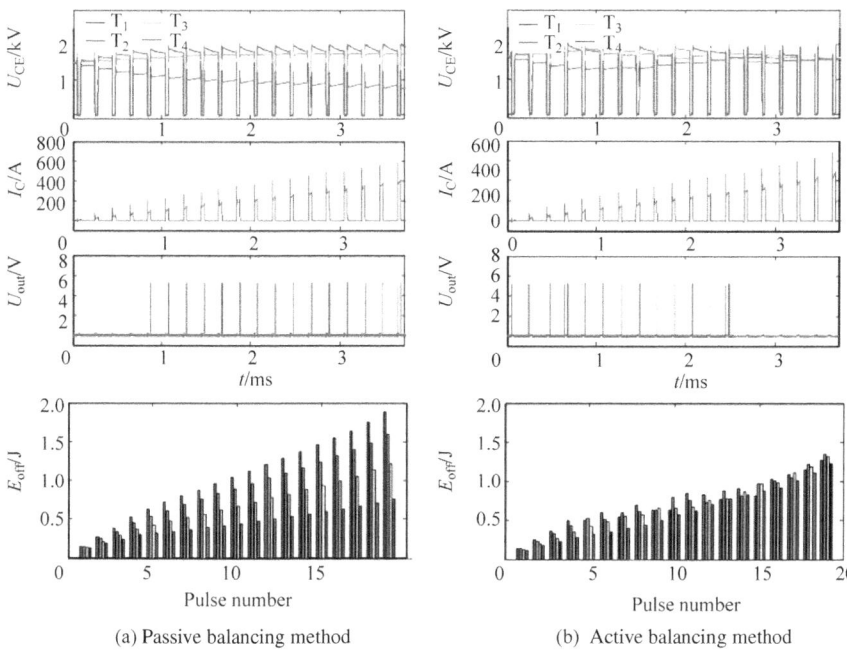

Fig. 8.30 Experimental comparison of the passive and active balancing control for four IGBTs in series connection ($U_{dc} = 6.6$ kV)

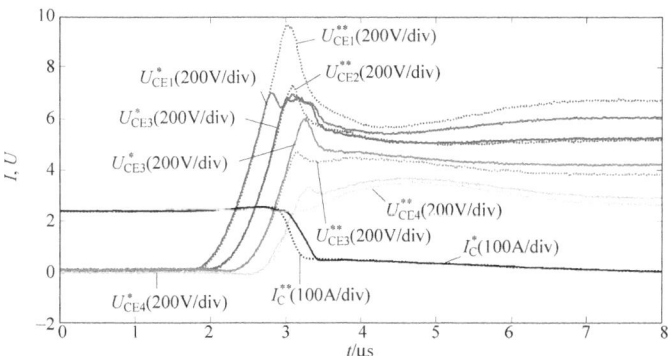

Fig. 8.31 Suppression of the IGBT voltage spike in the modulation process when four IGBTs are in series connection

In addition, the active balancing integrated with the main-power circuit has been tested in various topologies and working conditions, without finding any system fault or switch damage. This concludes that the chain-structure balancing circuit and distributive balancing module can secure the real-time performance of the balancing control in complex topology and multi-switch in series connection. Furthermore it saves software and hardware resources, resolves the potential conflicts caused by the centralized balancing control in complex topology and multi-switch in series connection, and effectively enhances the feasibility and reliability of the active balancing system.

Chapter 9
Balance of Electromagnetic Energy in Transients

Energy conservation is one of most fundamental rules in physics, which is interpreted as the input energy of any system is equal to the sum of the energy loss, storage and output. For power electronic converters where the electromagnetic energy conversion is the basic need, such rule is naturally applied to the analysis and control of electromagnetic transients.

As stated before, the main goal of the power electronic converter is to realize the effective conversion of the electromagnetic waveforms and characteristics. To reach the expected conversion performance, majority of power electronic converters adopt PWM control methods, which output the electromagnetic energy in the form of pulses and pulse sequences. Effective pulse or pulse sequences are basics of power electronic converters, the fundamental form of the transient energy transformation with the time constant of ns–μs. Such short-timescale transients are decisive for the reliability of power electronic converters. On one hand they are the fundamental of the waveform transformation. On the other hand, without the effective control they will result in the device failure and the system damage. Besides, as pointed in Chap. 2, various time-scaled transients co-exist in the power electronic converter with different time constants, which form multi-level timescale transient processes. Regardless of the complexity of transients, all of them need comply with the energy conservation and no sudden energy change.

As a multi-timescale energy conversion system, a power electronic converter needs balance the energy at different time constants to realize the system optimization. At a timescale far less than a switching period, energy imbalance in the transient process might cause the destructive energy local cluster. With the time constant of such energy conversion far less than a control period, only hardware instead of the software can be counted on to solve such issue, e.g., the conventional snubber circuit and soft-switching circuits to suppress the transient energy imbalance. For any energy conversion with timescale larger than one control period, the control algorithm can be optimized for the energy balance. Therefore the analysis and control of the energy balancing is multi-timescale and multi-layer.

© Tsinghua University Press and Springer Nature Singapore Pte Ltd. 2019
Z. Zhao et al., *Electromagnetic Transients of Power Electronics Systems*,
https://doi.org/10.1007/978-981-10-8812-4_9

In addition, the analysis and control of power electronics is usually targeting the current of inductive components, such as the motor current and grid current, or the voltage across capacitive components, such as the DC-bus voltage. In both the dynamic process and steady state, the control variable needs closely follow the reference. From the energy perspective, the control variable can be translated as the transient energy in energy storage components. Therefore in the dynamic process it is required that the energy storage needs rapidly follow the target energy while in the steady state the energy needs stay around the reference. When multiple energy storage components or various types of control variables (voltage and current) mix together, the conventional voltage or current closed-loop control is hard to coordinate them. From the energy perspective, voltage or current control can be replaced with the energy control. Therefore the energy balancing control is multi-objective.

Based on the distribution and flow of the transient energy, this chapter details the concept and theory of the energy balancing control. A three-level VSI PWM rectifier and a back-to-back four-quadrant PWM converter are taken as examples to present the conflict between the dynamic response and over modulation of the conventional outer-loop voltage control, establish the control method to balance the electromagnetic transient energy, and analyze the stability and robustness of such energy balancing control.

9.1 Balancing and Modelling of the Electromagnetic Energy

The power electronic converter essentially is an energy-conversion apparatus, which transforms the energy from one form to the other based upon the expectation. From the energy perspective, both inductor and capacitor are energy storage components, which can store and exchange the energy. The resistor is a lossy component, which consumes the energy. The power switch varies the energy flowing direction and form, while the ideal switch does not store or consume the energy. As a summary, the current and voltage in the circuit are both related to the energy, i.e., the inductor current is related to the magnetic energy and the capacitor voltage is related to the electric energy. Each component in the power electronic converter becomes an energy component. Voltage and current control then are translated into the energy control. The dynamic performance of a converter is the embodiness of the energy transforming process between input and output within different components. The stead-state performance, however, indicates the capability of electric components sustaining the reference energy. Therefore the analysis from the energy perspective reveals the internal characteristics of the converter, while the control from the energy point of view will enhance the conversion performance.

9.1.1 Balance of the Electromagnetic Transient Energy

Based on the energy conservation, the power electronic system needs balance the energy at any time interval, i.e., the balance among the energy input, output, loss and storage. Take one control period T_s as one example. The energy distribution and flow are shown in Fig. 9.1. The energy flow within one control period can be divided into four parts, i.e., loss of resistors and equivalent resistors, magnetic energy increment in inductive components, electric energy increment in capacitive components and the output to the load. Here all the energy input, output and energy storage are bidirectional.

Assume m sets of inductive components, n sets of capacitive components and k sets of equivalent resistors exist in the converter. Based on the description of energy balancing above, we have

$$\begin{aligned} E_{\mathrm{in}} = {} & E_{R1} + \cdots + E_{Rk} \\ & + \Delta W_{L1} + \cdots + \Delta W_{Lm} \\ & + \Delta W_{C1} + \cdots + \Delta W_{Cn} \\ & + E_{\mathrm{out}} \end{aligned} \tag{9.1}$$

Here E_{in} and E_{out} are the input and output energy of the converter within one control period. E_{Ri} ($i = 1 \sim k$) are the energy consumed by the equivalent resistor within one control period. ΔW_{Li} ($i = 1 \sim m$) are the magnetic energy increment of inductive components within one control period. ΔW_{Ci} ($i = 1 \sim n$) are the electric energy increment of capacitive components within one control period.

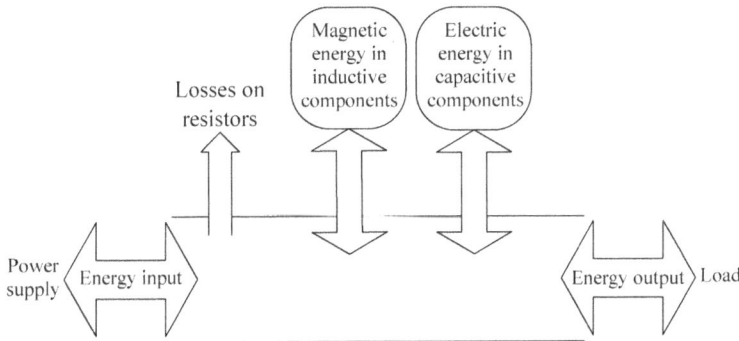

Fig. 9.1 Energy distribution and flow of the converter

9.1.2 Modelling of the Transient Energy Balancing Control

The common closed-loop control variables of power electronic converters are the voltage and current. All energy variables in Eq. (9.1) can be related to the voltage and current as below.

$$E_{in} = \int_0^{T_s} p_{in} \cdot dt = \int_0^{T_s} u_{in} i_{in} \cdot dt \tag{9.2}$$

$$E_{out} = \int_0^{T_s} p_{out} \cdot dt = \int_0^{T_s} u_{out} i_{out} \cdot dt \tag{9.3}$$

$$E_{Ri} = \int_0^{T_s} p_{Ri} \cdot dt = \int_0^{T_s} u_{Ri} i_{Ri} \cdot dt$$

$$= \int_0^{T_s} i_{Ri}^2 R_i \cdot dt = \int_0^{T_s} \frac{u_{Ri}^2}{R_i} \cdot dt \tag{9.4}$$

$$W_{Li} = \frac{1}{2} L_i i_{Li}^2 \tag{9.5}$$

$$W_{Ci} = \frac{1}{2} C_i u_{Ci}^2 \tag{9.6}$$

For physical variables above, the inductor current i_{Li} and capacitor voltage u_{Ci} are system state variables, which are usually used as closed-loop control objectives. The input voltage u_{in}, input current i_{in} or the input power p_{in} are control variables altered through switching actions. The closed-loop control needs calculate next-period control variables based on present state variables and control targets then further derive next-period switching states. Assume the transient energy balancing control is to reach and stay at the steady state within one control period T_s. The control equation is derived as

$$\begin{aligned} p_{in} = {} & \frac{E^*_{R1}}{T_s} + \cdots + \frac{E^*_{Rk}}{T_s} \\ & + \frac{W^*_{L1} - W_{L1}}{T_s} + \cdots + \frac{W^*_{Lm} - W_{Lm}}{T_s} \\ & + \frac{W^*_{C1} - W_{C1}}{T_s} + \cdots + \frac{W^*_{Cn} - W_{Cn}}{T_s} \\ & + p_{out} \end{aligned} \tag{9.7}$$

Right-side items without stars represent actual values, which can be sampled through transducers. Variables with stars are reference values (steady states), part of which are known control targets of the converter while others need be calculated based upon the balanced steady-state energy of the converter.

When reaching the steady state, energy storage components in the power electronics converter see no energy change any more. Within one control period, the input energy, output energy and the dissipated energy on resistors are well balanced, i.e.

$$p^*_{\text{in}} = \frac{E^*_{R1}}{T_{\text{s}}} + \cdots + \frac{E^*_{Rk}}{T_{\text{s}}} + p_{\text{out}} \qquad (9.8)$$

Based upon such an energy balancing equation, those unknown target values could be further calculated, which are employed to calculate next-period input variables of the converter. Switching signals then can be generated.

9.2 Transient Energy Balancing Based Control Strategy

Power electronic converters need balance the energy at different time scales. For the time scale, i.e., time constant, far less than a control period, the energy balancing relies on the hardware. For any transients with the time constant larger than a control period, the energy balancing depends on the software control.

9.2.1 Conventional Voltage Control Strategies

Take the three-level voltage-source PWM rectifier as an example with the main circuit topology shown in Fig. 9.2. The main circuit contains two sets of energy storage components, one of which is three grid-side inductors (EI_1), the other of which is two DC-bus capacitors (EI_2). The whole rectifier has two control objectives, i.e., DC-bus voltage and grid-side current/power. For a dual-closed-loop control structure, the outer loop controls the DC-bus voltage and calculates the input value of the inner loop. The inner loop is the current or power loop, controlling the grid-side current or the rectifier input power and generating switch gate signals through the bottom-level PWM control or the switch table.

For various conventional three-level PWM control strategies, the inner loop control is quite diverse with different pros and cons. However, the outer loop is very similar, i.e., employing the DC-bus reference voltage and actual feedback to provide the active power or current through a PI controller. From the analysis of the transient power flow in Sect. 9.1, it can be seen that the active power or current reference calculated by such a voltage-control loop only considers the energy stored in DC-bus capacitors (EI_2). The loss caused by the grid-inductor ESR, the DC-bus output power, and energy stored in grid-side inductors have little influence on the active power and current. Therefore the conventional voltage control strategy has the error of the active power and current control.

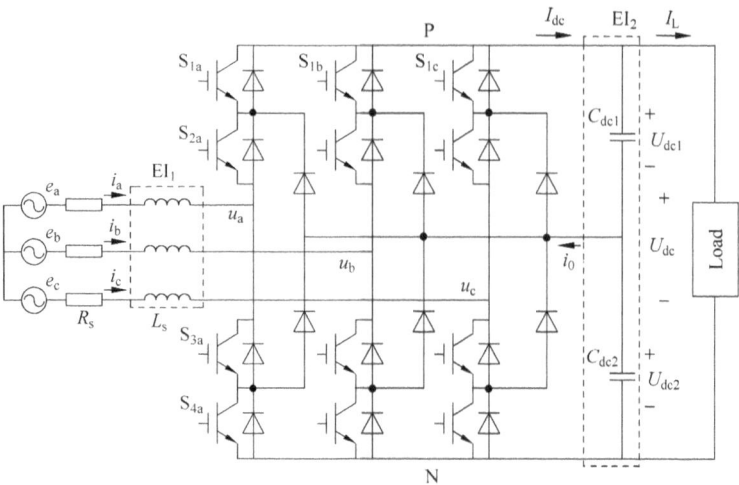

Fig. 9.2 Three-level voltage-source PWM rectifier

To eliminate such error brought by the DC-bus voltage control in the steady state, a PI controller is introduced for the voltage control,

$$\begin{cases} u(k) = u(k-1) + k_i[U_{dc}^*(k) - U_{dc}(k)] \\ p_{in}(k+1) = k_p[U_{dc}^*(k) - U_{dc}(k)] + u(k) \end{cases} \tag{9.9}$$

Such strategy counts on the integral to erase the steady-state error. However, the memory effect of the integral will create the over modulation during the dynamics. In the process of increasing the DC-bus voltage, the actual DC-bus voltage is different from the reference. The error between the actual and the reference is continually accumulated in the integral part, which gradually increases the active power and current. When the input power exceeds the output power and the equivalent-resistor loss, i.e., $p_{in} > p_{Rs} + p_{out}$, the energy balancing equation of Eq. (9.1) indicates that

$$E_{in} - E_{Rs} - E_{out} = \Delta W_{EI1} + \Delta W_{EI2} > 0 \tag{9.10}$$

Excessive input energy will compensate the difference between the actual and reference energy inside EI_1 and EI_2, which amplifies related energy characteristics, i.e., DC-bus voltage and the grid-side current. When the energy stored in EI_1 and EI_2 reaches the steady state, the input power of an ideal rectifier should immediately drop to $p_{in} = p_{Rs}^* + p_{out}^*$ to keep the three-level PWM rectifier at the targeted steady state ($\Delta W_{EI1} = \Delta W_{EI2} = 0$). This further keeps the DC-bus voltage around the reference value. However, due to the existence of the integral part of the PI controller, the input power and current reference only gradually reduce. So does the actual input active power. Before the actual active power drops to the target energy balancing state, i.e.,

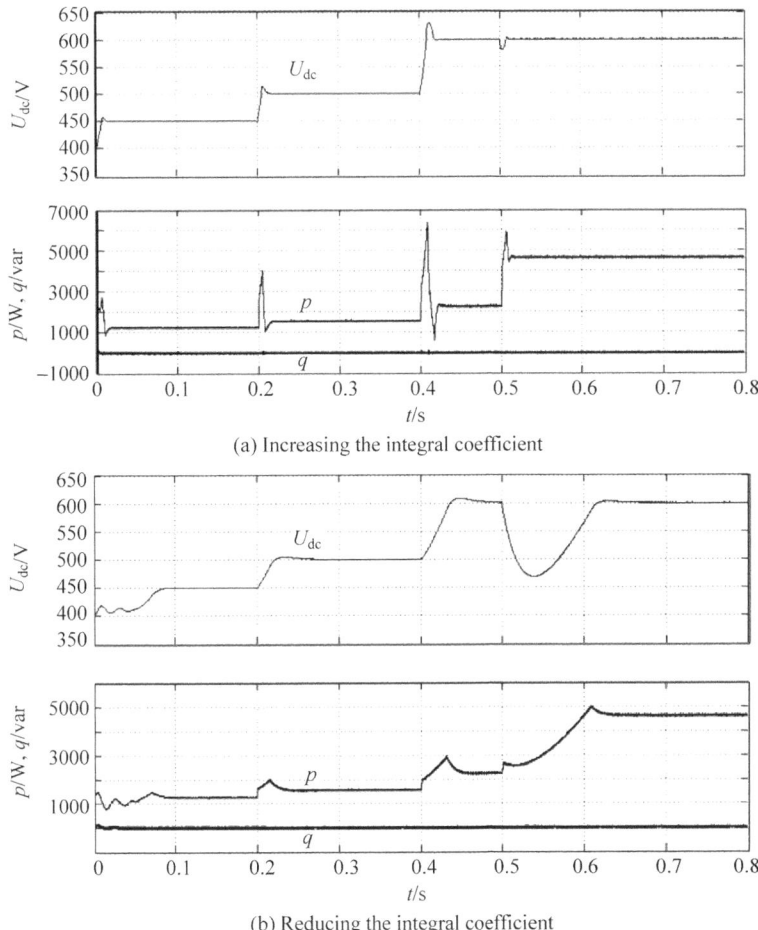

(a) Increasing the integral coefficient

(b) Reducing the integral coefficient

Fig. 9.3   Simulation waveform of the conventional voltage control strategy

$p_{Rs}^{*} + p_{out}^{*}$, the excessive input energy continually increases the energy stored in EI_1 and EI_2, which creates the over modulation of the DC-bus voltage.

Enlarging the integral coefficient k_i of the PI controller in Eq. (9.9) can expedite the dynamic process of the active power thereby dynamic response of the DC-bus voltage, but will at the same time worsen the memory effect of the controller thereby increasing the over modulation of the DC-bus voltage. On the contrary, reducing k_i alleviates the DC-bus over modulation but slows down the dynamic response. Furthermore, the DC-bus voltage drop caused by a load sudden change becomes more obvious, as simulated in Fig. 9.3. Therefore the conventional voltage control in the three-level PWM rectifier brings the inevitable conflict between the dynamic response and the over modulation.

9.2.2   Control Strategy Based on the Transient Energy Balancing

For a three-level PWM rectifier, based on Eq. (9.7), the transient energy balancing equation is derived as

$$p_{in} = p_{Rs}^* + \frac{W^*_{EI1} - W_{EI1} + W^*_{EI2} - W_{EI2}}{T_s} + p_{out} \tag{9.11}$$

Here the instantaneous energy stored in grid-side inductors (W_{EI1}) and DC-bus capacitors (W_{EI2}) can be calculated based on the sampling values as

$$
\begin{aligned}
W_{EI1} &= \frac{1}{2}L_s i_a^2 + \frac{1}{2}L_s i_b^2 + \frac{1}{2}L_s i_c^2 \\
&= \frac{1}{2}L_s i_\alpha^2 + \frac{1}{2}L_s i_\beta^2 \tag{9.12}
\end{aligned}
$$

$$W_{EI2} = \frac{1}{2}C_{dc1}U_{dc1}^2 + \frac{1}{2}C_{dc2}U_{dc2}^2 \tag{9.13}$$

Assume the upper and lower DC-bus voltage is identical. The reference energy stored in the DC-bus capacitor W^*_{EI2} can be calculated based on the reference DC-bus voltage as

$$
\begin{aligned}
W^*_{EI2} &= \frac{1}{2}C_{dc1}U_{dc1}^{*\,2} + \frac{1}{2}C_{dc2}U_{dc2}^{*\,2} \\
&= \frac{1}{8}(C_{dc1} + C_{dc2})U_{dc}^{*\,2} \tag{9.14}
\end{aligned}
$$

The output power can be calculated by samplings within two consecutive control periods, i.e.,

$$p_{out} = p_{in} - p_{Rs} - \frac{\Delta W_{EI1} + \Delta W_{EI2}}{T_s} \tag{9.15}$$

The equivalent resistive loss p_{Rs}^* and the reference energy stored in inductors W^*_{EI1} in the steady state can be calculated based on the ultimate target values using Eq. (9.8), i.e.,

$$
\begin{aligned}
p_{Rs}^* &= R_s(i_\alpha^{*2} + i_\beta^{*2}) \\
&= R_s\left(\frac{\sqrt{3}E - \sqrt{3E^2 - 4R_s p_{out}}}{2R_s}\right)^2 \tag{9.16}
\end{aligned}
$$

$$W_{E11}^* = \frac{1}{2}L_s i_\alpha^{*2} + \frac{1}{2}L_s i_\beta^{*2}$$

$$= \frac{1}{2}L_s \left(\frac{\sqrt{3}E - \sqrt{3E^2 - 4R_s p_{out}}}{2R_s} \right)^2 \qquad (9.17)$$

α-β coordinates are utilized to calculate variables. E is the grid-voltage RMS value. Based on the calculation procedure above for the active power control, a transient-power balancing based control diagram for the three-level PWM rectifier is shown in Fig. 9.4.

Figure 9.5 provides a head-to-head simulation comparison of the active and reactive power between the conventional voltage control with optimized PI parameters and the transient-energy balancing based control, when the DC-bus voltage command is changed instantaneously. Compared to the conventional control strategy, the transient-energy balancing based control strategy adjusts the power more rapidly and accurately. The DC-bus voltage has much faster dynamic response with nearly zero over modulation.

Figures 9.6 and 9.7 are the simulation and experimental comparison between the conventional voltage control and the transient-energy balancing based control, respectively, when the load power is changed abruptly. With the transient-energy balancing based control, the control of the active power is more rapid with the DC-bus voltage drop being much smaller and recovering faster. From experimental results, the transient-energy balancing based control strategy has 1/3 of the voltage drop as before, effectively stabilizing the DC-bus voltage.

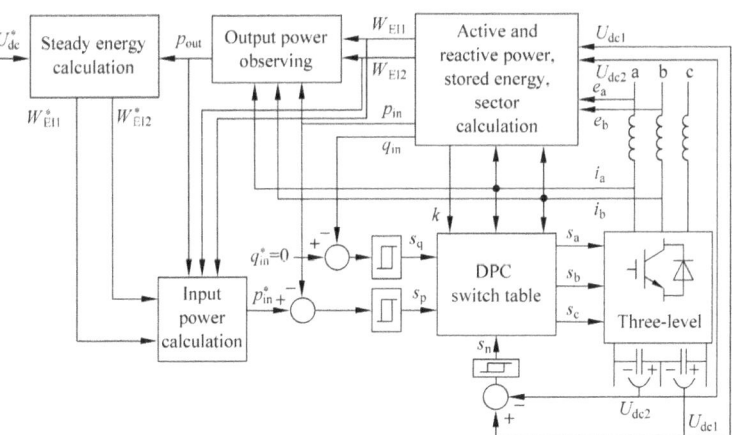

Fig. 9.4 The control diagram of the transient energy balancing based control for the three-level PWM rectifier

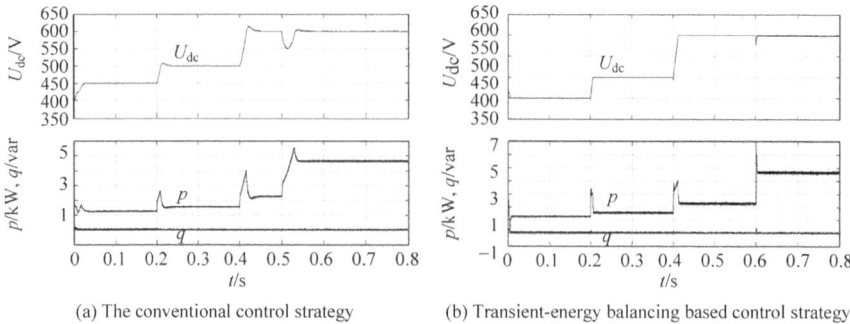

(a) The conventional control strategy   (b) Transient-energy balancing based control strategy

Fig. 9.5 Simulation comparison when the DC-bus voltage command is changed instantaneously for a three-level PWM rectifier

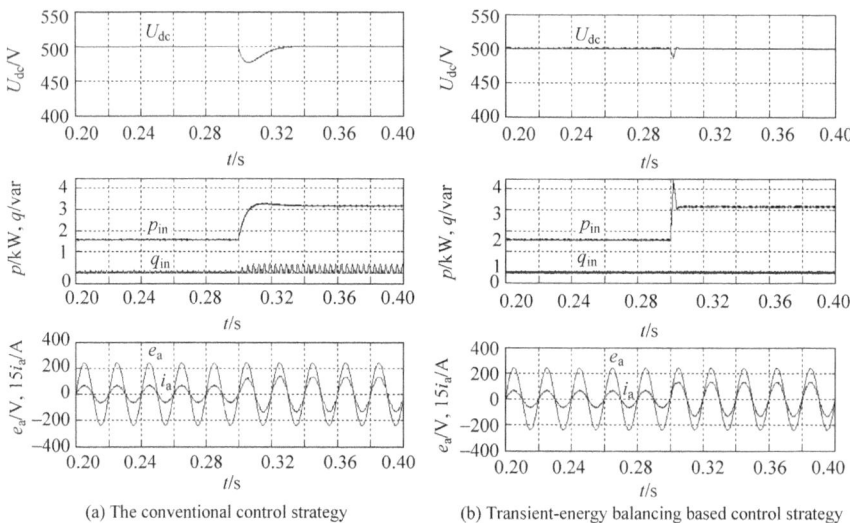

(a) The conventional control strategy   (b) Transient-energy balancing based control strategy

Fig. 9.6 Simulation comparison when the load is suddenly changed for a three-level PWM rectifier

9.3   Energy Balancing Control for a Back-to-Back Converter

The proposed transient-energy balancing based control can be extended to much complex converters, such as the back-to-back converter. Voltage-source three-phase dual inverters combine the PWM rectifier and inverter together to realize the bidirectional energy flow. Due to its symmetry, it is often called as back-to-back converter. Compared to the conventional AC-DC-AC converter, i.e., diode rectifier + PWM inverter, it has following merits.

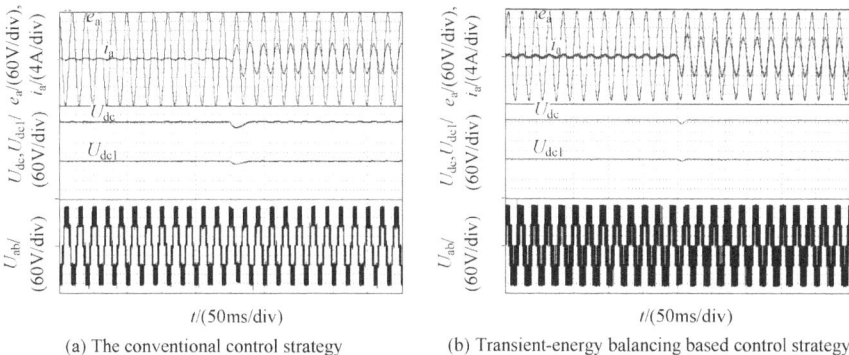

(a) The conventional control strategy   (b) Transient-energy balancing based control strategy

Fig. 9.7 Experimental comparison when the load is suddenly changed for a three-level PWM rectifier

(1) Controllable grid-side power factor

The PWM rectifier decouples the active and reactive power of the grid thereby realizing the accurate control of the power factor. To reduce the local reactive power and furthermore the grid transmission loss, the grid-side power factor is usually controlled as unity.

(2) Low grid current harmonics

The current control techniques of the PWM rectifier trim the grid current close to sinusoidal, which greatly reduces the low-order harmonics of the input current thereby alleviating the pollution to the grid.

(3) Bidirectional energy flow with high efficiency

The inverter side works at the four-quadrant mode, which can both generate and regenerate the energy.

(4) Controllable DC-bus voltage and smaller capacitance

The grid-side PWM rectifier resembles the DC-DC boost converter, making the DC-bus voltage higher than the peak value of the grid line-line voltage. A higher DC-bus voltage expedites the dynamic response of the load, adapts the system for a higher-voltage load thereby saving the cost and loss of an extra step-up transformer, and in addition reduces the DC-bus capacitance with the coordination of the rectifier and inverter, which stabilizes the DC-bus voltage and enhances the system reliability.

However, the increment of the system complexity brings the more severe impact of non-linear factors, such as the switch time delay and PWM distortion in the main circuit, yielding the conventional control vulnerable especially when controlling multiple objectives. Such conflict can be resolved by using the energy-balancing control.

For a dual-PWM-inverter variable speed system, the energy balancing processes can be categorized into two types based on the time scale versus the control period, i.e., (1) energy balancing within one control period, and (2) energy balancing during the system dynamics, which takes multiple control periods.

The first category of the energy balancing has the switch duty cycle follows the control command, which controls the input and output power. Based on the energy conservation, the input energy of the grid within one control period is the sum of the output energy and the increment of the energy stored inside energy storage components, such as filtering inductors, DC-bus capacitors, electromagnetic field inside the motor and motor mechanical rotatory system. Such rule allows us to predict the energy distribution in the next switching period thereby selecting the right command.

The second type of energy balancing processes is when large-time-constant energy variables reach the steady state. In this process all energy storage components gradually approach to the target and then keep constant. The relocation speed of the energy determines the system dynamic performance while holding the energy inside each component as constant reveals the system steady-state characteristics.

9.3.1 Energy-Balancing Model of Dual PWM Inverters

The main-circuit topology of dual PWM inverters is shown as Fig. 9.8. To facilitate the analysis, the grid is assumed to be Y type. L_g in the figure is the grid-side filtering inductor, R_g is the ESR of the inductor, C_{dc} is the DC-bus capacitor, R_{dc} is the discharging resistor of the DC-bus voltage, M is the load of the induction motor, $e_a \sim e_c$ are the grid phase voltage, $i_{ra} \sim i_{rc}$ are the grid phase current, $u_{ra} \sim u_{rc}$ are the equivalent phase voltage of the rectifier, u_{dc} is the DC-bus voltage, i_{rdc} is the current flowing from the rectifier towards the inverter DC-bus, i_{Cdc} is the current flowing into the DC-bus capacitor, i_{Rdc} is the current of the discharging resistor, i_{idc} is the current flowing from the DC-bus capacitor to the inverter, $u_{ia} \sim u_{ic}$ are the inverter equivalent phase voltage, i.e., the motor phase voltage, and $i_{ia} \sim i_{ic}$ are the inverter phase current, i.e., the motor phase current. Positive directions of the voltage and current are defined in Fig. 9.8.

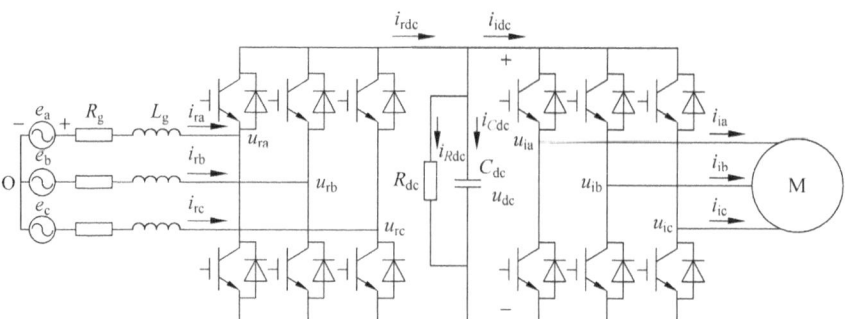

Fig. 9.8 The main circuit of dual PWM inverters

We need first study the energy variation within some specific periods. Based on the roles of components playing in the process of the energy transmission, all components can be categorized into three types, i.e., energy storage components (including L_g and C_{dc}), energy consuming components (including R_g, R_{dc} and the induction motor) and energy conversion components (including the three-phase rectifier and inverter).

With positive directions of all variables defined in Fig. 9.8, a positive energy flow happens when the grid provides energy to the mechanical load, while a negative energy flow happens when the mechanical load regenerates the energy back to the grid. For simplicity, we define the positive energy flow of dual PWM inverters as "rectifying mode" and the negative energy flow as "inverting mode". Based on the transient energy-balancing model proposed in Sect. 9.2, we can illustrate the energy balancing relationship within a time period of T, as shown in Fig. 9.9. Here the energy balancing equation is

$$E_g = E_R + \Delta E_{Lg} + \Delta E_{Cdc} + E_{inv} \qquad (9.18)$$

E_g is the grid input energy, E_R is the energy consumed by the inductor ESR, DC-discharging resistor and other related lossy components, ΔE_{Lg} is the energy increment of the inductor, ΔE_{Cdc} is the energy increment of the DC-bus capacitor and E_{inv} is the inverter output energy to the motor, including the electromagnetic-field energy, electromagnetic loss, mechanical loss, rotatory energy and load energy. Since the energy fluctuation of the system is caused by the imbalance between the system input and output active power, in this chapter synchronous rotatory coordinates (d-q coordinates) are used to mathematically formulate all energy in Eq. (9.18) as below.

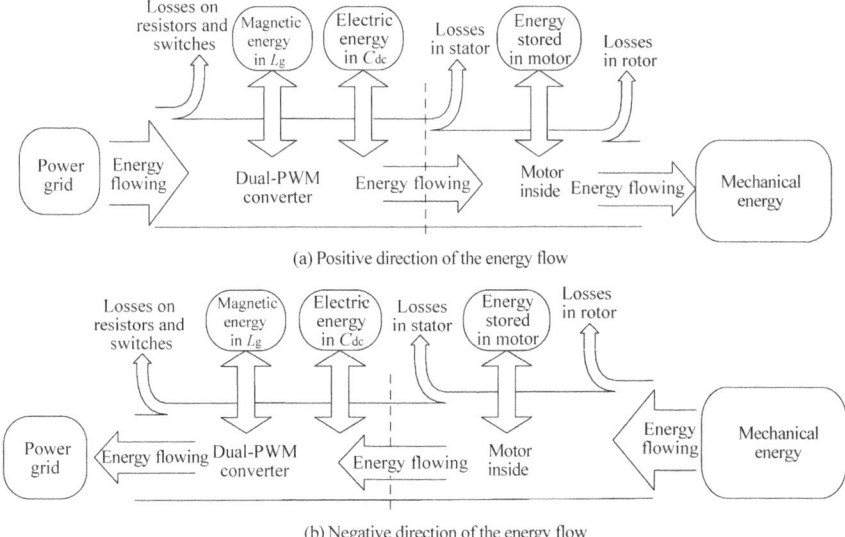

(a) Positive direction of the energy flow

(b) Negative direction of the energy flow

Fig. 9.9 The energy flow of all components inside dual-PWM inverters

$$E_g = \int_t^{t+T} (P_g)dt = \frac{3}{2} \int_t^{t+T} (e_d i_{rd})dt \tag{9.19}$$

$$E_R = \int_t^{t+T} (P_R)dt = \frac{3}{2} \int_t^{t+T} \left[\frac{3}{2} R_g \left(i_{rd}^2 + i_{rq}^2 \right) + \frac{u_{dc}^2}{R_{dc}} \right] dt \tag{9.20}$$

$$\Delta E_{Lg} = \frac{3}{4} L_g \left(i_{rd}(t+T)^2 - i_{rd}(t)^2 + i_{rq}(t+T)^2 - i_{rq}(t)^2 \right) \tag{9.21}$$

$$\Delta E_{Cdc} = \frac{1}{2} C_{dc} \left(u_{dc}(t+T)^2 - u_{dc}(t)^2 \right) \tag{9.22}$$

$$E_{inv} = \int_t^{t+T} (P_{inv})dt = \frac{3}{2} \int_t^{t+T} \left(u_{iD} i_{iD} + u_{iQ} i_{iQ} \right) dt \tag{9.23}$$

In the real practice, the unity power factor of the grid usually is required, i.e., $i_{rq} = 0$. In addition, the power loss of the DC-bus discharging resistor remains nearly unchanged with the input power. Therefore we can merge this related loss into the input power P_{inv}, regarding $u_{dc}^2/R_{dc} \approx 0$ while only counting this part of the loss in when calculating the inverter output power. Meanwhile to reduce the system loss as much as possible, the DC-discharging resistance is usually large, resulting in a minor discharging power. For 55 kW/380 V dual-PWM inverters, the loss of this DC-discharging resistor is only 52.1 W. Therefore, even if this loss calculation has some error, the estimation error of the output power is negligible.

9.3.2 Analysis of the DC-Bus-Capacitor Energy Oscillation in Dual PWM Inverters

Due to the existence of grid-side inductors the grid current can't change suddenly. Under the normal condition the amplitude and frequency of the grid voltage are constant, yielding no sudden change of the grid input power. Meanwhile due to the motor leakage inductance, the output current of dual PWM inverters cannot change abruptly either. The inverter output voltage, however, is controlled by switches, i.e., the system output voltage can be changed with the command. Therefore the system output power, i.e., the input power of the induction motor can be changed instantaneously.

The energy of the DC-bus capacitor will change as below in the dynamic process. Consider the fastest change of the output power, which step jumps from P_0 to P_1 at $t = t_0$. In this case, the system input power, output power and the DC-bus-capacitor energy are illustrated in Fig. 9.10. Here E_{CdcN} is the overall energy stored in the DC-bus capacitor in the steady state and P_{Rg} is the power loss of AC equivalent resistors.

Assume the system reaches the steady state before the power jump. At $t = t_0$ the transient energy balancing equation of the inverter is

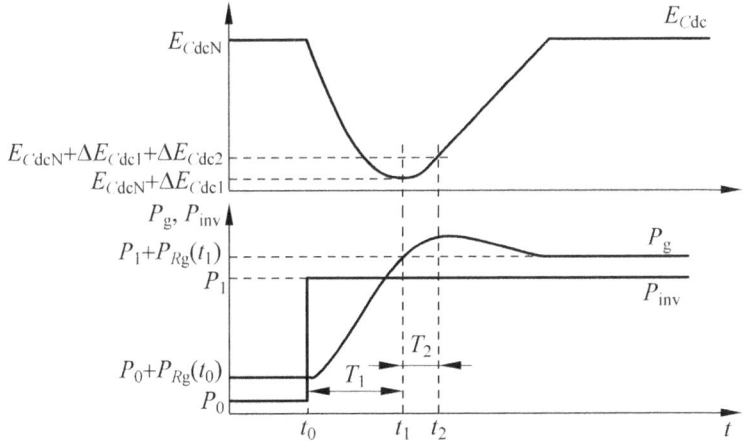

Fig. 9.10 The step response to the output power jump

$$\frac{3}{2}e_d i_{rd}(t_0) = \frac{3}{2}R_g i_{rd}(t_0)^2 + P_0 \tag{9.24}$$

Assume $P_1 > P_0 > 0$. Right after the step jump of the output power command, with the constant input power, the DC-bus voltage will slide given the input power is less than the sum of the output power and the loss. Therefore, to maintain a constant DC-bus voltage, the increment of the input power is must. Assume at $t = t_1$ the input power is equal to the sum of the output power and the loss, i.e.,

$$\frac{3}{2}e_d i_{rd}(t_1) = \frac{3}{2}R_g i_{rd}(t_1)^2 + P_1 \tag{9.25}$$

Based on the balancing of the transient energy, the energy stored in filtering inductors does not change any more given the grid AC current does not change. Neither does the energy in the DC-bus capacitor given the DC-bus voltage keeps constant after t_1. Between t_0 and t_1, the difference between the input energy and the consumed energy along with the energy increment in inductors is compensated by the DC-bus capacitor. The accumulative output energy of the DC-bus capacitor reaches the maximum at $t = t_1$, which can be expressed as

$$\Delta E_{Cdc1} = \int_{t_0}^{t_1} [P_g - P_1] dt - \frac{3}{4}L_g\left(i_{rd}(t_1)^2 - i_{rd}(t_0)^2\right) - \int_{t_0}^{t_1} \frac{3}{2}R_g i_{rd}(t)^2 dt \tag{9.26}$$

Here $i_{rd}(t)$ is the grid-side d-axis current, which is a function of time. Given the ESR of the filtering inductor R_g is very small, and the input-current variation can be done within a very short period to reduce the DC-bus voltage oscillation, we can assume the grid current linearly increases within $[t_0, t_1]$, i.e.,

$$i_{rd}(t) = \frac{i_{rd}(t_1) - i_{rd}(t_0)}{t_1 - t_0}(t - t_0) + i_{rd}(t_0), \ t \in [t_0, t_1] \tag{9.27}$$

Furthermore, Eq. (9.26) is simplified as

$$\Delta E_{Cdc1} = -\frac{3}{4}(i_{rd}(t_1) - i_{rd}(t_0))\big[e_d T_1 + L_g(i_{rd}(t_1) + i_{rd}(t_0))\big] \tag{9.28}$$

Here $T_1 = t_1 - t_0$, which is determined by the slew rate of the grid-current increment.

With very small R_g, the input power of the system in the steady state is approximately proportional to the output power, i.e., when $P_1 > P_0 > 0$, $i_{rd}(t_1) > i_{rd}(t_0) > 0$. Given positive values of the grid-voltage amplitude e_d and inductance L_g, for any time interval $T_1 > 0$, Eq. (9.28) is negative. When $P_0 > P_1 > 0$, a similar analysis yields $i_{rd}(t_0) > i_{rd}(t_1) > 0$, i.e., for any time interval $T_1 > 0$, Eq. (9.28) is positive. This means when the system energy flows from the grid to the motor, i.e., when dual PWM inverters are working at the rectifying mode, the instantaneous energy oscillation of the DC-bus capacitor cannot be controlled to zero. That being said, the oscillation amplitude can be reduced when T_1 shrinks.

On the other hand, when $P_0 < P_1 < 0$, we have $i_{rd}(t_0) < i_{rd}(t_1) < 0$. Similarly $P_1 < P_0 < 0$ results in $i_{rd}(t_1) < i_{rd}(t_0) < 0$. In this case the item of $e_d T_1$ in Eq. (9.28) has a different sign from $L_g[i_{rd}(t_0) + i_{rd}(t_1)]$, which allows Eq. (9.28) to be zero through adjusting T_1. Therefore, when the energy is flowing from the motor to the grid, i.e., dual PWM inverters work at the "inverting mode", theoretically the DC-bus voltage can be maintained constant through controlling the input-current changing rate.

At the rectifying mode, even though at $t = t_1$ the input power matches the consumed power, the DC-bus voltage has not reached the reference value yet, which requires the grid to compensate the energy gap of the DC-bus capacitor. For $P_1 > P_0 > 0$, given the energy increment of the DC-bus capacitor within $[t_0, t_1]$ is negative, the input current needs further increase to recover the DC-bus voltage. Assume the input current linearly increases for a time interval T_2 before $t = t_2$, i.e., $T_2 = t_2 - t_1$. With small R_g we can approximate the energy increment of the DC-bus capacitor during this time interval as

$$\Delta E_{Cdc2} = \int_{t_1}^{t_2} [P_g - P_1]dt - \frac{3}{4}L_g\big(i_{rd}(t_2)^2 - i_{rd}(t_1)^2\big) - \int_{t_1}^{t_2} \frac{3}{2}R_g i_{rd}(t)^2 dt$$

$$= \frac{3}{4}(i_{rd}(t_2) - i_{rd}(t_1))\big[e_d T_2 - L_g(i_{rd}(t_2) + i_{rd}(t_1))\big] \tag{9.29}$$

A rapid increment of the grid current during T_2 can lead to $e_d T_2 < L_g[i_{rd}(t_2) + i_{rd}(t_1)]$, resulting in Eq. (9.29) less than zero. Thus the DC-bus capacitor does not increase but decrease the energy stored. Such a positive feedback within this time interval will create a large oscillation of the DC-bus voltage and even system instability. When $P_0 > P_1 > 0$, similarly the DC-bus voltage can be further increased.

From Eqs. (9.28) to (9.29), the filtering inductor vary the stored energy when the grid current changes, which is related to the square of the input current. When the energy flows from the grid to the load, the input energy needs first fulfill the energy storage in inductors then the DC-bus capacitor. With too rapid change of the grid current, the inductors will absorb too much energy larger than the grid capability, choke part of the DC-capacitor energy, and reduce the DC-bus voltage. Similarly when the input current drops, without feeding energy back to the grid in time, the grid inductors will release part of the energy to the DC-bus capacitor, yielding an opposite voltage changing direction to the anticipation.

To prevent the DC-bus voltage oscillation from being further amplified after $t = t_1$, a limitation on the grid-current changing rate needs be imposed for the DC-bus voltage compensation. Based on the analysis above, such a limitation on the current changing rate within T_2 is

$$e_d T_2 - L_g(i_{rd}(t_2) + i_{rd}(t_1)) > 0 \qquad (9.30)$$

After $t = t_1$, as long as the grid-current changing rate within any T_2 meets Eq. (9.30), the energy stored in the DC-bus capacitor will keep the same trend as the grid current, i.e., the maximum grid-energy oscillation happens at $t = t_1$ with the oscillation amplitude of ΔE_{Cdc1}. It can be seen that not one but multiple switching periods are needed to balance all the energy.

9.3.3 Energy Balancing Strategy Based on the Stepwise Compensation

The conventional load-current feed-forward control is shown as Fig. 9.11, where the rectifier side adopts the grid voltage oriented vector control and the inverter side employs the rotor field oriented vector control. All controllers are PI or PID based. The system output power is fed forward to the active-current command to reduce the computation load of the DC-bus-voltage controller.

Different from conventional feed-forward control methods, the stepwise compensation based energy balancing control is shown in Fig. 9.12. To secure the control performance of the induction motor drive, the control method at the inverter side is the same as other conventional ones, and the rectifier side still uses the grid-voltage oriented synchronous rotatory coordinates to decouple the active and reactive current. However, the outer voltage control employs the adjustable energy regulator (AER). Based on the energy balancing, the DC-bus capacitor energy can be regulated at two levels, one within one switching period and the other is within the whole system dynamic process. In addition, to accurately follow the current command generated by the outer loop, a dead-beat prediction control is used for the inner current loop, which secures the actual current of the next switching period to follow the command.

1. **Outer-loop energy control**

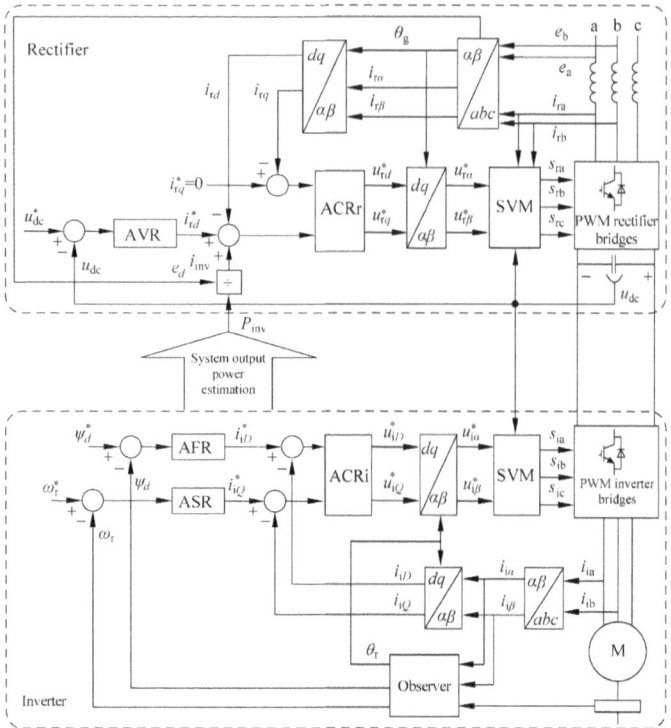

Fig. 9.11 The conventional load-current feed-forward control

Based on Eq. (9.28), at the rectifying mode, a fastest pace of the input power follow-ing the output will minimize the DC-bus voltage oscillation. However, as shown in Eq. (9.29), compensating the DC-bus energy within an extremely short time interval will create the DC-bus voltage oscillation and even yields poor controllability. Given the inductor energy increment is proportional to the current changing rate and the current amplitude, such scenario becomes worse with the power increasing.

Therefore the whole system energy needs be divided into two. One is the energy consumed by the lossy component, including the motor input power and system loss. This part of the energy is proportional to time and needs be supplied on time. We need control the input power equal to the consumed power. Since the DC-bus loss has been included in the input power, the control equation can be written as

$$P_{g1}^* = \frac{3}{2} R_g i_{rd}^2 + P_{inv} \qquad (9.31)$$

The other is energy stored in storage components, mainly filtering inductors and the DC-bus capacitor. The related energy demand is only determined by the initial and final states of the system. At the same time, the rectifier input voltage is employed

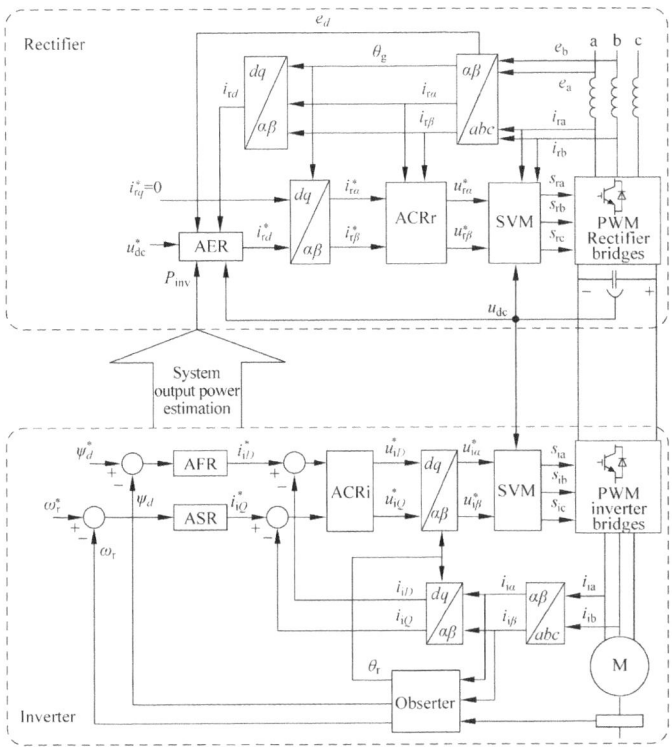

Fig. 9.12 Stepwise compensation based energy balancing control method

to control the input current rate, i.e., the input power. Therefore, within transients the energy distribution among inductors and capacitors is determined by the physical model, instead of any random values. For the sake of the system stability this part of the energy only meets the steady-state requirements, not considering transients. The system input energy within some time interval is equal to this part energy, with the control equation as

$$E_{g2}^* = \frac{3}{4}L_g\left(i_{2rd}^{*} - i_{rd}^2\right) + \frac{1}{2}C_{dc}\left(u_{dc}^{*2} - u_{dc}^2\right) \tag{9.32}$$

Since the grid voltage is constant, the active-current setting can reflect the energy control objective. Hence the setting of the grid d-axis current can be divided into two, i.e.,

$$i_{rd}^* = i_{rd1}^* + i_{rd2}^* \tag{9.33}$$

Here i_{rd1}^* is used to compensate the first part energy and i_{rd2}^* is used to compensate the second part energy in the whole dynamic process.

The first part energy compensates the consumed energy, which is related to the current command as below.

$$\frac{3}{2} e_d i_{rd1}^* = \frac{3}{2} R_g i_{rd1}^{*2} + P_{inv} \tag{9.34}$$

Here in order to enhance the system dynamic response, we adopt the ultimate current command to calculate the loss of equivalent resistors. Based on the previous analysis, P_{inv} contains the loss of the DC-bus discharging resistor, i.e., u_{dc}^2/R_{dc}. Two roots exist for the equation above.

$$\begin{cases} i_{rd11}^* = \frac{e_d}{2R_g} + \sqrt{\left(\frac{e_d}{2R_g}\right)^2 - 6R_g P_{inv}} \\ i_{rd12}^* = \frac{e_d}{2R_g} - \sqrt{\left(\frac{e_d}{2R_g}\right)^2 - 6R_g P_{inv}} \end{cases} \tag{9.35}$$

Since R_g is close to zero, these two roots are both real, making i_{rd11}^* close to infinity. This means most of the input power is consumed by the ESR of filtering inductors, which is certainly unrealistic. The other root i_{rd12}^* is close to $2\,P_{inv}/(3e_d)$, a reasonable real root. The transient-energy balancing equation in the steady state is shown as Eq. (9.34), which means i_{rd1}^* is the grid input current in the steady state.

The second part energy is expected to be compensated after n control periods, during which the energy compensation is shown as below.

$$\frac{3}{2} e_d n T_s i_{rd2}^* = \frac{3}{4} L_g \left(i_{rd1}^{*2} - i_{rd}^2\right) + \frac{1}{2} C_{dc} \left(u_{dc}^{*2} - u_{dc}^2\right) \tag{9.36}$$

Here T_s is the system control period.

Based on Eqs. (9.33)–(9.36), the energy regulator for the outer loop is obtained. Since R_g is very small, such control equation is close to be

$$i_{rd}^* = \frac{R_g}{e_d} \left(\frac{2 P_{inv}}{3e_d}\right)^2 + \frac{2 P_{inv}}{3e_d} + \frac{2}{3e_d n T_s} \left\{\frac{3}{4} L_g \left[\left(\frac{2 P_{inv}}{3e_d}\right)^2 - i_{rd}^2\right] + \frac{1}{2} C_{dc} \left(u_{dc}^{*2} - u_{dc}^2\right)\right\} \tag{9.37}$$

Such control method assumes the DC-bus voltage is stabilized within n control periods after the step power.

It is worthwhile pointing out that the control equation above is derived at the system rectifying mode. Based on Eq. (9.11), ideally the current tracking time $T_1 = 0$. With the same step input power, the DC-bus voltage variation should be the same for both the rectifying and inverting modes. Though theoretically other algorithms could zero out the DC-bus oscillation at the inverting mode, dual PWM inverter + induction motor mostly work at the rectifying mode. Plus the DC-bus capacitance is determined by the maximum energy oscillation. Therefore only reducing the DC-bus voltage oscillation at the inverting mode does not help much for the component selection. At the inverting mode ($P_0 < 0$, $P_1 < 0$) or when the system is switching between inverting and rectifying modes (P_0 and P_1 have different signs), we can still apply the same control equation.

Besides, the output voltage contains one control-period delay. To closely follow the output power, we need predict the output power of next period and substitute it into the energy regulator for the potential compensation. At the same time the constant loss in the system needs be compensated to reduce the prediction error. The final expression is

$$P_{inv} = \frac{3}{2}\left(u_{iD}^* \hat{i}_{iD}(k+1) + u_{iQ}^* \hat{i}_{iQ}(k+1)\right) + \frac{u_{dcN}^2}{R_{dc}} \tag{9.38}$$

Here u_{iD}^* and u_{iQ}^* are the inverter voltage commands calculated at the present period. Due to the reloading mechanism of digital signal processors, these two values will be implemented to the inverter in the next control period. Here $\hat{i}_{iD}(k+1)$ and $\hat{i}_{iQ}(k+1)$ are the estimated stator current in the next control period, which can be predicted using Euler method based upon the current-loop control bandwidth. u_{dcN} is the rated DC-bus voltage.

2. Deadbeat prediction control for the inner current loop

Because the outer energy loop calculates the current command based on the accurate system model, the inner current loop adopts the deadbeat control to accurately follow the current command. The deadbeat control is an accurate method based on the physical model, which is very good at tracking AC variables. Therefore there is no need for the Park transformation during the current control but only using the two-phase static coordinates, i.e., α-β coordinates. A linearized control equation of the current loop can be derived as below based on the mathematical model of the rectifier.

$$u_{rx}^* = e_x - \frac{L_g}{T_s}\left(i_{rx}^* - i_{rx}\right) - R_g i_{rx} \tag{9.39}$$

Here subscripts α and β represent Phases α and β, respectively. i_{rx}^* means the reference current. u_{rx}^* means the inverter voltage command.

Theoretically, the grid input current can reach the reference value within one control period. However, the digital signal system contains one-period delay, i.e., the voltage command calculated in this control period can only be implemented in the next control period. To accurately track the reference current, we need utilize the sampled current in the present period to predict the actual current in the next period, which can be expressed as

$$i_{rx}(k+1) = \frac{T_s}{L_g}\left(e_x - u_{rx}^*(k-1) - R_g i_{rx}\right) + i_{rx} \tag{9.40}$$

Here k means the present period. The grid voltage can be treated as constant in two adjacent control periods. The control-delay compensation based deadbeat current prediction is then further derived as below.

$$u_{rx}^* = e_x - \frac{L_g}{T_s}\left(i_{rx}^* - i_{rx}(k+1)\right) - R_g i_{rx}(k+1) \tag{9.41}$$

With such control theoretically the grid current will reach the reference value in the next control period.

In addition the outer energy loop calculates the current references i_{rd}^* and i_{rq}^* under d-q coordinates based on the grid active and reactive power, respectively. To generate the unity power factor, $i_{rq}^* = 0$. Such references need be converted to $i_{r\alpha}^*$ and $i_{r\beta}^*$ under α-β coordinates through inverse Park transform, then substituted into Eqs. (9.40) and (9.41).

9.3.4 Minimization of the DC-Bus Voltage Oscillation Based on the Energy Balancing Control

In the stepwise compensation based energy balancing control, the control objective is divided into two. i_{rd1}^* is used to compensate the real-time system loss and i_{rq}^* is used to compensate the energy in storage components within n periods. Here n determines the control performance of the DC-bus voltage. A reasonable selection of n can minimize the DC-bus voltage oscillation while securing the system stability. We need first calculate the theoretical minimal oscillation of the DC-bus voltage. Secondly we can select the appropriate n to let the maximum oscillation of the DC-bus voltage at specific control algorithm approach the theoretical minimum. Lastly, we need study the difference between the maximum and minimum of the DC-bus voltage oscillation.

1. **The theoretical minimum oscillation of the DC-bus voltage**

Based on the previous analysis, when the power steps from P_0 to P_1, without considering the system loss the theoretical minimum DC-bus energy oscillation occurs when the input power tracks the output power at the fastest pace. The energy is the function of the output power, i.e.,

$$\Delta E_{Cdcmin} = \left| \frac{(P_1 - P_0)T_{min}}{2} + \frac{L_g\left(P_1^2 - P_0^2\right)}{3e_d^2} \right| \tag{9.42}$$

Here T_{min} is the minimum time for the current variation, which in the digital control system has the minimum value as one control period T_s. T_{min} is also restrained by system parameters, which can be expressed as

$$T_{\min} = \frac{2L_g|P_1 - P_0|}{3e_d|e_d - u_{rd}|} \tag{9.43}$$

Since the rectifier input range is between positive and negative peaks of the grid voltage while the d-axis grid voltage is always a constant positive, the grid-current decrement will take much longer time than increment.

In addition, the energy oscillation of the DC-bus capacitor ΔE_{Cdc} has the relationship with DC-bus voltage oscillation Δu_{dc} shown as below.

$$\Delta E_{Cdc} = \frac{1}{2}C_{dc}(u_{dc0} + \Delta u_{dc})^2 - \frac{1}{2}C_{dc}u_{dc0}^2 \tag{9.44}$$

Here u_{dc0} is the DC-bus initial voltage. With a step power, the theoretical minimum variation of the DC-bus voltage is

$$C_{dc}\left|u_{dcN}\Delta u_{dcmin} + \frac{1}{2}\Delta u_{dcmin}^2\right| = \left|\frac{(P_1 - P_0)T_{\min}}{2} + \frac{L_g\left(P_1^2 - P_0^2\right)}{3e_d^2}\right| \tag{9.45}$$

When the DC-bus voltage oscillation is not large, the second-order item can be neglected, yielding the theoretical minimum oscillation of the DC-bus voltage as

$$\Delta u_{dcmin} = \frac{1}{C_{dc}u_{dcN}}\left|\frac{(P_1 - P_0)T_{\min}}{2} + \frac{L_g\left(P_1^2 - P_0^2\right)}{3e_d^2}\right| \tag{9.46}$$

Equation (9.46) indicates the minimum DC-bus voltage oscillation is determined by two factors. One is the speed of the input power tracking the output power, the other is the energy variation of AC filtering inductors.

In the actual system the output power can continually change. Since the feedforward control variable in dual PWM inverters, i.e., the induction motor power is an average value calculated within one control period, the continually changing output power can be piecewise linearized as power steps within multiple control periods. The amplitude after each power step is the average power of the related period, which can be used to calculate the theoretical minimum value of the DC-bus voltage oscillation.

2.   The energy-loop design to minimize the DC-bus oscillation

With a sudden change of the output power, to minimize the DC-bus oscillation and secure the rapid regulation speed, an appropriate n is needed. Firstly, in order not to violate preconditions of previous control equations, the DC-bus voltage should keep the same changing trend as the input current after n periods. An inequality can be derived based on Eq. (9.30) as below.

$$n > \frac{L_g(i_{rd}(t_2) + i_{rd}(t_1))}{T_s e_d} \tag{9.47}$$

where $i_{rd}(t_1)$ is the grid steady-state current and $i_{rd}(t_2)$ is the maximum current needed to compensate the energy of the DC-bus capacitor and grid inductors. To calculate the minimum n, $i_{rd}(t_1)$ and $i_{rd}(t_2)$ can be selected as the grid ratings, i.e.,

$$n > \frac{2L_g I_{rN}}{T_s E_N} \tag{9.48}$$

Here I_{rN} is the grid-side rated current of dual-PWM inverters. E_N is the rated peak value of the grid phase voltage.

Secondly, when the output power is subject to an instantaneous change, after n control periods the DC-bus voltage should approach the setting. Assume the power step from P_0 to P_1 occurs at $t = t_0$, before which the system is in the steady state. The DC-bus voltage $u_{dc}(t_0) = u_{dc}{}^*$, with the steady-state power balancing equation shown as Eq. (9.24). After the system reaches the steady state again, the grid current is i_{rds}, same as $i_{rd}(t_0)$ shown in Eq. (9.25). Based on control rules in this chapter, the DC-bus voltage variation after n control periods when ignoring R_g is

$$\Delta E_{Cdc} = \frac{3}{2} e_d i_{rd}^* n T_s - \frac{3}{4} L_g \left(i_{rd}^{*2} - i_{rd}(t_0)^2 \right) - P_1 n T_s$$

$$= \frac{3}{4} L_g \left\{ i_{rds}^2 - \left[i_{rds} + \frac{L_g \left(i_{rds}^2 - i_{rd}(t_0)^2 \right)}{2 e_d n T_s} \right]^2 \right\} \tag{9.49}$$

With actual parameters substituted in the equation above, we can build the relationship between n and ΔE_{Cdc} then solve the value of n to minimize the DC-bus oscillation.

Two sets of dual PWM inverters are used to demonstrate how to select n, with system parameters shown in Table 9.1. Substituting parameters of the 2.2 kW system into Eq. (9.48) results in $n > 1.07$. When the input power steps from 0 to ± 2.2 kW, Δu_{dc} versus n is calculated by Eq. (9.49) and illustrated in Fig. 9.13. It can be seen the larger the value of n the closer the DC-bus voltage to the reference value. When n is small, the DC-bus voltage after n control periods is still distant from the reference value, which violates the presumption of control rules, yields a large DC-bus voltage oscillation and even makes the whole system unstable. On the other hand, a too large n will result in a long modulation time. Therefore n should be located at the turning point of the curve in Fig. 9.13, which in this example is 5. For 55 kW dual PWM inverters, we select $n > 9.75$ based on Eq. (9.48). When the power steps from 0 to ± 55 kW, Δu_{dc} versus n is calculated by Eq. (9.49) and illustrated in Fig. 9.14. Here $n = 10$.

3. Analysis of the DC-bus voltage variation under specific control strategies

Based on control equations and the mathematical model of the actual system, we can calculate the maximum DC-bus oscillation under any specific control strategy. Such value can be further compared with the theoretical minimum variation to verify settings of the outer energy loop.

Table 9.1 Key parameters of dual PWM inverters based motor drive systems

Parameter	2.2 kW inverter	55 kW inverter
Rated power (kW)	2.2	55
Grid line-line voltage (V)	380	380
Grid frequency (Hz)	50	50
DC-bus rated voltage (V)	700	700
Filter inductance (mH)	5.5	2.0
Equivalent resistance (Ω)	0.015	0.010
DC-bus capacitance (μF)	110	4700
DC-bus discharge resistance (kΩ)	400	9.4
Switching frequency (kHz)	6.4	6.4

In the dynamic process, the system loss does not contribute much to the DC-bus variation. Therefore the system loss can be ignored, i.e., $R_g = 0$ and $R_{dc} = \infty$. With the energy balancing control, the energy stored in the DC-bus capacitor can be calculated by the end of each control period based on energy outer-loop control equations and the system energy balancing equation. Assume the output power step occurs at $t = t_0$, when the power jumps from P_0 to P_1. The DC-bus capacitor energy within each period after the power step is

$$E_{Cdc}(t + T_s) = E_{Cdc}(t) + \frac{3}{4}e_d(i_{rd}(t) + i_{rd}(t + T_s))T_s$$
$$- \frac{3}{4}L_g\left(i_{rd}(t + T_s)^2 - i_{rd}(t)^2\right) - P_1 T_s \qquad (9.50)$$

At the initial moment t_0, $E_{Cdc}(t_0) = E_{CdcN}$ and $i_{rd}(t_0) = 2P_0/(3e_d)$. Here E_{CdcN} is the energy stored in the capacitor at the rated DC-bus voltage.

Since the control strategy has already compensated the one-period delay, the next-period grid current values are calculated references within the present period based on Eq. (9.37), i.e.,

Fig. 9.13 DC-bus voltage variation versus n at the step power of a 2.2 kW system

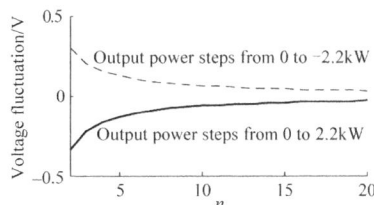

Fig. 9.14 DC-bus voltage
variation versus n at the step
power of a 55 kW system

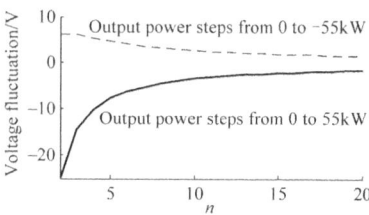

$$i_{rd}(t + T_s) = \frac{2P_1}{3e_d} + \frac{2}{3e_d n T_s} \left\{ \frac{3}{4} L_g \left[\left(\frac{2P_1}{3e_d} \right)^2 - i_{rd}(t)^2 \right] + E_{CdcN} - E_{Cdc}(t) \right\}$$

(9.51)

However, due to the limitation of the grid inductor, grid voltage and DC-bus voltage, the grid current can only change at its fastest pace even when the reference current has a big step. The next-period grid current is

$$i_{rd}(t + T_s) = \frac{(e_d - u_{rd}) T_s}{L_g} + i_{rd}(t)$$

(9.52)

Thus the recurrence formula between the DC-bus capacitor energy and the input current can be derived. Meanwhile, the relationship between the energy variation of the DC-bus capacitor and the DC-bus voltage change can be obtained through Eq. (9.44). When the DC-bus voltage change is small, the second-order item of Δu_{dc} can be ignored. Together with Eqs. (9.50) and (9.51) the recurrence formula on the oscillation amplitude of the DC-bus voltage is

$$\Delta u_{dc}(t + T_s) = \Delta u_{dc}(t)$$
$$+ \frac{1}{u_{dcN} C_{dc}} \left[\frac{3}{4} e_d (i_{rd}(t) + i_{rd}(t + T_s)) T_s - \frac{3}{4} L_g \left(i_{rd}(t + T_s)^2 - i_{rd}(t)^2 \right) - P_1 T_s \right]$$

(9.53)

$$i_{rd}(t + T_s) = \begin{cases} \frac{2P_1}{3e_d} + \frac{2}{3e_d n T_s} \left\{ \frac{3}{4} L_g \left[\left(\frac{2P_1}{3e_d} \right)^2 - i_{rd}(t)^2 \right] - C_{dc} \Delta u_{dc}(t) u_{dcN} \right\} \\ \frac{(e_d - u_{rd}) T_s}{L_g} + i_{rd}(t) \end{cases}$$

(9.54)

Here $\Delta u_{dc}(t_0) = 0$ and $i_{rd}(t_0) = 2P_0/(3e_d)$. Limited by actual system parameters, $i_{rd}(t + T_s)$ should be one of two values shown in Eq. (9.54) closer to $i_{rd}(t)$. When the change of the output power is large, the limitation of i_{rd2}^* needs be considered as well. Through the iteration, the discrete maximum value of the DC-bus voltage oscillation can be derived based on the energy-balancing control algorithm. With a shorter control period, the actual maximum energy oscillation of the DC-bus capacitor is similar to this value. Note the maximum DC-bus oscillation refers to the maximum change compared to the steady-state value. When the output power rapidly increases Δu_{dc} will be negative.

Table 9.2 Comparison of the DC-bus voltage variation

Rated power(kW)	Output power (kW)		Theoretical minimum values of DC-bus voltage oscillation (V)	Maximum values of the DC-bus voltage oscillation by the energy balancing control (V)
	From	To		
2.2	0	2.2	3.43	2.2
2.2	2.2	0	5.15	2.2
2.2	−2.2	2.2	4.46	2.2
55	0	55	10.29	55
55	55	0	27.41	55
55	−55	55	13.06	55

In addition, Eq. (9.19) indicates that the expected energy compensation within one control period is $1/n$ of the overall needed energy. When $P_1 > P_0$, i.e., the input power exceeds the output power, the needed extra energy reduces, yielding the reduction of both the grid current and grid-inductor energy. Meanwhile the DC-bus energy increases, and vice versa. Therefore the maximum DC-bus energy oscillation must happen around this moment. With the input current tracking references, 1~2 iterations are sufficient to calculate the maximum variation of the DC-bus voltage.

Back to 2.2 kW and 55 kW dual PWM inverters, their theoretical maximum values of the DC-bus voltage oscillation calculated by the energy balancing control method are compared to the theoretical minimum values in Table 9.2. It can be seen that in the 2.2 kW system, these two values are very close to each other, while in the 55 kW system there is some gap. This is because in the high-power system the power stepping to the rated value yields a large grid current, i.e., a large $i_{rd}(t_2) + i_{rd}(t_1)$. These two systems have the same grid voltage while the grid-side inductance does not proportionally reduce with the power rating. Therefore the high-power system needs longer energy-restoring time to regulate the DC-bus voltage towards the right direction, i.e., a large T_2. On the other hand, to secure the dynamic response and implement the compensation of the loss-estimation error, the system energy-restoring time cannot be too long, i.e., n cannot be too big. Therefore the high-power system design requires the maximum oscillation of the DC-bus voltage to be slightly higher than the theoretical minimum thereby shortening the system regulation time. When the output power suddenly drops, the rectifier input voltage is limited by the DC-bus voltage, yielding a small current decreasing rate. This results in that the theoretical minimum value of the DC-bus voltage variation is higher than that under the power-increasing scenario. Therefore in the high-power system, increasing the DC-bus voltage and reducing the filtering inductance help enhance the system dynamic response.

Overall the energy balancing control can restrain the DC-bus voltage oscillation close to the theoretical minimum value. This validates such control strategy fully utilizes the system hardware performance to minimize the DC-bus variation. From another perspective, a limited DC-bus variation Δu_{dc} reduces the DC-bus capacitance to minimum, as shown below.

$$C_{dcmin} = \frac{1}{u_{dcN}\Delta u_{dc}} \left| \frac{(P_1 - P_0)T_{min}}{2} + \frac{L_g\left(P_1^2 - P_0^2\right)}{3e_d^2} \right| \tag{9.55}$$

9.4   Analysis of Energy Balancing Control

9.4.1   Small-Signal Model of the Control System

Due to the existence of nonlinear elements in the transient energy balancing control, a small-signal model is needed to analyze the system stability and robustness. Firstly all state variables are divided into steady-state operational points (large signals) and small errors (small signals), as formulated below.

$$e_d = \overline{E}_d + \tilde{e}_d \tag{9.56}$$

$$i_{rd} = \overline{I}_{rd} + \tilde{i}_{rd} \tag{9.57}$$

$$u_{rd} = \overline{U}_{rd} + \tilde{u}_{rd} \tag{9.58}$$

$$u_{dc} = \overline{U}_{dc} + \tilde{u}_{dc} \tag{9.59}$$

$$P_{inv} = \overline{P}_{inv} + \tilde{p}_{inv} \tag{9.60}$$

Here $\overline{E}_d$, $\overline{I}_{rd}$, $\overline{U}_{rd}$, $\overline{U}_{dc}$ and $\overline{P}_{inv}$ represent related variables of steady-state operational points. $\tilde{e}_d$, $\tilde{i}_{rd}$, $\tilde{u}_{rd}$, $\tilde{u}_{dc}$ and $\tilde{p}_{inv}$ are corresponding small signals. For the convenience of analysis, the system mathematical model and control equations can be transformed into d-q coordinates based on the grid-voltage vector orientation. Given the reactive power is controlled as zero and the d-axis current control is identical to the q-axis current control, for the analysis below we only consider the d-axis component. Besides, the grid voltage can be treated as constant, i.e., $\tilde{e}_d \equiv 0$.

Substituting Eqs. (9.56)–(9.60) to d-q coordinates based differential Eqs. (9.61) and (9.18) of dual PWM motor drive systems yields

$$\begin{cases} L_g \dfrac{di_{rd}}{dt} = -R_g i_{rd} + \omega L_g i_{rq} + e_d - u_{rd} \\[2mm] L_g \dfrac{di_{rq}}{dt} = -\omega L_g i_{rd} - R_g i_{rq} + e_q - u_{rq} \\[2mm] C_{dc} \dfrac{du_{dc}}{dt} = -i_{dc} - \dfrac{u_{dc}}{R_{dc}} + \dfrac{3}{2u_{dc}}\left(u_{rd} i_{rd} + u_{rq} i_{rq}\right) \end{cases} \tag{9.61}$$

Accordingly state-variable equations of the d-axis current and the DC-bus voltage are

$$L_g \frac{d\left(\overline{I}_{rd} + \tilde{i}_{rd}\right)}{dt} = -R_g\left(\overline{I}_{rd} + \tilde{i}_{rd}\right) + \omega L_g i_{rq} + \overline{E}_d - \left(\overline{U}_d + \tilde{u}_{rd}\right) \tag{9.62}$$

$$\frac{3}{2}\overline{E}_d\left(\overline{I}_{rd} + \tilde{i}_{rd}\right) = R_g\left(\overline{I}_{rd} + \tilde{i}_{rd}\right)^2 + \frac{3}{4}L_g\left(\overline{I}_{rd} + \tilde{i}_{rd}\right)\frac{d\left(\overline{I}_{rd} + \tilde{i}_{rd}\right)}{dt}$$
$$+ \frac{1}{2}C_{dc}\frac{d\left(\overline{U}_{dc} + \tilde{u}_{dc}\right)^2}{dt} + \left(P_{inv} + \tilde{p}_{inv}\right) \tag{9.63}$$

Given the grid q-axis current is always zero while the large-signal model meets the energy balancing in the steady state, i.e.,

$$\overline{E}_d - \overline{U}_d = R_g\overline{I}_{rd} \tag{9.64}$$

$$\frac{3}{2}\overline{E}_d\overline{I}_{rd} = P_{inv} + \frac{3}{2}R_g\overline{I}_{rd}^2 \tag{9.65}$$

The system small-signal state equations are

$$L_g\frac{d\tilde{i}_{rd}}{dt} = -R_g\tilde{i}_{rd} - \tilde{u}_{rd} \tag{9.66}$$

$$\frac{3}{2}\overline{E}_d\tilde{i}_{rd} = 3R_g\overline{I}_{rd}\tilde{i}_{rd} + \frac{3}{2}R_g\tilde{i}_{rd}^2 + \frac{3}{4}L_g\left(\overline{I}_{rd} + \tilde{i}_{rd}\right)\frac{d\tilde{i}_{rd}}{dt} + C_{dc}\left(\overline{U}_{dc} + \tilde{u}_{dc}\right)\frac{d\tilde{u}_{dc}}{dt} + \tilde{p}_{inv} \tag{9.67}$$

With the z-transform of Eqs. (9.66) and (9.67) and overlooking the second-order item of small signals for further linearization of the system model, the z-domain discrete state equations is

$$\tilde{i}_{rd} = G_i(z)\tilde{u}_{rd} \tag{9.68}$$

$$\tilde{u}_{dc} = G_{dc}(z)\tilde{i}_{rd} - G_p(z)\tilde{p}_{inv} \tag{9.69}$$

Here $G_i(z)$, $G_{dc}(z)$ and $-G_p(z)$ are discrete small-signal transfer functions of the grid d-axis current over the rectifier d-axis voltage, the DC-bus voltage over the grid d-axis current, and the DC-bus voltage over the rectifier output power, respectively, as formulated below.

$$G_i(z) = \frac{-1}{\frac{L_g}{T_s}(z-1) + R_g} \tag{9.70}$$

$$G_{dc}(z) = \frac{T_s}{C_{dc}\overline{U}_{dc}(z-1)}\left[\frac{3}{4}\overline{E}_d(z+1) - \frac{3}{2}L_g\overline{I}_{rd}\frac{(z-1)}{T_s} - \frac{3}{2}R_g\overline{I}_{rd}(z+1)\right] \tag{9.71}$$

$$G_p(z) = \frac{T_s(z+1)}{2C_{dc}\overline{U}_{dc}(z-1)} \tag{9.72}$$

Substituting Eqs. (9.56)–(9.60) into (9.41) and (9.37) with the d-q transformation results in system control equations as below.

$$
\overline{U}_{\mathrm{rd}}^{*} + \tilde{u}_{\mathrm{rd}}^{*} = \left(2 - \frac{\hat{R}_{\mathrm{g}} T_{\mathrm{s}}}{\hat{L}_{\mathrm{g}}}\right)\overline{E}_{\mathrm{d}} - \frac{\hat{L}_{\mathrm{g}}}{T_{\mathrm{s}}}\left(\overline{I}_{\mathrm{rd}}^{*} + \tilde{i}_{\mathrm{rd}}^{*} - \overline{I}_{\mathrm{rd}} - \tilde{i}_{\mathrm{rd}}\right)
$$
$$
- \left(2 - \frac{\hat{R}_{\mathrm{g}} T_{\mathrm{s}}}{\hat{L}_{\mathrm{g}}}\right)\hat{R}_{\mathrm{g}}\left(\overline{I}_{\mathrm{rd}} + \tilde{i}_{\mathrm{rd}}\right) - \left(1 - \frac{\hat{R}_{\mathrm{g}} T_{\mathrm{s}}}{\hat{L}_{\mathrm{g}}}\right)\left(\overline{U}_{\mathrm{rd}}^{*}(k-1) + \tilde{u}_{\mathrm{rd}}^{*}(k-1)\right)
$$

(9.73)

$$
\overline{I}_{\mathrm{rd}}^{*} + \tilde{i}_{\mathrm{rd}}^{*} = \frac{\hat{R}_{\mathrm{g}}}{\overline{E}_{\mathrm{d}}}\left(\frac{2\widehat{\overline{P}}_{\mathrm{inv}} + 2\widehat{\tilde{p}}_{\mathrm{inv}}}{3\overline{E}_{\mathrm{d}}}\right)^{2} + \frac{2\left(\widehat{\overline{P}}_{\mathrm{inv}} + \widehat{\tilde{p}}_{\mathrm{inv}}\right)}{3\overline{E}_{\mathrm{d}}}
$$
$$
+ \frac{\hat{L}_{\mathrm{g}}}{2\overline{E}_{\mathrm{d}} n T_{\mathrm{s}}}\left[\left(\frac{2\widehat{\overline{P}}_{\mathrm{inv}} + 2\widehat{\tilde{p}}_{\mathrm{inv}}}{3\overline{E}_{\mathrm{d}}}\right)^{2} - \left(\overline{I}_{\mathrm{rd}} + \tilde{i}_{\mathrm{rd}}\right)^{2}\right]
$$
$$
+ \frac{\hat{C}_{\mathrm{dc}}}{3\overline{E}_{\mathrm{d}} n T_{\mathrm{s}}}\left[\left(\overline{U}_{\mathrm{dc}}^{*} + \tilde{u}_{\mathrm{dc}}^{*}\right)^{2} - \left(\overline{U}_{\mathrm{dc}} + \tilde{u}_{\mathrm{dc}}\right)^{2}\right]
$$

(9.74)

Here $\hat{C}_{\mathrm{dc}}$, $\hat{L}_{\mathrm{g}}$ and $\hat{R}_{\mathrm{g}}$ are estimated values of C_{dc}, L_{g} and R_{g} in the control system. $\widehat{\overline{P}}_{\mathrm{inv}}$ and $\widehat{\tilde{P}}_{\mathrm{inv}}$ are calculated values of the output power. Similarly, large signals meet the energy balancing in steady states, with the system reference operating points the same as actual working points. Given the system equivalent resistance is small, we have

$$
\begin{cases}
\widehat{\overline{P}}_{\mathrm{inv}} = \overline{P}_{\mathrm{inv}} = \dfrac{3}{2}\overline{E}_{\mathrm{d}}\overline{I}_{\mathrm{rd}} - R_{\mathrm{g}}\overline{I}_{\mathrm{rd}}^{2} \approx \dfrac{3}{2}\overline{E}_{\mathrm{d}}\overline{I}_{\mathrm{rd}} \\
\overline{I}_{\mathrm{rd}}^{*} = \overline{I}_{\mathrm{rd}},\ \overline{U}_{\mathrm{dc}}^{*} = \overline{U}_{\mathrm{dc}},\ \overline{U}_{\mathrm{rd}}^{*} = \overline{U}_{\mathrm{rd}}^{*}(k-1)
\end{cases}
$$

(9.75)

In addition, to linearize the system model we can ignore second-order items of small signals, which generates the small-signal model based control equations as

$$
\begin{cases}
\tilde{u}_{\mathrm{rd}}^{*} + \left(1 - \dfrac{\hat{R}_{\mathrm{g}} T_{\mathrm{s}}}{\hat{L}_{\mathrm{g}}}\right)\tilde{u}_{\mathrm{rd}}^{*}(k-1) = \dfrac{\hat{L}_{\mathrm{g}}}{T_{\mathrm{s}}}\left(\tilde{i}_{\mathrm{rd}}^{*} - \tilde{i}_{\mathrm{rd}}\right) - \left(2 - \dfrac{\hat{R}_{\mathrm{g}} T_{\mathrm{s}}}{\hat{L}_{\mathrm{g}}}\right)\hat{R}_{\mathrm{g}}\tilde{i}_{\mathrm{rd}} \\
\tilde{i}_{\mathrm{rd}}^{*} = \dfrac{4\hat{R}_{\mathrm{g}}\overline{I}_{\mathrm{rd}}\widehat{\overline{P}}_{\mathrm{inv}}}{3\overline{E}_{\mathrm{d}}^{2}} + \dfrac{2\widehat{\tilde{p}}_{\mathrm{inv}}}{3\overline{E}_{\mathrm{d}}} + \dfrac{\hat{L}_{\mathrm{g}}\overline{I}_{\mathrm{rd}}}{\overline{E}_{\mathrm{d}} n T_{\mathrm{s}}}\left(\dfrac{2\widehat{\tilde{p}}_{\mathrm{inv}}}{3\overline{E}_{\mathrm{d}}} - \tilde{i}_{\mathrm{rd}}\right) + \dfrac{2\hat{C}_{\mathrm{dc}}\overline{U}_{\mathrm{dc}}}{3\overline{E}_{\mathrm{d}} n T_{\mathrm{s}}}\left(\tilde{u}_{\mathrm{dc}}^{*} - \tilde{u}_{\mathrm{dc}}\right)
\end{cases}
$$

(9.76)

Furthermore, we can implement z-transform to Eq. (9.76). Since the system switching frequency is multiple kHz, equivalent resistance is $\sim$mΩ and the filtering inductance is $\sim$mH, we have $\hat{R}_{\mathrm{g}} T_{\mathrm{s}}/\hat{L}_{\mathrm{g}} \ll 1$, which allows us to further simplify Eq. (9.76) thereby getting z-domain discrete control equations as

$$
\tilde{u}_{\mathrm{rd}}^{*} = G_{\mathrm{i1}}(z)\tilde{i}_{\mathrm{rd}}^{*} - G_{\mathrm{i2}}(z)\tilde{i}_{\mathrm{rd}}
$$

(9.77)

$$\tilde{i}^*_{\mathrm{rd}} = G_{\mathrm{dc1}}(z)\big(\tilde{u}^*_{\mathrm{dc}} - \tilde{u}_{\mathrm{dc}}\big) - G_{\mathrm{dc2}}(z)\tilde{i}_{\mathrm{rd}} + G_{\mathrm{dc3}}(z)\hat{p}_{\mathrm{inv}} \qquad (9.78)$$

Here $G_{\mathrm{i1}}(z)$, $-G_{\mathrm{i2}}(z)$, $G_{\mathrm{dc1}}(z)$, $-G_{\mathrm{dc2}}(z)$ and $G_{\mathrm{dc3}}(z)$ are discrete small-signal transfer functions of the rectifier d-axis voltage over the d-axis current, the rectifier d-axis reference voltage over the d-axis sampled current, the grid d-axis reference current over the error between the DC-bus reference and actual voltage, the grid d-axis reference current over the d-axis actual current, and the grid d-axis reference current over the estimated power of the inverter, respectively. All functions above are formulated as below.

$$G_{\mathrm{i1}}(z) = -\frac{\hat{L}_{\mathrm{g}}}{(z+1)T_{\mathrm{s}}} \qquad (9.79)$$

$$G_{\mathrm{i2}}(z) = -\frac{1}{z+1}\left(\frac{\hat{L}_{\mathrm{g}}}{T_{\mathrm{s}}} + 2\hat{R}_{\mathrm{g}}\right) \qquad (9.80)$$

$$G_{\mathrm{dc1}}(z) = \frac{2\hat{C}_{\mathrm{dc}}\overline{U}_{\mathrm{dc}}}{3\overline{E}_{\mathrm{d}}nT_{\mathrm{s}}} \qquad (9.81)$$

$$G_{\mathrm{dc2}}(z) = \frac{\hat{L}_{\mathrm{g}}\overline{I}_{\mathrm{rd}}}{\overline{E}_{\mathrm{d}}nT_{\mathrm{s}}} \qquad (9.82)$$

$$G_{\mathrm{dc3}}(z) = \frac{2}{3\overline{E}_{\mathrm{d}}nT_{\mathrm{s}}}\left(\frac{\hat{L}_{\mathrm{g}}\overline{I}_{\mathrm{rd}}}{\overline{E}_{\mathrm{d}}} + nT_{\mathrm{s}} + \frac{2\hat{R}_{\mathrm{g}}nT_{\mathrm{s}}\overline{I}_{\mathrm{rd}}}{\overline{E}_{\mathrm{d}}}\right) \qquad (9.83)$$

The closed-loop small-signal model with the energy balancing control can be built as Fig. 9.15, based on Eqs. (9.68), (9.77) & (9.78). z^{-1} means the output voltage reference of the rectifier has a one-control-period delay. $G_{\mathrm{pre}}(z)$ is the open-loop transfer function of the output power prediction. Since the predicted output power for the next control period is used in the control system, $G_{\mathrm{pre}}(z)$ is formulated as

$$G_{\mathrm{pre}}(z) = z \qquad (9.84)$$

$H_{\mathrm{i}}(z)$ is the closed-loop transfer function of the inner current loop, which is

$$H_{\mathrm{i}}(z) = \frac{G_{\mathrm{i}}(z)G_{\mathrm{i1}}(z)}{z + G_{\mathrm{i}}(z)G_{\mathrm{i2}}(z)} \qquad (9.85)$$

9.4.2   System Stability Analysis

To analyze the impact of parameters on the system stability, the characteristic equation of the closed-loop system can be derived based on Fig. 9.15, i.e.,

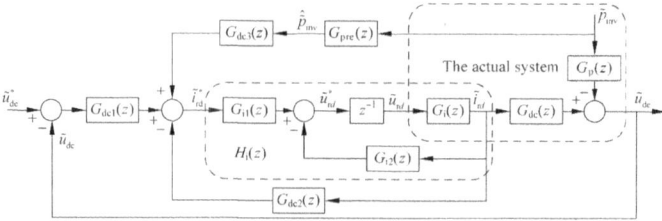

Fig. 9.15 The system small-signal model based on the energy balancing control

Fig. 9.16 z-domain root
locus when $n = 5$, $0 \leq \hat{R}_g$
$\leq 10R_g$

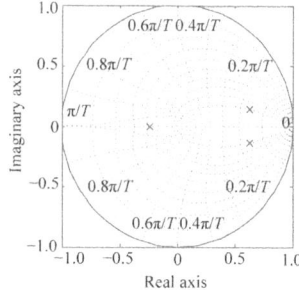

$$\lambda(z) = \left\{ \left[z^2 L_g + \hat{L}_g - L_g + R_g T_s \left(z^2 + z \right) \right] \overline{E}_d n T_s + \left(\hat{L}_g + 2\hat{R}_g T_s \right) \hat{L}_g \overline{I}_{rd} \right\} (z - 1)$$
$$+ \left[\frac{1}{2} \overline{E}_d T_s (z + 1) - L_g \overline{I}_{rd} (z - 1) - R_g T_s \overline{I}_{rd} (z + 1) \right] \left(\hat{L}_g + 2\hat{R}_g T_s \right) \frac{\hat{C}_{dc}}{C_{dc}}$$
$$\tag{9.86}$$

Furthermore zeros and poles can be derived. For the 2.2 kW system, the switching
frequency is 6.4 kHz. Assume the system is running with the rated power at the
rectifying mode. Based on Sect. 9.3, $n = 5$ is selected. The distribution of poles is
given in Fig. 9.16, where the ESR of the filtering inductor is changed from 0 to 10
times of the actual value. It can be seen that with a small R_g, the pole distribution
is nearly unchanged even when the estimated R_g varies in a wide range. The root
locus with the AC inductance and DC-bus capacitance is given in Fig. 9.17 and 9.18,
respectively. With observed values increasing, the trend of poles is marked with
arrows. $0.2L_g \leq \hat{L}_g \leq 1.7 L_g$ or $0.2C_{dc} \leq \hat{C}_{dc} \leq 2C_{dc}$ result in all poles locate inside
the circle, indicating the energy balancing control enhances the system robustness.

The impact of the n value on the system stability is analyzed below. When n
increases from 1 to 40, the system root locus is shown in Fig. 9.19. It can be seen that
when $n = 1$ the system root is outside the unity circle, i.e., system loses the stability
when using the fastest pace to compensate the capacitor energy. With n increasing
the stability margin is reinforced. For $n \geq 5$, damping coefficients of system poles
are close to each other. When $n = 2$, the system root locus is shown in Figs. 9.20 and
9.21, when estimated values of the AC inductance and the DC capacitance increase,
respectively. Compared to Figs. 9.17 and 9.18, the system robustness is significantly

Fig. 9.17 z-domain root locus when $n = 5$, $0.2L_g$ $\leq \hat{L}_g \leq 1.7\,L_g$

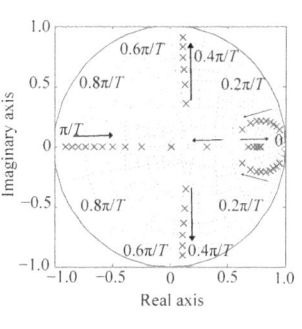

Fig. 9.18 z-domain root locus when $n = 5$, $0.2C_{dc}$ $\leq \hat{C}_{dc} \leq 2C_{dc}$

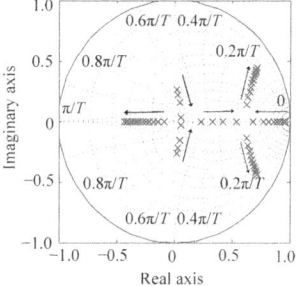

Fig. 9.19 z-domain root locus when $1 \leq n \leq 40$

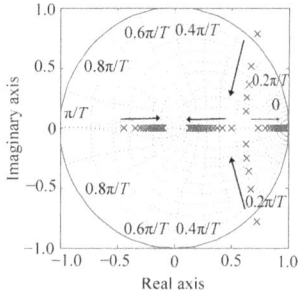

weakened. When the estimated inductance is above 1.6 times of the actual value and the estimated capacitance is beyond 1.5 times of the actual value, some poles are located outside the unity circle already, i.e., the system is not stable any more. Therefore appropriately prolonging the time to compensate the capacitor energy helps enhance the system robustness.

In addition, to study the impact of steady-state operating points on the small-signal model, Fig. 9.22 shows the root locus when $n = 5$ and steady-state working points shift from negative twice of the rated power (inverting mode) to positive twice of the rated power (rectifying mode). It can be seen that the steady-state operating points nearly have no impact on the pole distribution. Given a larger gap between the large-signal and small-signal absolute values means a more accurate small-signal model,

Fig. 9.20 z-domain root locus when $n = 2, 0.2L_g$ $\leq \widehat{L}_g \leq 1.7L_g$

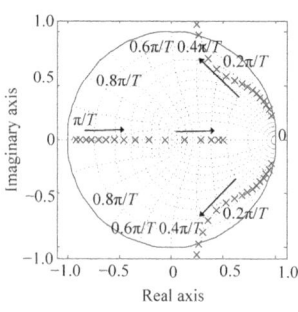

Fig. 9.21 z-domain root locus when $n = 2, 0.2C_{dc}$ $\leq \widehat{C}_{dc} \leq 2C_{dc}$

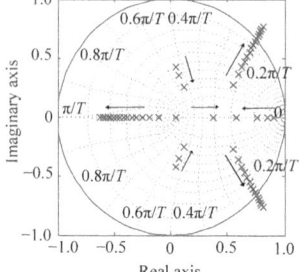

Fig. 9.22 z-domain root locus when $n = 5, -2I_{rN}$ $\leq \widehat{I}_{rd} \leq 2I_{rN}$

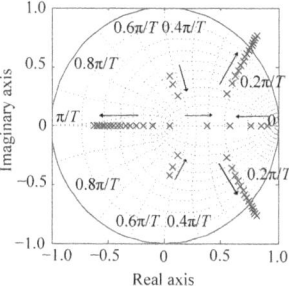

we select the rated-power point as the steady-state operating point in the dynamic analysis below.

9.4.3   Analysis of the System Dynamic Performance

Assume the ultimate goal of the control strategy is to minimize the DC-bus voltage oscillation when the output power changes suddenly. In Fig. 9.15, $\widetilde{u}_{dc}^* = 0$. Then the discrete closed-loop transfer function of the DC-bus voltage over the system output power is

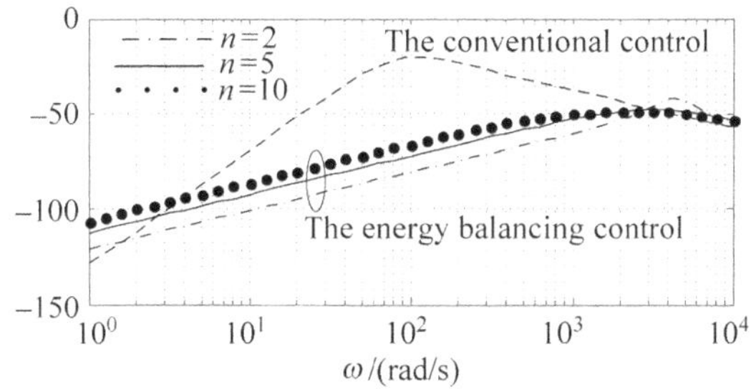

Fig. 9.23 Amplitude-frequency response of the DC-bus voltage over the system output power with different values of n

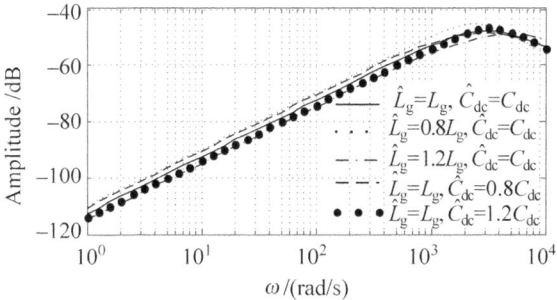

Fig. 9.24 Amplitude-frequency response of the DC-bus voltage over the system output power with inaccurate estimation of system parameters

$$\frac{\tilde{u}_{dc}}{\tilde{p}_{inv}} = \frac{G_{dc}(z)H_i(z)G_{dc3}(z)z - G_p(z)(1 + H_i(z)G_{dc2}(z))}{1 + H_i(z)G_{dc2}(z) + G_{dc}(z)H_i(z)G_{dc1}(z)} \tag{9.87}$$

Based on Eq. (9.87), the amplitude-frequency response of the DC-bus voltage over the system output is shown in Fig. 9.23, when $n = 2, 5, 10$, respectively. With n decreasing, the system low-frequency gain gradually reduces. However, when n is too small, e.g., 2, some spike emerges in the high-frequency domain. Compared to the conventional load-current feed-forward control, the system gain in the medium-frequency range, which is also the area with the maximum gain, drops significantly. When $n = 5$ and estimated values of the DC-bus capacitance and the filtering inductance are $\pm 20\%$ off, the amplitude-frequency response of the DC-bus voltage over the system output power is shown in Fig. 9.24. It can be seen that even with $\pm 20\%$ estimation error, the system gain is nearly the same as that under the accurate estimation, indicating strong robustness of the energy balancing control. With different switching frequencies and n, the amplitude-frequency response of the DC-bus voltage over the system output power is shown in Fig. 9.25. Tuning the value of n at higher switching frequency can yield better control performance.

Besides, in order to analyze the capability of the actual DC-bus voltage tracking the reference, the discrete closed-loop transfer function of the actual DC-bus voltage over the reference value can be derived as below, based on Fig. 9.15 when the output power is constant, i.e., $\widehat{P}_{inv} = \tilde{P}_{inv} = 0$.

Fig. 9.25 Amplitude-frequency response of the DC-bus voltage over the system output power with different switching frequency

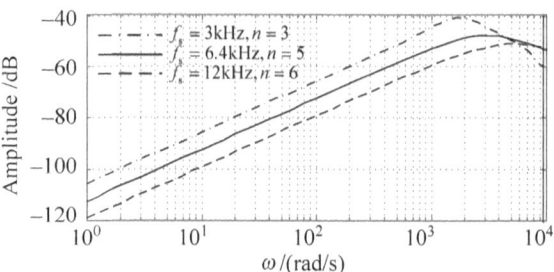

Fig. 9.26 Amplitude-frequency response of the DC-bus voltage over the reference with inaccurate estimation of system parameters

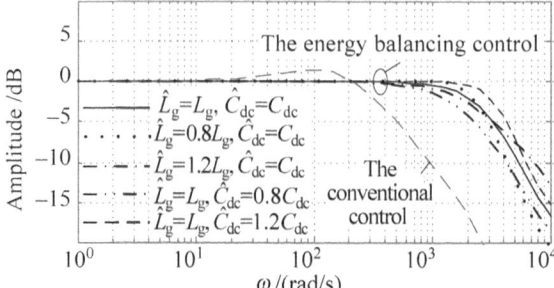

$$\frac{\tilde{u}_{dc}}{\tilde{u}_{dc}^{*}} = \frac{G_{dc}(z)H_i(z)G_{dc1}(z)}{1 + H_i(z)G_{dc2}(z) + G_{dc}(z)H_i(z)G_{dc1}(z)} \qquad (9.88)$$

With Eq. (9.88) the amplitude-frequency response of the actual DC-bus voltage over the reference value is shown in Fig. 9.26. Here $n = 5$ and the estimated DC-bus capacitance and the AC filtering inductance are $\pm 20\%$ off. The gain of the conventional control at the turning point is greater than 0 dB, indicating the over modulation with the step input. Meanwhile the system tends to amplify sampling errors within such frequency range, resulting in larger harmonics. For the energy balancing control, the gain at the low-frequency range is less than 0 dB without amplifying errors. Its control bandwidth is wider than the conventional control, indicating a faster tracking speed to the DC-bus reference than the conventional control.

9.4.4 Analysis of the System Static Error

The static error of the DC-bus voltage of dual PWM inverters can be derived based on Fig. 9.23, i.e.,

$$e_{ss} = \lim_{k \to \infty} \left(\tilde{u}^*_{dc}(kT_s) - \tilde{u}_{dc}(kT_s) \right)$$

$$= \lim_{z \to 1} \left[\left(1 - z^{-1}\right) \left(\tilde{u}^*_{dc}(z) - \tilde{u}_{dc}(z) \right) \right]$$

$$= \lim_{z \to 1} \left[\left(1 - z^{-1}\right) \left(\frac{nT_s}{\hat{C}_{dc}\overline{U}_{dc}} + \frac{\hat{L}_g \overline{I}_{rd}}{\overline{E}_d \hat{C}_{dc}\overline{U}_{dc}} \right) \left(\tilde{p}_{inv} - \hat{p}_{inv} \right) \right] \tag{9.89}$$

To simplify the analysis, the ESR of the inductor R_g has been ignored. Through Eq. (9.89), with an accurately estimated output power, no static error exists. With the power estimation error as a step function, i.e.,

$$\tilde{p}_{inv} - \hat{p}_{inv} = \frac{z}{z-1} \Delta p_{inv} \tag{9.90}$$

The static error of the DC-bus voltage will be

$$e_{ss} = \left(\frac{nT_s}{\hat{C}_{dc}\overline{U}_{dc}} + \frac{\hat{L}_g \overline{I}_{rd}}{\overline{E}_d \hat{C}_{dc}\overline{U}_{dc}} \right) \Delta p_{inv} \tag{9.91}$$

Thus the system static error is proportional to the accuracy of the predicted power and meanwhile related to estimated system parameters, static working point, switching frequency and value of n. Assume 2.2 kW dual PWM inverters are working at the rated point with the predicted power having 10% error. This results in the DC-bus voltage measurement has 2.47 V error. Therefore it is necessary to precisely estimate the system loss to further reduce the static error of the DC-bus voltage.

9.4.5 Simulation and Experimental Analysis

The system simulation and experiments were finished on the same 2.2 kW dual PWM inverters. To further verify the scalability of the proposed control, same experiments were repeated in a 55 kW system. Main parameters of inverters and motors are shown in Table 9.1.

1. Simulation analysis

In the 2.2 kW system, to effectively test both the steady state and dynamic response, the control bandwidth for the conventional load-current feed-forward control is set as 400 Hz for the current loop and 20 Hz for the voltage loop. With the energy balancing control when $n = 5$, the simulated steady-state waveforms are shown in Fig. 9.27. It can be seen that when the q-axis reference current is 0, the grid current is aligned with the voltage, i.e., the power factor is 1 with the current THD of 3.33%. Under the conventional control strategy, the grid current THD is 3.36%, indicating the energy balancing control is quite similar to the conventional control in the steady state.

Figure 9.28 shows simulation waveforms of the conventional control and the energy balancing control when the motor runs at no load and the speed steps from 1200 to 1500 r/min. Since the inverter side employs the same control algorithm, the motor speed with these two different controls is the same, as shown in Fig. 9.28a. With the energy balancing control, when the system output power, i.e., input power of the induction motor changes suddenly, the system input power can rapidly follow the output power, as shown in Fig. 9.28c. After that, a relatively small power is used to compensate the energy in the DC-bus capacitor to effectively reduce the DC-bus oscillation. The comparison of the DC-bus voltage is shown in Fig. 9.28b. Besides, the system output power ramps from 0 to 4 kW, which in theory yields the minimum DC-bus oscillation of 8.0 V. As shown in Fig. 9.28b, at $t = 0.5$ s the DC-bus voltage under the energy balancing control dropped ~8.0 V, close to the theoretical value. When running at the rated speed, the motor changed its torque from the positive rated value (in the motor mode) to the negative rated value (in the generator mode) at $t = 0.55$ s and switched back to the positive rated torque at $t = 0.58$ s. The related simulation waveforms are shown in Fig. 9.29. Similar to the scenario where the motor speed suddenly changed, the DC-bus voltage oscillation can be effectively suppressed as well through the energy balancing control.

After the DC-bus capacitance is changed to 45 μF, under the same test condition as Fig. 9.28, the dynamic response of the DC-bus voltage under the energy balancing control algorithm is simulated as Fig. 9.30, where the DC-bus voltage oscillation is the same as the conventional control with 110 μF DC-bus capacitance. Thus the energy balancing control can reduce the DC-bus capacitance by ~60%.

With the same simulation condition as Fig. 9.29, the system robustness based on the energy balancing control can be analyzed. Shown in Table 9.3 are the grid-current THD at the rated current and the DC-bus maximum oscillation when the output power switches between the positive and negative rated values. Here the estimated

Fig. 9.27 The simulated steady-state waveforms under the energy balancing control

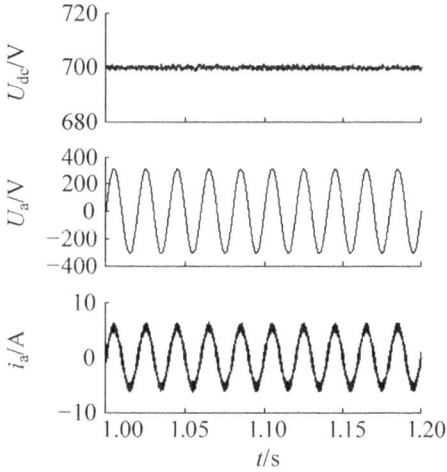

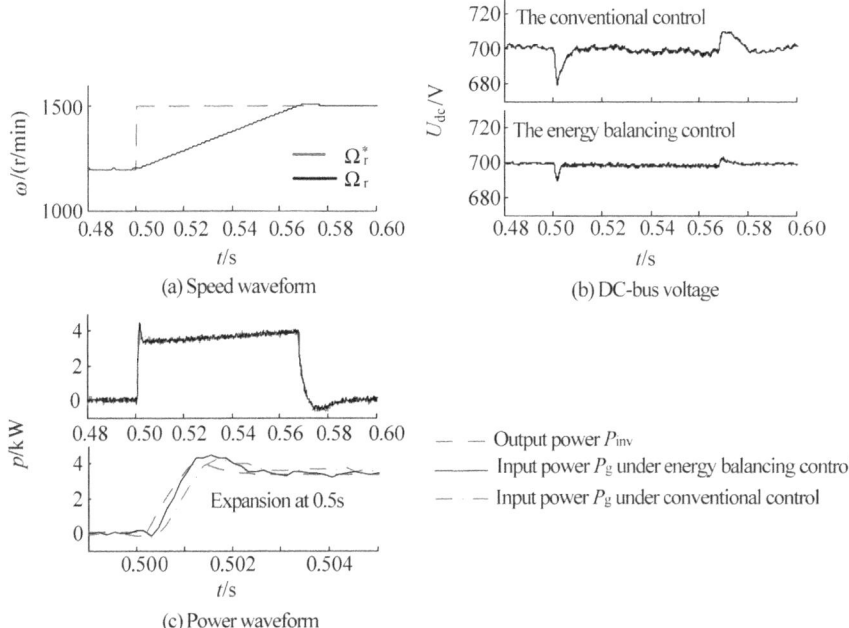

Fig. 9.28 Simulated dynamic process when the motor was given a step-speed command

Table 9.3 Simulated energy balancing control when estimated values are inaccurate

$\dfrac{\hat{C}_{dc}}{C_{dc}}$	$\dfrac{\hat{L}_g}{L_g}$	Grid current THD (%)	DC-bus maximum oscillation
1.0	1.0	3.33	+4.77 V, −6.50 V
0.8	1.0	3.32	+5.04 V, −7.07 V
1.2	1.0	3.36	+4.50 V, −6.10 V
1.0	0.8	3.38	+4.88 V, −7.10 V
1.0	1.2	3.33	+4.10 V, −5.65 V

inductance and capacitance are ±20% off. Even so the control performance is nearly unaffected.

To further analyze the DC-bus-voltage tracking performance, Fig. 9.31 exhibits the simulation comparison between these two control algorithms when the DC-bus voltage has a step reference. Compared to the conventional control, the energy balancing control has much faster dynamic response with nearly no over modulation of the DC-bus voltage.

To compare the DC-bus oscillation caused by the sudden change of the output power and the grid-voltage asymmetry, respectively, we let one phase voltage drop by 30%. The simulated waveform under the energy balancing control is shown in Fig. 9.32, with the same simulation condition as Fig. 9.29. It can be seen that there is

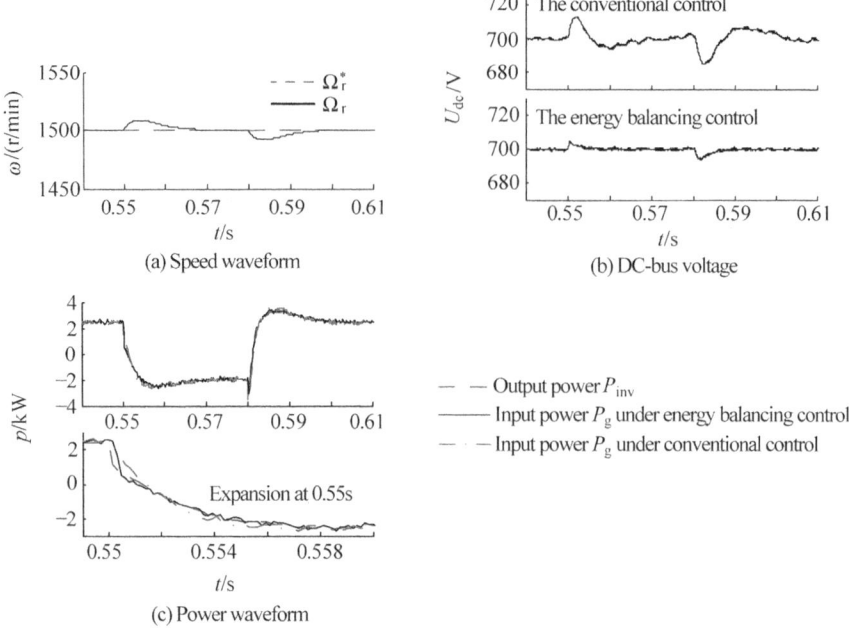

Fig. 9.29 Simulated dynamic process when the motor was given a step-torque command

Fig. 9.30 The simulated dynamic process of the energy balancing control with a step motor speed. Here the DC-bus capacitance is 45 μF

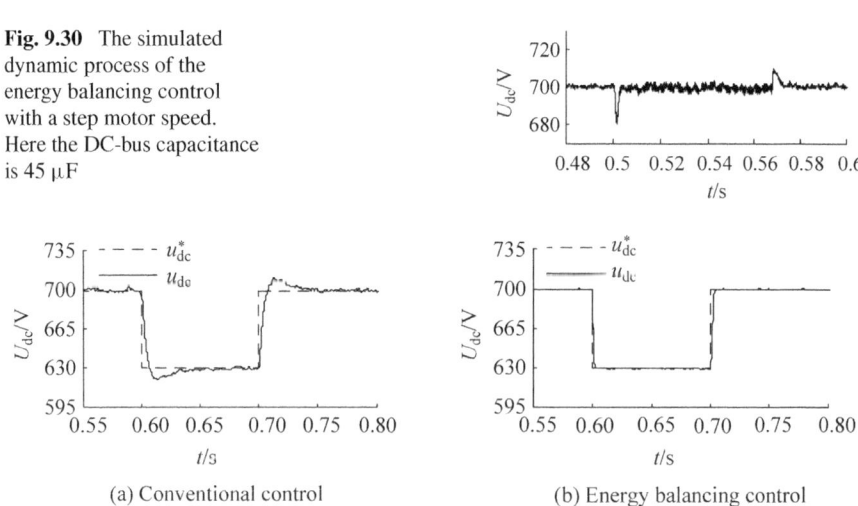

Fig. 9.31 Simulated dynamic process when the DC-bus reference voltage is a step function

some 100 Hz component in the DC-bus voltage, caused by the grid-voltage negative-sequence component. The amplitude of such oscillation is ±2 V, far less than the DC-bus vibration caused by the step power.

Fig. 9.32 Simulated
waveforms when one phase
voltage drops by 30%

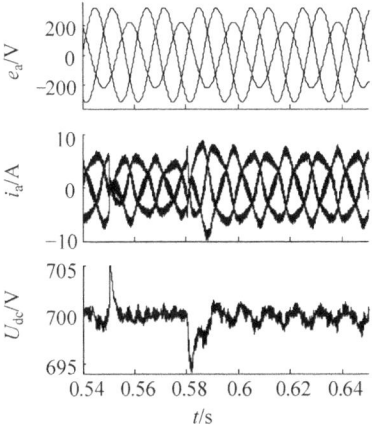

2. Experimental results

The same control strategy was adopted in experiments as the simulation for the 2.2 kW dual-PWM-inverter setup. The dynamic response of both the conventional control and the energy balancing control is shown in Fig. 9.33, when the speed command of a fully loaded motor jumped from 1200 to 1500 r/min. Similar to the simulation result, the speed waveform is the same for two control strategies, as shown in Fig. 9.33a. Both have the steady-state power factor equal to 1, as shown in Fig. 9.33b. In the dynamic process the energy balancing control effectively reduced the DC-bus voltage variation. When the motor torque stepped from the rated to 0 at the rated speed, experimental results of two control algorithms were shown in Fig. 9.34, indicating the effective suppression of the DC-bus voltage variation by the energy balancing control. In addition, as shown in Fig. 9.34a, with a sudden drop of the load, the inverter output power began to drop, lasting for 20 ms. The static power before power dropping is 2.64 kW and the minimum power is −0.44 kW, which can be used to further calculate the minimum DC-bus oscillation, i.e., 4.8 V. As shown in Fig. 9.34c, the actual DC-bus oscillation was 4.9 V, aligned with the theoretical calculation.

For a 55 kW system, $n = 10$ based on the analysis of Sect. 9.3. Experimental results of the conventional control and the energy balancing control were given in Fig. 9.35, when the system was running at the full power while the DC-bus voltage reference jumped from 700 to 650 V. Similar to simulation results, the DC-bus voltage can quickly track the reference value when using the energy balancing control, without any over modulation in the dynamic process.

At the rated torque when the motor speed reference stepped from 1200 to 1500 r/min, the system dynamic output power is shown in Fig. 9.36. Under the conventional control and the energy balancing control, the dynamic DC-bus voltage and the grid current are shown in Fig. 9.37, with all results summarized as Table 9.4. The energy balancing control effectively suppressed the DC-bus voltage variation to 13.9 V. Besides, it can be seen in Fig. 9.36 that the DC-bus theoretical minimum vari-

Fig. 9.33 Experimental
results of a 2.2 kW system
when the motor speed
reference has a step
command

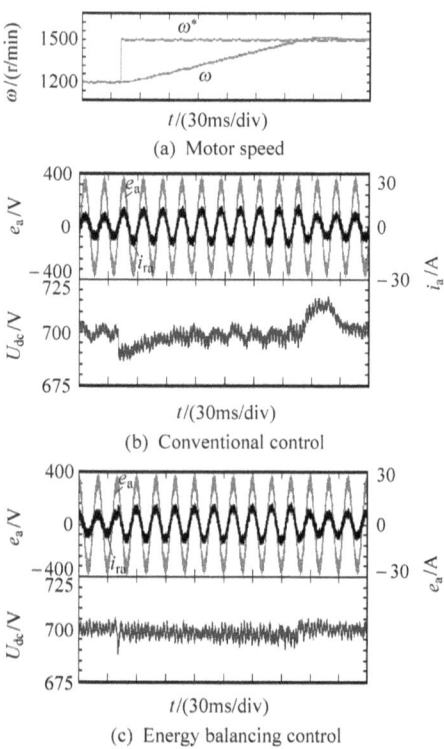

(a) Motor speed

(b) Conventional control

(c) Energy balancing control

Table 9.4 Experimental results of a 55 kW system using the energy balancing control when estimated parameters are inaccurate

$\frac{\hat{C}_{dc}}{C_{dc}}$	$\frac{\hat{L}_g}{L_g}$	Grid current THD (%)	DC-bus maximum oscillation
1.0	1.0	2.33	13.9
0.8	1.0	2.28	14.1
1.2	1.0	2.33	14.4
1.0	0.8	2.36	13.8
1.0	1.2	2.35	14.5

ation is 12.1 V when the system output power stepped from 42.0 to 81.5 kW, which
is 1.8 V less than the actual maximum variation. Such difference mainly attributes
to errors of the loss estimation and ADC sampling.

The distribution and flow of the transient energy in power electronic converters
comply with no instantaneous energy change and energy conservation, which is the
foundation of the energy balancing control. The implementation of such control
strategy can be summarized as: based on the distribution, flow and balance of the
transient energy in the converter tracking the steady-state energy associated to control
objectives in the fastest pace with the secured system reliability.

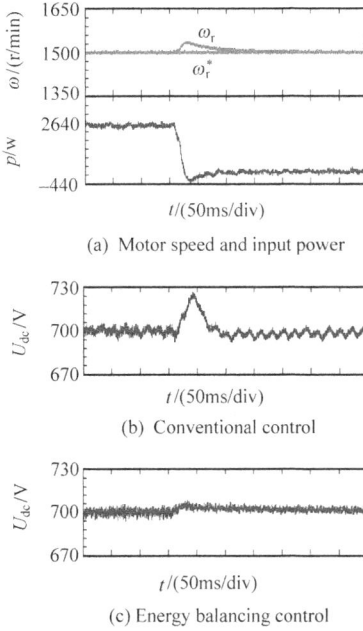

(a) Motor speed and input power

(b) Conventional control

(c) Energy balancing control

Fig. 9.34 Experimental results of a 2.2 kW system when the motor torque reference has a step command

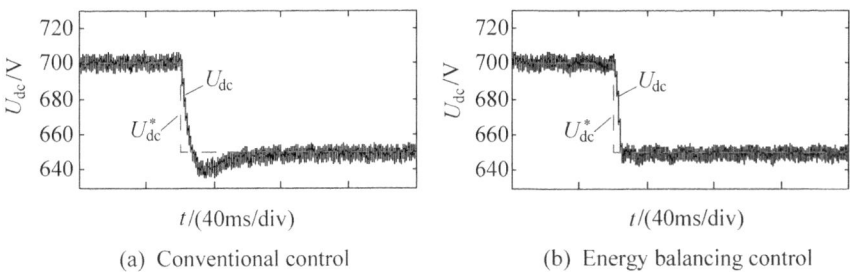

(a) Conventional control   (b) Energy balancing control

Fig. 9.35 Experimental results of a 55 kW system when the DC-bus voltage reference has a step command

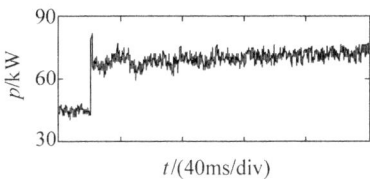

Fig. 9.36 Experimental dynamic process of the output power in a 55 kW system

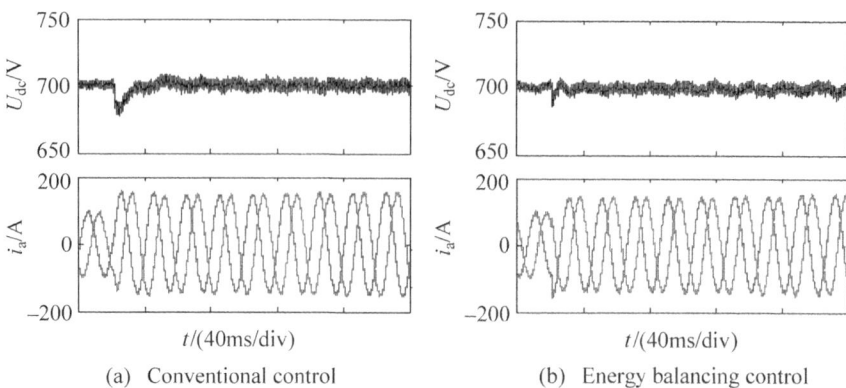

(a) Conventional control   (b) Energy balancing control

Fig. 9.37 Experimental results of a 55 kW system when the motor speed reference has a step command

The equation of the energy balancing control was derived from the balance of the transient energy in the power electronic converter, which is related to the input energy, output energy, loss and energy storage increment. Each item has a clear physical meaning. When applying the energy balancing control in a multi-control-objective power electronics converter, we need convert current and voltage related variables into the transient energy stored in the inductor and capacitor, respectively, relate all elements through energy balancing thereby considering the relationship among all control variables together. Since the control equation of the energy balancing was strictly derived through the transient energy with the energy flow and distribution of all related variables taken into account, it is expected for such control to have a high accuracy. Given the target is to control the converter energy to reach the steady state as soon as possible, only states of the present and previous control periods are involved, avoiding the memory effect of the conventional control thereby yielding the fast dynamic response. With all energy storage components approaching the energy steady state, control variables will be updated quickly by the energy balancing control, which stops the further energy exchange among components thereby eliminating the potential over modulation.

The energy balancing control in the power electronic converter needs adjust the mathematical model, parameter selection, control objectives and control algorithms based upon the timescale of the transient energy. With the energy balancing model and control targets varying with transient timescales, we can find the relationship among all energy balancing models to realize the full-time-domain full-system calculation, analysis and control.

Chapter 10
Applications of Transient Analysis in Power Converters

The analysis methods of electromagnetic transients have been widely used, especially in high-power and high-performance power electronics converters. This chapter will emphasize the application of transient analysis in series connected HV IGBTs with dynamic balancing circuit and SiC device based power amplifiers.

10.1 Electromagnetic Transient Analysis in Series Connected HV IGBTs Based Converter

Series connecting LV-rating switches can yield HV switching modules-based high-power converters, which effectively overcomes the limitation of switching voltage ratings. Such technology is found very promising in HV converters, but it is restrained by the voltage imbalance across switches during the switching transients. Previous chapters adopted the closed-loop active balancing control to improve the performance of series connected switches, while this chapter will focus on transient processes and characteristics of HV IGBTs in series connection.

10.1.1 The Transient Mechanism Model of HV IGBTs in Series Connection

1. Voltage imbalance in the current tailing process

Two IGBTs in series connection are shown in Fig. 10.1, where each IGBT package is made of one IGBT module and its accessory circuit. The accessory circuit is for the snubber and voltage balancing of the IGBT, which is tightly connected to the IGBT module through bus bars.

© Tsinghua University Press and Springer Nature Singapore Pte Ltd. 2019

Z. Zhao et al., *Electromagnetic Transients of Power Electronics Systems*,

https://doi.org/10.1007/978-981-10-8812-4_10

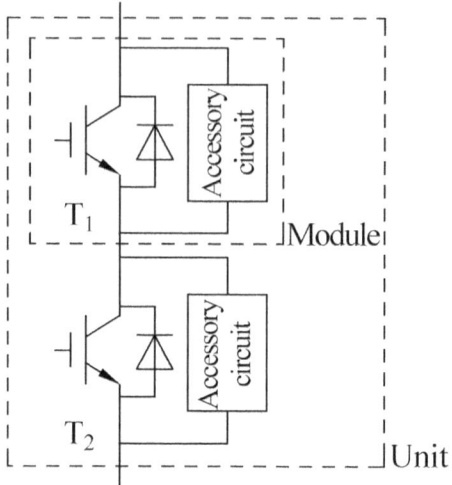

Fig. 10.1 IGBT package with two switches in series connection

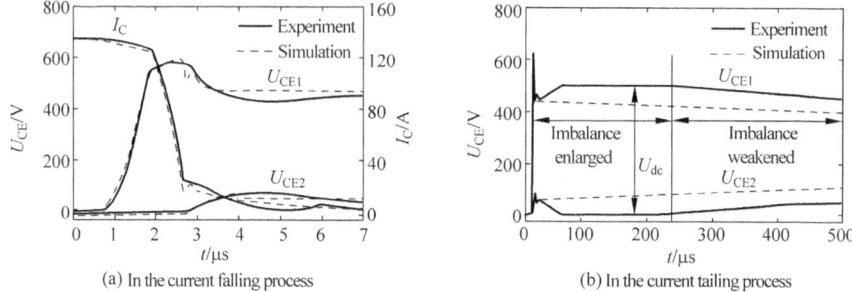

Fig. 10.2 Simulation of the turn-off process of series connected IGBTs using the conventional model

The transient analysis of switches in series connection is highly related to the connection between components. Certain applicability exists for the device model. For example, when the conventional IGBT model is directly used in series connected IGBTs, simulated and experimental switching-off transients are shown in Fig. 10.2. The experimental waveform was measured at certain DC-bus voltage. The accessory circuit contains the steady balancing resistor and the RC snubber circuit. The voltage imbalance was caused by longer turn-off delay of T_2 than T_1. In the current falling process before the tailing stage, simulation results are well aligned with the experiment, accurately revealing the voltage imbalance. Once in the current tailing stage, simulation results exhibited obvious errors.

The experimental waveform indicated that the voltage U_{CE} in the current tailing stage could be divided into two parts, the first of which had enlarged voltage imbalance, where $\Delta U_{CE}(U_{CE1} - U_{CE2})$ increased and could even reach the whole

Fig. 10.3 IGBT modules
with the internal stray
inductance

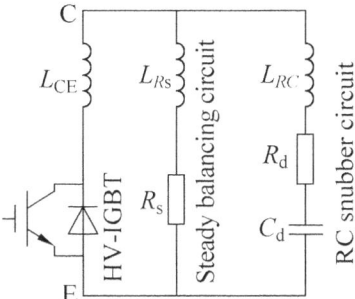

DC-bus voltage in the worst case. The second part had the voltage imbalance gradually weakened. The simulation results, however, showed that the voltage imbalance at the current tailing stage was always attenuating. The voltage spike at the current tailing stage could potentially breakdown the IGBT when the DC-bus voltage was high, which could further cause system damages. Therefore, it will be rewarding to design a model of series connected HV IGBTs that accurately describes voltage distribution in the whole process.

2. **Influential factors of the voltage imbalance during the current tailing stage**

It is very difficult to determine influential factors of the voltage imbalance at the current tailing stage through direct measurement. Firstly, the HV IGBT converters have very compact assembly, which cannot fit probes to directly measure most of variables. Secondly, high-bandwidth wide-range current probes provide poor accuracy for the small current measurement, and thus cannot accurately measure the tailing current. Thirdly, some inner variables of the IGBT, such as the carrier distribution in the base, are inaccessible due to missing appropriate equipment. Therefore, in analyzing transient processes, those influential factors, especially main impact factors of the voltage imbalance in the current tailing stage, need be comprehended. The influential factors can be categorized into two types. One is external factors, mainly the stray inductance inside the package. The other is IGBT itself, mainly the voltage imbalance at the current falling stage, which further causes a different tailing current.

The equivalent circuit of the IGBT package with the internal stray inductance is shown in Fig. 10.3, which includes the stray inductance of the IGBT modules L_{CE}, stray inductance of the steady balancing resistor L_{Rs}, and stray inductance of the RC snubber circuit L_{RC}. The LCR bridge or the partial element equivalent circuit (PEEC) method can be used to approximate the stray inductance.

In general, the module and accessory circuit are connected through compact bus bars, resulting in the small stray inductance L_{RC} (less than 100 nH) and even smaller L_{Rs} and L_{CE} (less than 50 nH). Simulation results of w/n stray inductance in the current tailing stage are given in Fig. 10.4. We can see that the stray inductance at the current tailing stage does not affect much thereby not being considered as the main contributor to the voltage imbalance.

Fig. 10.4 Simulation results
when considering the
internal stray inductance

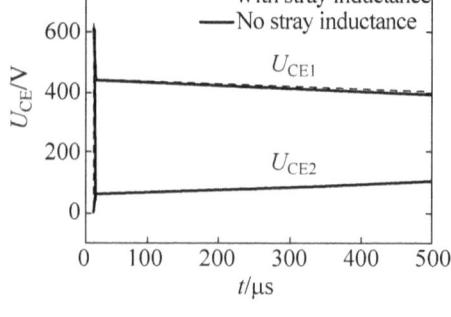

Fig. 10.5 The equivalent
circuit of two series
connected IGBTs in the
current tailing process

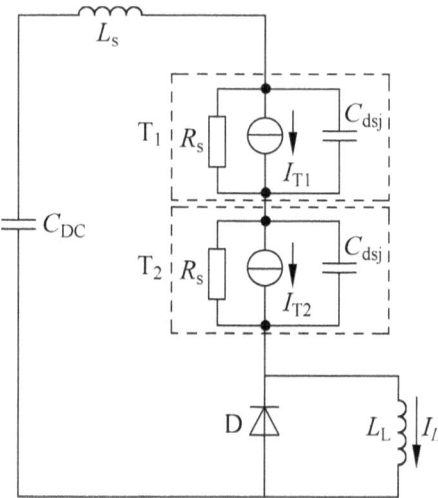

In the current tailing process the IGBT can be treated as a parallel circuit made of the junction capacitor C_{dsj}, off-state resistor R_s, and equivalent current source I_T of the tailing current. Without considering the accessory circuit, the equivalent circuit of series connected IGBTs is shown in Fig. 10.5. At the time the freewheeling diode is conducted, the voltage drop is $U_d = 0$. Since current decrement in the tailing stage is very slow, the voltage drop across the stray inductance can be approximated to 0. For T_1 and T_2, we have

$$U_{dc} = U_{CE1} + U_{CE2} \tag{10.1}$$

Since the off-state resistance R_s is very large, the current difference through C_{dsj} of T_1 and T_2 is

$$\Delta I_{cds}(t) = I_{cds1}(t) - I_{cds2}(t) = I_{T2}(t) - I_{T1}(t) \tag{10.2}$$

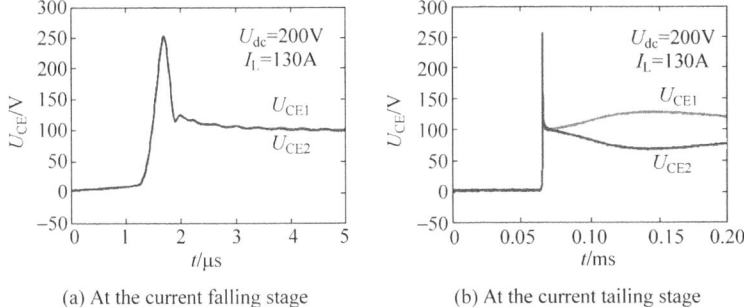

(a) At the current falling stage   (b) At the current tailing stage

Fig. 10.6 Experimental turn-off transients of two series connected IGBTs without RC snubber circuits

The difference of the tailing current I_{T1} and I_{T2} will cause the voltage imbalance of T_1 and T_2 during the current tailing stage. Such voltage imbalance ΔU_{CEt} (voltage increment of T_1 to T_2) is

$$\Delta U_{CEt}(t) = \frac{\int_0^t \Delta I_{cds}(t)dt}{C_{dsj}} \qquad (10.3)$$

Therefore, with the understanding of the tailing-current difference, as shown in Fig. 10.2, the voltage imbalance in this process can be fully explained. However, the conventional IGBT model cannot reveal such tailing-current difference. Assume the initial voltage difference ΔU_{CE} between T_1 and T_2 is U_{dif}. In the conventional model ΔU_{CE} is

$$\Delta U_{CE}(t) = U_{dif} \exp\left(-\frac{t}{R_s C_{dsj}}\right) \qquad (10.4)$$

Namely, the IGBT voltage imbalance decays in an exponential manner at the current-tailing stage, as shown in the simulation waveform of Fig. 10.2. This is misaligned with the experiments. Mark ΔU_{CE} in Eq. (10.4) as ΔU_{CEd}, the voltage differential in the discharging process through RC circuits. In reality, $\Delta U_{CE} = \Delta U_{CEd} + \Delta U_{CEt}$. Therefore, the key of building a model applicable to the series-connected IGBTs is to accurately exhibit the difference of the tailing current based on the conventional IGBT model.

Figure 10.6 gives experimental results of two series connected IGBTs without the dynamic RC circuit. Two major features are exhibited in terms of the voltage imbalance at the current tailing stage.

(1) Significant impact of the tailing current on the IGBT voltage imbalance. In the current falling process, voltage of two series connected IGBTs, U_{CE1} and U_{CE2}, only have very small difference. This means in the beginning of the tailing current stage the differential of the tailing current of two IGBTs is very small.

Fig. 10.7 Two-dimension
base structure of the IGBT

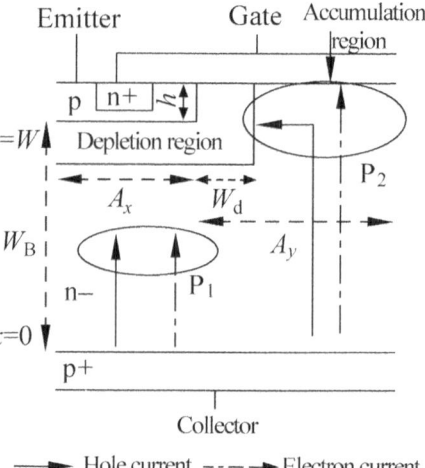

——— Hole current ----→ Electron current

However, a major difference between U_{CE1} and U_{CE2} emerges in the current
tailing process. This indicates that even a tiny differential of the tailing current
between two switches can cause a big voltage imbalance, as shown in Eq. (10.3).

(2) Long influential time of the tailing current. As shown in Figs. 10.2b and 10.6b,
    it took nearly 100 μs from the end of the current falling stage to the initiation
    of ΔU_{CE} shrinking, which means the tailing current of the IGBT acted for
    ~100 μs. After that ΔU_{CEd} is dominant over ΔU_{CEt}, which weakened the voltage
    imbalance. The study to the turn-off process of a single IGBT is usually focused
    on the timescale of several microseconds with the current tailing time confined
    to ~μs level. For IGBTs in series connection, the tailing time should not be
    limited to ~μs level, but at a larger timescale.

3. **Transient model of series connected HV IGBTs**

A two-dimension distributive model of the IGBT base is mainly used to study IGBT
static characteristics and merits of the enhanced-gate IGBT. The model structure is
highly complex and is mainly to model the carrier distribution in the IGBT base in the
steady state. It is not suitable for studying the transient behavior of series connected
IGBTs. To establish the two-dimension model of the IGBT base for the transient
analysis, the cell structure shown in Fig. 10.7 is recommended.

The IGBT base consists of P_1 and P_2 regions. The current of P_1 is directly injected
to the P region in the emitter through the base, with the related area of A_x. The current
of P_2 flows into the P region in the emitter through the side area h, with the related
area of A_y. Given the h/l_x is small, the hole current is mainly concentrated in P_1.
However, due to the existence of the electron layer below the gate, electrons in P_2
can flow towards the emitter through such an electron layer and MOSFET channel,
resulting in electron currents existing in both P_1 and P_2. The electron and hole current
meets

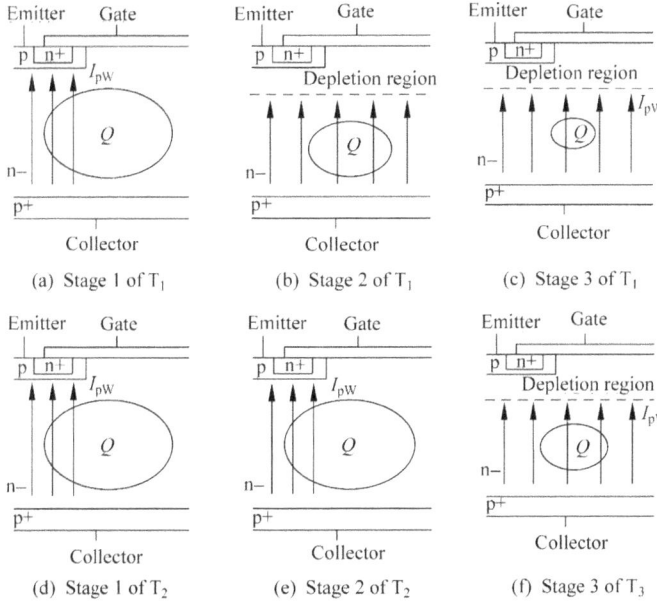

Fig. 10.8 Base carrier distribution when the turn-off signal of T_2 falls behind T_1

$$I_C = A\left(J_n + k J_p\right) \tag{10.5}$$

where I_C is the overall current, J_n and J_p are the density of the electron current and the hole current, respectively, A is the area of the whole cell, and k is the percentage of P_1 area over the whole cell area.

$$k = \frac{A_x}{A_x + A_y} \tag{10.6}$$

k is determined by the gate area and cell area, which is ~ 1/2 for the HV IGBT.

When the turn-off signal of T_2 falls behind T_1, the base current carriers of two switches are shown in Fig. 10.8. The whole process of the carrier distribution can be divided into three stages.

Stage 1: Before U_{CE1} and U_{CE2} rise, the base carrier distribution of T_1 and T_2 are identical, i.e., the current carriers in the steady state.

Stage 2: U_{CE1} rises first, widening the depletion region of T_1. The hole-current area is enlarged, the extracting current I_{pW} in the base is rapidly increasing and the current carriers in the base are greatly reduced. Since the impact of the current-carrier reduction on I_{pW} is offset by the increment of the conduction area, the base carriers can be further reduced.

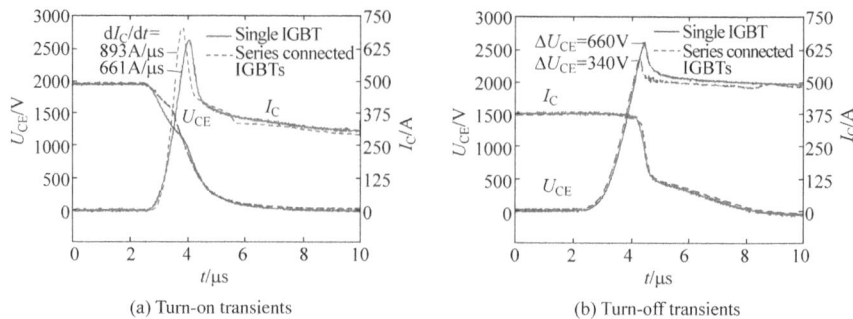

Fig. 10.9 Experimental comparison of switching behaviors for a single IGBT and fully balanced series connected IGBTs

Stage 3: After a while, U_{CE2} begins to rise. The base carries of T_2 begin to reduce. Since the amount of the current carriers of T_1 is less than T_2, the tailing current of T_1 is less than T_2.

The lifetime of current carriers has a high impact on the tailing current. Assume the carrier lifetime is a constant 4 μs based on the turn-off time constant of a single IGBT, the influential time of the tailing current should be no more than 10 μs. However, the influential time of the tailing current in experiments shown in Figs. 10.2 and 10.6 is ~100 μs. This reveals the limitation of using the carrier lifetime to model series connected IGBTs. If we consider the carrier lifetime versus the concentration, the lower the carrier concentration the longer the lifetime, thereby the longer the influential time of the tailing current.

10.1.2 Analysis of the Transient Behavior of Series Connected IGBTs

1. Transient behavior with fully balanced voltage

The comparison of switching transients for a single IGBT and series connected IGBTs when the voltage is fully balanced is shown in Fig. 10.9, with certain DC-bus voltage and the load current. At the turn-on moment, the series connected IGBTs undertake the DC-bus voltage of 4 kV and the load current of 400 A, while the single IGBT undertakes the DC-bus voltage of 2 kV and the load current of 400 A. For the turn-off test, series connected IGBTs undertake the DC-bus voltage of 4 kV and the load current of 375 A, while the single IGBT undertakes the DC-bus voltage of 2 kV and the load current of 375 A. The bus-bar parasitics and freewheeling diodes are the same in two cases.

For turn-on transients, test results of these two scenarios are shown in Fig. 10.9a. Series connected IGBTs have performance similar to single IGBT, indicating that the performance of N IGBTs in series connection at the DC-bus voltage U_{dc} is similar

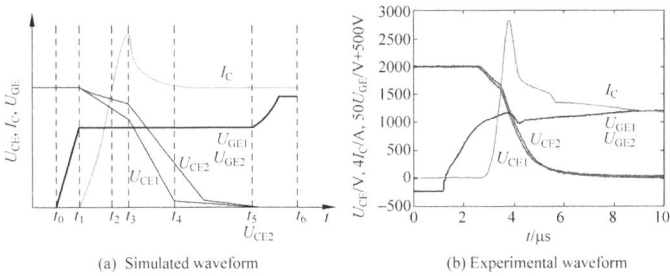

(a) Simulated waveform (b) Experimental waveform

Fig. 10.10 The voltage imbalance during the turn-on process with different dU_{CE}/dt

to one single IGBT at the DC-bus voltage U_{dc}/N, as long as the voltage is fully balanced. The dI_C/dt of series connected IGBTs is 893 A/μs, larger than that of the single IGBT (661 A/μs).

Similarly, for the turn-off transient shown in Fig. 10.9b, N IGBTs in series connection with fully balanced voltage at the DC-bus voltage U_{dc} are similar to one single IGBT at the DC-bus voltage U_{dc}/N. The difference is that the voltage spike of each series connected IGBT is 340 V, only half of that in the single-switch test bench. The difference is caused by two IGBTs undertaking the voltage spike together. Therefore, if a single switch undertakes the voltage spike of ΔU_{CE}, each switch in the N-switch-series string will only undertake $\Delta U_{CE}/N$.

2. Turn-on behavior with unbalanced voltage

The voltage imbalance during the turn-on transient is mainly caused by different dU_{CE}/dt and different initial time when U_{CE} begins to change. Simulated and experimental turn-on transients with different dU_{CE}/dt are shown in Fig. 10.10. Since the conduction current of series connected IGBTs is the same while U_{GE1} and U_{GE2} are identical, yielding the same current flowing through gates of IGBTs, dU_{CE}/dt follows

$$\frac{U_{G\_on} - U_{GE}}{R_{G1}} + \frac{C_{GD1,1V}}{\sqrt{U_{CE1}}}\frac{dU_{CE1}}{dt} = \frac{U_{G\_on} - U_{GE}}{R_{G2}} + \frac{C_{GD2,1V}}{\sqrt{U_{CE2}}}\frac{dU_{CE2}}{dt} \quad (10.7)$$

If series connected IGBTs have different R_G or $C_{GD,1V}$, the voltage imbalance in the turn-on process is mainly due to different dU_{CE}/dt. Beginning from $t = t_1$, U_{CE} of different IGBTs drops with different slope. With very small parameter diversity, e.g., the gate resistance R_G has the tolerance of ±1% and device manufacturers can secure the same C_{GD}, the voltage imbalance caused by different dU_{CE}/dt is usually not a major concern. A significantly different dU_{CE}/dt was detected in Fig. 10.10b during one test of many series connected IGBTs.

The simulated and experimental U_{CE} versus time of series connected IGBTs during the turn-on stage are given in Fig. 10.11. The different U_{CE} is caused by the turn-on delay of series connected IGBTs, which is the time interval between the rising edge of the control signal generated by the control board and the moment when the gate voltage climbs up to the turn-on threshold U_T. Such a turn-on delay

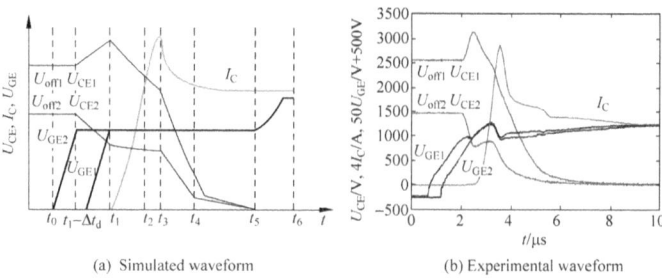

(a) Simulated waveform   (b) Experimental waveform

Fig. 10.11   The voltage imbalance in the turn-on process with different U_{CE}

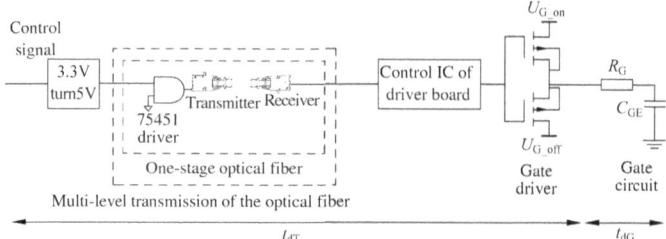

Fig. 10.12   the propagation of the IGBT control signal

can be divided into two processes. One is the interval between generation of the turn-on signal from the control board and the moment when the gate voltage U_G increases from U_{G_off} to U_{G_on}. This delay is defined as the propagation delay t_{dT}. The other is when U_G increases from U_{G_on} to the moment when $U_{GE} = U_T$, defined as t_{dG}. As shown in Fig. 10.12, the control signal passes through a 3.3 V → 5 V voltage-level transition IC, the driver of the optical fiber, transmitter of the optical fiber, the receiver of the optical fiber, the control IC of the driver board, and the gate-drive circuit. For two series connected IGBTs, one-stage optical fiber is enough. For multiple series connected IGBTs, multi-level transmission of the optical fiber is needed. The transmission delay might be subject to change due to diversity of ICs and optical fibers. It can be varied by levels of optical fibers as well. For instance, the commonly used HFBR-1512 and HFBR-2521 have the propagation delay of <100 ns, i.e., a 100 ns propagation delay between one IGBT using one-stage optical fibers and the other IGBT using two-stage optical fibers. The first type of the delay is defined as Δt_{dT} and the second type of the delay is

$$\Delta t = R_G C_{GE} \ln \frac{U_{G\_on} - U_{G\_off}}{U_{G\_on} - U_T} \tag{10.8}$$

Therefore, the overall differential of the turn-on time delay is

$$\Delta t_d = \Delta t_{dT} + R_{G1} C_{GE1} \ln \frac{U_{G\_on} - U_{G\_off}}{U_{G\_on} - U_{T1}} - R_{G2} C_{GE2} \ln \frac{U_{G\_on} - U_{G\_off}}{U_{G\_on} - U_{T2}} \tag{10.9}$$

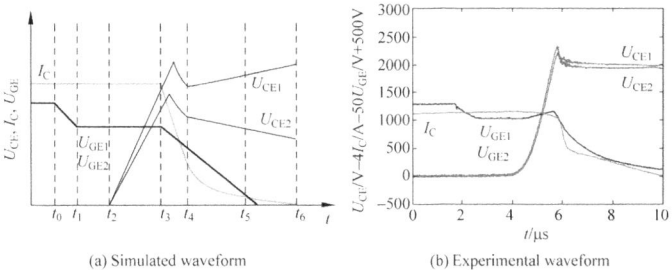

(a) Simulated waveform   (b) Experimental waveform

Fig. 10.13 The voltage imbalance caused by different dU_{CE}/dt during turn-off transients

As shown in Fig. 10.11, when the turn-on delay of T_1 is longer than that of T_2, U_{GE2} will climb to U_T earlier than U_{GE1}. During this period there is no current flowing through the main circuit. U_{GE2} will be held at U_T while U_{CE2} will drop in order to maintain U_{GE2} constant. With different turn-on delay time, the turn-off delay usually is different as well, resulting in different voltage across series connected IGBTs in the off state. Assuming voltage across switches at the turn-on initial moment has the difference, i.e., U_{off1} and U_{off2}, the voltage spike of U_{CE1} at $t = t_1 - \Delta t_d \sim t_1$ can be calculated as

$$U_{CE1\_max} = U_{dc} - \left(\sqrt{U_{off2}} - \frac{(U_{G\_on} - U_T)\Delta t_d}{2R_G C_{GD,1V}} \right)^2 \qquad (10.10)$$

At $t = t_1$, $U_{GE1} = U_T$, I_C begin to increase and U_{CE1} & U_{CE2} drop simultaneously. At $t = t_1 - t_4$, U_{CE1} & U_{CE2} both comply with Eq. (10.10). In this process, since U_{CE1} and U_{CE2} are different, their dU_{CE}/dt is different even with the same R_G and $C_{GD,1V}$. The voltage imbalance at the turn-on moment with $\Delta t_d = 500$ ns is shown in Fig. 10.11b, indicating that Δt_d has a big impact on the turn-on voltage imbalance. Therefore, electromagnetic transients within Δt_d are the main reason of the voltage imbalance.

3. **Turn-off behavior with unbalanced voltage**

The turn-off voltage imbalance exhibits mainly in three parts, i.e., different dU_{CE}/dt, different initial time of U_{CE}, and different voltage at the current tailing stage. Figure 10.13 shows simulated and experimental switching-off transients with different dU_{CE}/dt. Here U_{GE1} is identical as U_{GE2}. In the voltage rising process, the current flowing through the gate resistance of series connected IGBTs is identical as well. Therefore dU_{CE}/dt complies with

$$\frac{U_{G\_off} - U_{GE}}{R_{G1}} + \frac{C_{GD1,1V}}{\sqrt{U_{CE1}}} \frac{dU_{CE1}}{dt} = \frac{U_{G\_off} - U_{GE}}{R_{G2}} + \frac{C_{GD2,1V}}{\sqrt{U_{CE2}}} \frac{dU_{CE2}}{dt} \qquad (10.11)$$

We can see that R_G and $C_{GD,1V}$ also influence dU_{CE}/dt during the turn-off transients. With the diversified R_G and $C_{GD,1V}$ for series connected IGBTs, the voltage imbalance caused by different dU_{CE}/dt emerges. Such voltage imbalance initiates from $t = t_1$. The larger R_G or $C_{GD,1V}$, the smaller dU_{CE}/dt. Experimental waveform of the voltage imbalance caused by different dU_{CE}/dt is presented in Fig. 10.13b, revealing that such voltage imbalance is not severe.

Shown in Fig. 10.14 is the voltage imbalance caused by the different initial moment when U_{CE} begins to change, which in turn is caused by the diversity of the IGBT turn-off delay. Such a delay is the time interval between the control signal generation and the moment when U_{CE} rapidly increases. Different from the turn-on moment, the time interval of t_0-t_2 can be formulated as

$$\Delta t = R_G(C_{GE} + C_{oxd}) \ln \frac{U_{G\_on} - U_{G\_off}}{U_{ml} - U_{G\_off}} + \frac{R_G C_{oxd} U_{lim}}{3(U_{ml} - U_{G\_off})} \qquad (10.12)$$

Note: in Fig. 10.14b I_C has been multiplied by 4. U_{GE} has been multiplied by 50 and added an offset of 500 V.

The first item of the right side of Eq. (10.12) is the time for U_G to drop from U_{G_on} to the miller plateau U_{ml}. The second item represents the gradually rising process of U_{CE} before the depletion region fully covers the gate. The turn-off delay can be formulated as

$$\Delta t_d = \Delta t_{dT} + \Delta t_{02,T1} - \Delta t_{02,T2} \qquad (10.13)$$

When the time delay of T_2 is larger than that of T_1, U_{CE1} rises prior to U_{CE2}. After Δt_d, U_{CE2} begins to rise. At $t = t_2$, $U_{CE1} + U_{CE2} = U_{dc}$. Assume Δt_d is negligible compared to $t_3 - t_2$. Then U_{CE1} and U_{CE2} are solved as

$$U_{CE1}(t_3) = \frac{U_{dc}}{2} + \sqrt{\frac{U_{dc}}{2} \frac{U_{ml} - U_{G\_off}}{R_G C_{GD,1V}} \frac{\Delta t_d}{2}} \qquad (10.14)$$

$$U_{CE2}(t_3) = \frac{U_{dc}}{2} - \sqrt{\frac{U_{dc}}{2} \frac{U_{ml} - U_{G\_off}}{R_G C_{GD,1V}} \frac{\Delta t_d}{2}} \qquad (10.15)$$

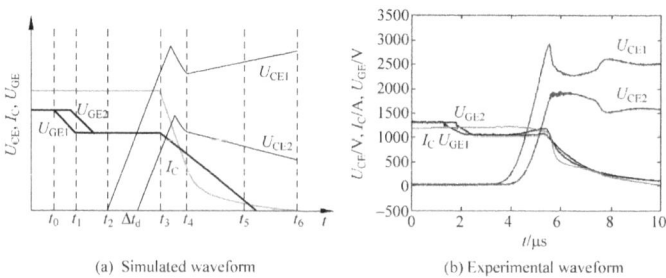

(a) Simulated waveform   (b) Experimental waveform

Fig. 10.14 The voltage imbalance caused by the U_{CE} initial moment during turn-off transients

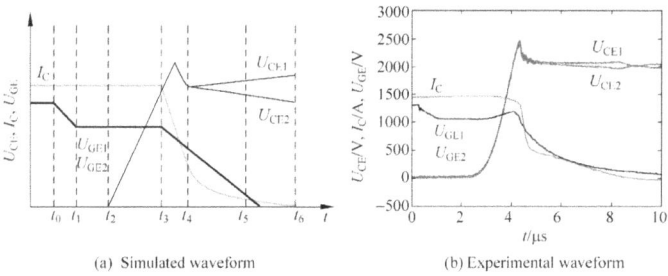

(a) Simulated waveform   (b) Experimental waveform

Fig. 10.15 The voltage imbalance in the current tailing process

At $t = t_3 - t_4$, I_C begins to drop, where U_{CE1} and U_{CE2} maintain nearly the same trend. Therefore the miller capacitance of the IGBTs undertakes the same current. In this process, dU_{CE}/dt is inversely proportional to C_{GD} and proportional to $\sqrt{U_{CE}}$ The voltage distribution across T_1 and T_2 can be approximated as

$$U_{CE1\_max} = U_{CE1}(t_3) + \frac{\sqrt{U_{CE1}(t_3)}\left(U_{CE\_max} - U_{dc}\right)}{\sqrt{U_{CE1}(t_3)} + \sqrt{U_{CE2}(t_3)}} \tag{10.16}$$

$$U_{CE2\_max} = U_{CE2}(t_3) + \frac{\sqrt{U_{CE2}(t_3)}\left(U_{CE\_max} - U_{dc}\right)}{\sqrt{U_{CE1}(t_3)} + \sqrt{U_{CE2}(t_3)}} \tag{10.17}$$

The voltage imbalance during the turn-off transient with $\Delta t_d = 500$ ns is given in Fig. 10.14b. It indicates that Δt_d plays an important role in the voltage imbalance.

As described in previous section, the voltage imbalance during the current falling stage caused by different dU_{CE}/dt or different moments when U_{CE} begins to change will certainly cause difference of the tailing current, which will be further detailed in Sect. 10.1.3. In this section we will only cover the voltage imbalance during the current tailing stage even with the voltage totally balanced in the current falling process, as shown in Fig. 10.15.

Note: in Fig. 10.15b I_C has been multiplied by 4. U_{GE} has been multiplied by 50 and added an offset of 500 V.

The difference of the tailing current is caused by the diversity of the parameters in the IGBT base, even though the switch voltage is totally balanced during the current falling stage. However, the diversity of the IGBT base is usually minor, as shown in Fig. 10.15b, the experiments of the voltage imbalance caused by the different tailing current. Such a voltage imbalance is negligible.

4. **Influential factors of the voltage imbalance of series connected IGBTs**

With the analysis above of the voltage imbalance during switching transients of series connected IGBTs with related mathematical expressions, all influential factors of the voltage imbalance can be summarized as

(1) In the control-signal propagating loop, t_{dT};
(2) In the gate-drive loop, R_G, U_{G_on}, U_{G_off} and C_{GE};

Table 10.1 The influential factors of various types of the voltage imbalance

	dU_{CE}/dt differences	Differences of U_{CE} changing moment	Voltage imbalance in the tailing stage
Switching-on	(2), (3), (5)	(1), (2), (5)	–
Switching-off	(2), (3), (5)	(1), (2), (3), (5)	(4)

(3) In the gate, U_{lim}, N_B, C_{oxd} and A_{GD};
(4) In the base, A, N_B, W_B, C_p, τ_t, Q_{by} and λ;
(5) In the MOSFET channel, k_p and U_T.

The impact of all the influential factors above on the voltage imbalance during switching transients is summarized as Table 10.1.

5. **The impact of thermal characteristics on the voltage imbalance of series connected IGBTs**

Assume the junction temperature of two series connected IGBTs is T_{j1} and T_{j2}. Furthermore, the power loss of two IGBTs is defined as $P_{loss1}(T_{j1}, T_{j2})$ and $P_{loss2}(T_{j1}, T_{j2})$, respectively. Here T_1 and T_2 are equipped with different heatsinks. Thus, the thermal dissipation of T_1 and T_2 can be expressed as $P_{D1}(T_{j1})$ and $P_{D2}(T_{j2})$, respectively. Therefore, the junction temperature at the IGBT working point should comply with

$$P_{D1}\left(T_{j1}\right) = P_{loss1}\left(T_{j1}, T_{j2}\right), \quad P_{D2}\left(T_{j2}\right) = P_{loss2}\left(T_{j1}, T_{j2}\right) \tag{10.18}$$

If such a working point is stable, it should also comply with

$$\frac{\partial P_{D1}}{\partial T_{j1}} > \frac{\partial P_{loss1}}{\partial T_{j1}}, \quad \frac{\partial P_{D2}}{\partial T_{j2}} > \frac{\partial P_{loss2}}{\partial T_{j2}} \tag{10.19}$$

Obviously if the voltage is balanced between two series connected IGBTs, their junction temperature should be identical. With the voltage imbalance caused by the turn-off delay, at the same junction temperature, the early turned-off IGBT has higher switching-off loss, i.e., $P_{loss1}(T_{j1}, T_{j2}) \neq P_{loss2}(T_{j1}, T_{j2})$. While the heatsink design for IGBTs is usually uniformed, i.e., $P_{D1}(T_{j1}) = P_{D2}(T_{j2})$ when $T_{j1} = T_{j2}$, Eq. (10.18) indicates that the junction temperatures of two IGBTs at the steady state are different.

Assume T_1 has higher switching loss than T_2 at the same junction temperature. Then the junction temperature of T_1 increases, which will affect T_1 in two aspects. One is that the conduction loss increases, and the other is that the increased junction temperature will slow down dU_{CE}/dt, thereby weakening the voltage imbalance and decreasing its transient loss. In addition, the temperature increment will enlarge the tailing current and increase its transient loss. Given the conduction loss increases greatly with the temperature, usually $\partial P_{loss}/\partial T_j > 0$.

Shown in Fig. 10.16 is loss and thermal dissipation versus the junction temperature. The thermal dissipation can be approximated with straight lines, with the slope

Fig. 10.16 The junction temperature of the devices in steady state

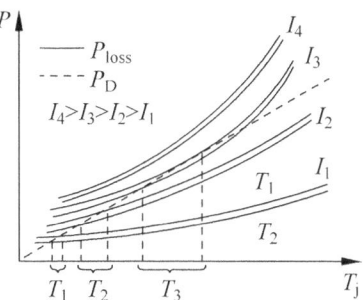

of which as the thermal resistance. The loss curve is a concave curve and related to the rms value of the load current. Due to the voltage imbalance, the loss curve of T_1 is higher than T_2. Under smaller current such as I_1 and I_2, the intersectional points of the loss and thermal dissipating curves are $(T_{j1,1}, T_{j2,1})$ and $(T_{j1,2}, T_{j2,2})$, complying with stability condition of Eq. (10.19). With the load current increasing to I_3, T_1, the switch with higher loss between two series connected IGBTs reaches the critical stability at $T_{j1,3}$ of the thermal dissipation and the loss. At this working point the junction temperatures of T_1 and T_2 are $T_{j1,3}$ and $T_{j2,3}$, respectively, with $T_{j1,3} > T_{j2,3}$, though the system is still stable. Of course if the voltage balance is realized, the loss curve of T_1 will move downwards and that of T_2 will move upwards, which makes both T_1 and T_2 work at some junction temperature between $T_{j1,3}$ and $T_{j2,3}$. In this case the load current can still be increased. In another word, even though the voltage imbalance among series connected IGBTs will not lead to uncontrollable junction temperature, it does shrink the utilization of the whole system.

10.1.3 Analysis of the Transients at the Current Tailing Stage

In Sect. 10.1.1, we have pointed out that the voltage imbalance in the current falling process results in different tailing current, which further causes more severe voltage imbalance. Quantifying such voltage imbalance through modelling this stage is of importance. However, majority of the models can only provide numerical solutions. Simplifying such models can generate relatively simple analytical equations. Analysis of the transient voltage imbalance during the current falling stage in Sect. 10.1.2 indicates that the difference of turn-on/off delay is the main influential factor. Therefore, in this section we will mainly focus on the analytical solution of the switching delay versus the voltage imbalance in the current tailing process.

Given majority of the converters are working at the constant DC-bus voltage, e.g., 5 kV for two HV IGBTs in series connection in this chapter, the voltage imbalance at the current falling stage is mainly due to the turn-off current, i.e., the load current I_L and the turn-off delay difference Δt_d. Furthermore, the voltage imbalance at this stage determines the difference of the tailing current and its related voltage imbalance.

Fig. 10.17 Equivalent circuits of two series connected IGBTs

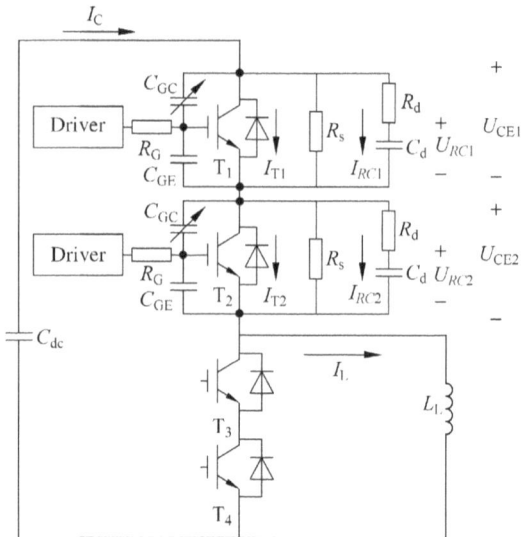

Here we will quantify the relationship between the voltage imbalance in the current tailing process and I_L, Δt_d.

For simplicity of the analysis, we assume:

(1) The tailing current is an exponential function of time with the time constant of τ_1. The difference of the tailing current is caused entirely by the initial current difference of the tailing stage;

(2) the base carriers are Q instead of Q_H and Q_L. The lifetime of current carriers is τ;

(3) the influence of the non-quasi-steady-state hypothesis is ignorable;

(4) the effective width of the base W is the same as the base width W_B.

1. The relationship between the tailing current and its voltage imbalance

The equivalent circuit of two series connected IGBTs is shown in Fig. 10.17. Shown in Fig. 10.18 is the experimental result where the DC-bus voltage for two series connected IGBTs is 5 kV, the load current is 600 A, and the difference of the turn-on delay is 500 ns. At $t = t_0$, U_{CE} rapidly increases. At $t = t_1$, I_C drops to the tailing current. The tailing process starts. At $t = t_2$, the difference between U_{CE1} and U_{CE2} does not increase any more, which means the impact of the tailing current diminishes. In the current tailing process, I_C mainly contains two parts, i.e., the equivalent tailing current of the IGBT I_T and the RC snubber current I_{RC}.

$$I_C = I_{T1} + I_{RC1} = I_{T2} + I_{RC2} \tag{10.20}$$

Difference of I_{RC} induces the gap of the voltage across the snubber capacitor C_d, i.e.,

Fig. 10.18 Experimental voltage imbalance in the current tailing process

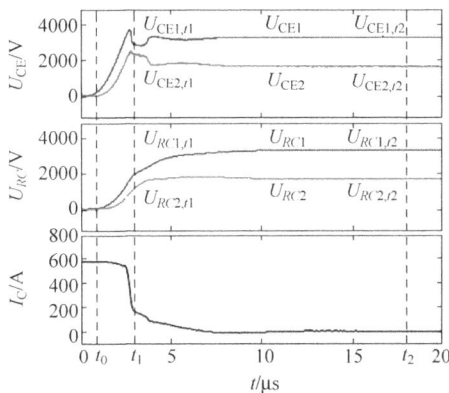

$$\Delta U_{RC} = \Delta U_{RC,t2} - \Delta U_{RC,t1} = \frac{1}{C_d} \int_{t_1}^{t_2} (I_{T2} - I_{T1}) dt \tag{10.21}$$

Assume the tailing current is

$$I_T(t) = I_{T,t1} \exp\left(-\frac{t - t_1}{\tau_i}\right) \tag{10.22}$$

Define $I_{T2} - I_{T1}$ at $t = t_1$ as $\Delta I_{T,t1}$. When the snubber resistance R_d is small, U_{CE} can be formulated as

$$\Delta U_{CEt} = \Delta U_{CE,t2} - \Delta U_{CE,t1} = \frac{\tau_i - R_d C_d}{C_d} \Delta I_{T,t1} \tag{10.23}$$

Thus acquiring the difference of the initial tailing current through two series connected IGBTs will determine the voltage imbalance in the current tailing process.

2. **The change of current carriers in the current falling process**

A two-dimension distribution of the base current carriers has been modelled in Sect. 10.1.1. In such a model, when the IGBT conducts the current, the area of the hole current is $A - A_{gd}$. With the gate being covered by the depletion region, the area of the hole current becomes A. The depletion region covers the gate in a much faster pace than the change of current carriers. Therefore such a process can be simplified into two stages, as shown in Fig. 10.19. One stage is without the depletion region. The other stage is when depletion region fully covers the gate. In each stage, the change of current carriers complies with following equations, respectively.

$$\frac{dQ}{dt} = \frac{D_n A}{D_n A + D_p(A - A_{gd})} I_c - \frac{4 D_n D_p(A - A_{gd}) k_t Q}{D_n A + D_p(A - A_{gd})} - \lambda_Q Q^2 - \frac{Q}{\tau} \tag{10.24}$$

$$\frac{dQ}{dt} = \frac{D_n}{D_n + D_p} I_c - \frac{4 D_n D_p k_t Q}{D_n + D_p} - \lambda_Q Q^2 - \frac{Q}{\tau} \tag{10.25}$$

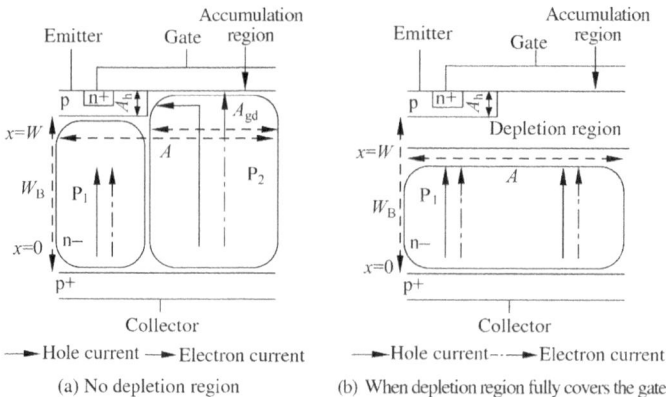

Fig. 10.19 Two stages of the two-dimension base

After neglecting the non-quasi-steady-state hypothesis and assuming the effective width of the base to be W_B, we can conclude that dp/dx is proportional to Q. Here k_t is the proportion coefficient, which is related to the base width and carrier lifetime. λ_Q is not the same as λ, but can be calculated through λ. Assigning $dQ/dt = 0$ in Eq. (10.24), we can calculate the amount of carriers when the IGBT is on:

$$Q_{t0} = \sqrt{k_1^2 + k_2^2 I_L} - k_1 \tag{10.26}$$

where

$$\begin{cases} k_1 = \dfrac{4 D_n D_p (A - A_{gd}) \tau k_t + (D_n A + D_p (A - A_{gd}))}{2 \lambda_Q (D_n A + D_p (A - A_{gd})) \tau} \\[4mm] k_2 = \sqrt{\dfrac{D_n A}{\lambda_Q (D_n A + D_p (A - A_{gd}))}} \end{cases} \tag{10.27}$$

With small I_L, Q_{t0} can be expanded with first-order Taylor Series. When I_L is large, k_1 can be ignored. Therefore, a piece-wise function can be employed to express Q_{t0} as follows.

$$Q_{t0} = \begin{cases} \dfrac{k_2^2}{2 k_1} I_L & I_L < \dfrac{4 k_1^2}{k_2^2} \\[4mm] k_2 \sqrt{I_L} & I_L \geq \dfrac{4 k_1^2}{k_2^2} \end{cases} \tag{10.28}$$

When the depletion region fully covers the gate, the carrier amount Q reduces. With the assumption of $dQ/dt = 0$ in Eq. (10.25), the steady-state value of Q can be derived as

$$Q_{tf} = \sqrt{k_{1f}^2 + k_{2f}^2 I_L} - k_{1f} \tag{10.29}$$

where

$$k_{1f} = \frac{4 D_n D_p \tau k_t + (D_n + D_p)}{2\lambda_Q (D_n + D_p)\tau}, \quad k_{2f} = \sqrt{\frac{D_n}{\lambda_Q (D_n + D_p)}} \tag{10.30}$$

The time constant for Q decaying is

$$\tau_c = \frac{1}{2\lambda_Q \sqrt{k_{1f}^2 + k_{2f}^2 I_L}} \tag{10.31}$$

Such a time constant is relatively large compared to $t_{10} = t_1 - t_0$. For simplicity of the calculation, we can approximately attribute the decrement of Q to the increment of the hole current I_{pW} at $x = W$, i.e.,

$$\frac{dQ}{dt} = -\Delta I_{pW} \tag{10.32}$$

Furthermore, such an increment of I_{pW} is due to the area growth of the hole current, i.e.,

$$\Delta I_{pW} = \alpha (I_L + k_t D_n Q_{t0}) \tag{10.33}$$

where

$$\alpha = \frac{D_n D_p A_{gd}}{(D_n + D_p)(D_n A + D_p (A - A_{gd}))} \tag{10.34}$$

Using Eqs. (10.28) and (10.33), we can formulate the carrier amount in the base at the current tailing moment as

$$Q_{t1} = \begin{cases} \left[\frac{k_2^2}{2k_1}(1 - \alpha k_t D_n t_{10}) - \alpha t_{10} \right] I_L & I_L < \frac{4k_1^2}{k_2^2} \\ k_2(1 - \alpha k_t D_n t_{10})\sqrt{I_L} - \alpha t_{10} I_L & I_L \geq \frac{4k_1^2}{k_2^2} \end{cases} \tag{10.35}$$

where the initial current of the tailing process is

$$I_{T,t1} = 4 D_p k_t Q_{t1} \tag{10.36}$$

Based on the analysis above, the change of the base current carriers at the turn-off transient can be illustrated in Fig. 10.20. Meanwhile, Eqs. (10.35) and (10.36) bridge the initial tailing current with the load current and employ the piece-wise function to

Fig. 10.20 The base current carriers during the turn-off transient

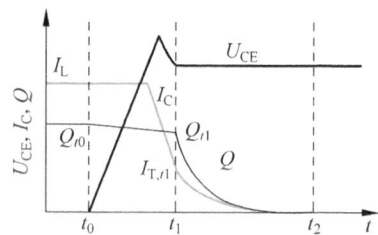

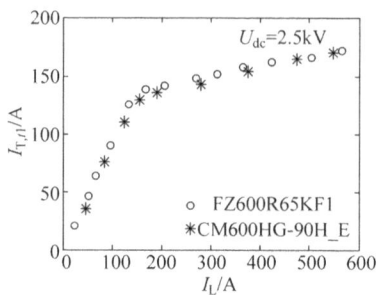

Fig. 10.21 The initial tailing current versus the load current

formulate the tailing current, which has been tested and verified with several different IGBTs under the DC-bus voltage of 2.5 kV, as shown in Fig. 10.21.

For series connected IGBTs, the one turned off early will have its U_{CE} climbing up first; i.e., its t_{10} is longer than the one turned-off late. Assume the time difference of the turn-off delay between series connected IGBTs is Δt_d; at the initial moment of the tailing current, the difference of the carrier amount is

$$\Delta Q_{t1} = \begin{cases} \left(\frac{k_2^2}{2k_1}k_t D_n + 1\right)\alpha I_L \Delta t_d & I_L < \frac{4k_1^2}{k_2^2} \\ \left(k_2 k_t D_u \sqrt{I_L} + I_L\right)\alpha \Delta t_d & I_L \geq \frac{4k_1^2}{k_2^2} \end{cases} \tag{10.37}$$

If we bring together Eqs. (10.21), (10.22) and (10.36), we can formulate the voltage difference across of snubber capacitors in the current tailing process as

$$C_d \Delta U_{RC} = 4 D_p \tau_t k_t \Delta Q_{t1} \tag{10.38}$$

When we further combine the above with Eq. (10.37) and (10.38), we can formulate ΔU_{RC} formulate as

$$C_d \Delta U_{RC} = \begin{cases} (a I_L + I_L)c\Delta t_d & I_L < \left(\frac{b}{a}\right)^2 \\ (b\sqrt{I_L} + I_L)c\Delta t_d & I_L \geq \left(\frac{b}{a}\right)^2 \end{cases} \tag{10.39}$$

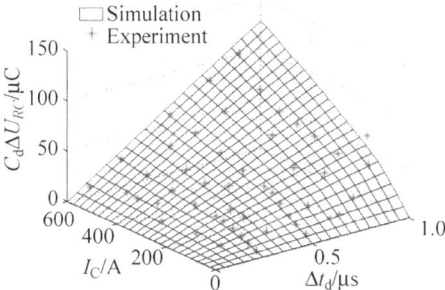

Fig. 10.22 The increment of voltage imbalance in the current tailing process versus the load current and the turn-off-delay difference

where a, b and c are all related to the IGBT structure.

3. **Experimental validation**

The voltage imbalance across the snubber capacitor, ΔU_{RC} versus the turn-off-delay difference and the load current has been formulated in Eq. (10.39). For one specific IGBT, $a = 2.09$, $b = 22.9$ and $c = 0.11$. Simulation and experimental results of the voltage imbalance in the current tailing process are compared in Fig. 10.22. Here the experimental curves are shown in Fig. 10.18. ΔU_{RC} is the voltage difference of snubber capacitors between t_1 and t_2. It shows that the simulation and experimental results are perfectly aligned with each other. The effectiveness of using Eq. (10.39) to approximate the voltage imbalance at the current tailing stage has been verified.

With $\Delta t_d = 250$ ns and 500 ns, respectively, the voltage imbalance versus the load current in the current tailing process is presented in Fig. 10.23a, where a clear piece-wise function is also shown. With $I_L = 175/a$ and 580 A, respectively, the voltage imbalance versus the turn-off-delay difference in the current tailing process is shown in Fig. 10.23b, showing strong linear relationship. Both can be explained by Eq. (10.39).

10.2 SiC Device Based High-Frequency Converters

Wide-bandgap (WBG) materials such as silicon carbide (SiC) have come into focus due to their superior electrical and physical characteristics. We have witnessed recent development of devices using those materials, with attempts to utilize such devices to power electronics systems so as to significantly improve system performance. In theory, SiC material has higher breakdown dielectric and thermal conductivity than Si. In high-frequency, high-power-density, and high-temperature applications SiC devices are more promising.

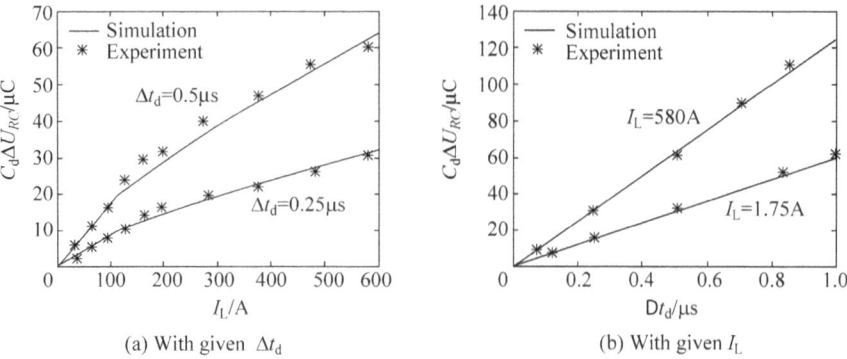

Fig. 10.23 The increment of voltage imbalance versus Δt_d and I_L in the current tailing process

10.2.1 Analysis and Modelling of Switching Transients

In high-frequency power electronic converters, transient commutations caused by switching actions will highly affect switching loss, switching speed, pulse distortion, EMC, and etc. Therefore, analyzing and modelling such switching transient is of high importance. Since transient characteristics of the commutating loop of SiC devices are different from those of Si devices, in this section, we will first discuss the characteristics of transient commutations based on SiC devices and then analyze and model the turn-on and turn-off transient process quantitatively by experimental results.

1. Switch transient characteristics with its basic hypothesis

For the MOSFET based VSCs, regardless of the topology variety, switching transients of a single device are always the same as the one shown in Fig. 10.24. Therefore the analysis, simulation, and experiment of such circuit can be excellent reference, with its related Pspice model shown in Fig. 10.25. Here R_s and L_s are the transient damping resistance and stray inductance, respectively. C_{p1} and C_{p2} are the parasitic capacitance of probes. L_1 is the stray inductance of the load connection and C_1 is the parasitic capacitance of the load inductor. The gate-drive voltage of the SiC MOSFET is $-4/+20$ V. The external turn-on and turn-off gate resistance is 4 and 2.5 Ω, respectively.

The current commutations of SiC MOSFET and diodes, in addition to sharing some common features with Si devices, such as the stray inductance and switching modes, have their own unique features as follows:

(1) Nearly no reverse recovery process due to absence of minor-wiping-out process in SiC diode. This significantly reduces the current spike in the reverse recovery process when the diode is hard turned off, which is related to the hard turn-on of the complementary MOSFET. However, the current spike still exists because reverse biasing the diode leads to charging the diode junction capacitance. On

Fig. 10.24 The test setup of one single switch

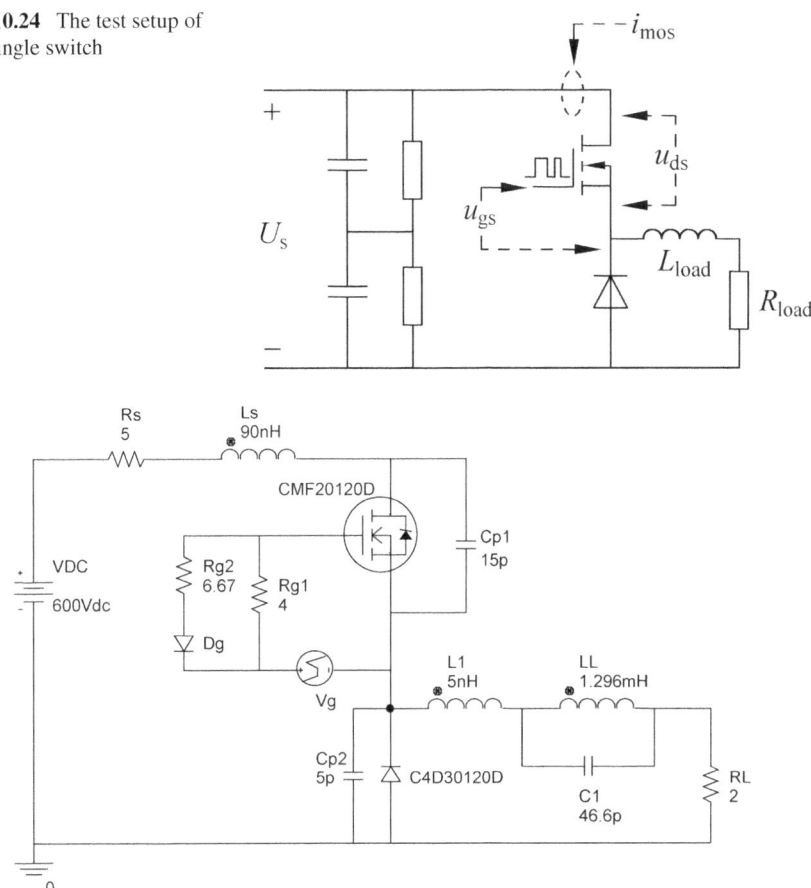

Fig. 10.25 Pspice simulation model of the single-switch setup

one hand, such current spike will add extra switching-on loss of the MOSFET. On the other hand, compared to the current spike caused by the reverse recovery process of the conventional diode, that of SiC diode has shorter time constant, e.g., higher equivalent frequency thereby more severe EMI.

(2) Higher doping concentration and narrower depletion region of the SiC diode, due to higher breakdown electric strength. The setback, however, is its higher junction capacitance than Si devices. This not only creates higher current spike described in (1) but also intensifies oscillation in switching transients. Besides, the high doping concentration and narrow depletion region reduces its damping resistance, and the SiC MOSFET has lower on-state resistance than Si devices at the same rating. These features make the electric spike and oscillation during the current commutation of SiC devices worse than Si ones.

Fig. 10.26 Parasitic capacitance of the power MOSFET

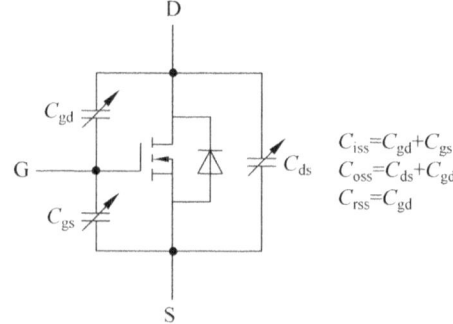

Fig. 10.27 Parasitic capacitance of one MOSFET versus the voltage drop

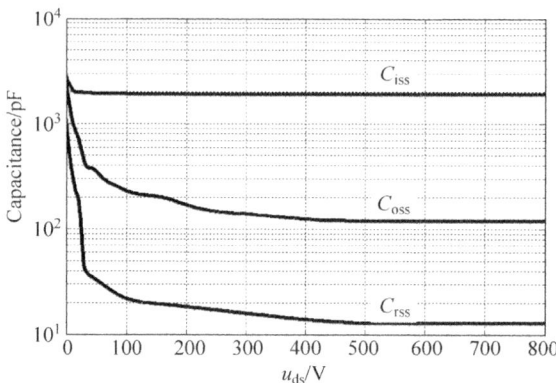

(3) Negligible miller effect. For regular MOSFETs, the voltage drop across switches will rapidly change during switching actions. The gate current is mainly charging/discharging the miller capacitance, resulting in a plateau on the gate voltage, i.e., the miller plateau. For SiC MOSFETs, the related C_{gd} is relatively small. With sufficient gate-drive capability, the charging/discharging process of C_{gd} does not greatly influence the gate voltage. Therefore, the miller effect can be ignored.

(4) The power MOSFET has three critical parasitic capacitance, as shown in Fig. 10.26. All three capacitance along with the diode junction capacitance affect the transient commutating process. As shown in Fig. 10.27, all these capacitance has strong non-linearity. So does the diode junction capacitance. On the other hand, when MOSFETs are working at different states, such parasitic capacitance has different effects. For instance, it has little impact on low-frequency or slow-switching applications, which can thus be simplified and even neglected. However, for SiC based high-frequency converters, accurate analysis and prediction of switching transients is a must. Therefore, such capacitance must be considered in the transient analysis.

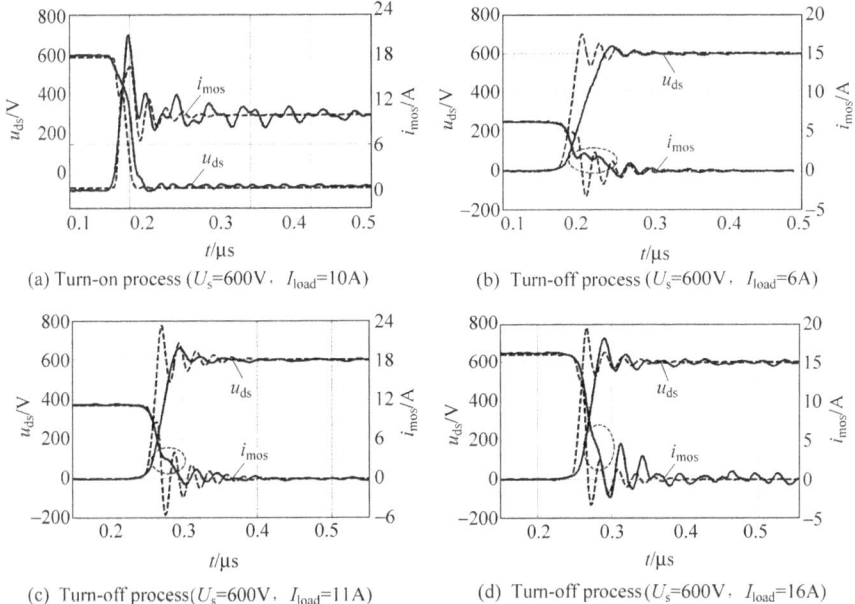

Fig. 10.28 Simulated and experimental switching transients without considering the nonlinearity of the junction capacitance

It is worth pointing out that the conventional Spice model of SiC MOSFETs only details C_{gd} while using constant values for C_{gs} and C_{ds}. Since the nonlinearity of C_{gs} is relatively weak, using the constant value for C_{gs} is acceptable. However, without considering the nonlinearity of C_{ds}, significant discrepancies will be exhibited between experimental results and simulation. Same thing also happens to the diode junction capacitance. Shown in Fig. 10.28 is the transient waveform of the SiC MOSFET based on the Pspice model, including one turn-on waveform and three turn-off waveforms at different load current. The simulation circuit is shown in Fig. 10.25. To emphasize the importance of nonlinear characteristics of C_{ds}, both C_{ds} and diode junction capacitance C_D are set as constants, the values of which can be determined by the steady high-voltage zone. In the figures, dashed lines are the simulation while solid lines are the experiments. Note all experimental waveforms were filtered by a Gauss low-pass filter with the cutoff frequency of 150 MHz.

As shown in Fig. 10.28, at least three major differences exist between the simulation and experiments.

(1) In the turn-on process the simulated current spike i_{mos} is much smaller than the one in experiments;
(2) In the turn-off process, the simulated rising rate of u_{ds} is much higher than the one in experiments;
(3) For turn-off waveforms shown in Fig. 10.28b–d, a very obvious plateau exists in the falling process of i_{mos}, which shrinks with the current reduction and even

Fig. 10.29 The equivalent
circuit of a single-switch test
bench

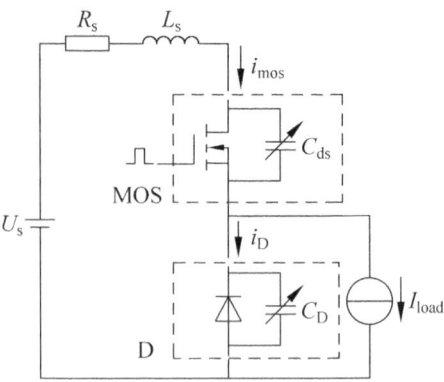

evolves into a reverse spike or downwards oscillation in Fig. 10.28b. This will
create negative impact on the turn-off loss and time. However, such phenomenon
cannot be accurately revealed by simulation results in Fig. 10.28.

Obviously, the discrepancies above will distort accuracy of the calculation and
simulation of related transient features, such as switching loss, switching time and
voltage, and current stress. As explained in the analysis below, after taking full
consideration of nonlinear parasitic capacitance, we can explain and eliminate all
these discrepancies.

To analyze and model switching behaviors in the test bench shown in Fig. 10.24,
we can simplify its related transient model shown in Fig. 10.25 into the one shown
Fig. 10.29.

Several assumptions and clarifications are needed for such modeling.

(1) In switching transients, the DC-bus voltage is equivalent to a constant DC volt-
    age source and the load current is equivalent to a constant DC current source;
(2) All parasitic inductance in the main-power loop is merged into one lumped
    inductance L_s. Meanwhile, R_s represents the high-frequency damping resistance
    in the transient process of the whole main-power circuit;
(3) The diode reverse recovery process is ignored, which is acceptable for SiC
    diodes;
(4) With small external gate resistance and sufficient gate-drive power, the miller
    effect is negligible;
(5) For the theoretical analysis, the parasitic capacitance of probes is negligible;
(6) Given the important effect of the MOSFET parasitic capacitance C_{ds} and the
    diode junction capacitance C_D on the transient commutating process of the
    main-power loop, we have particularly emphasized these two capacitance in
    Fig. 10.29.

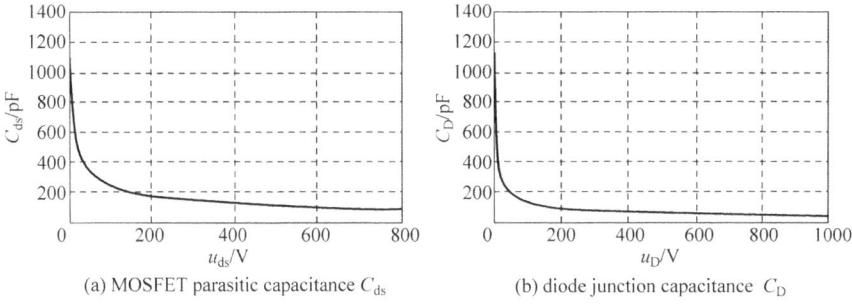

Fig. 10.30 Curve fitting of C_{ds} and C_D

Fig. 10.31 Cross-section area of an exemplary power MOSFET

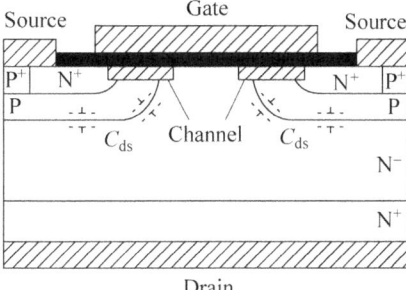

Specifically, we will emphasize the following two aspects of the nonlinearity of C_{ds} and C_D.

(1) The capacitance varies with the voltage. Shown in Fig. 10.30 is the capacitance curve fit through the data in the switch datasheet. We can see that in the low-voltage region, both capacitance values drop sharply with the voltage increasing. In the high-voltage region, the capacitance changes slowly with the voltage;

(2) The parasitic capacitance of the MOSFET, C_{ds}, is mainly induced by the reverse biased PN junction between the drain and the source, as shown in Fig. 10.31. When the MOSFET is on, C_{ds} will be discharged through low resistance of the MOSFET channel. Meanwhile, the channel will vary distribution of holes and electrons in the PN junction to some extent, thereby reducing C_{ds}. Therefore, the role of C_{ds} is significantly weakened in the on-state until the switch is turned off with the conduction channel gradually diminishing. Note in Figs. 10.27 and 10.30a, C_{ds} values are all based on $u_{gs} = 0$.

2. Analysis and modelling of the turn-on transient

The experimental turn-on waveform at $U_s = 600$ V and $I_{load} = 10$ A is shown in Fig. 10.32, based on which we can analyze and model the switching transient of a single-switch circuit. Beginning from the moment when the MOSFET gate voltage reaches the turn-on threshold, the whole turn-on process can be divided into three stages.

Stage 1 (t_1–t_2), which starts from the moment when the gate voltage reaches the turn-on threshold U_{th}, has its transient equivalent circuit shown as Fig. 10.33a. Within this stage the voltage drop u_{ds} is still quite high. Therefore, the MOSFET works in the saturation region, acting as a voltage controlled current source with its channel current i_{ch} controlled by the gate voltage, i.e.,

$$i_{ch}(t) = g_{fs}[u_{gs}(t) - U_{th}] \tag{10.40}$$

where g_{fs} is the trans-conductance of the MOSFET.

At $t = t_1$, i_{ch} is fully utilized to discharge C_{ds}. With u_{ds} dropping, L_s begins to undertake the positive voltage. Externally we can observe i_{mos} increasing. From now on, i_{ch} is split into two parts: one is the device current i_{mos} which can be observed externally, and the other is internal discharging current of C_{ds}. Meanwhile, during this whole stage, the freewheeling diode is always on, clamping the bridge output voltage to zero. Therefore, the rising rate of i_{mos} is solely determined by the voltage drop across the MOSFET, u_{ds}. When using u_{ds} and i_{mos} as state variables, we can set up state space equations of this stage as

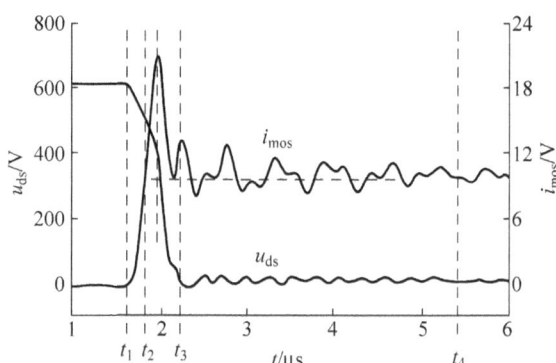

Fig. 10.32 Experimental turn-on waveform of one SiC MOSFET

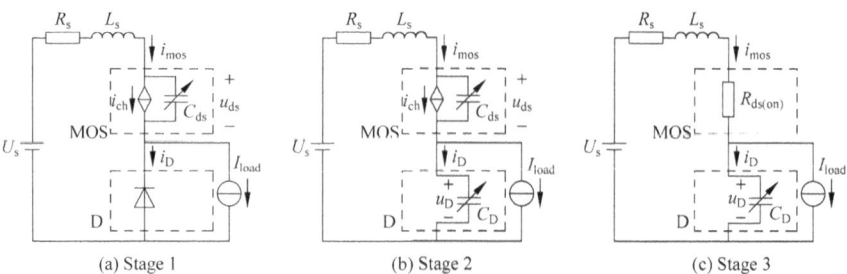

(a) Stage 1   (b) Stage 2   (c) Stage 3

Fig. 10.33 Equivalent circuits of all stages in the turn-on process

$$\begin{cases} U_S - u_{ds}(t) = R_s i_{mos}(t) + L_s \dfrac{di_{mos}(t)}{dt} \\ i_{mos}(t) - i_{ch}(t) = C_{ds}(u_{ds}) \dfrac{du_{ds}(t)}{dt} \end{cases} \tag{10.41}$$

where C_{ds} is the function of u_{ds}.

It is worth pointing out that with u_{ds} decreasing, i_{mos} is always increasing rapidly in an exponential manner. However, based on the previous assumption, i_{ch} is only linearly increasing with u_{gs}. Therefore, the discharging current of C_{ds} is quite limited. In addition, C_{ds} does increase with u_{ds} sliding, which is not very obvious. In this stage, the dropping rate of u_{ds} is relatively low. At some points, the voltage dropping trend even slows down.

When $i_{mos} = I_{load}$, the stage 1 ends.

Stage 2 (t_2–t_3): when $i_{mos} = I_{load}$, the freewheeling diode becomes reverse biased. The corresponding equivalent circuit of this transient process is shown in Fig. 10.33b. Without wiping out minors, the diode can be modelled as the junction capacitance charged with the current as $i_D = i_{mos} - I_{load}$. Therefore, to complete the charging of the diode junction capacitance, i_{mos} has to continue increasing, which eventually induces the turn-on current spike shown in Fig. 10.32. At this stage, u_{ds} usually complies with

$$u_{ds} > u_{gs} - U_{th} \tag{10.42}$$

Therefore, the MOSFET is still saturated, with the equivalent circuit of the commutating process in the upper leg similar to the Stage 1. Adding another extra state variable u_D, we can set up the state space equations of this stage as

$$\begin{cases} U_S - u_{ds}(t) - u_D(t) = R_s i_{mos}(t) + L_s \dfrac{di_{mos}(t)}{dt} \\ i_{mos}(t) - i_{ch}(t) = C_{ds}(u_{ds}) \dfrac{du_{ds}(t)}{dt} \\ i_{mos}(t) - I_{load} = C_D(u_D) \dfrac{du_D(t)}{dt} \end{cases} \tag{10.43}$$

Similar to Eq. (10.41), C_D is also a function of u_D. Note once i_{mos} reaches the peak, it will drop sharply, which makes i_{ch} quickly charge C_{ds}. At the same time, with the switch being turned on, the formation of the conduction channel weakens C_{ds}. This process expedites the decrement of u_{ds}, as shown in the period of t_2–t_3 in Fig. 10.32.

At $t = t_3$, u_{ds} is too low to comply with the precondition of Eq. (10.42). Now, the MOSFET can be assumed as fully turned on, with Stage 2 ended.

Stage 3 (t_3–t_4): In this stage, the MOSFET is fully turned on and equivalent to the constant resistance R_{ds_on}. The equivalent circuit of this transient is shown in Fig. 10.33c, a second-order resonance circuit made of L_s and C_D. Eventually, the

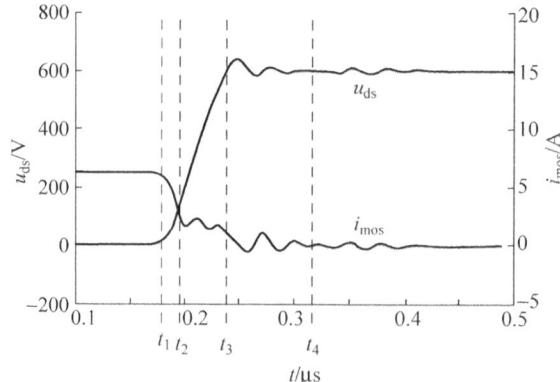

Fig. 10.34 The experimental waveform of the SiC MOSFET in the turn-off process

circuit reaches the steady state due to the damping effect of R_{ds_on} and R_s, with the state space equation as

$$\begin{cases} U_S - u_D(t) = (R_s + R_{ds\_on})i_{mos}(t) + L_s \dfrac{\mathrm{d}i_{mos}(t)}{\mathrm{d}t} \\ i_{mos}(t) - I_{load} = C_D(u_D)\dfrac{\mathrm{d}u_D(t)}{\mathrm{d}t} \end{cases} \tag{10.44}$$

In real practice, due to the parasitic capacitance of the load inductor, some oscillation of the load current will appear when the bridge output voltage changes. Such oscillation will be further introduced to i_{mos} and, along with the bridge loop, will further create the irregular oscillation of i_{mos}, as shown in Fig. 10.32.

In some extreme cases, the voltage drop of the MOSFET, u_{ds}, might be low enough before $t = t_2$, contrary to the precondition of Eq. (10.42). This will result in the absence of Stage 2. The MOSFET will fully turn on in Stage 1, form a RL circuit made of R_s, R_{ds_on} and L_s, charge the circuit with U_s, and enter Stage 3 directly after $i_{mos} = I_{load}$.

Simulation waveforms of Fig. 10.28a indicate that right at the beginning of Stage 2, the actual value of C_D at such low voltage is much higher than that in the simulation. This means the charging process of C_D is much slower and u_D is harder to rise. Therefore, i_{mos} climbs faster under U_s, creating a higher current spike. This explains why the simulated current spike in Fig. 10.28a is much smaller than the actual one.

3.   Analysis and modelling of the turn-off transient

In the turn-off process, to focus on the current plateau, we select the actual turn-off waveforms at low current ($U_s = 600$ V, $I_{load} = 6$ A), as shown in Fig. 10.34. With the current dropping moment as the initial point, the turn-off process can similarly be divided into three stages.

Stage 1 (t_1–t_2): At $t = t_1$, the MOSFET gate voltage begins to drop, weakening its conduction channel thereby increasing its resistance. Meanwhile, u_{ds} begins to increase and meet the precondition of Eq. (10.42). The switch transitions from the constant channel resistance to the saturation region, turning the switch into a voltage controlled current source. On the other hand, the rise of u_{ds} leads to a negative voltage across the leakage inductance, causing i_{mos} to start dropping. Given that $i_{mos} < I_{load}$, the diode junction capacitance C_D also begins to discharge. The transient equivalent circuit is shown in Fig. 10.35a. It is worth mentioning that the channel still exists and plays an important role, making the behavior of C_{ds} unable to be fully exhibited. Therefore, the voltage drop u_{ds} is mostly determined by the external circuit. The state space equations of this stage are

$$
\begin{cases}
U_S - u_{ds}(t) - u_D(t) = R_s i_{mos}(t) + L_s \dfrac{di_{mos}(t)}{dt} \\[2mm]
i_{mos}(t) - I_{load} = C_D(u_D)\dfrac{du_D(t)}{dt}
\end{cases}
\tag{10.45}
$$

where

$$
i_{mos}(t) = i_{ch}(t) = g_{fs}[u_{gs}(t) - U_{th}]
\tag{10.46}
$$

Stage 2 (t_2–t_3): With the switch being further turned off, the channel disappears, making C_{ds} become dominant, as shown in Fig. 10.35b. This stage can be equivalent to a third-order resonance made of L_s, C_{ds}, and C_D.

At this stage, C_D continues to be discharged. C_{ds} begins to be charged by the current of $i_{mos} - i_{ch}$. Since the channel is nearly gone, i_{ch} is small enough to be ignored. Therefore, we can approximate the charging current of C_{ds} as i_{mos}. The state space equations of this stage are

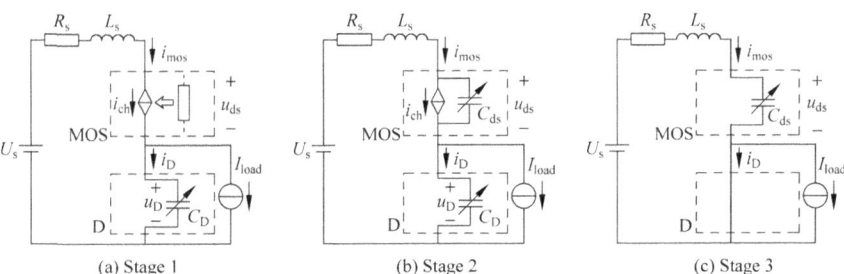

(a) Stage 1   (b) Stage 2   (c) Stage 3

Fig. 10.35 Equivalent circuits of three stages in the turn-off transients

$$\begin{cases} U_{\text{s}} - u_{\text{ds}}(t) - u_{\text{D}}(t) = R_{\text{s}}i_{\text{mos}}(t) + L_{\text{s}}\dfrac{di_{\text{mos}}(t)}{dt} \\[2mm] i_{\text{mos}}(t) = C_{\text{ds}}(u_{\text{ds}})\dfrac{du_{\text{ds}}(t)}{dt} \\[2mm] i_{\text{mos}}(t) - I_{\text{load}} = C_{\text{D}}(u_{\text{D}})\dfrac{du_{\text{D}}(t)}{dt} \end{cases} \tag{10.47}$$

The sum of the voltage across C_{D} and C_{ds} determines the voltage of R_{s} and L_{s} and further affects i_{mos}. Once the following scenario becomes true,

$$u_{\text{ds}} + u_{\text{D}} + R_{\text{s}}i_{\text{mos}} < U_{\text{s}} \tag{10.48}$$

i_{mos} will increase instead, which is the cause of the current plateau or even the inverse oscillation. We will further explain this phenomenon after discussing Stage 3.

Stage 3(t_3–t_4): At $t = t_3$, C_{D} finishes discharging, when the freewheeling diode is completely forward biased. When the diode voltage drop is ignored, the bridge output voltage is clamped to zero. Since the MOSFET is fully turned off, the commutation turns to the second-order resonance made of L_{s} and C_{ds}, which is damped by R_{s} to reach the steady state. The equivalent circuit of such transients is shown in Fig. 10.35c, with state space equations shown as below:

$$\begin{cases} U_{\text{s}} - u_{\text{ds}}(t) = R_{\text{s}}i_{\text{mos}}(t) + L_{\text{s}}\dfrac{di_{\text{mos}}(t)}{dt} \\[2mm] i_{\text{mos}}(t) = C_{\text{ds}}(u_{\text{ds}})\dfrac{du_{\text{ds}}(t)}{dt} \end{cases} \tag{10.49}$$

The current plateau in Stage 2 can be further explained based on following two aspects:

(1) There is a reciprocal relationship between the charging current of C_{ds} and discharging current of C_{D}. The sum of the two current is always I_{load};
(2) If the voltage drop across the high-efficiency damping resistance R_{s} is ignored, i_{mos} will increase if $u_{\text{ds}} + u_{\text{D}} < U_{\text{s}}$, and vice versa.

With the relatively large load current, such as 11 and 16 A in Fig. 10.28c and d, respectively, the value of i_{mos} in Stage 2 is also large, which charges C_{ds} quickly and makes its voltage reach U_{s} before discharging C_{D}. Therefore, throughout Stage 2, it is not possible to have $u_{\text{ds}} + u_{\text{D}} < U_{\text{s}}$. i_{mos} continually decreases. However, at the late Stage 2, the discharging current of C_{D} increases, and in turn the discharging process expedites. This will weaken the negative voltage drop across L_{s} to some extent, thereby slowing down the decrement of i_{mos}, which forms the current plateau in Fig. 10.28. With the load current increasing, such a plateau will fade into a region where current slowly drops and becomes invisible.

When the load current is relatively small, e.g., 6 A in Fig. 10.28b, the small i_{mos} charges C_{ds} slowly, which makes u_{ds} unable to reach U_{s} when the discharging process of C_D is finished, i.e., the end of Stage 2. In such a case, $u_{ds} + u_D < U_s$, which will introduce the reverse rise of i_{mos}. Once i_{mos} begins to increase, the charging process of C_{ds} will expedite while the discharging process of C_D will slow down. This process will quickly result in $u_{ds} + u_D > U_s$, thereby reducing i_{mos} again. If the discharging of C_D finishes before the second reverse rise of i_{mos}, the transient directly enters Stage 3, leaving only one reverse rise for the current. However, if the load current is small and Stage 2 lasts long, multiple reverse current rises might be observed as an oscillation. If during the oscillation, C_{ds} and C_D are treated as constant, theoretically, the average value of the oscillating current is a constant. The current begins to fall upon the completion of Stage 2. In reality, such capacitance has strong nonlinearity as shown in Fig. 10.30, i.e., large capacitance at low voltage and small capacitance at high voltage. During the oscillation, C_{ds} is continually being charged with its capacitance value being reduced. The charging efficiency, i.e., the charging speed at the same current, gradually increases. Thus, the actual charging speed is faster while the discharging speed is slower than anticipated, leading to an overall negative voltage across L_s during the oscillation. Therefore, i_{mos} gradually declines, as shown in Fig. 10.28b. Such a trend will be more visible if the load current is smaller.

To summarize, analysis of both the transition from Stage 1 to Stage 2 and various current plateaus in Stage 2 employs the nonlinear characteristics of capacitance. Such current transients cannot be accurately simulated if the capacitance nonlinearity is not taken into account. In addition, in the majority part of the turn-off transient, the actual value of C_{ds} is higher than the one in the simulation setting in Fig. 10.28. Therefore, the actual rising rate of u_{ds} is relatively slower.

Note the transition from Stage 1 to Stage 2 reveals the diminishing process of the conduction channel and the enforcing process of C_{ds}. This in real practice is a continuous process.

10.2.2 Analysis of Electromagnetic Transients in the High-Frequency Converter

The output-pulse distortion of the high-frequency converter includes the quantization distortion in ideal cases and those distortion caused by nonlinear factors. The non-ideal power pulses generated in switching transients are one typical example. With the switching frequency increasing, the pulse width becomes narrower and the time delay and distortion in switching transients become more prominent. On the other hand, with the switching frequency and the power level increasing, the switching loss and EMI during switching transients become more severe. When the switching frequency increases to a certain level, the switching loss will become dominant among all power losses. With the switching frequency and its power capacity further increasing, the system efficiency declines and the demand on the thermal dissipation increases. If inappropriately designed, the switch can be destroyed due to exceeding its power rating. In the high-power and fast-switching process, the induced spike, oscillation, du/dt, and di/dt will create severe EMI to weak-electricity circuits, such as gate-drive circuit and control circuit. Here the large spike and oscillation can even impact the output-power waveform, thereby causing additional distortion.

Here we will introduce a passive transient damping circuit for a topology called parallel frequency-multiplication topology based on SiC MOSFET, although the theory of such topology will not be discussed in detail here. Such a method has effective suppression on spikes and oscillations without slowing down the switching speed or increasing the steady-state loss. It can enhance system reliability, reduce switching loss to some extent, and alleviate output distortion. Some dynamic processes are inevitable during switching transitions, e.g., the time delay of current and voltage along with their rising and falling processes, which however have already become quite prominent after application of the above method.

1. **Existing problems in switching transients.**

The frequency-multiplication topology used in this test bench is shown in Fig. 10.36. The fundamental structure is an H-bridge. Each switching array is made of three paralleled SiC MOSFETs. One extra SiC diode is anti-paralleled. The single-switch test circuit of such topology is shown in Fig. 10.37, where Q_1 and Q_2 are switching modules made of SiC MOSFETs. For the purpose of comparison, this circuit is used to test switches without parallel (hereinafter referred to as "no-parallel") and with three in parallel (hereinafter referred to as "three-parallel"). Q_2 is always maintained off while only the switch in the Q_1 module is switched on. C_1 and C_2 represent the sum of the capacitance of paralleled switches and the anti-paralleled diode.

As described in Sect. 10.2.1, during current commutation of SiC devices, their damping coefficient is smaller than that of Si devices at the same rating. Therefore, their related spike, oscillation, and other voltage/current transients become more visible. In general, the junction capacitance of all paralleled devices should get charged and discharged simultaneously, causing the equivalent junction capacitance to increase by multiple times. This will further worsen spikes and oscillations during switching transients, thus increasing switching loss, transition time, and EMI.

Exemplary switching waveforms at no-parallel and three-parallel cases are shown in Fig. 10.38. Solid lines are experiments while dashed lines are simulation wave-

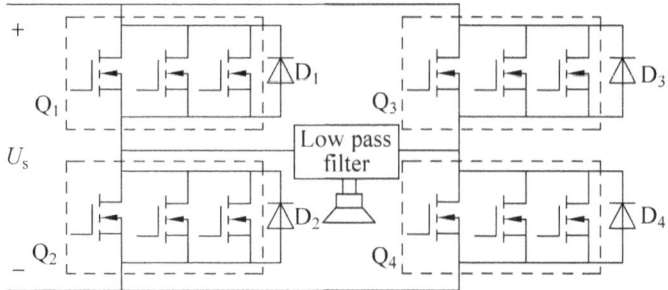

Fig. 10.36 Double-frequency topology with three-paralleled devices

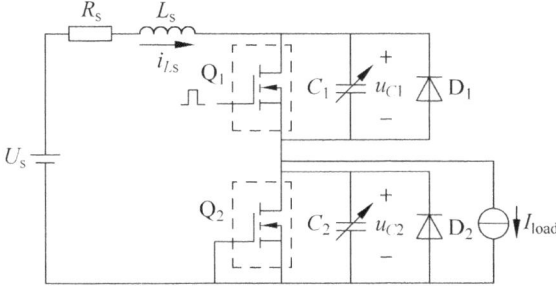

Fig. 10.37 Single-switch test circuit for the double-frequency topology using paralleled switches

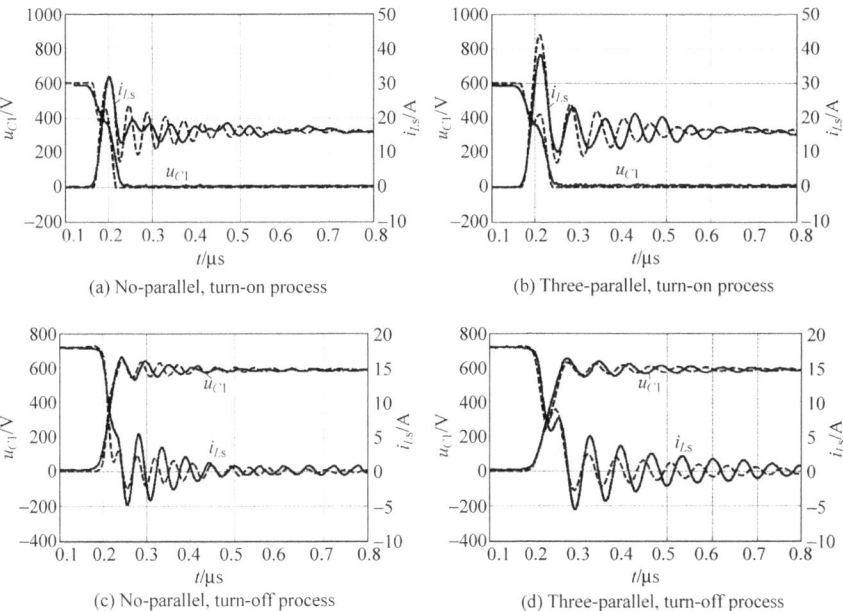

Fig. 10.38 Exemplary switching waveforms in cases of no-parallel and three-parallel

forms, with the enhanced model in Sect. 10.2.1. Given that the measured current flows through the whole switching array instead of one single switch, here we use i_{Ls} to replace i_{mos} in the previous section. To further differentiate the voltage of upper and lower switches, we use u_{C1} to replace the previous u_{ds}.

From figures above, several harmful influential factors require special attention during switching transients.

(1)   Turn-on current spike, which is caused by the charging process of the junction capacitance of the freewheeling diode (C_2, including the junction capacitance of Q_2). This current spike has small time constant and high equivalent frequency range, and is prone to disturbing the control circuit, gate-drive circuit, and the

measurement equipment. In addition, u_{ds} during this current spike has not fully dropped to zero, which creates extra switching-on loss. It is worth pointing out that for the three-parallel case, the junction capacitance of three paralleled MOSFETs (C_{ds}) and one SiC diode needs be charged together, and the overall charge is much larger than the no-parallel case. Therefore, the current spike becomes more severe, as shown in Fig. 10.38a and b. Such a large turn-on current spike will greatly affect the system reliability.

(2) Current oscillation in the switching processes, which also includes the current plateau in the turn-off process, as mentioned in Sect. 10.2.1. Usually the frequency range of such current oscillation is from more than ten MHz to several tens of MHz, which tends to create high-frequency EMI. Without damping such oscillation sufficiently before the next switching action, the extra switching loss is expected. Therefore, such oscillations in turn increases the switching transition time and limits the switching frequency. As described before, the SiC device based main-power circuit has a relatively small damping coefficient, making it difficult to decay the current oscillation. For the three-parallel case, the resonant capacitance is made of several junction capacitances in parallel, with larger capacitance and smaller damping resistance. Therefore, the resonant frequency is lower but the damping period is longer.

(3) Voltage-drop slowing down during the turn-on process.

According to the analysis in Sect. 10.2.1, the decline of u_{ds} is slow in the first part of the turn-on process and becomes even slower until the current passes the peak value. This has been verified in Fig. 10.38. Such a behavior will certainly exacerbate the turn-on loss.

To achieve better switching transients, zero-voltage switching (ZVS) turn-on is recommended, i.e., utilizing extra resonance components with the switch junction capacitance to turn on the anti-parallel diode before the switch is turned on. This will greatly reduce the turn-on loss. However, such technique creates extra current due to the resonance, resulting in larger average and RMS values of the on-state current than the conventional hard-switching technique. Therefore, ZVS turn-on not only increases the on-state loss of the converter, but also increases the requirement of power ratings, which devalues its applicability in the high-power converter. At the same time, various snubber circuits can be used to reduce the switching loss, some of which can suppress the spike and oscillation in switching processes. However all of these circuits slow down du/dt and di/dt in the switching process via prolonging transient processes, which is not desirable in the high-power converter. In addition, adding relatively complex snubber circuits will require extra components, and thus further increases cost and complexity.

From another perspective, if the damping resistance of the commutating loop can be increased appropriately, the spike and oscillation will be effectively suppressed. Its demerit lies in the extra power loss. This is why the passive transient damping circuit was introduced. Such circuit only consists of three elements, i.e., resistor, inductor, and capacitor, and is very easy to implement, adjust, and optimize. The main lossy component is the damping resistor. Through taking advantage of the big gap between the equivalent frequency range in steady states and transient states, the damping resistor will only function in switching transients, thereby adding the transient damping of the related commutating loop. This will effectively suppress current spikes and oscillation, and meanwhile barely create any extra loss in the steady-state. In addition, such method will expedite the voltage drop of the switch in the turn-on process thereby reducing the turn-on loss. With appropriate allocation of parameters, we can expect to greatly improve the transient characteristics of power switches, lower system EMI, enhance system reliability, and reduce switching loss of the converter.

2. Analysis and implementation of the transient damping circuit

Shown in Fig. 10.39 is the referred transient damping circuit, where the right half of the circuit is the same as Fig. 10.37, except that three passive components were added on the DC-bus side, i.e., L_0, R_{damp} and C_{damp}. During switching transients, the large du/dt and di/dt represent much wider frequency range than the steady states, yielding significantly different impedance of each component. Selecting right parameters and especially enlarging L_0 appropriately (compared to L_s) should meet following critical preconditions:

(1) in switching transients

$$Z(L_0)|_{transient} >> [Z(R_{damp}) + Z(C_{damp})]|_{transient} \qquad (10.50)$$

(2) in steady states

$$Z(L_0)|_{steady} << [Z(R_{damp}) + Z(C_{damp})]|_{steady} \qquad (10.51)$$

Therefore, the transient current is mainly provided by the branch made of R_{damp} and C_{damp}, as it reduces the spike and oscillation with the damping effect of R_{damp}. In the steady state, the damping branch is bypassed by the low-impedance L_0 and DC-bus capacitance. Consequently, the steady-state current is mainly flowing through L_0 instead of R_{damp}.

To further verify the function of such transient damping circuit, the simulation waveform is given in Fig. 10.40 for the three-parallel case. Since the transient damping circuit is mainly affecting the turn-on process, here we demonstrate its function

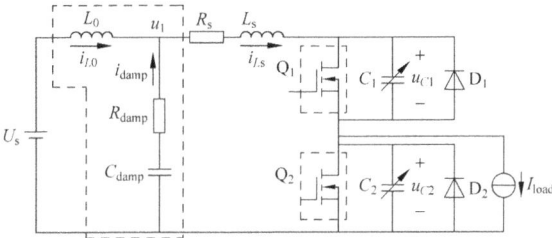

Fig. 10.39 the transient damping circuit

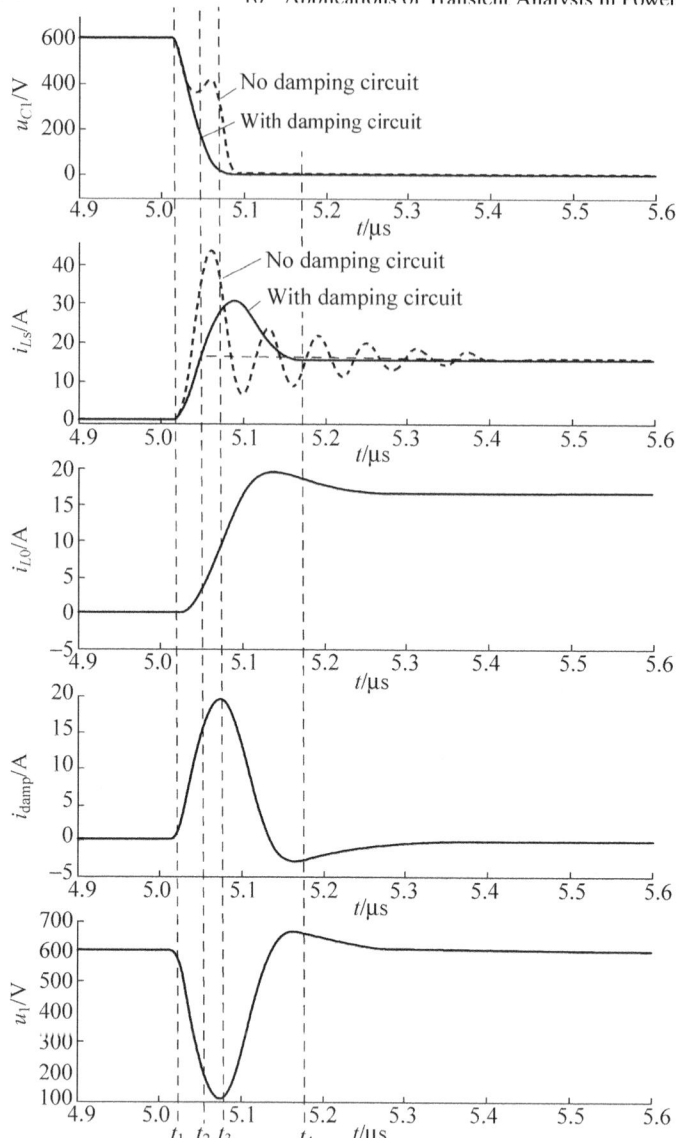

Fig. 10.40 The simulation waveform of turn-on transients in the three-parallel topology w/n the transient damping circuit

only during turn-on transients, i.e., solid lines in the figures. The dashed lines are the switch voltage drop and main-circuit current without the damping circuit.

Similar to Sect. 10.2.1, the turn-on process can be divided into several stages based upon several critical moments, shown as below.

$$\begin{cases} t_1 : u_{gs} = U_{th} \\ t_2 : i_{Ls} = I_L \\ t_3 : u_{C1} = u_{gs} - U_{th} \\ t_4 : \text{steady state} \end{cases} \tag{10.52}$$

The current commutations in this bridge are the same as what has been described in Sect. 10.2.1, which we will not repeat here. As shown in the waveforms, a large L_0, e.g., 1700 nH slows down i_{L0}, resulting in i_{L0} unable to catch up with the switch turn-on current, i_{Ls} when $t < t_2$ and even when $t = t_2 - t_3$. Therefore, part of i_{Ls} can only be compensated by i_{damp} from the damping branch. With the damping effect of R_{damp}, the turn-on current spike is reduced. After $t = t_3$, i_{L0} climbs high enough to reach the load current I_L, when i_{damp} begins to decline towards zero. On the other hand, due to the relatively large time constant caused by L_0, the high-frequency resonant current cannot form the loop through the DC-bus capacitor any more. The only resonant circuit is the damping branch. Simulation results show that the current oscillation has been effectively suppressed as the whole loop is at the critical damping state. Such a theory also holds true for the turn-off transient.

Based on the analysis above, the functions of the three passive components are summarized below:

R_{damp}: damping the current spike and oscillation during switching transients;

C_{damp}: undertaking the DC-bus voltage in the steady state and providing the transient current in switching processes;

L_0: acting as a frequency-selective switch. It passes the low-frequency current, the steady state current, and blocks the high-frequency current, the fast changing current in switching processes. As the key component in the transient damping branch, its value should be selected to slow down i_{L0} in early and middle stages of switching transients so that most of the transient current is provided by the damping circuit, and to allow i_{L0} to reach I_L or 0 as soon as the switching transition is completed so that the static loss of the damping circuit is minimal.

An extra benefit of such a damping circuit branch is that the damping current i_{damp} creates the voltage drop across R_{damp} during the turn-on transient, which yields an instantaneous voltage drop of u_1, as shown in the last plot of Fig. 10.40. Such a voltage drop greatly expedites the voltage dropping rate, thereby reducing the turn-on loss of the MOSFET. Besides, the voltage drop enhances linearity of the voltage pulse and facilitates compensation and trimming of power-pulse edges.

In order to validate the analysis above, the transient damping circuit was added to the double-frequency test bench, with three switches in parallel. The experimental switching transients are shown in Fig. 10.41. As a comparison, the experimental and simulation results are plotted together with solid lines and dashed line in dark color, respectively, and the experimental results without the damping circuit are shown in the solid lines in light color. Aligned with the analysis above, both the current spike and oscillation during turn-on transients are effectively reduced with the transient damping circuit. Meanwhile, the region where the turn-on voltage drops slowly is totally gone.

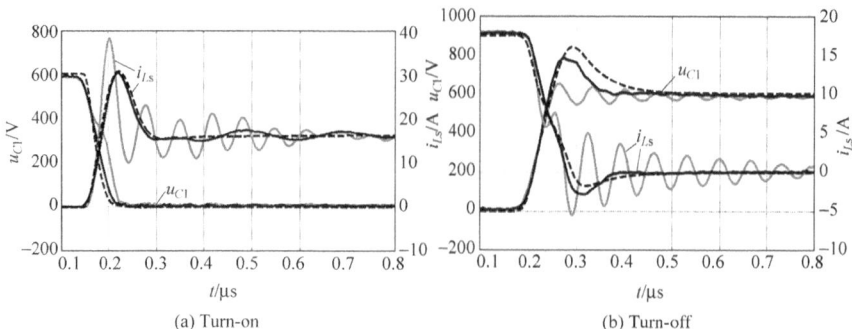

Fig. 10.41 Comparison of the switching waveforms in the three-parallel double-frequency circuit w/n the transient damping circuit

Of course, as a universal approach to improve switching transient performance, the transient damping circuit can be applied in a variety of power electronic converters. Given that problems caused by switching transients are more severe in the double-frequency topology with paralleled devices, this section mainly focuses on the simulation and experiments of the type of topology.

It is worth clarifying several differences between the transient damping circuit and the regular snubber circuit. Firstly, in theory, the inductor and the capacitor in regular snubber circuit are mostly used to reduce the switch di/dt or du/dt, which helps reduce the switching loss and EMI. The inductor of the transient damping circuit, however, acts as the frequency-selective switch without directly influencing the current changing trend of the bridge. Secondly, the capacitor of the damping circuit has the same purpose as the DC-bus capacitors. The only component suppressing the current spike and oscillation is the damping resistor. Thirdly, as described previously, the regular snubber circuit slows down di/dt and du/dt, which increases the switching time, thereby obstructing performance enhancement of the high-frequency converter. The transient damping circuit, however, does not prolong the switching time. It actually expedites the voltage declination, thereby reducing the switching time and loss.

3. Design and parameter optimization of the transient damping circuit

Based on the analysis above, the principle of the transient damping circuit is not complex. The critical part is how to design and optimize related parameters to maximize its effectiveness and minimize its negativity.

Among the three passive components in the transient damping circuit, the selection of C_{damp} is relatively simple. The circuit performance remains insensitive to C_{damp} as well. As described above, the function of C_{damp} is to provide enough current during switching transients without altering its voltage too much. If we take into account the cost and size of the circuit, the smaller the value of C_{damp}, the better. Since the circuit will be working at high frequency, capacitors with excellent high-frequency performance are preferred, e.g., high-frequency ceramic or film capacitors.

L_0 and R_{damp} need be carefully selected for the reason that both the merits and demerits of the circuit are highly related to these two parameters. Therefore, we should quantify the relationship between performance indexes and these two parameters, based on which we can further fine-tune and optimize these two parameters in the real application.

In the turn-on process of the three-parallel circuit equipped with the transient damping branch, the state space equations of each stage can be derived as follows, in reference to the methodology in Sect. 10.2.1. Here each stage is shown in Fig. 10.40 and Eq. (10.52).

1. Stage 1 $(t_1\text{--}t_2)$:

$$\begin{cases} U_S - u_1(t) = L_0\dfrac{di_{L0}(t)}{dt} \\[2mm] u_1(t) - u_{C1}(t) = R_s i_{Ls}(t) + L_s\dfrac{di_{Ls}(t)}{dt} \\[2mm] i_{Ls}(t) - i_{ch}(t) = C_1(u_{C1})\dfrac{du_{C1}(t)}{dt} \end{cases} \tag{10.53}$$

2. Stage 2 $(t_2\text{--}t_3)$:

$$\begin{cases} U_S - u_1(t) = L_0\dfrac{di_{L0}(t)}{dt} \\[2mm] u_1(t) - u_{C1}(t) - u_{C2}(t) = R_s i_{Ls}(t) + L_s\dfrac{di_{Ls}(t)}{dt} \\[2mm] i_{Ls}(t) - i_{ch}(t) = C_1(u_{C1})\dfrac{du_{C1}(t)}{dt} \\[2mm] i_{Ls}(t) - I_L = C_2(u_{C2})\dfrac{du_{C2}(t)}{dt} \end{cases} \tag{10.54}$$

3. Stage 3 $(t_3\text{--}t_4)$:

$$\begin{cases} U_S - u_1(t) = L_0\dfrac{di_{L0}(t)}{dt} \\[2mm] u_1(t) - u_{C2}(t) = (R_s + R_{ds(on)})i_{Ls}(t) + L_s\dfrac{di_{Ls}(t)}{dt} \\[2mm] i_{Ls}(t) - I_L = C_2(u_{C2})\dfrac{du_{C2}(t)}{dt} \end{cases} \tag{10.55}$$

where

$$u_1(t) = U_S - i_{damp}(t)R_{damp} = U_S - [i_{Ls}(t) - i_{L0}(t)]R_{damp} \tag{10.56}$$

Similarly, we can curve fit C_1 and C_2 versus the voltage. Given that C_1 and C_2 are the sum of the junction capacitances of paralleled MOSFETs and anti-paralleled diode, we will assume the above equations can also be applied to the junction capac-

Fig. 10.42 Comparison of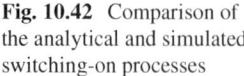
the analytical and simulated
switching-on processes

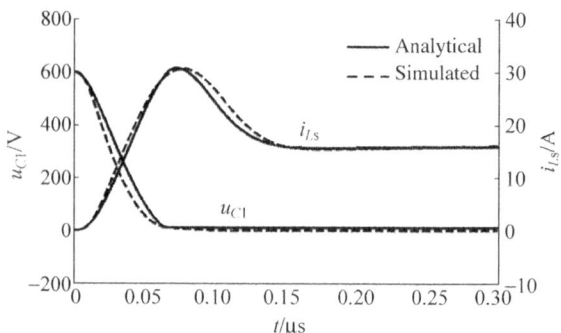

itance of the SiC diode. Since the curve of diode junction capacitance versus voltage
has a trajectory similar to that of the switch junction capacitance, such assumption
is feasible.

Using the analytical equations above, we are able to roughly calculate the turn-
on waveform at different working points, with related characteristic parameters
extracted. Simulation and analytical results are shown in Fig. 10.42. Both exhibit
an excellent alignment.

Based on the analysis above, the major characteristic parameters of the turn-on
process include the current peak value and the overall turn-on loss, both of which
can be calculated with different R_{damp} and L_0 using the equations above. Here we
will ignore the suppressing effect of oscillation caused by different parameters.

In the turn-off process, the major characteristics parameters include the turn-
off peak voltage and the overall turn-off loss, both of which we can adopt similar
approaches as in Sect. 10.2.1. to analytically calculate with the equations above.
We will not repeat it here. In real practice, the extraction of parameters above can
be further simplified. Without the transient damping circuit, the stray inductance of
the main circuit L_s does not induce obvious voltage spike. Therefore, the voltage
spike with the transient damping branch is assumably caused at the peak voltage of
u_1. From another aspect, as shown in Fig. 10.41b, the plateau of i_{Ls} in the falling
process is drastically weakened with the transient damping branch. For purpose of
simplicity, the current can be linearized, with its slew rate calculated by simulation or
experiments. For instance, the related slope in Fig. 10.41b is -0.2 A/ns. Accordingly,
the voltage waveform and peak of u_1 can be calculated through Eq. (10.57), i.e.,

$$\begin{cases} U_S - u_1(t) = L_0 \dfrac{di_{L0}(t)}{dt} \\ u_1(t) = U_S - i_{\text{damp}} R_{\text{damp}} = U_S - [i_{Ls}(t) - i_{L0}(t)] R_{\text{damp}} \end{cases} \tag{10.57}$$

Besides, we can calculate the transient waveform of i_{damp} with the above equations,
and thereby further calculate the loss of R_{damp}. In actual design, the overall switching
loss of switch modules and damping resistor within one switching period should be
employed as one objective for parameter optimization.

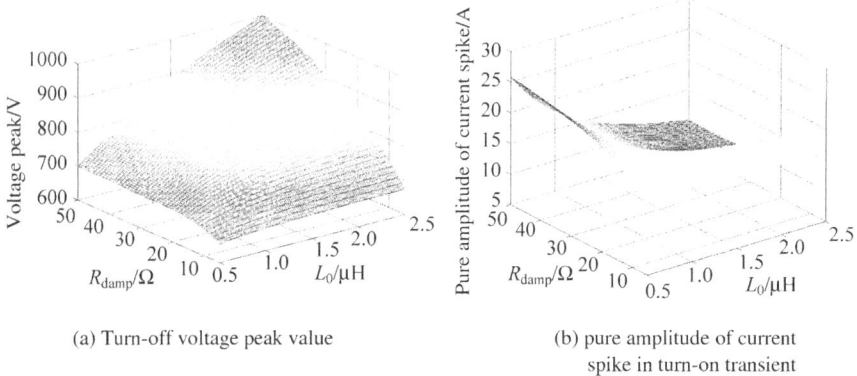

(a) Turn-off voltage peak value

(b) pure amplitude of current spike in turn-on transient

Fig. 10.43 The transient characteristics versus L_0 and R_{damp}

According to the analysis above, we can optimize L_0 and R_{damp} based on actual performance requirement and constraints. For a direct illustration, three major characteristics, i.e., turn-on current peak, turn-off voltage peak and overall switching loss of switch modules and damping resistor are calculated with different L_0 and R_{damp}, as shown in Fig. 10.43. Here, $U_s = 600$ V and $I_L = 16$ A. The switching-off loss is neglected. We can see that a larger L_0 will make the inductor more frequency sensitive, i.e., more adequately meeting the precondition of Eq. (10.50), thereby achieving better damping effect and reducing the turn-on current spike. However, with L_0 increasing, energy accumulation and release during switching transients significantly increase the turn-off voltage spike and the overall switching loss. Therefore, L_0 shouldn't be too large. On the other hand, a larger R_{damp} yields a better damping effect, which further lowers the current spike and the switching loss, but it also causes the higher turn-off voltage spike. Besides, a further increment of R_{damp} does not show better improvement of system performance, especially in terms of suppressing the turn-on current spike. Specific tradeoffs must be processed for the selection of R_{damp} based on actual needs (Fig. 10.43).

10.3 Summary

Electromagnetic transients of power electronics systems exhibit special characteristics, such as quasi-discrete electromagnetic pulses, pulse sequences, and large electromagnetic energy variation within short timescales. All these characteristics are rooted in power switches, conversion circuit, and switching control, which in turn have tremendous impact on power electronic circuits and systems. The focus of this book is rules and analytical methodologies of electromagnetic transients in typical power electronic systems, which however is far from covering connotation and denotation of power electronics. The study of electromagnetic transients in power electronics systems is still ongoing, but conventional analysis and methods are inadequate for us to peek into the core of electromagnetic transients. As we all know, the size of the material has entered the nanometer level, so has the related material revolution been brought to the material domain. We can well imagine that as the research of electromagnetic transients enters the nanosecond range, consequently, related theories, analytical methodologies and applications will bring revolutions as well, such as:

(1) Characterization of ultra-short-timescale transients. The analyzing angle will shift from the macroscopic level such as "$P = U*I$" to the microscopic view of charged particle dynamics, which will remove constraints caused by the ideal switch model and the lumped parameters. More accurate switch model and distributive parameter adopting small-space-scale based circuit model are to be built, with non-ideal characteristics and non-linear mechanism. These all help us to more accurately calculate and control distortion of energy pulses and to further understand the operation and failure mechanism of the device and system.

(2) Characterization of intensity, direction, and shape of electromagnetic energy transmission. Traveling waves based on diffusion equations in electromagnetics will be applied to energy transmission of power electronic systems, which helps us to understand the in-depth movement of carriers inside switches and free electrons inside transmission lines. The energy function then can be extended from a one-dimension time function to a four-dimension function of time and space. The energy flow and distribution of power electronic systems can be reproduced precisely at any time and location;

(3) Achievement of arbitrary transformation of any electromagnetic variables, including the voltage and current. The high-level realm of power electronics is to achieve arbitrary transformation of any waveforms, while the electromagnetic transient energy pulse and pulse sequence are the fundamental. With rapid development of computer science and engineering, we have witnessed various virtual realities in our lives. While electromagnetic variables simulated by computer science and technology is at the information level, we need convert such information signals into power signals with the same waveform to drive the real world, i.e., the power electronic amplifier. The present large-scale power system is based on the transformer. The future electric world will be based on high-power arbitrary waveform power electronic converters.

(4) Based on the pulsed power transient theory of power electronics, the conventional control strategies, component selection, system design and cooling design can all be converted to mathematical optimization and physical energy conversion at different timescales. This will make the design of power electronic systems more scientific, programmable, and visible.

(5) Repetitive energy pulses and pulse sequences will become a new research branch of electromagnetics. The pulsed power in high-voltage research is concentrated on the single pulsing process, which is generated by multiple capacitors and

not easy to control. The pulsed power in power electronics, however, is sequential, repetitive, and controllable. Such transients are based on specific control strategies with specific switches and topologies. With further development and breakthrough of the electromagnetic transient analysis and control theory, power electronic energy pulses will replace high-voltage pulses and embrace wider applicabilities.

Therefore, the study of pulsed power transients of power electronics will provide important scientific guidance to and dramatically benefit research of energy distribution and variation, implementation of high-efficiency, high-reliability arbitrary energy conversion, analysis and effective simulation of device failure mechanism, and accurate evaluation of switch characteristics, such as the switching loss, thermal dissipation, component placement, EMC, and reduction of the switch faulty turn-on and faulty protections. Their scientific significance is also highly related to their applications. Based on the research of electromagnetic transients, the effective non-linear analytical model of power switches can be established, which can then accurately measure switch performance under different working conditions. Based on the functionality of the electromagnetic energy in the loop of the high-power converter, along with its optimal solution, the structure of the power electronic system can be optimized. Further based on the transient energy balancing of power electronic converters, the snubber circuit can be effectively designed to suppress du/dt and di/dt, thereby precisely controlling electromagnetic transients. By combining the PWM control and switching transients, we can obtain optimal settings of the switching frequency, minimum pulse width, dead band, and modulation index through analysis of electromagnetic pulsed power transients.

References

1. Zhao, Zhengming, Liqiang Yuan, Ting Lu, and Fanbo He. 2015. Overview of the developments on high power electronic technologies and applications in China. *Journal of Electrical Engineering* 10 (4): 16–24.
2. Qian, Zhaoming, Junming Zhang, and Kuang Sheng. 2014. Status and development of power semiconductor devices and its applications. *Proceedings of the CSEE* 34 (29): 5149–5161.
3. Zhao, Zhengming, Fanbo He, Liqiang Yuan, and Ting Lu. 2014. Techniques and applications of electromagnetic transient analysis in high power electronic systems. *Proceedings of the CSEE* 34 (18): 3013–3019.
4. Palmour, J.W., L. Cheng, V. Pala, E.V. Brunt, D.J. Lichtenwalner, G.Y. Wang, J. Richmond, M. O' Loughlin, S. Ryu, S.T. Allen, A.A. Burk, and C. Scozzie. 2014. Silicon carbide power MOSFETs: breakthrough performance from 900V up to 15kV. In *26th International Symposium on Power Semiconductor Devices and ICs(ISPSD), 2014*, 79–82.
5. Bauer, Friedhelm, Iulian Nistor, Andrei Mihaila, Marina Antoniou, and Florin Udrea 2012. SuperJunction IGBTs: an evolutionary step of silicon power devices with high impact potential. In *International Semiconductor Conference (CAS), 2012*, 27–36.
6. Rao, Hong, Qiang Song, Yu. Wenhuang Liu, Shukai Xu Luo, and Xiaolin Li. 2013. Optimized design solutions for multi-terminal VSC-HVDC system using modular multilevel converters and their comparison. *Automation of Electric Power System* 37 (15): 103–108.
7. Hartmann, Samuel, Venkatesh Sivasubramaniam, David Guillon, David E. Hajas, Rolf Schuetz, Dominik Truessel, and Charalampos Papadopoulos. 2014. Packaging technology platform for next generation high power IGBT modules. In *International Exhibition and Conference for Power Electronics, Intelligent Motion, Renewable Energy and Energy Management (PCIM), 2014*, 1–7.
8. Zhao, Zhengming, and Liqiang Yuan. 2009. *Integrated analysis of power electronics and motor drive system.* China Machine Press.
9. Yuan, Liqiang, Zhengming Zhao, Gaosheng Song, and Zhengyuan Wang. 2011. *The principle and application of power semiconductor devices.* China Machine Press.
10. Zhao, Zhengming, Hua Bai, and Liqiang Yuan. 2007. Transient of power pulse and its sequence in power electronics. *Science in China, Series E: Technological Sciences* 50 (3): 351–360.

© Tsinghua University Press and Springer Nature Singapore Pte Ltd. 2019
Z. Zhao et al., *Electromagnetic Transients of Power Electronics Systems*,
https://doi.org/10.1007/978-981-10-8812-4

11. Zhao, Zhengming, Hua Bai, and Liqiang Yuan. 2008. Analysis of cutting-edge techniques in the high voltage and high power adjustable speed drive systems. *Science in China, Series E: Technological Sciences* 52 (2): 442–449.

12. Zhao, Zhengming, Jian Chen, and Xiaoying Sun. 2012. *Maximum power point tracking technology for photovoltaic power generation.* Beijing: Publishing House of Electronics Industry.

13. On semiconductor. *MC33153 single IGBT gate driver data sheet.* On Semiconductor.

14. Infineon technologies AG. *FF300R17ME4 IGBT data sheet.* Infineon Technologies AG.

15. Zhao, Zhengming, Hua Bai, and Liqiang Yuan. 2007. Transient process and sequence of electromagnetic pulsed power in power electronics. *Science China Series E Technological Sciences* 37 (01): 60–69.

16. Wang, Pengcheng, Jianyong Zheng, and Jun You. 2009. Improved design and simulation of snubber circuits for IGBT inverters. *Electrical Measurement and Instrumentation* 46 (10): 67–71.

17. Zhang, Quanzhu, Chengyu Huang, and Yonghong Deng. 2009. Matlab simulation and research for the IGBT absorbing circuits of inverter. *Electric Drive Automation* 31 (6): 27–31.

18. Zou, Gaoyu, Zhengming Zhao, and Liqiang Yuan. 2013. Study on DC busbar structure considering stray inductance for the back-to-back IGBT-based converter. In *Applied Power Electronics Conference and Exposition (APEC), 2013*, 1213–1218.

19. Jiao, Yang, and Fred C. Lee. 2015. LCL filter design and inductor current ripple analysis for a three-level NPC grid interface converter. *IEEE Transactions on Power Electronics* 30 (9): 4659–4668.

20. Wang, Xuesong, Zhengming Zhao, and Liqiang Yuan. 2011. Systematic safe operating area of IGBT based converters. *J T singhua Univ (Sci & Tech)* 51 (07): 914–920.

21. Chen, Zheng, Yiying Yao, Dushan Boroyevich, Khai Ngo, Paolo Mattavelli, and Kaushik Rajashekara. 2013. A 1200V 60A SiC MOSFET multi-chip phase-leg module for high-temperature, high-frequency applications. *Applied Power Electronics Conference and Exposition (APEC)* 2013: 608–615.

22. Chen, Kainan. 2014. *Research on the key technologies of high performance and wide bandwidth power electronic amplifiers.* PhD diss., Department of Electrical Engineering, Tsinghua University.

23. Bai, Hua, and Chris Mi. 2011. *Transients of modern power electronics.* America: Wiley.

24. Luo, Xiaoshu. 2012. *Nonlinear dynamics behavior and chaotic control in DC-DC converters.* Beijing: Science Press.

25. Pathak, A.D. 2001. *MOSFET/IGBT drivers theory and applications.* Milpitas, USA: IXYS Corporation.

26. Rashid, M.H. 2001. *Power electronics handbook*, 407–429. America: Academic Press.

27. Bai, Hua, Zhengming Zhao, Hu Xian, and Xin Chen. 2006. The experimental analysis of DC pre-excitation for 3-level inverter-motor system. *Proceedings of the CSEE* 26 (3): 159–163.

28. Dziech, Andrzej. 1993. *Random pulse streams and their applications*, 105–122. Polish scientific publishers PWN Ltd.

29. Bai, Hua. 2007. *Research on the transient process of electromagnetic pulsed power in power electronic converter.* PhD diss., Department of Electrical Engineering, Tsinghua University.

30. Yuan, Liqiang. 2006. *Design and optimization of the IGCT-based three-level voltage source converter.* Postdoctoral thesis, Department of Electrical Engineering, Tsinghua University.

31. Yuan, Liqiang. 2004. *IGCT-based multilevel converters for medium voltage drive system.* PhD diss., Department of Electrical Engineering, Tsinghua University.

32. Yuan, Liqiang, Zhengming Zhao, and Hua Bai. 2004. The functional model of IGCTs for the circuit simulation of high voltage converters. *Proceedings of the CSEE* 24 (6): 65–69.

33. Wang, Xuesong. 2011. *Systematic safe operating area of the power electronic converters and its application.* PhD diss., Department of Electrical Engineering, Tsinghua University.

34. Kevin, Motto. 2000. *Application of high-power snubberless semiconductor switches in high-frequency PWM converters.* MS thesis, Virginia Polytechnic Institute and State University, UAS.

35. Yuan, Liqiang, Zhengming Zhao, Hua Bai, Chongjian Li, and Yaohua Li. 2003. The IGCT test platform for voltage source inverters. *IEEE Proceedings of PEDS* 2003 (1): 1291–1294.

36. Yuan, Liqiang, Zhengming Zhao, Jianzheng Liu, and Bing Li. 2005. Experiment and analysis for single IGCT equipped in MV three-level inverter. *Conference Record of IAS2005* 2: 825–829.

37. Sun, Qiang, Xueru Wang, and Yuelong Cao. 2004. Study of the current balances of IGBTs in paralleling. *China Academic Journal Electronic Publishing* 9: 84–85, 89.

38. Zhao, Hongtao, Wu Jun, and Wensen Chang. 2007. Research on the feasibility of balancing on-state current for paralled IGBTs by controlling gate voltage. *Power Electronics* 41 (9): 101–103.

39. Zhao, Zhengyuan, and Jihua Xie. 2008. Research and simulation on characteristics of paralleled IGBTs. *Electrotechnical Application* 27 (20): 64–67.

40. Zha, Shensen, Jianyong Zheng, Su Lin, Wu Hengrong, and Jun Chen. 2005. Research on current balancing of parallel IGBTs. *Electric Power Automation Equipment* 25 (7): 32–34.

41. Qiao, Ermin, Xuhui Wen, and Xin Guo 2006. Development of high power intelligent module based on paralleled IGBTs. *Transactions of China Electrotechnological Society* 21 (10): 90–93, 100.

42. Letor, Romeo. 1992. Static and dynamic behavior of paralleled IGBTs. *IEEE Transactions on Industry Application* 28 (2): 395–402.

43. Hofer-Noser, Patrick, Nick Karrer, and Christian Gerster. 1996. Paralleling intelligent IGBT power modules with active gate-controlled current balancing. *IEEE PESC* 2: 1312–1316.

44. Hofer-Noser, Patrick, and Nicolas Karrer. 1999. Monitoring of paralleled IGBT/diode modules. *IEEE Transactions on Power Electronics* 4 (3): 438–444.

45. Miyazaki, Hideki, Hideshi Fukumoto, Shigeru Sugiyama, Makoto Tachikawa, and Noboru Azusawa. 2000. Neutral-point-clamped inverter with parallel driving of IGBTs for industrial applications. *IEEE Transactions on Industry Application* 36 (1): 146–151.

46. Azar, Ramy, Florin Udrea, Wai Tung Ng, Francis Dawson, and William Findlay. 2008. The current sharing optimization of paralleled IGBTs in a power module tile using a PSpice frequency dependent impedance model. *IEEE Transactions on Power Electronics* 23 (1): 206–217.

47. Bortis, Dominik, Juergen Biela, and Johann Kolar. 2008. Active gate control for current balancing of parallel-connected IGBT modules in solid-state modulators. *IEEE Transactions on Plasma Science* 36 (5): 2632–2637.

48. Wang, Xuesong, Zhengming Zhao, and Liqiang Yuan. 2010. Current sharing of IGBT modules in parallel with thermal imbalance. In *IEEE ECCE*, 234–238.

49. Yuan, Shoucai. 2007. *Introduction of the IGBT field effect semiconductor power devices.* Beijing: Science Press.

50. Infineon Technologies AG. 2007. *IGBT-modules FF450R12ME3 datasheet.* Germany: Infineon Technologies AG.

51. Nabae, Akira, Isao Takahashi, and Hirofumi Akagi. 1981. A new neutral point clamped PWM inverter. *IEEE Transactions on Industry Application* IA–15 (5): 518–523.

52. Bhagwat, Pradeep M. 1999. Generailzed structure of a multilevel PWM inverter. *IEEE Transactions on Industry Application* IA–19 (6): 1057–1069.

53. Lai, Jin-Sheng, and Fang Zheng Peng. 1996. Multilevel converter-a new breed of power converters. *IEEE Transactions on Industry Application* 32 (3): 509–517.

54. Yi, Rong, and Zhengming Zhao. 2007. Research on the turn-off characteristic of IGCT influenced by the stray inductance in high power inverters. *Proceedings of the CSEE* 27 (31): 115–120.

55. Yi, Rong, Zhengming Zhao, and Liqiang Yuan. 2008. Busbar optimization design for high power converters. *Transactions of China Electrotechnical Society* 23 (8): 94–100.

56. Marquardt, Rainer. 2011. Modular multilevel converter topologies with DC-short circuit current limitation. *Power Electronics and ECCE Asia (ICPE & ECCE)* 2011: 1425–1431.

57. Solas, Estíbaliz, Gonzalo Abad, Jon Andoni Barrena, Sergio Aurtenetxea, Ainhoa Cárcar, and Ludwik Zając. 2013. Modular multilevel converter with different submodule concepts-Part I: capacitor voltage balancing method. *IEEE Transactions on Industrial Electronics* 60 (10): 4525–4535.

58. Glinka, Martin, and Rainer Marquardt. 2005. A new AC/AC multilevel converter family. *IEEE Transactions on Industrial Electronics* 52 (3): 662–669.

59. Li, Xiaoqian, Wenhua Liu, Qiang Song, Hong Rao, and Shukai Xu. 2013. An enhanced MMC topology with DC fault ride-through capability. In *39th Annual Conference of the IEEE Industrial Electronics Society IECON 2013*, 6182–6188.

60. Hiller, Marc, Dietmar Krug, Rainer Sommer, and Steffen Rohner. 2009. A new highly modular medium voltage converter topology for industrial drive applications.In *13th European Conference on Power Electronics and Applications EPE '09*, 1–10.

61. Chen, Chingchi. 2003. Characterization of power electronics EMI emission. *IEEE International Symposium on Electromagnetic Compatibility* 2: 553–557.

62. Qian, Zhaoming, and Henglin Chen. 2007. State of art of electromagnetic compatibility research on power electronic equipment. *Transactions of China Electrotechnical Society* 22 (7): 1–9.

63. Liu, Qian, Fred Wang, and Dushan Boroyevich. 2007. Conducted-EMI prediction for AC converter systems using an equivalent modular-terminal-behavioral (MTB) source model. *IEEE Transactions on Industry Applications* 43 (5): 1360–1370.

64. Wang, Xuesong, Zhengming Zhao, and Liqiang Yuan. 2010. Conducted EMI reduction in IGBT-based converters. In *IEEE APEMC*, 230–234.

65. Zhong, Yulin. 2005. *High frequency equivalent circuit of grounding net*. MS thesis, Department of Electrical Engineering, Tsinghua University, Beijing.

66. Zou, Gaoyu. 2013. *Design and analysis of the power electronics converters based on systematic safe operating area*. PhD diss., Department of Electrical Engineering, Tsinghua University, Beijing.

67. Ma, Weiming, Lei Zhang, and Jin Meng. 2007. *Electromagnetic compatibility of independent power system and power electronics*. Beijing: Science Press.

68. Zou, Gaoyu, Zhengming Zhao, Liqiang Yuan, Xuesong Wang, and Ting Lu. 2011. Optimal design of the back-to-back IGBT-based converter with the concept of systematic safe operating area. In *International Conference on Electrical Machines and Systems (ICEMS)*, 1–5.

69. Zou, Gaoyu, Zhengming Zhao, Liqiang Yuan, and Lu Yin. 2014. Design of main circuit parameters for high-performance three-phase back-to-back converter. *Electric Power Automation Equipment* 01: 72–79.

70. Zou, Gaoyu, Zhengming Zhao, Liqiang Yuan, and Xuesong Wang. 2014. Systematic safe operating area of dual PWM converter and its application. *Electric Power Automation Equipment* 03: 82–88.

71. Zou, Gaoyu, Zhengming Zhao, Liqiang Yuan, and Lu Yin. 2013. Losses in IGBT modules in dual-PWM converters. *J Tsinghua Univ (Sci & Technol)* (07): 1011–1018.

72. Stevanovic, Ljubisa D., Richard A. Beaupre, Eladio C. Delgado, and Arun V. Gowda. 2010. Low inductance power module with blade connector. In *Proceedings of IEEE Applied Power Electronics Conference and Exposition (APEC)*, 1603–1609.

73. Infineon Technologies AG. 2011. *IGBT-modules FF300R12ME3 datasheet*. Germany: Infineon Technologies AG.

74. Wang, Jun, Bingjian Yang, Xu Zhixin, Yan Deng, Rongxiang Zhao, and Xiangning He. 2010. Configuration of low inductive laminated bus bar in 750 kVA NPC three-level universal converter module of high power density. *Proceedings of the CSEE* 30 (18): 47–54.

75. Bai, Hua, and Zhengming Zhao. 2007. Research on starting strategies in the three-level high voltage high power inverters. *Transactions of China Electrotechnological Society* 22 (11): 91–97.

76. Mao, Peng, Shajun Xie, and Xu Zegang. 2010. Switching transient model and loss analysis of IGBT module. *Proceedings of the CSEE* 30 (15): 40–47.

77. Bierhoff, M.H., and F.W. Fuchs. 2004. Semiconductor losses in voltage source and current source IGBT converters based on analytical derivation. *Power Electronics Specialists Conference (PESC)* 2836–2842.

78. Jing, Wei, Guojun Tan, and Zongbin Ye. 2011. Losses calculation and heat dissipation analysis of high-power three-level converters. *Transactions of China Electrotechnological Society* 26 (2): 134–140.

79. Wang, Xuesong, Hua Bai, Zhengming Zhao, and Liqiang Yuan. 2011. Mathematical models of the system-level safe operational areas of power electronic converters in plug-in hybrid electric vehicles. *IEEE Transactions on Vehicular Technology* 60 (9): 4288–4298.

80. Yuan, Liqiang, Zhengming Zhao, Fanbo He, and Eltawil M. 2007. Safe operating area of high power three-level neutral point clamped voltage source inverters equipped with IGCTs. In *2007 International Conference on Electrical Machines and Systems, ICEMS*, 64–68.

81. Zhao, Zhengming, Hua Bai, and Qiangyuan Li. 2009. Analysis of cutting-edge techniques in the high voltage and high power adjustable speed drive systems. *Science China Series E Technological Sciences* 39 (3): 394–401.

82. Bai, Hua, Chunting Mi, and Sonya Gargies. 2008. The short-time-scale transient processes in high-voltage and high-power isolated bidirectional DC-DC converters. *IEEE Transactions on Power Electronics* 23 (6): 2648–2656.

83. Akdag, Alper. 2006. SOA in high power semiconductors. *IEEE Industry Applications Conference* 3: 1473–1477.

84. Abbate, C., G. Busatto, R. Manzo, L. Fratelli, B. Cascone, G. Giannini, and F. Iannuzzo. 2004. Experimental optimisation of high power IGBT modules performances working at the edges of their safe operating area. In *IEEE PESC*, 2588–2592.

85. Li, Ming, Xiong Fang, Yue Wang, Leqiang Zhang, Zhaoan Wang, Gang Liu, and Weizheng Yao. 2009. Overvoltage protection of high power IGBTs in wind power converters under short circuit. In *IEEE IPEMC*, 2288–2290.

86. Tianjin Electric Drive Design Institute. 2005. *Manual of electric drive automation technology*. Beijing: China Machine Press.

87. Lu, Ting. 2010. *Characteristics and control methods of power pulses in high power electronic conversion*. PhD diss., Department of Electrical Engineering, Tsinghua University, Beijing.

88. Wang, Xuesong, Zhengming Zhao, and Liqiang Yuan. 2012. Protection scheme for converter based on the systematic safe operating area. *J T singhua Univ (Sci & Tech)* 52 (8): 1029–1034.

89. Guangshu, Hu. 1997. *Digital signal processing*. Beijing: Tsinghua University Press.

90. Rodriguez, Jose, Jih-sheng Lai, and Fangzheng Peng. 2002. Multilevel inverters: a survey of topologies, controls, and applications. *IEEE Transactions on Industrial Electronics* 49 (4): 724–738.

91. Lu, Ting, Zhengming Zhao, Yingchao Zhang, and Liqiang Yuan. 2008. General-purpose control platform based on dual DSPs for power electronic converters. *J T singhua Univ (Sci & Tech)* 48 (10): 1541–1544.

92. Texas Instruments Incorporated. 2005. *CPU and peripherals of TMS320C28x*. Beijing: Tsinghua University Press.

93. Lu, Ting. 2012. R*esearch on electromagnetic transient process of high power electronic conversion*. Postdoctoral thesis, Department of Electrical Engineering, Tsinghua University, Beijing.
94. Hu, Sideng. 2011. *Research on control strategy of adjustable speed driver considering non-ideal character and specific operation*. PhD diss., Department of Electrical Engineering, Tsinghua University, Beijing.
95. Yingchao, Zhang. 2008. *Research on the control techniques of neutral point clamped three-level dual-PWM converter*. PhD diss., Department of Electrical Engineering, Tsinghua University, Beijing.
96. Wei, Lixiang. 2000. *Research on dual-PWM three-level flux orientation induction motor control system*. PhD diss., Department of Electrical Engineering, Tsinghua University, Beijing.
97. Pengsheng, Su, and Lianwei Jiao. 2003. *Automatic control theory*. Beijing: Publishing House of Electronics Industry.
98. Chongwei, Zhang, and Zhang Xing. 2003. *PWM rectifier and its control technology*. Beijing: China Machine Press.
99. Kuo, B.C. 1995. *Automatic Control Systems*, 7th ed. New Jersey: Prentice Hall.
100. Akagi, Hirofumi, Yoshihira Kanazawa, and Akira Nabae. 1984. Instantaneous reactive power compensator comprising switching devices without energy storage components. *IEEE Transactions on Industry Application* 20 (3): 625–630.
101. Ohnishi, Tokuo. 1991. Three phase PWM converter/inverter by means of instantaneous active and reactive power control. In *IECON'91*, 819–824.
102. Malinowski, Mariusz, Marian P. Kazmierkowski, and Andrzej M. Trzynadlowski. 2003. A comparative study of control techniques for PWM rectifier in ac adjustable speed drives. *IEEE Transactions on Power Electronics* 18 (6): 1390–1396.
103. Zhang, Yingchao, Zhengming Zhao, Mohammed Eltawil, and Liqiang Yuan. 2008. Performance evaluation of three control strategies for three-level neutral point clamped PWM rectifier. In *IEEE APEC'08*, 259–264.
104. Zhang, Yingchao, Zhengming Zhao, Liqiang Yuan, Ting Lu, and Yongchang Zhang. 2008. Direct power control for three-Level PWM rectifier. *Transactions of China Electrotechnological Society* 23 (5): 62–68.
105. Qian, Jixin, Jun Zhao, and Xu Zuhua. 2007. *Predictive control*. Beijing: Chemical Industry Press.
106. Aurtenechea, Sergio, Miguel Angel Rodriguez Vidal, Estanis Oyarbide, and Jos Ramn Torrealday Apraiz. 2007. Predictive control strategy for dc/ac converters based on direct power control. *IEEE Transactions on Industry Electronics* 54 (3): 1261–1271.
107. Antoniewicz, P., M.P. Kazmierkowski, S. Aurtenechea, and M.A. Rodriguez. 2008. Comparative study of two predictive direct power control algorithms for three-phase ac/dc converters. In *SIBIRCON'08*, 159–163.
108. Zhang, Haitao. 2006. *Research on the control system of three level inverters based on IGCT*. PhD diss., Department of Electrical Engineering, Tsinghua University, Beijing.
109. Zhang, Yongchang. 2009. *Research on the control technologies of neutral point clamped three-level dual-PWM converter*. PhD diss., Department of Electrical Engineering, Tsinghua University, Beijing.
110. Zhao, Zhengming, Ting Lu, Liqiang Yuan, Fanbo He, and Lu Yin. 2014. Power electronic converter control strategy based on transient electromagnetic energy balance. Delta Power Electronics Seminar.
111. Ferreira, J.A., and J. Daan Van Wyk. 2001. Electromagnetic energy propagation in power electronic converters: toward future electromagnetic integration. *Proceedings of IEEE* 89 (6): 879–889.
112. Yin, Lu, Zhengming Zhao, Ting Lu, Fanbo He, Liqiang Yuan. 2011. A predictive dc voltage control scheme for back-to-back converters based on energy balance modeling. In *ICEMS'11*.

113. He, Fanbo, Zhengming Zhao, Ting Lu, Liqiang Yuan. 2010. Predictive dc voltage control for three-phase grid-connected PV inverters based on energy balance modeling. In *IEEE PEDG'10*, 516–519.

114. He, Fanbo. 2012. *Design and control of large-scale grid-connected photovoltaic systems*. PhD diss., Department of Electrical Engineering, Tsinghua University, Beijing.

115. Wang, Cheng. 2006. *Improving the active power filter performance with control delay and dead-time effect*. MS thesis, College of Electrical Engineering and Information Technology, Sichuan University, Chengdu.

116. Gutierrez, Monica, Giri Venkataramanan, and Ashok Sundaram. 1999. Performance characterization of integrated gate commutated thyristors. *Thirty-Fourth IAS Annual Meeting* 1999: 359–363.

117. Ji, Shiqi. 2015. *Modeling and active voltage balancing control for series-connected HV-IGBTs*. PhD diss., Department of Electrical Engineering, Tsinghua University, Beijing.

118. Odegard, B., and R. Ernst 2003. *Applying IGCT's*. ABB Semiconductors, Switzerland. March 2003, No. 5SYA 2032-01.

119. ABB. 2002. *Reverse Conducting Integrated Gate-Commutated Thyristor 5SHX 14H4502*. ABB Semiconductors, Switzerland. January 2002, No. 5SYA 1227-03.

120. ABB. 2002. *Reverse Conducting Integrated Gate-Commutated Thyristor 5SHX 08F4502*. ABB Semiconductors, ABB Switzerland Ltd. January 2002, No. 5SYA 1223-04.

Printed by Printforce, the Netherlands